Matilde Marcolli | Deepak Parashar (Eds.)

Quantum Groups and Noncommutative Spaces

Edited by Klas Diederich

The texts published in this series are intended for graduate students and all mathematicians who wish to broaden their research horizons or who simply want to get a better idea of what is going on in a given field. They are introductions to areas close to modern research at a high level and prepare the reader for a better understanding of research papers. Many of the books can also be used to supplement graduate course programs.

Vol. E 34: I. Lieb/J. Michael: The Cauchy-Riemann Complex

Vol. E 35: G. Harder: Lectures on Algebraic Geometry I

Vol. E 36: C. Hertling/M. Marcolli (Eds.): Frobenius Manifolds*

Vol. E 37: C. Consani/M. Marcolli (Eds.): Noncommutative Geometry and Number Theory*

Vol. E 38: S. Albeverio/M. Marcolli/S. Paycha/J. Plazas (Eds.): Traces in Number Theorey, Geometry and Quantum Fields

Vol. E 39: G. Harder: Lectures on Algebraic Geometry II

Vol. E 40: H. Abbaspour/M. Marcolli/T. Tradler (Eds.): Deformation Spaces*

Vol. E 41: M. Marcolli / D. Parashar (Eds.): Quantum Groups and Noncommutative Spaces*

*A Publication of the Max-Planck-Institute for Mathematics, Bonn

Matilde Marcolli | Deepak Parashar (Eds.)

Quantum Groups and Noncommutative Spaces

Perspectives on Quantum Geometry

A Publication of the Max-Planck-Institute for Mathematics, Bonn

VIEWEG+
TEUBNER

Bibliographic information published by the Deutsche Nationalbibliothek
The Deutsche Nationalbibliothek lists this publication in the Deutsche Nationalbibliografie; detailed bibliographic data are available in the Internet at http://dnb.d-nb.de.

Prof. Dr. Matilde Marcolli
Mathematics Department
California Institute of Technology
1200 E.California Blvd.
Pasadena, CA 91125
USA

matilde@caltech.edu

Prof. Dr. Klas Diederich (Series Editor)
Bergische Universität Wuppertal
Fachbereich Mathematik
Gaußstraße 20
42119 Wuppertal
Germany

klas@uni-wuppertal.de

Dr. Deepak Parashar
University of Cambridge
Cambridge Cancer Trials Centre
Department of Oncology
Addenbrooke's Hospital (Box 279)
Hills Road
Cambridge CB2 0QQ

Cambridge Hub in Trials Methodology Research
MRC Biostatistics Unit
University Forvie Site
Robinson Way
Cambridge CB2 0SR
UK

dp409@cam.ac.uk

Mathematics Subject Classification
17B37 Quantum groups (quantized enveloping algebras) and related deformations, **58B34** Noncommutative geometry (à la Connes) , **58B32** Geometry of quantum groups, **20G42** Quantum groups (quantized function algebras) and their representations, **16T05** Hopf algebras and their applications, **19D55** K-theory and homology; cyclic homology and cohomology, **81T75** Noncommutative geometry methods

Softcover re-print of the Hardcover 1st edition 2011

Editorial Office: Ulrike Schmickler-Hirzebruch

Vieweg+Teubner Verlag is a brand of Springer Fachmedien.
Springer Fachmedien is part of Springer Science+Business Media.
www.viewegteubner.de

Cover design: KünkelLopka Medienentwicklung, Heidelberg
Printed on acid-free paper

ISBN 978-3-8348-2689-3

Contents

Preface

The present volume is based on an activity organized at the Max Planck Institute for Mathematics in Bonn, during the days August 6–8, 2007, dedicated to the topic of Quantum Groups and Noncommutative Geometry. The main purpose of the workshop was to focus on the interaction between the many different approaches to the topic of Quantum Groups, ranging from the more algebraic techniques, revolving around algebraic geometry, representation theory and the theory of Hopf algebras, and the more analytic techniques, based on operator algebras and noncommutative differential geometry. We also focused on some recent developments in the field of Noncommutative Geometry, especially regarding spectral triples and their applications to models of elementary particle physics, where quantum groups are expected to play an important role.

The contributions to this volume are written, as much as possible, in a pedagogical and expository way, which is intended to serve as an introduction to this area of research for graduate students, as well as for researchers in other areas interested in learning about these topics.

The first contribution to the volume, by Brzezinski, deals with the important topic of Hopf-cyclic homology, which is the right cohomology theory in the context of Hopf algebras, playing a role, with respect to cyclic homology of algebras, similar to the cohomology of Lie algebras in the context of de Rham cohomology. The contribution in this volume focuses on the observation that anti-Yetter-Drinfeld contramodules can serve as coefficients for cyclic homology.

The second contribution, by Ćaćić, focuses on recent developments in particle physics models based on noncommutative geometry. In particular, the paper describes a general framework for the classification of Dirac operators on the finite geometries involved in specifying the field content of the particle physics models. These Dirac operators have interesting moduli spaces, which are analyzed extensively in this paper.

The paper by Fioresi deals with supergeometry aspects. More precisely, it describes how one can treat the general linear supergroup from the point of view of group schemes and Hopf algebras.

Fioresi and Gavarini contributed a paper on a generalization of the quantum duality principle to quantizations of projective quantum homogeneous spaces. The procedure is illustrated completely explicitly in the important case of the quantum Grassmannians.

The paper by Goswami considers the problem of finding an analogue in Noncommutative Geometry of the isometry group in Riemannian geometry. The noncommutative analog of Riemannian manifolds is provided by spectral triples, hence the replacement is provided by a compact quantum group, which acts on the spectral triple.

Kassel's paper deals with the geometry of Hopf Galois extensions. Hopf Galois extensions can be constructed from Hopf algebras, whose product is twisted with a cocycle. The algebra obtained in this way is a flat deformation over a central subalgebra. This paper presents a construction of elements in this commutative subalgebra. It also shows that an integrality condition is satisfied by all finite-dimensional Hopf algebras generated by grouplike and skew-primitive elements. Explicit computations are given for the case of the Hopf algebra of a cyclic group.

Mukherjee's paper gives a survey or recent results on the quantization of the moduli space of stable parabolic Higgs bundles of rank two over a Riemann surface of genus at least two. This is obtained via the deformation quantization of the Poisson structure associated to a natural holomorphic symplectic structure. The choice of a projective structure on the Riemann surface induces a canonical star product over a Zariski open dense subset of the moduli space.

Van Daele's paper discusses the Radford formula expressing the forth power of the antipode in terms of modular operators. It is first shown how the formula simplifies in the case of compact and discrete quantum groups. Then the setting of locally compact quantum groups is recalled and it is shown that the square of the antipode is an analytical generator of the scaling group of automorphisms.

A paper dealing with the idea of Hopf monads over arbitrary categories was contributed by Wisbauer, as a generalization to arbitrary categories of the notion of Hopf algebras in module categories.

The last paper in the volume, by Zampini, deals with the important topic of covariant differential calculus on quantum groups. The example of the quantum Hopf fibration on the standard Podleś sphere is analysed in full details. It is shown then how one obtains from the differential calculus gauged Laplacians on associated line bundles and a Hodge star operator on the total space and base space of the Hopf bundle. The paper includes an explicit review of the ordinary differential calculus on SU(2) based on the classisal geometry of the Hopf fibration, so that the comparison with the quantum groups case becomes more transparent.

We are grateful to the numerous referees for their expertise in ensuring a high standard of the contributions, and to all speakers and participants for a very lively interaction during the workshop. Finally, we wish to thank the MPIM, Bonn, for financial support for the activity and for hosting the workshop, and Vieweg Verlag for publishing this volume.

Matilde Marcolli and Deepak Parashar

Hopf-cyclic homology with contramodule coefficients

Tomasz Brzeziński

ABSTRACT. A new class of coefficients for the Hopf-cyclic homology of module algebras and coalgebras is introduced. These coefficients, termed *stable anti-Yetter-Drinfeld contramodules*, are both modules and *contramodules* of a Hopf algebra that satisfy certain compatibility conditions.

1. Introduction

It has been demonstrated in [**8**], [**9**] that the Hopf-cyclic homology developed by Connes and Moscovici [**5**] admits a class of non-trivial coefficients. These coefficients, termed *anti-Yetter-Drinfeld modules* are modules and comodules of a Hopf algebra satisfying a compatibility condition reminiscent of that for cross modules. The aim of this note is to show that the Hopf-cyclic (co)homology of module coalgebras and module algebras also admits coeffcients that are modules and *contramodules* of a Hopf algebra with a compatibility condition.

All (associative and unital) algebras, (coassociative and counital) coalgebras in this note are over a field k. The coproduct in a coalgebra C is denoted by Δ_C, and counit by ε_C. A Hopf algebra H is assumed to have a bijective antipode S. We use the standard Sweedler notation for coproduct $\Delta_C(c) = c_{(1)}\otimes c_{(2)}$, $\Delta_C^2(c) = c_{(1)}\otimes c_{(2)}\otimes c_{(3)}$, etc., and for the left coaction ${}^N\varrho$ of a C-comodule N, ${}^N\varrho(x) = x_{(-1)}\otimes x_{(0)}$ (in all cases summation is implicit). $\mathrm{Hom}(V, W)$ denotes the space of k-linear maps between vector spaces V and W.

2. Contramodules

The notion of a *contramodule* for a coalgebra was introduced in [**6**], and discussed in parallel with that of a comodule. A *right contramodule* of a coalgebra C is a vector space M together with a k-linear map $\alpha : \mathrm{Hom}(C, M) \to M$ rendering the following diagrams commutative

$$\begin{array}{ccccc}
\mathrm{Hom}(C, \mathrm{Hom}(C, M)) & & \xrightarrow{\mathrm{Hom}(C,\alpha)} & & \mathrm{Hom}(C, M) \\
\Big\downarrow{\scriptstyle \Theta} & & & & \Big\downarrow{\scriptstyle \alpha} \\
\mathrm{Hom}(C\otimes C, M) & \xrightarrow{\mathrm{Hom}(\Delta_C,M)} & \mathrm{Hom}(C, M) & \xrightarrow{\alpha} & M,
\end{array}$$

2000 *Mathematics Subject Classification.* 19D55.

$$\begin{array}{ccc} \mathrm{Hom}(k,M) & \xrightarrow{\mathrm{Hom}(\varepsilon_C,M)} & \mathrm{Hom}(C,M) \\ & {\scriptstyle\simeq}\searrow \quad \swarrow{\scriptstyle\alpha} & \\ & M, & \end{array}$$

where Θ is the standard isomorphism given by $\Theta(f)(c\otimes c') = \Theta(f)(c)(c')$. *Left contramodules* are defined by similar diagrams, in which Θ is replaced by the isomorphism $\Theta'(f)(c\otimes c') = f(c')(c)$ (or equivalenty, as right contramodules for the co-opposite coalgebra C^{op}). Writing blanks for the arguments, and denoting by matching dots the respective functions α and their arguments, the associativity and unitality conditions for a right C-contramodule can be explicitly written as, for all $f \in \mathrm{Hom}(C\otimes C, M)$, $m \in M$,

$$\dot{\alpha}\left(\ddot{\alpha}\left(f\left(\dot{-}\otimes\ddot{-}\right)\right)\right) = \alpha\left(f\left((-)_{(1)}\otimes(-)_{(2)}\right)\right), \qquad \alpha\left(\varepsilon_C(-)m\right) = m.$$

With the same conventions the conditions for left contramodules are

$$\dot{\alpha}\left(\ddot{\alpha}\left(f\left(\ddot{-}\otimes\dot{-}\right)\right)\right) = \alpha\left(f\left((-)_{(1)}\otimes(-)_{(2)}\right)\right), \qquad \alpha\left(\varepsilon_C(-)m\right) = m.$$

If N is a left C-comodule with coaction ${}^N\varrho : N \to C\otimes N$, then its dual vector space $M = N^* := \mathrm{Hom}(N,k)$ is a right C-contramodule with the structure map

$$\alpha : \mathrm{Hom}(C,M) \simeq \mathrm{Hom}(C\otimes N, k) \to \mathrm{Hom}(N,k) = M, \qquad \alpha = \mathrm{Hom}({}^N\varrho, k).$$

Explicitly, α sends a functional f on $C\otimes N$ to the functional $\alpha(f)$ on N,

$$\alpha(f)(x) = f(x_{(-1)}\otimes x_{(0)}), \qquad x \in N.$$

The dual vector space of a right C-comodule N with a coaction $\varrho^N : N \to N\otimes C$ is a left C-contramodule with the structure map $\alpha = \mathrm{Hom}(\varrho^N, k)$. The reader interested in more detailed accounts of the contramodule theory is referred to [**1**], [**15**].

3. Anti-Yetter-Drinfeld contramodules

Given a Hopf algebra H with a bijective antipode S, anti-Yetter-Drinfeld contramodules are defined as H-modules and H-contramodules with a compatibility condition. Similarly to the case of anti-Yetter-Drinfeld modules [**7**] they come in four different flavours.

(1) A *left-left anti-Yetter-Drinfeld contramodule* is a left H-module (with the action denoted by a dot) and a left H-contramodule with the structure map α, such that, for all $h \in H$ and $f \in \mathrm{Hom}(H,M)$,

$$h\cdot\alpha(f) = \alpha\left(h_{(2)}\cdot f\left(S^{-1}(h_{(1)})(-)h_{(3)}\right)\right).$$

M is said to be *stable*, provided that, for all $m \in M$, $\alpha(r_m) = m$, where $r_m : H \to M$, $h \mapsto h\cdot m$.

(2) A *left-right anti-Yetter-Drinfeld contramodule* is a left H-module and a right H-contramodule, such that, for all $h \in H$ and $f \in \mathrm{Hom}(H,M)$,

$$h\cdot\alpha(f) = \alpha\left(h_{(2)}\cdot f\left(S(h_{(3)})(-)h_{(1)}\right)\right).$$

M is said to be *stable*, provided that, for all $m \in M$, $\alpha(r_m) = m$.

(3) A *right-left anti-Yetter-Drinfeld contramodule* is a right H-module and a left H-contramodule, such that, for all $h \in H$ and $f \in \mathrm{Hom}(H,M)$,

$$\alpha(f)\cdot h = \alpha\left(f\left(h_{(3)}(-)S(h_{(1)})\right)\cdot h_{(2)}\right).$$

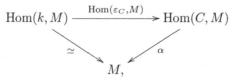

M is said to be *stable*, provided that, for all $m \in M$, $\alpha(\ell_m) = m$, where $\ell_m : H \to M$, $h \mapsto m{\cdot}h$.

(4) A *right-right anti-Yetter-Drinfeld contramodule* is a right H-module and a right H-contramodule, such that, for all $h \in H$ and $f \in \mathrm{Hom}(H, M)$,

$$\alpha(f){\cdot}h = \alpha\left(f\left(h_{(1)}(-)S^{-1}(h_{(3)})\right){\cdot}h_{(2)}\right).$$

M is said to be *stable*, provided that, for all $m \in M$, $\alpha(\ell_m) = m$.

In a less direct, but more formal way, the compatibility condition for left-left anti-Yetter-Drinfeld contramodules can be stated as follows. For all $h \in H$ and $f \in \mathrm{Hom}(H, M)$, define k-linear maps $\ell\ell_{f,h} : H \to M$, by

$$\ell\ell_{f,h} : h' \mapsto h_{(2)}{\cdot}f\left(S^{-1}(h_{(1)})h'h_{(3)}\right).$$

Then the main condition in (1) is

$$h{\cdot}\alpha(f) = \alpha\left(\ell\ell_{f,h}\right), \qquad \forall h \in H,\ f \in \mathrm{Hom}(H, M).$$

Compatibility conditions between action and the structure maps α in (2)–(4) can be written in analogous ways.

If N is an anti-Yetter-Drinfeld module, then its dual $M = N^*$ is an anti-Yetter-Drinfeld contramodule (with the sides interchanged). Stable anti-Yetter-Drinfeld modules correspond to stable contramodules. For example, consider a right-left Yetter-Drinfeld module N. The compatibility between the right action and left coaction ${}^N\varrho$ thus is, for all $x \in N$ and $h \in H$,

$${}^N\varrho(x{\cdot}h) = S(h_{(3)})x_{(-1)}h_{(1)}\otimes x_{(0)}h_{(2)}.$$

The dual vector space $M = N^*$ is a left H-module by $h\otimes m \mapsto h \cdot m$,

$$(h{\cdot}m)(x) = m(x{\cdot}h),$$

for all $h \in H$, $m \in M = \mathrm{Hom}(N, k)$ and $x \in N$, and a right H-contramodule with the structure map $\alpha(f) = f \circ {}^N\varrho$, $f \in \mathrm{Hom}(H\otimes N, k) \simeq \mathrm{Hom}(H, M)$. The space $\mathrm{Hom}(H\otimes N, k)$ is a left H-module by $(h{\cdot}f)(h'\otimes x) = f(h'\otimes x{\cdot}h)$. Hence

$$(h{\cdot}\alpha(f))(x) = \alpha(f)(x{\cdot}h) = f\left({}^N\varrho(x{\cdot}h)\right),$$

and

$$\begin{aligned}\alpha\left(h_{(2)}{\cdot}f\left(S(h_{(3)})(-)h_{(1)}\right)\right)(x) &= h_{(2)}{\cdot}f\left(S(h_{(3)})x_{(-1)}h_{(1)}\otimes x_{(0)}\right)\\ &= f\left(S(h_{(3)})x_{(-1)}h_{(1)}\otimes x_{(0)}{\cdot}h_{(2)}\right).\end{aligned}$$

Therefore, the compatibility condition in item (2) is satisfied. The k-linear map $r_m : H \to M$ is identified with $r_m : H\otimes N \to k$, $r_m(h\otimes x) = m(x{\cdot}h)$. In view of this identification, the stability condition comes out as, for all $m \in M$ and $x \in N$,

$$m(x) = \alpha(r_m)(x) = r_m(x_{(-1)}\otimes x_{(0)}) = m(x_{(0)}{\cdot}x_{(-1)}),$$

and is satisfied provided N is a stable right-left anti-Yetter-Drinfeld module. Similar calculations establish connections between other versions of anti-Yetter-Drinfeld modules and contramodules.

4. Hopf-cyclic homology of module coalgebras

Let C be a left H-module coalgebra. This means that C is a coalgebra and a left H-comodule such that, for all $c \in C$ and $h \in H$,

$$\Delta_C(h\cdot c) = h_{(1)}\cdot c_{(1)} \otimes h_{(2)}\cdot c_{(2)}, \qquad \varepsilon_C(h\cdot c) = \varepsilon_H(h)\varepsilon_C(c).$$

The multiple tensor product of C, $C^{\otimes n+1}$, is a left H-module by the diagonal action, that is

$$h\cdot(c^0 \otimes c^1 \otimes \ldots \otimes c^n) := h_{(1)}\cdot c^0 \otimes h_{(2)}\cdot c^1 \otimes \ldots \otimes h_{(n+1)}\cdot c^n.$$

Let M be a stable left-right anti-Yetter-Drinfeld contramodule. For all positive integers n, set $C_n^H(C,M) := \mathrm{Hom}_H(C^{\otimes n+1}, M)$ (left H-module maps), and, for all $0 \le i,j \le n$, define $d_i : C_n^H(C,M) \to C_{n-1}^H(C,M)$, $s_j : C_n^H(C,M) \to C_{n+1}^H(C,M)$, $t_n : C_n^H(C,M) \to C_n^H(C,M)$, by

$$\begin{aligned}
d_i(f)(c^0,\ldots,c^{n-1}) &= f(c^0,\ldots,\Delta_C(c^i),\ldots,c^{n-1}), \qquad 0 \le i < n,\\
d_n(f)(c^0,\ldots,c^{n-1}) &= \alpha\left(f\left({c^0}_{(2)}, c^1,\ldots,c^{n-1},(-)\cdot {c^0}_{(1)}\right)\right),\\
s_j(f)(c^0,\ldots,c^{n+1}) &= \varepsilon_C(c^{j+1}) f(c^0,\ldots,c^j,c^{j+2},\ldots,c^{n+1}),\\
t_n(f)(c^0,\ldots,c^n) &= \alpha\left(f\left(c^1,\ldots,c^n,(-)\cdot c^0\right)\right).
\end{aligned}$$

It is clear that all the maps s_j, d_i, $i < n$, are well-defined, i.e. they send left H-linear maps to left H-linear maps. That d_n and t_n are well-defined follows by the anti-Yetter-Drinfeld condition. To illustrate how the anti-Yetter-Drinfeld condition enters here we check that the t_n are well defined. For all $h \in H$,

$$\begin{aligned}
t_n(f)(h\cdot(c^0,\ldots,c^n)) &= t_n(f)(h_{(1)}\cdot c^0,\ldots,h_{(n+1)}\cdot c^n)\\
&= \alpha\left(f\left(h_{(2)}\cdot c^1,\ldots,h_{(n+1)}\cdot c^n,(-)h_{(1)}\cdot c^0\right)\right)\\
&= \alpha\left(f\left(h_{(2)}\cdot c^1,\ldots,h_{(n+1)}\cdot c^n,h_{(n+2)}S(h_{(n+3)})(-)h_{(1)}\cdot c^0\right)\right)\\
&= \alpha\left(h_{(2)}\cdot f\left(c^1,\ldots,c^n,S(h_{(3)})(-)h_{(1)}\cdot c^0\right)\right)\\
&= h\cdot\alpha\left(f\left(c^1,\ldots,c^n,(-)\cdot c^0\right)\right) = h\cdot t_n(f)(c^0,\ldots,c^n),
\end{aligned}$$

where the third equation follows by the properties of the antipode and counit, the fourth one is a consequence of the H-linearity of f, while the anti-Yetter-Drinfeld condition is used to derive the penultimate equality.

THEOREM 1. *Given a left H-module coalgebra C and a left-right stable anti-Yetter-Drinfeld contramodule M, $C_*^H(C,M)$ with the d_i, s_j, t_n defined above is a cyclic module.*

Proof. One needs to check whether the maps d_i, s_j, t_n satisfy the relations of a cyclic module; see e.g. [**12**, p. 203]. Most of the calculations are standard, we only display examples of those which make use of the contramodule axioms. For example,

$$\begin{aligned}
(t_{n-1}\circ d_{n-1})(f)(c^0,\ldots,c^{n-1}) &= \alpha\left(d_{n-1}(f)\left(c^1,\ldots,c^{n-1},(-)\cdot c^0\right)\right)\\
&= \alpha\left(f\left(c^1,\ldots,c^{n-1},\Delta_C\left((-)\cdot c^0\right)\right)\right)\\
&= \alpha\left(f\left(c^1,\ldots,c^{n-1},(-)_{(1)}\cdot {c^0}_{(1)},(-)_{(2)}\cdot {c^0}_{(2)}\right)\right)\\
&= \dot\alpha\left(\ddot\alpha\left(f\left(c^1,\ldots,c^{n-1},(\dot-)\cdot {c^0}_{(1)},(\ddot-)\cdot {c^0}_{(2)}\right)\right)\right)\\
&= \alpha\left(t_n(f)\left({c^0}_{(2)},c^1,\ldots,c^{n-1},(-)\cdot {c^0}_{(1)}\right)\right)\\
&= (d_n\circ t_n)(f)(c^0,\ldots,c^{n-1}),
\end{aligned}$$

where the third equality follows by the module coalgebra property of C, and the fourth one is a consequence of the associative law for contramodules. In a similar way, using compatibility of H-action on C with counits of H and C, and that $\alpha(\varepsilon_C(-)m) = m$, for all $m \in M$, one easily shows that $d_{n+1} \circ s_n$ is the identity map on $C_n^H(C, M)$. The stability of M is used to prove that t_n^{n+1} is the identity. Explicitly,

$$\begin{aligned} t_n^{n+1}(f)(c^0, \ldots, c^n) &= \alpha^{n+1}(f((-)\cdot c^0, \ldots, (-)\cdot c^n)) \\ &= \alpha(f((-)_{(1)}\cdot c^0, \ldots, (-)_{(n+1)}\cdot c^n)) = \alpha(r_{f(c^0,\ldots,c^n)}) = f(c^0, \ldots, c^n), \end{aligned}$$

where the second equality follows by the n-fold application of the associative law for contramodules, and the penultimate equality is a consequence of the H-linearity of f. The final equality follows by the stability of M. □

Let N be a right-left stable anti-Yetter-Drinfeld module, and $M = N^*$ be the corresponding left-right stable anti-Yetter-Drinfeld contramodule, then

$$C_n^H(C, M) = \mathrm{Hom}_H(C^{\otimes n+1}, \mathrm{Hom}(N, k)) \simeq \mathrm{Hom}(N \otimes_H C^{\otimes n+1}, k).$$

With this identification, the cyclic module $C_n^H(C, N^*)$ is obtained by applying functor $\mathrm{Hom}(-, k)$ to the cyclic module for N described in [**8**, Theorem 2.1].

5. Hopf-cyclic cohomology of module algebras

Let A be a left H-module algebra. This means that A is an algebra and a left H-module such that, for all $h \in H$ and $a, a' \in A$,

$$h\cdot(aa') = (h_{(1)}\cdot a)(h_{(2)}\cdot a), \qquad h\cdot 1_A = \varepsilon_H(h)1_A.$$

LEMMA 1. *Given a left H-module algebra A and a left H-contramodule M, $\mathrm{Hom}(A, M)$ is an A-bimodule with the left and right A-actions defined by*

$$(a\cdot f)(b) = f(ba), \qquad (f\cdot a)(b) = \alpha\left(f\left(((-)\cdot a)\, b\right)\right),$$

for all $a, b \in A$ and $f \in \mathrm{Hom}(A, M)$.

Proof. The definition of left A-action is standard, compatibility between left and right actions is immediate. To prove the associativity of the right A-action, take any $a, a', b \in A$ and $f \in \mathrm{Hom}(A, M)$, and compute

$$\begin{aligned} ((f\cdot a)\cdot a')(b) &= \dot\alpha\left(\ddot\alpha\left(f\left(\left((\ddot{-})\cdot a\right)\left((\dot{-})\cdot a'\right) b\right)\right)\right) \\ &= \alpha\left(f\left(\left((-)_{(1)}\cdot a\right)\left((-)_{(2)}\cdot a'\right) b\right)\right) \\ &= \alpha\left(f\left(((-)\cdot(aa'))\, b\right)\right) = ((aa')\cdot f)(b), \end{aligned}$$

where the second equality follows by the definition of a left H-contramodule, and the third one in a consequence of the module algebra property. The unitality of the right A-action follows by the triangle diagram for contramodules and the fact that $h\cdot 1_A = \varepsilon_H(h)1_A$. □

For an H-module algebra A, $A^{\otimes n+1}$ is a left H-module by the diagonal action

$$h\cdot(a^0 \otimes a^1 \otimes \ldots \otimes a^n) := h_{(1)}\cdot a^0 \otimes h_{(2)}\cdot a^1 \otimes \ldots \otimes h_{(n+1)}\cdot a^n.$$

Take a stable left-left anti-Yetter-Drinfeld contramodule M, set $C_H^n(A, M)$ to be the space of left H-linear maps $\mathrm{Hom}_H(A^{\otimes n+1}, M)$, and, for all $0 \le i, j \le n$, define

$\delta_i : C^{n-1}_H(A,M) \to C^n_H(A,M)$, $\sigma_j : C^{n+1}_H(A,M) \to C^n_H(A,M)$, $\tau_n : C^n_H(A,M) \to C^n_H(A,M)$, by

$$
\begin{aligned}
\delta_i(f)(a^0,\dots,a^n) &= f(a^0,\dots,a^{i-1},a^i a^{i+1},a^{i+2},\dots,a^n), \qquad 0 \le i < n,\\
\delta_n(f)(a^0,\dots,a^n) &= \alpha\left(f\left(((-)\cdot a^n)\,a^0,a^1,\dots,a^{n-1}\right)\right),\\
\sigma_j(f)(c^0,\dots,c^n) &= f(a^0,\dots,a^j,1_A,a^{j+1},\dots,a^n),\\
\tau_n(f)(a^0,\dots,a^n) &= \alpha\left(f\left((-)\cdot a^n,a^0,a^1,\dots,a^{n-1}\right)\right).
\end{aligned}
$$

Similarly to the module coalgebra case, the above maps are well-defined by the anti-Yetter-Drinfeld condition. Explicitly, using the aformentioned condition as well as the fact that the inverse of the antipode is the antipode for the co-opposite Hopf algebra, one computes

$$
\begin{aligned}
\tau_n(f)(h\cdot(a^0,\dots,a^n)) &= \alpha\left(f\left(((-)h_{(n+1)})\cdot a^n, h_{(1)}\cdot a^0, h_{(2)}\cdot a^1,\dots,h_{(n)}\cdot a^{n-1}\right)\right)\\
&= \alpha\left(f\left((h_{(2)}S^{-1}(h_{(1)})(-)h_{(n+3)})\cdot a^n, h_{(1)}\cdot a^0, h_{(2)}\cdot a^1,\dots,h_{(n+2)}\cdot a^{n-1}\right)\right)\\
&= \alpha\left(h_{(2)}\cdot f\left((S^{-1}(h_{(1)})(-)h_{(3)})\cdot a^n, a^0,\dots,a^{n-1}\right)\right) = h\cdot\tau_n(f)(a^0,\dots,a^n).
\end{aligned}
$$

Analogous calculations ensure that also δ_n is well-defined.

THEOREM 2. *Given a left H-module algebra A and a stable left-left anti-Yetter-Drinfeld contramodule M, $C^*_H(A,M)$ with the δ_i, σ_j, τ_n defined above is a (co)cyclic module.*

Proof. In view of Lemma 1 and taking into account the canonical isomorphism $\mathrm{Hom}(A^{\otimes n+1},M) \simeq \mathrm{Hom}(A^{\otimes n},\mathrm{Hom}(A,M))$,

$$
\mathrm{Hom}(A^{\otimes n+1},M) \ni f \mapsto \left[a^1\otimes a^2\otimes\dots\otimes a^n \mapsto f\left(-,a^1,a^2,\dots,a^n\right)\right],
$$

the simplicial part comes from the standard A-bimodule cohomology. Thus only the relations involving τ_n need to be checked. In fact only the equalities $\tau_n \circ \delta_n = \delta_{n-1}\circ\tau_{n-1}$ and $\tau_n^{n+1} = \mathrm{id}$ require one to make use of definitions of a module algebra and a left contramodule. In the first case, for all $f \in C^n_H(A,M)$,

$$
\begin{aligned}
(\tau_n\circ\delta_n)(f)(a^0,\dots,a^n) &= \dot\alpha\left(\ddot\alpha\left(f\left(\left((\ddot{-})\cdot a^{n-1}\right)\left((\dot{-})\cdot a^n\right),a^0,\dots,a^{n-2}\right)\right)\right)\\
&= \alpha\left(f\left(\left((-)_{(1)}\cdot a^{n-1}\right)\left((-)_{(2)}\cdot a^n\right),a^0,\dots,a^{n-2}\right)\right)\\
&= \alpha\left(f\left((-)\cdot\left(a^{n-1}a^n\right),a^0,\dots,a^{n-2}\right)\right)\\
&= (\delta_{n-1}\circ\tau_{n-1})(f)(a^0,\dots,a^n),
\end{aligned}
$$

where the second equality follows by the associative law for left contramodules and the third one by the definition of a left H-module algebra. The equality $\tau_n^{n+1} = \mathrm{id}$ follows by the associative law of contramodules, the definition of left H-action on $A^{\otimes n+1}$, and by the stability of anti-Yetter-Drinfeld contramodules. □

In the case of a contramodule M constructed on the dual vector space of a stable right-right anti-Yetter-Drinfeld module N, the complex described in Theorem 2 is the right-right version of Hopf-cyclic complex of a left module algebra with coefficients in N discussed in [**8**, Theorem 2.2].

6. Anti-Yetter-Drinfeld contramodules and hom-connections

Anti-Yetter-Drinfeld modules over a Hopf algebra H can be understood as comodules of an H-coring; see [**2**] for explicit formulae and [**4**] for more information about corings. These are corings with a group-like element, and thus their comodules can be interpreted as modules with a flat connection; see [**2**] for a review. Consequently, anti-Yetter-Drinfeld modules are modules with a flat connection (with respect to a suitable differential structure); see [**10**].

Following similar line of argument anti-Yetter-Drinfeld contramodules over a Hopf algebra H can be understood as contramodules of an H-coring. This is a coring of an entwining type, as a vector space built on $H\otimes H$, and its form is determined by the anti-Yetter-Drinfeld compatibility conditions between action and contra-action. The coring $H\otimes H$ has a group-like element $1_H\otimes 1_H$, which induces a differential graded algebra structure on tensor powers of the kernel of the counit of $H\otimes H$. As explained in [**3**, Section 3.9] contramodules of a coring with a group-like element correspond to *flat hom-connections*. Thus, in particular, anti-Yetter-Drinfeld contramodules are flat hom-connections. We illustrate this discussion by the example of right-right anti-Yetter-Drinfeld contramodules.

First recall the definition of hom-connections from [**3**]. Fix a differential graded algebra ΩA over an algebra A. A *hom-connection* is a pair (M, ∇_0), where M is a right A-module and ∇_0 is a k-linear map from the space of right A-module homomorphisms $\mathrm{Hom}_A(\Omega^1 A, M)$ to M, $\nabla_0 : \mathrm{Hom}_A(\Omega^1 A, M) \to M$, such that, for all $a \in A$, $f \in \mathrm{Hom}_A(\Omega^1 A, M)$,

$$\nabla_0(f\cdot a) = \nabla_0(f)\cdot a + f(da),$$

where $f\cdot a \in \mathrm{Hom}_A(\Omega^1 A, M)$ is given by $f\cdot a : \omega \mapsto f(a\omega)$, and $d : \Omega^* A \to \Omega^{*+1} A$ is the differential. Define $\nabla_1 : \mathrm{Hom}_A(\Omega^2 A, M) \to \mathrm{Hom}_A(\Omega^1 A, M)$, by $\nabla_1(f)(\omega) = \nabla_0(f\cdot\omega) + f(d\omega)$, where, for all $f \in \mathrm{Hom}_A(\Omega^2 A, M)$, the map $f\cdot\omega \in \mathrm{Hom}_A(\Omega^1 A, M)$ is given by $\omega' \mapsto f(\omega\omega')$. The composite $F = \nabla_0 \circ \nabla_1$ is called the *curvature* of (M, ∇_0). The hom-connection (M, ∇_0) is said to be *flat* provided its curvature is equal to zero. Hom-connections are non-commutative versions of *right connections* or *co-connections* studied in [**13**, Chapter 4 § 5], [**16**], [**17**].

Consider a Hopf algebra H with a bijective antipode, and define an H-coring $\mathcal{C} = H\otimes H$ as follows. The H bimodule structure of $\mathcal{C}$ is given by

$$h\cdot(h'\otimes h'') = h_{(1)}h'S^{-1}(h_{(3)})\otimes h_{(2)}h'', \qquad (h'\otimes h'')\cdot h = h'\otimes h''h,$$

the coproduct is $\Delta_H\otimes\mathrm{id}_H$ and counit $\varepsilon_H\otimes\mathrm{id}_H$. Take a right H-module M. The identification of right H-linear maps $H\otimes H \to M$ with $\mathrm{Hom}(H, M)$ allows one to identify right contramodules of the H-coring $\mathcal{C}$ with right-right anti-Yetter-Drinfeld contramodules over H.

The kernel of the counit in $\mathcal{C}$ coincides with $H^+\otimes H$, where $H^+ = \ker\varepsilon_H$. Thus the associated differential graded algebra over H is given by $\Omega^n H = (H^+\otimes H)^{\otimes_H n} \simeq (H^+)^{\otimes n}\otimes H$, with the differential given on elements h of H and one-forms $h'\otimes h \in H^+\otimes H$ by

$$dh = 1_H\otimes h - h_{(1)}S^{-1}(h_{(3)})\otimes h_{(2)},$$

$$d(h'\otimes h) = 1_H\otimes h'\otimes h - h'_{(1)}\otimes h'_{(2)}\otimes h + h'\otimes h_{(1)}S^{-1}(h_{(3)})\otimes h_{(2)}.$$

Take a right-right anti-Yetter-Drinfeld contramodule M over a Hopf algebra H and identify $\mathrm{Hom}_H(\Omega^1 H, M)$ with $\mathrm{Hom}(H^+, M)$. For any $f \in \mathrm{Hom}(H^+, M)$, set $\bar{f} : H \to M$ by $\bar{f}(h) = f(h - \varepsilon_H(h)1_H)$, and then define

$$\nabla_0 : \mathrm{Hom}(H^+, M) \to M, \qquad \nabla_0(f) = \alpha(\bar{f}).$$

(M, ∇_0) is a flat hom-connection with respect to the differential graded algebra ΩH.

7. Final remarks

In this note a new class of coefficients for the Hopf-cyclic homology was introduced. It is an open question to what extent Hopf-cyclic homology with coefficients in anti-Yetter-Drinfeld contramodules is useful in studying problems arising in (non-commutative) geometry. The answer is likely to depend on the supply of (calculable) examples, such as those coming from the transverse index theory of foliations (which motivated the introduction of Hopf-cyclic homology in [**5**]). It is also likely to depend on the structure of Hopf-cyclic homology with contramodule coefficients. One can easily envisage that, in parallel to the theory with anti-Yetter-Drinfeld module coefficients, the cyclic theory described in this note admits cup products (in the case of module coefficients these were foreseen in [**8**] and constructed in [**11**]) or homotopy formulae of the type discovered for anti-Yetter-Drinfeld modules in [**14**]. Alas, these topics go beyond the scope of this short note. The author is convinced, however, of the worth-whileness of investigating them further.

References

[1] Böhm, G., Brzeziński, T., Wisbauer, R., *Monads and comonads in module categories,* arxiv:0804.1460 (2008).

[2] Brzeziński, T., *Flat connections and (co)modules,* in *New Techniques in Hopf Algebras and Graded Ring Theory,* S. Caenepeel and F. Van Oystaeyen (eds.), Universa Press, Wetteren, pp. 35–52, (2007).

[3] Brzeziński, T., *Non-commutative connections of the second kind,* arxiv:0802.0445, J. Algebra Appl., in press (2008).

[4] Brzeziński, T., Wisbauer, R., *Corings and Comodules,* Cambridge University Press, Cambridge (2003). Erratum: http://www-maths.swan.ac.uk/staff/tb/Corings.htm

[5] Connes, A., Moscovici, H., *Cyclic cohomology and Hopf algebra symmetry,* Lett. Math. Phys. 52, 1–28 (2000).

[6] Eilenberg, S., Moore, J.C., *Foundations of relative homological algebra,* Mem. Amer. Math. Soc. 55 (1965).

[7] Hajac, P.M., Khalkhali, M., Rangipour, B., Sommerhäuser, Y., *Stable anti-Yetter-Drinfeld modules,* C. R. Math. Acad. Sci. Paris, 338, 587–590 (2004).

[8] Hajac, P.M., Khalkhali, M., Rangipour, B., Sommerhäuser, Y., *Hopf-cyclic homology and cohomology with coefficients,* C. R. Math. Acad. Sci. Paris, 338, 667–672 (2004).

[9] Jara, P., Ştefan, D., *Cyclic homology of Hopf-Galois extensions and Hopf algebras,* Proc. London Math. Soc. 93, 138–174 (2006).

[10] Kaygun, A., Khalkhali, M., *Hopf modules and noncommutative differential geometry,* Lett. Math. Phys. 76, 77–91 (2006).

[11] Khalkhali, M., Rangipour, B., *Cup products in Hopf-cyclic cohomology,* C. R. Math. Acad. Sci. Paris 340, 9–14 (2005).

[12] Loday, J.-L., *Cyclic Homology* 2nd ed., Springer, Berlin (1998).

[13] Manin, Yu.I., *Gauge Field Theory and Complex Geometry,* Springer-Verlag, Berlin (1988).

[14] Moscovici, H., Rangipour, B., *Cyclic cohomology of Hopf algebras of transverse symmetries in codimension 1,* Adv. Math. 210, 323–374 (2007).

[15] Positselski, L., *Homological algebra of semimodules and semicontramodules,* arXiv:0708.3398 (2007).

[16] Vinogradov, M.M., *Co-connections and integral forms,* (Russian) Dokl. Akad. Nauk 338, 295–297 (1994); English translation in Russian Acad. Sci. Dokl. Math. 50, 229–233 (1995).

[17] Vinogradov, M.M., *Remarks on algebraic Lagrangian formalism,* Acta Appl. Math. 49, 331–338 (1997).

Department of Mathematics, Swansea University, Singleton Park, Swansea SA2 8PP, U.K.

E-mail address: T.Brzezinski@swansea.ac.uk

Moduli Spaces of Dirac Operators for Finite Spectral Triples

Branimir Ćaćić

ABSTRACT. The structure theory of finite real spectral triples developed by Krajewski and by Paschke and Sitarz is generalised to allow for arbitrary KO-dimension and the failure of orientability and Poincaré duality, and moduli spaces of Dirac operators for such spectral triples are defined and studied. This theory is then applied to recent work by Chamseddine and Connes towards deriving the finite spectral triple of the noncommutative-geometric Standard Model.

1. Introduction

From the time of Connes's 1995 paper [6], spectral triples with finite-dimensional $*$-algebra and Hilbert space, or *finite spectral triples*, have been central to the noncommutative-geometric (NCG) approach to the Standard Model of elementary particle physics, where they are used to encode the fermionic physics. As a result, they have been the focus of considerable research activity.

The study of finite spectral triples began in earnest with papers by Paschke and Sitarz [20] and by Krajewski [18], first released nearly simultaneously in late 1996 and early 1997, respectively, which gave detailed accounts of the structure of finite spin geometries, *i.e.* of finite real spectral triples of KO-dimension $0 \bmod 8$ satisfying orientability and Poincaré duality. In their approach, the study of finite spectral triples is reduced, for the most part, to the study of *multiplicity matrices*, integer-valued matrices that explicitly encode the underlying representation-theoretic structure. Krajewski, in particular, defined what are now called *Krajewski diagrams* to facilitate the classification of such spectral triples. Iochum, Jureit, Schücker, and Stephan have since undertaken a programme of classifying Krajewski diagrams for finite spectral triples satisfying certain additional physically desirable assumptions [12–14, 22] using combinatorial computations [17], with the aim of fixing the finite spectral triple of the Standard Model amongst all other such triples.

However, there were certain issues with the then-current version of the NCG Standard Model, including difficulty with accomodating massive neutrinos and the so-called fermion doubling problem, that were only to be resolved in the 2006 papers by Connes [7] and by Chamseddine, Connes and Marcolli [4], which use the Euclidean signature of earlier papers, and by Barrett [1], which instead uses Lorentzian signature; we restrict our attention to the Euclidean signature approach of [7] and [4], which has more recently been set forth in the monograph [8] of

2000 *Mathematics Subject Classification.* Primary 58J42; Secondary 58B34, 58D27, 81R60.

Connes and Marcolli. The finite spectral triple of the current version has KO-dimension 6 mod 8 instead of 0 mod 8, fails to be orientable, and only satisfies a certain modified version of Poincaré duality. It also no longer satisfies S^0-reality, another condition that holds for the earlier finite geometry of [6], though only because of the Dirac operator. Jureit, and Stephan [15, 16] have since adopted the new value for the KO-dimension, but further assume orientability and Poincaré duality. As well, Stephan [25] has proposed an alternative finite spectral triple for the current NCG Standard Model with the same physical content but satisfying Poincaré duality; it also just fails to be S^0-real in the same manner as the finite geometry of [4]; in the same paper, Stephan also discusses non-orientable finite spectral triples.

More recently, Chamseddine and Connes [2, 3] have sought a purely algebraic method of isolating the finite spectral triple of the NCG Standard Model, by which they have obtained the correct $*$-algebra, Hilbert space, grading and real structure using a small number of fairly elementary assumptions. In light of these successes, it would seem reasonable to try to view this new approach of Chamseddine and Connes through the lens of the structure theory of Krajewski and Paschke–Sitarz, at least in order to understand better their method and the assumptions involved. This, however, would require adapting that structure theory to handle the failure of orientability and Poincaré duality, yielding the initial motivation of this work.

To that end, we provide, for the first time, a comprehensive account of the structure theory of Krajewski and Paschke–Sitarz for finite real spectral triples of arbitrary KO-dimension, without the assumptions of orientability or Poincaré duality; this consists primarily of straightforward generalisations of the results and techniques of [20] and [18]. In this light, the main features of the approach presented here are the following:

(1) A finite real spectral triple with algebra $\mathcal{A}$ is to be viewed as an $\mathcal{A}$-bimodule with some additional structure, together with a choice of Dirac operator compatible with that structure.
(2) For fixed algebra $\mathcal{A}$, an $\mathcal{A}$-bimodule is entirely characterised by its multiplicity matrix (in the ungraded case) or matrices (in the graded case), which also completely determine(s) what sort of additional structure the bimodule can admit; this additional structure is then unique up to unitary equivalence.
(3) The form of suitable Dirac operators for an $\mathcal{A}$-bimodule with real structure is likewise determined completely by the multiplicity matrix or matrices of the bimodule and the choice of additional structure.

However, we do not discuss Krajewski diagrams, though suitable generalisation thereof should follow readily from the generalised structure theory for Dirac operators.

Once we view a real spectral triple as a certain type of bimodule together with a *choice* of suitable Dirac operator, it then becomes natural to consider moduli spaces of suitable Dirac operators, up to unitary equivalence, for a bimodule with fixed additional structure, yielding finite real spectral triples of the appropriate KO-dimension. The construction and study of such moduli spaces of Dirac operators first appear in [4], though the focus there is on the sub-moduli space of Dirac operators commuting with a certain fixed subalgebra of the relevant $*$-algebra. Our last point above almost immediately leads us to relatively concrete expressions for general moduli spaces of Dirac operators, which also appear here for the first time. Multiplicity matrices and moduli spaces of Dirac operators are then worked out for the bimodules appearing in the Chamseddine–Connes–Marcolli formulation of the NCG Standard Model [4,8] as examples.

Finally, we apply these methods to the work of Chamseddine and Connes [2,3], offering concrete proofs and some generalisations of their results. In particular, the choices determining the finite geometry of the current NCG Standard Model within their framework are made explicit.

This work, a revision of the author's qualifying year project (master's thesis equivalent) at the Bonn International Graduate School in Mathematics (BIGS) at the University of Bonn, is intended as a first step towards a larger project of investigating in generality the underlying noncommutative-geometric formalism for field theories found in the NCG Standard Model, with the aim of both better understanding current versions of the NCG Standard Model and facilitating the further development of the formalism itself.

The author would like to thank his supervisor, Matilde Marcolli, for her extensive comments and for her advice, support, and patience, Tobias Fritz for useful comments and corrections, and George Elliott for helpful conversations. The author also gratefully acknowledges the financial and administrative support of BIGS and of the Max Planck Institute for Mathematics, as well as the hospitality and support of the Department of Mathematics at the California Institute of Technology and of the Fields Institute.

2. Preliminaries and Definitions

2.1. Real C^*-algebras. In light of their relative unfamiliarity compared to their complex counterparts, we begin with some basic facts concerning real C^*-algebras.

First, recall that a *real $*$-algebra* is a real associative algebra $\mathcal{A}$ together with an *involution* on $\mathcal{A}$, namely an antihomomorphism $*$ satisfying $*^2 = \mathrm{id}$, and that the *unitalisation* of a real $*$-algebra $\mathcal{A}$ is the unital real $*$-algebra $\tilde{\mathcal{A}}$ defined to be $\mathcal{A} \oplus \mathbb{R}$ as a real vector space, together with the multiplication $(a, \alpha)(b, \beta) := (ab + \alpha b + \beta a, \alpha\beta)$ for a, $b \in \mathcal{A}$, α, $\beta \in \mathbb{R}$ and the involution $\star \oplus \mathrm{id}_{\mathbb{R}}$. Note that if $\mathcal{A}$ is already unital, then $\tilde{\mathcal{A}}$ is simply $\mathcal{A} \oplus \mathbb{R}$.

DEFINITION 2.1. A *real C^*-algebra* is a real $*$-algebra $\mathcal{A}$ endowed with a norm $\|\cdot\|$ making $\mathcal{A}$ a real Banach algebra, such that the following two conditions hold:

(1) $\forall a \in \mathcal{A}$, $\|a^*a\| = \|a\|^2$ (*C^*-identity*);
(2) $\forall a \in \tilde{\mathcal{A}}$, $1 + a^*a$ is invertible in $\tilde{\mathcal{A}}$ (*symmetry*).

The symmetry condition is redundant for complex C^*-algebras, but not for real C^*-algebras. Indeed, consider $\mathbb{C}$ as a real algebra together with the trivial involution $* = \mathrm{id}$ and the usual norm $\|\zeta\| = |\zeta|$, $\zeta \in \mathbb{C}$. Then $\mathbb{C}$ with this choice of involution and norm yields a real Banach $*$-algebra satisfying the C^*-identity but not symmetry, for $1 + i^*i = 0$ is certainly not invertible in $\tilde{\mathbb{C}} = \mathbb{C} \oplus \mathbb{R}$.

Now, in the finite-dimensional case, one can give a complete description of real C^*-algebras, which we shall use extensively in what follows:

THEOREM 2.2 (Wedderburn's theorem for real C^*-algebras [11]). *Let $\mathcal{A}$ be a finite-dimensional real C^*-algebra. Then*

$$\mathcal{A} \cong \bigoplus_{i=1}^{N} M_{n_i}(\mathbb{K}_i), \tag{2.1}$$

where $\mathbb{K}_i = \mathbb{R}$, $\mathbb{C}$, or $\mathbb{H}$, and $n_i \in \mathbb{N}$. Moreover, this decomposition is unique up to permutation of the direct summands.

Note, in particular, that a finite-dimensional real C^*-algebra is necessarily unital.

Given a finite-dimensional real C^*-algebra $\mathcal{A}$ with fixed *Wedderburn decomposition* $\oplus_{i=1}^{N} M_{n_i}(\mathbb{K}_i)$ we can associate to $\mathcal{A}$ a finite dimensional complex C^*-algebra $\mathcal{A}_\mathbb{C}$, the *complex form* of $\mathcal{A}$, by setting

$$\mathcal{A}_\mathbb{C} := \bigoplus_{i=1}^{N} M_{m_i}(\mathbb{C}), \tag{2.2}$$

where $m_i = 2n_i$ if $\mathbb{K}_i = \mathbb{H}$, and $m_i = n_i$ otherwise. Then $\mathcal{A}$ can be viewed as a real $*$-subalgebra of $\mathcal{A}_\mathbb{C}$ such that $\mathcal{A}_\mathbb{C} = \mathcal{A} + i\mathcal{A}$, that is, as a *real form* of $\mathcal{A}_\mathbb{C}$. Here, $\mathbb{H}$ is considered as embedded in $M_2(\mathbb{C})$ by

$$\zeta_1 + j\zeta_2 \mapsto \begin{pmatrix} \zeta_1 & \zeta_2 \\ -\overline{\zeta_2} & \overline{\zeta_1} \end{pmatrix},$$

for $\zeta_1, \zeta_2 \in \mathbb{C}$.

In what follows, we will consider only finite-dimensional real C^*-algebras with fixed Wedderburn decomposition.

2.2. Representation theory. In keeping with the conventions of noncommutative differential geometry, we shall consider $*$-representations of real C^*-algebras on complex Hilbert spaces. Recall that such a (left) representation of a real C^*-algebra $\mathcal{A}$ consists of a complex Hilbert space $\mathcal{H}$ together with a $*$-homomorphism $\lambda : \mathcal{A} \to \mathcal{L}(\mathcal{H})$ between real C^*-algebras. Similarly, a *right representation* of $\mathcal{A}$ is defined to be a complex Hilbert space $\mathcal{H}$ together with a $*$-*antihomomorphism* $\rho : \mathcal{A} \to \mathcal{L}(\mathcal{H})$ between real C^*-algebras. For our purposes, then, an $\mathcal{A}$-*bimodule* consists of a complex Hilbert space $\mathcal{H}$ together with a left $*$-representation λ and a right $*$-representation ρ that commute, *i.e.* such that $[\lambda(a), \rho(b)] = 0$ for all a, $b \in \mathcal{A}$. In what follows, we will consider only finite-dimensional representations and hence only finite-dimensional bimodules; since finite-dimensional C^*-algebras are always unital, we shall require all representations to be unital as well.

Now, given a left [right] representation $\alpha = (\mathcal{H}, \pi)$ of an algebra $\mathcal{A}$, one can define its *transpose* to be the right [left] representation $\alpha^T = (\mathcal{H}^*, \pi^T)$, where $\pi^T(a) := \pi(a)^T$ for all $a \in \mathcal{A}$. Note that for any left or right representation α, $(\alpha^T)^T$ can naturally be identified with α itself. In the case that $\mathcal{H} = \mathbb{C}^N$, we shall identify $\mathcal{H}^*$ with $\mathcal{H}$ by identifying the standard ordered basis on $\mathcal{H}$ with the corresponding dual basis on $\mathcal{H}^*$. The notion of the transpose of a representation allows us to reduce discussion of right representations to that of left representations.

Since real C^*-algebras are semisimple, any left representation can be written as a direct sum of irreducible representations, unique up to permutation of the direct summands, and hence any right representation can be written as a direct sum of transposes of irreducible representations, again unique up to permutation of the direct summands.

DEFINITION 2.3. The *spectrum* $\widehat{\mathcal{A}}$ of a real C^*-algebra $\mathcal{A}$ is the set of unitary equivalence classes of irreducible representations of $\mathcal{A}$.

Now, let $\mathcal{A}$ be a real C^*-algebra with Wedderburn decomposition $\oplus_{i=1}^{N} M_{k_i}(\mathbb{K}_i)$. Then

$$\widehat{\mathcal{A}} = \bigsqcup_{i=1}^{N} \widehat{M_{k_i}(\mathbb{K}_i)}, \tag{2.3}$$

where the embedding of $\widehat{M_{k_i}(\mathbb{K}_i)}$ in $\widehat{\mathcal{A}}$ is given by composing the representation maps with the projection of $\mathcal{A}$ onto the direct summand $M_{k_i}(\mathbb{K}_i)$. The building blocks for $\widehat{\mathcal{A}}$ are as follows:

(1) $\widehat{M_n(\mathbb{R})} = \{[(\mathbb{C}^n, \lambda)]\}$,
(2) $\widehat{M_n(\mathbb{C})} = \{[(\mathbb{C}^n, \lambda)], [(\mathbb{C}^N, \overline{\lambda})]\}$,
(3) $\widehat{M_n(\mathbb{H})} = \{[(\mathbb{C}^{2n}, \lambda)]\}$,

where $\lambda(a)$ denotes left multiplication by a and $\overline{\lambda}(a)$ denotes left multiplication by $\overline{a}$.

DEFINITION 2.4. Let $\mathcal{A}$ be a real C^*-algebra, and let $\alpha \in \widehat{\mathcal{A}}$. We shall call α *conjugate-linear* if it arises from the conjugate-linear irreducible representation $(a \mapsto \overline{a}, \mathbb{C}^{n_i})$ of a direct summand of $\mathcal{A}$ of the form $M_{n_i}(\mathbb{C})$; otherwise we shall call it *complex-linear*.

Thus, a representation α of the real C^*-algebra $\mathcal{A}$ extends to a $\mathbb{C}$-linear $*$-representation of $\mathcal{A}_{\mathbb{C}}$ if and only if α is the sum of complex-linear irreducible representations of $\mathcal{A}$.

Finally, for an individual direct summand $M_{k_i}(\mathbb{K}_i)$ of $\mathcal{A}$, let e_i denote its unit, n_i the dimension of its irreducible representations (which is therefore equal to $2k_i$ if $\mathbb{K}_i = \mathbb{H}$, and to k_i itself otherwise), $\mathbf{n}_i$ its complex-linear irreducible representation, and, if $\mathbb{K}_i = \mathbb{C}$, $\overline{\mathbf{n}}_i$ its conjugate-linear irreducible representation. We define a strict ordering $<$ on $\widehat{\mathcal{A}}$ by setting $\alpha < \beta$ whenever $\alpha \in \widehat{M_{n_i}(\mathbb{K}_i)}$, $\beta \in \widehat{M_{n_j}(\mathbb{K}_j)}$ for $i < j$, and by setting $\mathbf{n}_i < \overline{\mathbf{n}}_i$ in the case that $\mathbb{K}_i = \mathbb{C}$. Note that the ordering depends on the choice of Wedderburn decomposition, *i.e.* on the choice of ordering of the direct summands. Let S denote the cardinality of $\widehat{\mathcal{A}}$. We shall identify $M_S(\mathbb{R})$ with the real algebra of functions $\widehat{\mathcal{A}}^2 \to \mathbb{R}$, and hence index the standard basis $\{E_{\alpha\beta}\}$ of $M_S(\mathbb{R})$ by $\widehat{\mathcal{A}}^2$.

2.3. Bimodules and spectral triples. Let us now turn to spectral triples. Recall that we are considering only finite-dimensional algebras and representations (*i.e.* Hilbert spaces), so that we are dealing only with what are termed *finite* or *discrete* spectral triples.

Let $\mathcal{H}$ and $\mathcal{H}'$ be $\mathcal{A}$-bimodules. We shall denote by $\mathcal{L}_{\mathcal{A}}^{\mathrm{L}}(\mathcal{H}, \mathcal{H}')$, $\mathcal{L}_{\mathcal{A}}^{\mathrm{R}}(\mathcal{H}, \mathcal{H}')$, and $\mathcal{L}_{\mathcal{A}}^{\mathrm{LR}}(\mathcal{H}, \mathcal{H}')$ the subspaces of $\mathcal{L}(\mathcal{H}, \mathcal{H}')$ consisting of left $\mathcal{A}$-linear, right $\mathcal{A}$-linear, and left and right $\mathcal{A}$-linear operators, respectively. In the case that $\mathcal{H}' = \mathcal{H}$, we shall write simply $\mathcal{L}_{\mathcal{A}}^{\mathrm{L}}(\mathcal{H})$, $\mathcal{L}_{\mathcal{A}}^{\mathrm{R}}(\mathcal{H})$ and $\mathcal{L}_{\mathcal{A}}^{\mathrm{LR}}(\mathcal{H})$. If N is a subalgebra or linear subspace of a real or complex C^*-algebra, we shall denote by N_{sa} the real linear subspace of N consisting of the self-adjoint elements of N, and we shall denote by $\mathrm{U}(N)$ set of unitary elements of N. Finally, for operators A and B on a Hilbert space, we shall denote their anticommutator $AB + BA$ by $\{A, B\}$.

2.3.1. *Conventional definitions.* We begin by recalling the standard definitions for spectral triples of various forms. Since we are working with the finite case, all analytical requirements become redundant, leaving behind only the algebraic aspects of the definitions.

The following definition first appeared in a 1995 paper [5] by Connes:

DEFINITION 2.5. A *spectral triple* is a triple $(\mathcal{A}, \mathcal{H}, D)$, where:

- $\mathcal{A}$ is a unital real or complex $*$-algebra;
- $\mathcal{H}$ is a complex Hilbert space on which $\mathcal{A}$ has a left representation $\lambda : \mathcal{A} \to \mathcal{L}(\mathcal{H})$;
- D, the *Dirac operator*, is a self-adjoint operator on $\mathcal{H}$.

Moreover, if there exists a $\mathbb{Z}/2\mathbb{Z}$-grading γ on $\mathcal{H}$ (*i.e.* a self-adjoint unitary on $\mathcal{H}$) such that:

(1) $[\gamma, \lambda(a)] = 0$ for all $a \in \mathcal{A}$,
(2) $\{\gamma, D\} = 0$;

then the spectral triple is said to be *even*. Otherwise, it is said to be *odd*.

In the context of the general definition for spectral triples, a finite spectral triple necessarily has metric dimension 0.

In a slightly later paper [6], Connes defines the additional structure on spectral triples necessary for defining the noncommutative spacetime of the NCG Standard Model; indeed, the same paper also contains the first version of the NCG Standard Model to use the language of spectral triples, in the form of a reformulation of the so-called Connes-Lott model.

DEFINITION 2.6. A spectral triple $(\mathcal{A}, \mathcal{H}, D)$ is called a *real spectral triple of KO-dimension* $n \bmod 8$ if, in the case of n even, it is an even spectral triple, and if there exists an antiunitary $J : \mathcal{H} \to \mathcal{H}$ such that:

(1) J satisfies $J^2 = \varepsilon$, $JD = \varepsilon' DJ$ and $J\gamma = \varepsilon'' \gamma J$ (in the case of even n), where ε, ε', $\varepsilon'' \in \{-1, 1\}$ depend on $n \bmod 8$ as follows:

n	0	1	2	3	4	5	6	7
ε	1	1	−1	−1	−1	−1	1	1
ε'	1	−1	1	1	1	−1	1	1
ε''	1		−1		1		−1	

(2) The *order zero condition* is satisfied, namely $[\lambda(a), J\lambda(b)J^*] = 0$ for all a, $b \in \mathcal{A}$;
(3) The *order one condition* is satisfied, namely $[[D, \lambda(a)], J\lambda(b)J^*] = 0$ for all a, $b \in \mathcal{A}$.

Moreover, if there exists a self-adjoint unitary ϵ on $\mathcal{H}$ such that:

(1) $[\epsilon, \lambda(a)] = 0$ for all $a \in \mathcal{A}$;
(2) $[\epsilon, D] = 0$;
(3) $\{\epsilon, J\} = 0$;
(4) $[\epsilon, \gamma] = 0$ (even case);

then the real spectral triple is said to be S^0-*real*.

REMARK 2.7 (Krajewski [18, §2.2], Paschke–Sitarz [20, Obs. 1]). If $(\mathcal{A}, \mathcal{H}, D)$ is a real spectral triple, then the order zero condition is equivalent to the statement that $\mathcal{H}$ is an $\mathcal{A}$-bimodule for the usual left action λ and the right action $\rho : a \mapsto J\lambda(a^*)J^*$.

It was commonly assumed until fairly recently that the finite geometry of the NCG Standard Model should be S^0-real. Though the current version of the NCG Standard Model no longer makes such an assumption [4, 7], we shall later see that its finite geometry can still be seen as satisfying a weaker version of S^0-reality.

2.3.2. *Structures on bimodules.* In light of the above remark, the order one condition, the strongest algebraic condition placed on Dirac operators for real spectral triples, should be viewed more generally as a condition applicable to operators on bimodules [18, §2.4]. This then motivates our point of view that a finite real spectral triple $(\mathcal{A}, \mathcal{H}, D)$ should be viewed rather as an $\mathcal{A}$-bimodule with additional structure, together with a Dirac operator satisfying the order one condition that is compatible with that additional structure. We therefore begin by defining a suitable notion of "additional structure" for bimodules.

DEFINITION 2.8. A *bimodule structure* P consists of the following data:

- A set $\mathcal{P} = \mathcal{P}_\gamma \sqcup \mathcal{P}_J \sqcup \mathcal{P}_\epsilon$, where each set $\mathcal{P}_X$ is either empty or the singleton $\{X\}$, and where $\mathcal{P}_\epsilon$ is non-empty only if $\mathcal{P}_J$ is non-empty;
- If $\mathcal{P}_J$ is non-empty, a choice of *KO-dimension* n mod 8, where n is even if and only if $\mathcal{P}_\gamma$ is non-empty.

In particular, we call a structure P:

- *odd* if $\mathcal{P}$ is empty;
- *even* if $\mathcal{P} = \mathcal{P}_\gamma = \{\gamma\}$;
- *real* if $\mathcal{P}_J$ is non-empty and $\mathcal{P}_\epsilon$ is empty
- *S^0-real* if $\mathcal{P}_\epsilon$ is non-empty.

Finally, if P is a graded structure, we call γ the *grading*, and if P is real or S^0-real, we call J the *charge conjugation*.

Since this notion of KO-dimension is meant to correspond with the usual KO-dimension of a real spectral triple, we assign to each real or S^0-real structure P of KO-dimension n mod 8 constants ε, ε' and, in the case of even n, ε'', according to the table in Definition 2.6.

We now define the *structure algebra* of a structure P to be the real associative algebra with generators $\mathcal{P}$ and relations, as applicable,

$$\gamma^2 = 1, \quad J^2 = \varepsilon, \quad \epsilon^2 = 1; \gamma J = \varepsilon'' J\gamma, \quad [\gamma, \epsilon] = 0, \quad \{\epsilon, J\} = 0.$$

DEFINITION 2.9. An $\mathcal{A}$-bimodule $\mathcal{H}$ is said to have structure P whenever it admits a faithful representation of the structure algebra of P such that, when applicable, γ and ϵ are represented by self-adjoint unitaries in $\mathcal{L}_{\mathcal{A}}^{\mathrm{LR}}(\mathcal{H})$, and J is represented by an antiunitary on $\mathcal{H}$ such that

$$\forall a \in \mathcal{A}, \quad \rho(a) = J\lambda(a^*)J. \tag{2.4}$$

Note that a S^0-real bimodule can always be considered as a real bimodule, and a real bimodule of even [odd] KO-dimension can always be considered as an even [odd] bimodule. Note also that an even bimodule is simply a graded bimodule such that the algebra acts from both left and right by degree 0 operators, and the grading itself respects the Hilbert space structure; an odd bimodule is then simply an ungraded bimodule. We use the terms "even" and "odd" so as to keep the terminology consistent with that for spectral triples.

Note also that for a real or S^0-real structure P, the structure algebra of P is independent of the value of ε'. Thus the notions of real [S^0-real] $\mathcal{A}$-bimodule with KO-dimension 1 mod 8 and 7 mod 8 are identical, as are the notions of [S^0-real] $\mathcal{A}$-bimodule with KO-dimension 3 mod 8 and 5 mod 8; again, we make the distinction with an eye to the discussion of Dirac operators (and hence of spectral triples) later on.

Now, a *unitary equivalence* of $\mathcal{A}$-bimodules $\mathcal{H}$ and $\mathcal{H}'$ with structure P is a unitary equivalence of $\mathcal{A}$-bimodules (*i.e.* a unitary element of $\mathcal{L}_{\mathcal{A}}^{\mathrm{LR}}(\mathcal{H},\mathcal{H}')$) that effects unitary equivalence of the representations of the structure algebra of P. We denote the set of all such unitary equivalences $\mathcal{H} \to \mathcal{H}'$ by $\mathrm{U}_{\mathcal{A}}^{\mathrm{LR}}(\mathcal{H},\mathcal{H}';\mathcal{P})$. In particular, $\mathrm{U}_{\mathcal{A}}^{\mathrm{LR}}(\mathcal{H},\mathcal{H};\mathcal{P})$, which we denote by $\mathrm{U}_{\mathcal{A}}^{\mathrm{LR}}(\mathcal{H};\mathcal{P})$, is a subgroup of $\mathrm{U}_{\mathcal{A}}^{\mathrm{LR}}(\mathcal{H}) := \mathrm{U}(\mathcal{L}_{\mathcal{A}}^{\mathrm{LR}}(\mathcal{H}))$. In all such notation, we suppress the argument $\mathcal{P}$ whenever $\mathcal{P}$ is empty.

DEFINITION 2.10. Let $\mathcal{A}$ be a real C^*-algebra, and let P be a bimodule structure. The abelian monoid $(\mathrm{Bimod}(\mathcal{A},P),+)$ of $\mathcal{A}$-bimodules with structure P is defined as follows:

- $\mathrm{Bimod}(\mathcal{A},P)$ is the set of unitary equivalence classes of $\mathcal{A}$-bimodules with structure P;
- For $[\mathcal{H}]$, $[\mathcal{H}'] \in \mathrm{Bimod}(\mathcal{A},P)$, $[\mathcal{H}]+[\mathcal{H}'] := [\mathcal{H}\oplus\mathcal{H}']$.

For convenience, we shall denote $\mathrm{Bimod}(\mathcal{A},P)$ by:

- $\mathrm{Bimod}(\mathcal{A})$ if P is the odd structure;
- $\mathrm{Bimod}^{\mathrm{even}}(\mathcal{A})$ if P is the even structure;
- $\mathrm{Bimod}(\mathcal{A},n)$ if P is the real structure of KO-dimension n mod 8;
- $\mathrm{Bimod}^0(\mathcal{A},n)$ if P is the S^0-real structure of KO-dimension n mod 8.

These monoids will be studied in depth in the next section. In light of our earlier comment, we therefore have that

$$\mathrm{Bimod}(\mathcal{A},1)=\mathrm{Bimod}(\mathcal{A},7), \quad \mathrm{Bimod}(\mathcal{A},3)=\mathrm{Bimod}(\mathcal{A},5).$$

and

$$\mathrm{Bimod}^0(\mathcal{A},1)=\mathrm{Bimod}^0(\mathcal{A},7), \quad \mathrm{Bimod}^0(\mathcal{A},3)=\mathrm{Bimod}^0(\mathcal{A},5).$$

Finally, for the sake of completeness, we now define the notions of orientabilty and Poincaré duality in this more general context; in the case of a real spectral triple $(\mathcal{A},\mathcal{H},D,\gamma,J)$ of even KO-dimension, where the right action is given by $\rho(a) := J\lambda(a^*)J^*$, these definitions yield precisely the usual ones (*cf.* [18, §§2.2, 2.3]).

DEFINITION 2.11. We call an even $\mathcal{A}$-bimodule $(\mathcal{H},\gamma)$ *orientable* if there exist $a_1,\dots,a_k, b_1,\dots,b_k \in \mathcal{A}$ such that

$$\gamma = \sum_{i=1}^{k} \lambda(a_i)\rho(b_i). \tag{2.5}$$

DEFINITION 2.12. Let $\mathcal{A}$ be a real C^*-algebra, and let $(\mathcal{H}, \gamma)$ be an even $\mathcal{A}$-bimodule. Then the *intersection form* $\langle \cdot, \cdot \rangle : KO_0(\mathcal{A}) \times KO_0(\mathcal{A}) \to \mathbb{Z}$ associated with $(\mathcal{H}, \gamma)$ is defined by setting

$$\langle [e], [f] \rangle := \operatorname{tr}(\gamma\lambda(e)\rho(f)) \tag{2.6}$$

for projections $e, f \in \mathcal{A}$.

In the case that the intersection form is non-degenerate, we shall say that $(\mathcal{H}, \gamma)$ satisfies *Poincaré duality*.

The orientability assumption was used extensively in [20] and [18], as it leads to considerable algebraic simplifactions; we shall later define a weakened version of orientability that will yield precisely those simplifications.

2.3.3. *Bilateral spectral triples.* We now turn to Dirac operators on bimodules satisfying a generalised order one condition, and define the appropriate notion of compatibility with additional structure on the bimodule.

DEFINITION 2.13. A *Dirac operator* for an $\mathcal{A}$-bimodule $\mathcal{H}$ with structure P is a self-adjoint operator D on $\mathcal{H}$ satisfying the *order one condition*:

$$\forall a, b \in \mathcal{A}, \quad [[D, \lambda(a)], \rho(b)] = 0, \tag{2.7}$$

together with the following relations, as applicable:

$$\{D, \gamma\} = 0, \quad DJ = \varepsilon' JD, \quad [D, \epsilon] = 0.$$

We denote the finite-dimensional real vector space of Dirac operators for an an $\mathcal{A}$-bimodule $\mathcal{H}$ with structure P by $\mathcal{D}_0(\mathcal{A}, \mathcal{H}, \mathcal{P})$.

DEFINITION 2.14. A *bilateral spectral triple* with structure P is a triple of the form $(\mathcal{A}, \mathcal{H}, D)$, where $\mathcal{A}$ is a real C^*-algebra, $\mathcal{H}$ is an $\mathcal{A}$-bimodule with structure P, and D is a Dirac operator for $(\mathcal{H}, P)$.

We shall generally denote such a spectral triple by $(\mathcal{A}, \mathcal{H}, D; \mathcal{P})$, where $\mathcal{P}$ is the set of generators of the structure algebra; in cases where the presence or absence of a grading γ is immaterial, we will suppress the generator γ in this notation.

REMARK 2.15. In the case that P is a real [S^0-real] structure of KO-dimension $n \bmod 8$, a bilateral spectral triple with structure P is precisely a real [S^0-real] spectral triple of KO-dimension $n \bmod 8$.

More generally, an odd [even] bilateral spectral triple $(\mathcal{A}, \mathcal{H}, D)$ is equivalent to an odd [even] spectral triple $(\mathcal{A} \otimes \mathcal{A}^{\mathrm{op}}, \mathcal{H}, D)$ such that $[[D, \mathcal{A} \otimes 1], 1 \otimes \mathcal{A}^{\mathrm{op}}] = \{0\}$, an object that first appears in connection with S^0-real spectral triples [6]

A *unitary equivalence* of spectral triples $(\mathcal{A}, \mathcal{H}, D)$ and $(\mathcal{A}, \mathcal{H}', D')$ is then a unitary $U \in \mathrm{U}^{\mathrm{LR}}_{\mathcal{A}}(\mathcal{H}, \mathcal{H}')$ such that $D' = UDU^*$. This concept leads us to the following definition:

DEFINITION 2.16. Let $\mathcal{A}$ be a real C^*-algebra, and let $\mathcal{H}$ be an $\mathcal{A}$-bimodule with structure P. The *moduli space of Dirac operators* for $\mathcal{H}$ is defined by

$$\mathcal{D}(\mathcal{A}, \mathcal{H}, \mathcal{P}) := \mathcal{D}_0(\mathcal{A}, \mathcal{H}, \mathcal{P}) / \mathrm{U}^{\mathrm{LR}}_{\mathcal{A}}(\mathcal{H}, \mathcal{P}), \tag{2.8}$$

where $\mathrm{U}^{\mathrm{LR}}_{\mathcal{A}}(\mathcal{H}, \mathcal{P})$ acts on $\mathcal{D}_0(\mathcal{A}, \mathcal{H}, \mathcal{P})$ by conjugation.

If $\mathcal{C}$ is a central subalgebra of $\mathcal{A}$, we can form the subspace

$$\mathcal{D}_0(\mathcal{A}, \mathcal{H}, \mathcal{P}; \mathcal{C}) := \{D \in \mathcal{D}_0(\mathcal{A}, \mathcal{H}, \mathcal{P}) \mid [D, \lambda(\mathcal{C})] = [D, \rho(\mathcal{C})] = \{0\}\}. \tag{2.9}$$

and hence the sub-moduli space

$$\mathcal{D}(\mathcal{A},\mathcal{H},\mathcal{P};\mathcal{C}) := \mathcal{D}_0(\mathcal{A},\mathcal{H},\mathcal{P};\mathcal{C})/\operatorname{U}^{\mathrm{LR}}_{\mathcal{A}}(\mathcal{H},\mathcal{P}), \tag{2.10}$$

of $\mathcal{D}_0(\mathcal{A},\mathcal{H},\mathcal{P})$; the moduli space of Dirac operators studied by Chamseddine, Connes and Marcolli [4, §2.7],[8, §13.4] is in fact a sub-moduli space of this form.

Since $\mathcal{D}(\mathcal{A},\mathcal{H},\mathcal{P})$ $[\mathcal{D}(\mathcal{A},\mathcal{H},\mathcal{P};\mathcal{C})]$ is the orbit space of a smooth finite-dimensional representation of a compact Lie group, it is *a priori* locally compact Hausdorff, and is thus homeomorphic to a semialgebraic subset of $\mathbb{R}^d$ for some d [24]. The dimension of $\mathcal{D}(\mathcal{A},\mathcal{H},\mathcal{P})$ $[\mathcal{D}(\mathcal{A},\mathcal{H},\mathcal{P};\mathcal{C})]$ can then be defined as the dimension of this semialgebraic set. Such moduli spaces will be discussed in some detail.

2.3.4. *S^0-reality.* Following Connes [6], we now describe how to reduce the study of S^0-real bimodules of even [odd] KO-dimension to the study of even [odd] bimodules.

Let $(\mathcal{H},J,\epsilon)$ be an S^0-real $\mathcal{A}$-bimodule of even [odd] KO-dimension. Define mutually orthogonal projections P_i, P_{-i} in $\mathcal{L}^{\mathrm{LR}}_{\mathcal{A}}(\mathcal{H})$ by $P_{\pm i} = \frac{1}{2}(1 \pm \epsilon)$. Then, at the level of even [odd] bimodules, $\mathcal{H} = \mathcal{H}_i \oplus \mathcal{H}_{-i}$ for $\mathcal{H}_{\pm i} := P_{\pm i}\mathcal{H}$, where the left and right actions on $\mathcal{H}_{\pm i}$ are given by

$$\lambda_{\pm i}(a) := P_{\pm i}\lambda(a)P_{\pm i}, \quad \rho_{\pm i}(a) := P_{\pm i}\rho(a)P_{\pm i},$$

for $a \in \mathcal{A}$, and, in the case of even KO-dimension, the grading on $\mathcal{H}_{\pm i}$ is given by $\gamma_{\pm i} := P_{\pm i}\gamma P_{\pm i}$. Moreoever,

$$J = \begin{pmatrix} 0 & \varepsilon\tilde{J}^* \\ \tilde{J} & 0 \end{pmatrix},$$

where $\tilde{J} := P_{-i}JP_i$ is an antiunitary $\mathcal{H}_i \to \mathcal{H}_{-i}$, so that for $a \in \mathcal{A}$,

$$\lambda_{-i}(a) = \tilde{J}\rho_i(a^*)\tilde{J}^*, \quad \rho_{-i}(a) = \tilde{J}\lambda_i(a^*)\tilde{J}^*,$$

and in the case of even KO-dimension, $\gamma_{-i} = \varepsilon''\tilde{J}\gamma\tilde{J}^*$. Finally, note that $\tilde{J}$ can also be viewed as a unitary $\overline{\mathcal{H}_i} \to \mathcal{H}_{-i}$, where $\overline{\mathcal{H}_i}$ denotes the conjugate space of $\mathcal{H}$. Hence, for fixed KO-dimension, an S^0-real $\mathcal{A}$-bimodule $\mathcal{H}$ is determined, up to unitary equivalence, by the bimodule $\mathcal{H}_i$.

On the other hand, if $\mathcal{V}$ is an even [odd] $\mathcal{A}$-bimodule, we can construct an S^0-real $\mathcal{A}$-bimodule $\mathcal{H}$ for any even [odd] KO-dimension $n \bmod 8$ such that $\mathcal{H}_i = \mathcal{V}$, by setting $\mathcal{H} := \mathcal{H}_i \oplus \mathcal{H}_{-i}$ for $\mathcal{H}_i := \mathcal{V}$, $\mathcal{H}_{-i} := \overline{\mathcal{V}}$, defining $\tilde{J} : \mathcal{H}_i \to \mathcal{H}_{-i}$ as the identity map on $\mathcal{V}$ viewed as an antiunitary $\mathcal{V} \to \overline{\mathcal{V}}$, then using the above formulas to define J, γ (as necessary), λ, ρ, and finally setting $\epsilon = 1_{\mathcal{V}} \oplus (-1_{\overline{\mathcal{V}}})$. In the case that $\mathcal{V}$ is already $\mathcal{H}_i$ for some S^0-real bimodule $\mathcal{H}$, this procedure reproduces $\mathcal{H}$ up to unitary equivalence. We have therefore proved the following:

PROPOSITION 2.17. *Let $\mathcal{A}$ be a real C^*-algebra, and let $n \in \mathbb{Z}_8$. Then the map*

$$\operatorname{Bimod}^0(\mathcal{A},n) \to \begin{cases} \operatorname{Bimod}(\mathcal{A}), & \text{if } n \text{ is odd,} \\ \operatorname{Bimod}^{\mathrm{even}}(\mathcal{A}), & \text{if } n \text{ is even,} \end{cases}$$

defined by $[\mathcal{H}] \mapsto [\mathcal{H}_i]$ is an isomorphism of monoids.

Now, let $\mathcal{H}$ is an S^0-real $\mathcal{A}$-bimodule, and suppose that D is a Dirac operator for $\mathcal{H}$. We can define Dirac operators D_i and D_{-i} on $\mathcal{H}_i$ and $\mathcal{H}_{-i}$, respectively, by $D_{\pm i} := P_{\pm i}DP_{\pm i}$; then $D = D_i \oplus D_{-i}$ and, in fact, $D_{-i} = \varepsilon'\tilde{J}D_i\tilde{J}^*$. Thus, a Dirac operator D on $\mathcal{H}$ is completely determined by D_i; indeed, the map $D \mapsto D_i$ defines an isomorphism $\mathcal{D}_0(\mathcal{A},\mathcal{H},J,\epsilon) \cong \mathcal{D}_0(\mathcal{A},\mathcal{H})$.

Along similar lines, one can show that $U^{LR}_{\mathcal{A}}(\mathcal{H}, J) \cong U^{LR}_{\mathcal{A}}(\mathcal{H}_i)$ by means of the map $U \mapsto U_i := P_i U P_i$; this isomorphism is compatible with the isomorphism $\mathcal{D}_0(\mathcal{A}, \mathcal{H}, J, \epsilon) \cong \mathcal{D}_0(\mathcal{A}, \mathcal{H})$. Hence, the functional equivalence between $\mathcal{H}$ and $\mathcal{H}_i$ holds at the level of moduli spaces of Dirac operators:

PROPOSITION 2.18. *Let $\mathcal{H}$ be an S^0-real $\mathcal{A}$-bimodule. Then*

$$\mathcal{D}(\mathcal{A}, \mathcal{H}, J, \epsilon) \cong \mathcal{D}(\mathcal{A}, \mathcal{H}_i). \tag{2.11}$$

One can similarly show that for a central subalgebra $\mathcal{C}$ of $\mathcal{A}$,

$$\mathcal{D}(\mathcal{A}, \mathcal{H}, J, \epsilon; \mathcal{C}) \cong \mathcal{D}(\mathcal{A}, \mathcal{H}_i; \mathcal{C}).$$

Let us conclude by considering the relation between orientability and Poincaré duality for an S^0-real bimodule $\mathcal{H}$ of even KO-dimension and orientability and Poincaré duality, respectively, for the associated even bimodule $\mathcal{H}_i$.

PROPOSITION 2.19. *Let $\mathcal{H}$ be an S^0-real $\mathcal{A}$-bimodule of even KO-dimension. Then $\mathcal{H}$ is orientable if and only if there exist $a_1, \ldots, a_k, b_1, \ldots, b_k \in \mathcal{A}$ such that*

$$\gamma_i = \sum_{j=1}^{k} \lambda_i(a_j)\rho_i(b_j) = \varepsilon'' \sum_{j=1}^{k} \lambda_i(b_j^*)\rho_i(a_j^*). \tag{2.12}$$

PROOF. Let $a_1, \ldots, a_k$, $b_1, \ldots, b_k \in \mathcal{A}$, and set $T = \sum_{j=1}^{k} \lambda(a_j)\rho(b_j)$. Then

$$T_i := P_i T P_i = \sum_{j=1}^{k} \lambda_i(a_j)\rho_i(b_j),$$

while

$$T_{-i} := P_{-i} T P_{-i} = \sum_{j=1}^{k} \lambda_{-i}(a_j)\rho_{-i}(b_j) = \tilde{J}\left(\sum_{j=1}^{k} \lambda_i(b_j^*)\rho_i(a_j^*)\right)\tilde{J}^*.$$

Hence, $T_{-i} = \varepsilon'' \tilde{J} T_i \tilde{J}^*$ if and only if

$$\varepsilon'' \sum_{j=1}^{k} \lambda_i(b_j^*)\rho_i(a_j^*) = T_i = \sum_{j=1}^{k} \lambda_i(a_j)\rho_i(b_j).$$

Applying this intermediate result to a_j and b_j such that $\gamma = \sum_{j=1}^{k} \lambda(a_j)\rho(b_j)$, in the case that $\mathcal{H}$ is orientable, and then to a_j and b_j such that $\gamma_i = \sum_{j=1}^{k} \lambda_i(a_j)\rho_i(b_j)$, in the case that $\mathcal{H}_i$ is orientable, yields the desired result. □

Thus, orientability of an S^0-real bimodule $\mathcal{H}$ is equivalent to a stronger version of orientability on the bimodule $\mathcal{H}_i$.

Turning to Poincaré duality, we can obtain the following result:

PROPOSITION 2.20. *Let $\mathcal{H}$ be an S^0-real $\mathcal{A}$-bimodule of even KO-dimension with intersection form $\langle \cdot, \cdot \rangle$, and let $\langle \cdot, \cdot \rangle_i$ be the intersection form for $\mathcal{H}_i$. Then for any p, $q \in KO_0(\mathcal{A})$,*

$$\langle p, q \rangle = \langle p, q \rangle_i + \varepsilon'' \langle q, p \rangle_i \,.$$

PROOF. Let e, $f \in \mathcal{A}$ be projections. Then

$$\begin{aligned}
\langle [e],[f]\rangle &= \operatorname{tr}(\gamma\lambda(e)\rho(f)) \\
&= \operatorname{tr}(\gamma_i\lambda_i(e)\rho_i(f)) + \operatorname{tr}(\gamma_{-i}\lambda_{-i}(e)\rho_{-i}(f)) \\
&= \operatorname{tr}(\gamma_i\lambda_i(e)\rho_i(f)) + \varepsilon'' \operatorname{tr}(\tilde{J}\gamma_i\lambda_i(f)\rho_i(e)\tilde{J}^*) \\
&= \operatorname{tr}(\gamma_i\lambda_i(e)\rho_i(f)) + \varepsilon''\overline{\operatorname{tr}(\gamma_i\lambda_i(f)\rho_i(e))} \\
&= \langle [e],[f]\rangle_i + \varepsilon'' \langle [f],[e]\rangle_i ,
\end{aligned}$$

where we have used the fact that the intersection forms are integer-valued. □

Thus, Poincaré duality on an S^0-real bimodule $\mathcal{H}$ is equivalent to nondegeneracy of either the symmetrisation or antisymmetrisation of the intersection form on $\mathcal{H}_i$, as the case may be.

3. Bimodules and Multiplicity Matrices

We now turn to the study of bimodules, and in particular, to their characterisation by multiplicity matrices. We shall find that a bimodule admits, up to unitary equivalence, at most one real structure of any given KO-dimension, and that the multiplicity matrix or matrices of a bimodule will determine entirely which real structures, if any, it does admit.

In what follows, $\mathcal{A}$ will be a fixed real C^*-algebra.

3.1. Odd bimodules. Let us begin with the study of odd bimodules.

For $m \in M_S(\mathbb{Z}_{\geq 0})$, we define an $\mathcal{A}$-bimodule $\mathcal{H}_m$ by setting

$$\begin{aligned}
\mathcal{H}_m &:= \bigoplus_{\alpha,\beta\in\widehat{\mathcal{A}}} \mathbb{C}^{n_\alpha} \otimes \mathbb{C}^{m_{\alpha\beta}} \otimes \mathbb{C}^{n_\beta}, \\
\lambda_m(a) &:= \bigoplus_{\alpha,\beta\in\widehat{\mathcal{A}}} \lambda_\alpha(a) \otimes 1_{m_{\alpha\beta}} \otimes 1_{n_\beta}, \quad a \in \mathcal{A}, \\
\rho_m(a) &:= \bigoplus_{\alpha,\beta\in\widehat{\mathcal{A}}} 1_{n_\alpha} \otimes 1_{m_{\alpha\beta}} \otimes \lambda_\beta(a)^T, \quad a \in \mathcal{A}.
\end{aligned}$$

Here we use the convention that 1_n is the identity on $\mathbb{C}^n$, with $\mathbb{C}^0 := \{0\}$ and hence $1_0 := 0$.

PROPOSITION 3.1 (Krajewski [18, §3.1], Paschke–Sitarz [20, Lemmas 1, 2]). *The map* bimod : $M_S(\mathbb{Z}_{\geq 0}) \to \operatorname{Bimod}(\mathcal{A})$ *given by* $m \mapsto [\mathcal{H}_m]$ *is an isomorphism of monoids.*

PROOF. By construction, bimod is an injective morphism of monoids. It therefore suffices to show that $\text{bimod}^{\text{odd}}$ is also surjective.

Now, let $\mathcal{H}$ be an $\mathcal{A}$-bimodule. For $\alpha \in \widehat{\mathcal{A}}$ define projections P^L_α and P^R_α by

$$P^L_\alpha := \begin{cases} \lambda(e_i) & \text{if } \alpha = \mathbf{n}_i \text{ for } \mathbb{K}_i \neq \mathbb{C}, \\ \frac{1}{2}\left(\lambda(e_i) - i\lambda(ie_i)\right) & \text{if } \alpha = \mathbf{n}_i \text{ for } \mathbb{K}_i = \mathbb{C}, \\ \frac{1}{2}\left(\lambda(e_i) + i\lambda(ie_i)\right) & \text{if } \alpha = \overline{\mathbf{n}}_i \text{ for } \mathbb{K}_i = \mathbb{C}, \end{cases}$$

and

$$P_\alpha^R := \begin{cases} \rho(e_i) & \text{if } \alpha = \mathbf{n}_i \text{ for } \mathbb{K}_i \neq \mathbb{C}, \\ \frac{1}{2}\left(\rho(e_i) - i\rho(ie_i)\right) & \text{if } \alpha = \mathbf{n}_i \text{ for } \mathbb{K}_i = \mathbb{C}, \\ \frac{1}{2}\left(\rho(e_i) + i\rho(ie_i)\right) & \text{if } \alpha = \overline{\mathbf{n}}_i \text{ for } \mathbb{K}_i = \mathbb{C}, \end{cases}$$

respectively; by construction, $P_\alpha^L \in \lambda(\mathcal{A}) + i\lambda(\mathcal{A})$ and $P_\alpha^R \in \rho(\mathcal{A}) + i\rho(\mathcal{A})$, so that for α, $\beta \in \widehat{\mathcal{A}}$, P_α^L and P_β^R commute. We can therefore define projections $P_{\alpha\beta} := P_\alpha^L P_\beta^R$ for each α, $\beta \in \widehat{\mathcal{A}}$; it is then easy to see that each $\mathcal{H}_{\alpha\beta} := P_{\alpha\beta}\mathcal{H}$ is a sub-$\mathcal{A}$-bimodule of $\mathcal{H}$, and that $\mathcal{H} = \oplus_{\alpha,\beta\in\widehat{\mathcal{A}}}\mathcal{H}_{\alpha\beta}$.

Let α, $\beta \in \widehat{\mathcal{A}}$. As noted before, the left action of $\mathcal{A}$ on $\mathcal{H}_{\alpha\beta}$ must decompose as a direct sum of irreducible representations, but by construction of $\mathcal{H}_{\alpha\beta}$, those irreducible representations must all be α. Similarly, the right action on $\mathcal{H}_{\alpha\beta}$ must be a direct sum of copies of β. Since the left action and right action commute, we must therefore have that $\mathcal{H}_{\alpha\beta} \cong \mathcal{H}_{m_{\alpha\beta}E_{\alpha\beta}}$ for some $m_{\alpha\beta} \in \mathbb{Z}_{\geq 0}$. Taking the direct sum of the $\mathcal{H}_{\alpha\beta}$, we therefore see that $\mathcal{H}$ is unitarily equivalent to $\mathcal{H}_m$ for $m = (m_{\alpha\beta}) \in M_S(\mathbb{Z}_{\geq 0})$, that is, $[\mathcal{H}] = \operatorname{bimod}(m)$. □

We denote the inverse map $\operatorname{bimod}^{-1} : \operatorname{Bimod}(\mathcal{A}) \to M_S(\mathbb{Z}_{\geq 0})$ by mult.

DEFINITION 3.2. Let $\mathcal{H}$ be an $\mathcal{A}$-bimodule. Then the *multiplicity matrix* of $\mathcal{A}$ is the matrix $\operatorname{mult}[\mathcal{H}] \in M_S(\mathbb{Z}_{\geq 0})$.

From now on, without any loss of generality, we shall assume that an $\mathcal{A}$-bimodule $\mathcal{H}$ with multiplicity matrix m is $\mathcal{H}_m$ itself.

REMARK 3.3. Multiplicity matrices readily admit a K-theoretic interpretation [10]. For simplicity, suppose that $\mathcal{A}$ is a complex C^*-algebra and consider only complex-linear representations. Then for $\mathcal{H}$ an $\mathcal{A}$-bimodule, $\operatorname{mult}[\mathcal{H}]$ is essentially the *Bratteli matrix* of the inclusion $\lambda(\mathcal{A}) \hookrightarrow \rho(\mathcal{A})' \subset \mathcal{L}(\mathcal{H})$ (cf. [9, §2]), and can thus be interpreted as representing the induced map $K_0(\lambda(\mathcal{A})) \to K_0(\rho(\mathcal{A})')$ in complex K-theory. Likewise, $\operatorname{mult}[\mathcal{H}]^T$ can be interpreted as representing the map $K_0(\rho(\mathcal{A})) \to K_0(\lambda(\mathcal{A})')$ induced by the inclusion $\rho(\mathcal{A}) \hookrightarrow \lambda(\mathcal{A})' \subset \mathcal{L}(\mathcal{H})$. Similar interpretations can be made in the more general context of real C^*-algebras and KO-theory.

We shall now characterise left, right, and left and right $\mathcal{A}$-linear maps between $\mathcal{A}$-bimodules. Let $\mathcal{H}$ and $\mathcal{H}'$ be $\mathcal{A}$-bimodules with multiplicity matrices m and m', respectively, let $P_{\alpha\beta}$ be the projections on $\mathcal{H}$ defined as in the proof of Proposition 3.1, and let $P'_{\alpha\beta}$ be the analogous projections on $\mathcal{H}'$. Then any linear map $T : \mathcal{H} \to \mathcal{H}'$ is characterised by the components

$$T_{\alpha\beta}^{\gamma\delta} := P'_{\gamma\delta} T P_{\alpha\beta}, \tag{3.1}$$

which we view as maps $T_{\alpha\beta}^{\gamma\delta} : \mathbb{C}^{n_\alpha} \otimes \mathbb{C}^{m_{\alpha\beta}} \otimes \mathbb{C}^{n_\beta} \to \mathbb{C}^{n_\gamma} \otimes \mathbb{C}^{m'_{\gamma\delta}} \otimes \mathbb{C}^{n_\delta}$, or equivalently, as elements $T_{\alpha\beta}^{\gamma\delta} \in M_{n_\gamma \times n_\alpha}(\mathbb{C}) \otimes M_{m'_{\gamma\delta}\times m_{\alpha\beta}} \otimes M_{n_\delta\times n_\beta}(\mathbb{C})$. Thus we have an isomorphism

$$\operatorname{comp} : \mathcal{L}(\mathcal{H}, \mathcal{H}') \to \bigoplus_{\alpha,\beta,\gamma,\delta\in\widehat{\mathcal{A}}} M_{n_\gamma \times n_\alpha}(\mathbb{C}) \otimes M_{m'_{\gamma\delta}\times m_{\alpha\beta}} \otimes M_{n_\delta\times n_\beta}(\mathbb{C})$$

given by $\operatorname{comp}(T) := (T_{\alpha\beta}^{\gamma\delta})_{\alpha,\beta,\gamma,\delta\in\widehat{\mathcal{A}}}$. Note that when $\mathcal{H} = \mathcal{H}'$, T is self-adjoint if and only if $T_{\gamma\delta}^{\alpha\beta} = (T_{\alpha\beta}^{\gamma\delta})^*$ for all α, β, γ, $\delta \in \widehat{\mathcal{A}}$.

PROPOSITION 3.4 (Krajewski [18, §3.4]). *Let $\mathcal{H}$ and $\mathcal{H}'$ be $\mathcal{A}$-bimodules with multiplicity matrices m and m', respectively. Then*

$$\operatorname{comp}(\mathcal{L}^{\mathrm{L}}_{\mathcal{A}}(\mathcal{H},\mathcal{H}')) = \bigoplus_{\alpha,\beta,\delta\in\widehat{\mathcal{A}}} 1_{n_\alpha} \otimes M_{m'_{\alpha\delta}\times m_{\alpha\beta}}(\mathbb{C}) \otimes M_{n_\delta\times n_\beta}(\mathbb{C}), \tag{3.2}$$

$$\operatorname{comp}(\mathcal{L}^{\mathrm{R}}_{\mathcal{A}}(\mathcal{H},\mathcal{H}')) = \bigoplus_{\alpha,\beta,\gamma\in\widehat{\mathcal{A}}} M_{n_\gamma\times n_\alpha}(\mathbb{C}) \otimes M_{m'_{\gamma\beta}\times m_{\alpha\beta}}(\mathbb{C}) \otimes 1_{n_\beta}, \tag{3.3}$$

$$\operatorname{comp}(\mathcal{L}^{\mathrm{LR}}_{\mathcal{A}}(\mathcal{H},\mathcal{H}')) = \bigoplus_{\alpha,\beta\in\widehat{\mathcal{A}}} 1_{n_\alpha} \otimes M_{m'_{\alpha\beta}\times m_{\alpha\beta}}(\mathbb{C}) \otimes 1_{n_\beta}. \tag{3.4}$$

PROOF. Observe that $T \in \mathcal{L}(\mathcal{H},\mathcal{H}')$ is left, right, or left and right $\mathcal{A}$-linear if and only if each $T^{\gamma\delta}_{\alpha\beta}$ is left, right, or left and right $\mathcal{A}$. Thus, let α, β, γ and $\delta \in \widehat{\mathcal{A}}$ be fixed, and let $T \in M_{n_\gamma\times n_\alpha}(\mathbb{C}) \otimes M_{m'_{\gamma\delta}\times m_{\alpha\beta}} \otimes M_{n_\delta\times n_\beta}(\mathbb{C})$.

First, write $T = \sum_{i=1}^k A_i \otimes B_i$ for $A_i \in M_{n_\gamma\times n_\alpha}(\mathbb{C})$ and for linearly independent $B_i \in M_{m'_{\gamma\delta}\times m_{\alpha\beta}} \otimes M_{n_\delta\times n_\beta}(\mathbb{C})$. Then, for $a \in \mathcal{A}$,

$$(\lambda_\gamma(a) \otimes 1_{m'_{\gamma\delta}} \otimes 1_{n_\delta})T - T(\lambda_\alpha(a) \otimes 1_{m_{\alpha\beta}} \otimes 1_{n_\beta}) = \sum_{i=1}^k (\lambda_\gamma(a)A_i - A_i\lambda_\alpha(a)) \otimes B_i,$$

so that by linear independence of the B_i, T is left $\mathcal{A}$-linear if and only if each A_i intertwines the irreducible representations α and γ, and hence, by Schur's lemma, if and only if $\alpha = \gamma$ and each A_i is a constant multiple of 1_{n_α} or each $A_i = 0$. Thus,

$$\mathcal{L}^{\mathrm{L}}_{\mathcal{A}}(\mathbb{C}^{n_\alpha} \otimes \mathbb{C}^{m_{\alpha\beta}} \otimes \mathbb{C}^{n_\beta}, \mathbb{C}^{n_\gamma} \otimes \mathbb{C}^{m'_{\gamma\delta}} \otimes \mathbb{C}^{n_\delta}) = \begin{cases} 1_{n_\alpha} \otimes M_{m'_{\alpha\delta}\times m_{\alpha\beta}}(\mathbb{C}) \otimes M_{n_\delta\times n_\beta}(\mathbb{C}) & \text{if } \alpha = \gamma, \\ \{0\} & \text{otherwise.} \end{cases}$$

Analogously, one can show that

$$\mathcal{L}^{\mathrm{R}}_{\mathcal{A}}(\mathbb{C}^{n_\alpha} \otimes \mathbb{C}^{m_{\alpha\beta}} \otimes \mathbb{C}^{n_\beta}, \mathbb{C}^{n_\gamma} \otimes \mathbb{C}^{m'_{\gamma\delta}} \otimes \mathbb{C}^{n_\delta}) = \begin{cases} M_{n_\gamma\times n_\alpha}(\mathbb{C}) \otimes M_{m'_{\gamma\beta}\times m_{\alpha\beta}}(\mathbb{C}) \otimes 1_{n_\beta} & \text{if } \beta = \delta, \\ \{0\} & \text{otherwise,} \end{cases}$$

and then these first two results together imply that

$$\mathcal{L}^{\mathrm{LR}}_{\mathcal{A}}(\mathbb{C}^{n_\alpha} \otimes \mathbb{C}^{m_{\alpha\beta}} \otimes \mathbb{C}^{n_\beta}, \mathbb{C}^{n_\gamma} \otimes \mathbb{C}^{m'_{\gamma\delta}} \otimes \mathbb{C}^{n_\delta}) = \begin{cases} 1_{n_\alpha} \otimes M_{m'_{\alpha\beta}\times m_{\alpha\beta}}(\mathbb{C}) \otimes 1_{n_\beta} & \text{if } (\alpha,\beta) = (\gamma,\delta), \\ \{0\} & \text{otherwise,} \end{cases}$$

as was claimed. □

An immediate consequence is the following description of the group $\mathrm{U}^{\mathrm{LR}}_{\mathcal{A}}(\mathcal{H})$:

COROLLARY 3.5. *Let $\mathcal{H}$ be an $\mathcal{A}$-bimodule. Then*

$$\operatorname{comp}(\mathrm{U}^{\mathrm{LR}}_{\mathcal{A}}(\mathcal{H})) = \bigoplus_{\alpha,\beta\in\widehat{\mathcal{A}}} 1_{n_\alpha} \otimes \mathrm{U}(m_{\alpha\beta}) \otimes 1_{n_\beta} \cong \prod_{\alpha,\beta\in\widehat{\mathcal{A}}} \mathrm{U}(m_{\alpha\beta}),$$

with the convention that $\mathrm{U}(0) = \{0\}$ is the trivial group.

3.2. Even bimodules. We now turn to the study of even bimodules; let us begin by considering the decomposition of an even bimodule into its even and odd sub-bimodules.

Let $(\mathcal{H},\gamma)$ be an even $\mathcal{A}$-bimodule. Define mutually orthogonal projections P^{even} and P^{odd} by

$$P^{\text{even}} = \frac{1}{2}(1+\gamma), \quad P^{\text{odd}} = \frac{1}{2}(1-\gamma).$$

We can then define sub-bimodules $\mathcal{H}^{\text{even}}$ and $\mathcal{H}^{\text{odd}}$ of $\mathcal{H}$ by $\mathcal{H}^{\text{even}} = P^{\text{even}}\mathcal{H}$, $\mathcal{H}^{\text{odd}} = P^{\text{odd}}\mathcal{H}$; one has that $\mathcal{H} = \mathcal{H}^{\text{even}} \oplus \mathcal{H}^{\text{odd}}$ at the level of bimodules.

On the other hand, given $\mathcal{A}$-bimodules $\mathcal{H}_1$ and $\mathcal{H}_2$, we can construct an even $\mathcal{A}$-bimodule $(\mathcal{H},\gamma)$ such that $\mathcal{H}^{\text{even}} = \mathcal{H}_1$ and $\mathcal{H}^{\text{odd}} = \mathcal{H}_2$ by setting $\mathcal{H} = \mathcal{H}_1 \oplus \mathcal{H}_2$ and $\gamma = 1_{\mathcal{H}_1} \oplus (-1_{\mathcal{H}_2})$. If $\mathcal{H}_1$ and $\mathcal{H}_2$ are already $\mathcal{H}^{\text{even}}$ and $\mathcal{H}^{\text{odd}}$ for some $(\mathcal{H},\gamma)$, then this procedure precisely reconstructs $(\mathcal{H},\gamma)$. Since this procedure manifestly respects direct summation and unitary equivalence at either end, we have therefore proved the following:

PROPOSITION 3.6. *Let $\mathcal{A}$ be a real C^*-algebra. The map*

$$C : \operatorname{Bimod}^{\text{even}}(\mathcal{A}) \to \operatorname{Bimod}(\mathcal{A}) \times \operatorname{Bimod}(\mathcal{A})$$

given by

$$C([\mathcal{H}]) := ([\mathcal{H}^{\text{even}}],[\mathcal{H}^{\text{odd}}])$$

is an isomorphism of monoids.

One readily obtains a similar decomposition at the level of unitary groups:

COROLLARY 3.7. *Let $(\mathcal{H},\gamma)$ be an even $\mathcal{A}$-bimodule. Then*

$$\mathrm{U}^{\mathrm{LR}}_{\mathcal{A}}(\mathcal{H},\gamma) = \mathrm{U}^{\mathrm{LR}}_{\mathcal{A}}(\mathcal{H}^{\text{even}}) \oplus \mathrm{U}^{\mathrm{LR}}_{\mathcal{A}}(\mathcal{H}^{\text{odd}}).$$

Another immediate consequence is the following analogue of Proposition 3.1:

PROPOSITION 3.8. *Let $\mathcal{A}$ be a real C^*-algebra. The map*

$$\operatorname{bimod}^{\text{even}} : M_S(\mathbb{Z}_{\geq 0}) \times M_S(\mathbb{Z}_{\geq 0}) \to \operatorname{Bimod}^{\text{even}}(\mathcal{A})$$

defined by bimod$^{\text{even}} := C^{-1} \circ (\operatorname{bimod} \times \operatorname{bimod})$ *is an isomorphism of monoids.*

Just as in the odd case, we will find it convenient to denote $(\operatorname{bimod}^{\text{even}})^{-1} : \operatorname{Bimod}^{\text{even}}(\mathcal{A}) \to M_S(\mathbb{Z}_{\geq 0}) \times M_S(\mathbb{Z}_{\geq 0})$ by $\operatorname{mult}^{\text{even}}$. It then follows that $\operatorname{mult}^{\text{even}} = (\operatorname{mult} \times \operatorname{mult}) \circ C$.

DEFINITION 3.9. Let $(\mathcal{H},\gamma)$ be an even $\mathcal{A}$-bimodule. Then the *multiplicity matrices* of $(\mathcal{H},\gamma)$ are the pair of matrices

$$(\operatorname{mult}[\mathcal{H}^{\text{even}}], \operatorname{mult}[\mathcal{H}^{\text{odd}}]) = \operatorname{mult}^{\text{even}}[(\mathcal{H},\gamma)] \in M_S(\mathbb{Z}_{\geq 0}) \times M_S(\mathbb{Z}_{\geq 0}).$$

Let us now consider orientability of even bimodules.

LEMMA 3.10 (Krajewski [18, §3.4]). *Let $(\mathcal{H},\gamma)$ be an even $\mathcal{A}$-bimodule. Then $(\mathcal{H},\gamma)$ is orientable only if $\mathcal{L}^{\mathrm{LR}}(\mathcal{H}^{\text{even}},\mathcal{H}^{\text{odd}}) = \{0\}$.*

PROOF. Suppose that $(\mathcal{H},\gamma)$ is orientable, so that $\gamma = \sum_{i=1}^{k} \lambda(a_i)\rho(b_i)$ for some $a_1,\dots,a_k$, $b_1,\dots,b_k \in \mathcal{A}$. Now, let $T \in \mathcal{L}^{\mathrm{LR}}_{\mathcal{A}}(\mathcal{H}^{\text{even}},\mathcal{H}^{\text{odd}})$, and define $\tilde{T} \in \mathcal{L}^{\mathrm{LR}}_{\mathcal{A}}(\mathcal{H})$ by

$$\tilde{T} = \begin{pmatrix} 0 & T^* \\ T & 0 \end{pmatrix}.$$

Then, on the one hand, since $\gamma = 1_{\mathcal{H}^{\mathrm{even}}} \oplus (-1_{\mathcal{H}^{\mathrm{odd}}})$, $\tilde{T}$ anticommutes with γ, and on the other, since $\gamma = \sum_{i=1}^{k} \lambda(a_i)\rho(b_i)$, $\tilde{T}$ commutes with γ, so that $\tilde{T} = 0$. Hence, $T = 0$. □

This last result motivates the following weaker notion of orientability:

DEFINITION 3.11. An even $\mathcal{A}$-bimodule $(\mathcal{H}, \gamma)$ shall be called *quasi-orientable* whenever $\mathcal{L}_{\mathcal{A}}^{\mathrm{LR}}(\mathcal{H}^{\mathrm{even}}, \mathcal{H}^{\mathrm{odd}}) = \{0\}$.

The subset of $\mathrm{Bimod}^{\mathrm{even}}(\mathcal{A})$ consisting of the unitary equivalence classes of the quasi-orientable even $\mathcal{A}$-bimodules will be denoted by $\mathrm{Bimod}_q^{\mathrm{even}}(\mathcal{A})$.

We define the *support* of a real $p \times q$ matrix A to be the set

$$\mathrm{supp}(A) := \{(i,j) \in \{1,\ldots,p\} \times \{1,\ldots,q\} \mid A_{ij} \neq 0\}.$$

For $A \in M_S(\mathbb{R})$, we shall view $\mathrm{supp}(A)$ as a subset of $\widehat{\mathcal{A}}^2$ by means of the identification of $\{1,\ldots,S\}$ with $\widehat{\mathcal{A}}$ as ordered sets. We shall also find it convenient to associate to each matrix $m \in M_S(\mathbb{Z})$ a matrix $\widehat{m} \in M_N(\mathbb{Z})$ by

$$\widehat{m}_{ij} := \sum_{\alpha \in \widehat{M_{n_i}(\mathbb{K}_i)}} \sum_{\beta \in \widehat{M_{n_j}(\mathbb{K}_j)}} m_{\alpha\beta}. \tag{3.5}$$

One can check the map $M_S(\mathbb{Z}) \to M_N(\mathbb{Z})$ defined by $m \mapsto \widehat{m}$ is linear and respects transposes.

We can now offer the following characterisation of quasi-orientable bimodules:

PROPOSITION 3.12 (Krajewski [18, §3.3], Paschke–Sitarz [20, Lemma 3]). *Let $\mathcal{A}$ be a real C^*-algebra. Then*

$$\begin{aligned}(3.6)\quad &\mathrm{mult}^{\mathrm{even}}(\mathrm{Bimod}_q^{\mathrm{even}}(\mathcal{A})) \\ &\quad = \{(m^{\mathrm{even}}, m^{\mathrm{odd}}) \in M_S(\mathbb{Z}_{\geq 0})^2 \mid \mathrm{supp}(m^{\mathrm{even}}) \cap \mathrm{supp}(m^{\mathrm{odd}}) = \emptyset\}.\end{aligned}$$

PROOF. Let $(\mathcal{H}, \gamma)$ be an even $\mathcal{A}$-bimodule and let $(m^{\mathrm{even}}, m^{\mathrm{odd}})$ be its multiplicity matrices. Then by Proposition 3.4,

$$\mathcal{L}_{\mathcal{A}}^{\mathrm{LR}}(\mathcal{H}^{\mathrm{even}}, \mathcal{H}^{\mathrm{odd}}) \cong \bigoplus_{\alpha,\beta \in \widehat{\mathcal{A}}} M_{m_{\alpha\beta}^{\mathrm{odd}} \times m_{\alpha\beta}^{\mathrm{even}}}(\mathbb{C}),$$

whence the result follows immediately. □

We therefore define the *signed multiplicity matrix* of a quasi-orientable even $\mathcal{A}$-bimodule $(\mathcal{H}, \gamma)$, or rather, the unitary equivalence class thereof, to be the matrix

$$\mathrm{mult}_q[(\mathcal{H}, \gamma)] := \mathrm{mult}[\mathcal{H}^{\mathrm{even}}] - \mathrm{mult}[\mathcal{H}^{\mathrm{odd}}] \in M_S(\mathbb{Z}).$$

The map $\mathrm{Bimod}_q^{\mathrm{even}}(\mathcal{A}) \to M_S(\mathbb{Z})$ defined by

$$[(\mathcal{H}, \gamma)] \mapsto \mathrm{mult}_q[(\mathcal{H}, \gamma)]$$

is then bijective, and $\mathrm{mult}^{\mathrm{even}}[(\mathcal{H}, \gamma)]$ is readily recovered from $\mathrm{mult}_q[(\mathcal{H}, \gamma)]$. Indeed, if $(\mathcal{H}, \gamma)$ is a quasi-orientable even $\mathcal{A}$-bimodule with signed multiplicity matrix μ, then (cf. [20, Lemma 3],[18, 3.3])

$$\gamma = \bigoplus_{\alpha,\beta \in \widehat{\mathcal{A}}} \mu_{\alpha\beta} 1_{\mathcal{H}_{\alpha\beta}}. \tag{3.7}$$

These algebraic consequences of quasi-orientability, which were derived from the stronger condition of orientability in the original papers [20] and [18], are key to the formalism developed by Krajewski and Paschke–Sitarz, and hence to the later work by Iochum, Jureit, Schücker, and Stephan [12–14, 22].

We can now characterise orientable bimodules amongst quasi-orientable bimodules:

PROPOSITION 3.13 (Krajewski [18, §3.3]). *Let $(\mathcal{H},\gamma)$ be a quasi-orientable $\mathcal{A}$-bimodule with signed multiplicity matrix μ. Then $(\mathcal{H},\gamma)$ is orientable if and only if the following conditions all hold:*

(1) *For each $i \in \{1,\ldots,N\}$ such that $\mathbb{K}_i = \mathbb{C}$ and all $\beta \in \widehat{\mathcal{A}}$,*
$$\mu_{\mathbf{n}_i\beta}\mu_{\overline{\mathbf{n}}_i\beta} \geq 0;$$
(2) *For all $\alpha \in \widehat{\mathcal{A}}$ and each $j \in \{1,\ldots,N\}$ such that $\mathbb{K}_j = \mathbb{C}$,*
$$\mu_{\alpha\mathbf{n}_j}\mu_{\alpha\overline{\mathbf{n}}_j} \geq 0;$$
(3) *For all i, $j \in \{1,\ldots,N\}$ such that $\mathbb{K}_i = \mathbb{K}_j = \mathbb{C}$,*
$$\mu_{\mathbf{n}_i\overline{\mathbf{n}}_j}\mu_{\overline{\mathbf{n}}_i\mathbf{n}_j} \geq 0.$$

In particular, if $(\mathcal{H},\gamma)$ is orientable, then

$$\gamma = \sum_{i,j=1}^{N} \lambda(\operatorname{sgn}(\widehat{\mu}_{ij})e_i)\rho(e_j). \tag{3.8}$$

PROOF. First, suppose that $(\mathcal{H},\gamma)$ is indeed orientable, so that there exist $a_1,\ldots,a_n$, $b_1,\ldots,b_n \in \mathcal{A}$ such that $\gamma = \sum_{l=1}^{n}\lambda(a_l)\rho(b_l)$; in particular, then, for each α, $\beta \in \widehat{\mathcal{A}}$,

$$\operatorname{sgn}(\mu_{\alpha\beta})1_{n_\alpha} \otimes 1_{|\mu_{\alpha\beta}|} \otimes 1_{n_\beta} = \gamma^{\alpha\beta}_{\alpha\beta} = \sum_{l=1}^{n} \lambda_\alpha(a_l) \otimes 1_{|\mu_{\alpha\beta}|} \otimes \lambda_\beta(b_l)^T.$$

Now, let $i \in \{1,\ldots,N\}$ be such that $\mathbb{K}_i = \mathbb{C}$, and let $\beta \in \widehat{\mathcal{A}}$, and suppose that $\mu_{\mathbf{n}_i\beta}$ and $\mu_{\overline{\mathbf{n}}_i\beta}$ are both non-zero. It then follows that

$$\operatorname{sgn}(\mu_{\mathbf{n}_i\beta})1_{n_i}\otimes 1_{n_\beta} = \sum_{l=1}^{n}(a_l)_i \otimes \lambda_\beta(b_l)^T, \quad \operatorname{sgn}(\mu_{\overline{\mathbf{n}}_i\beta})1_{n_i}\otimes 1_{n_\beta} = \sum_{l=1}^{n}\overline{(a_l)_i}\otimes\lambda_\beta(b_l)^T,$$

where $(a_l)_i$ denotes the component of a_l in the direct summand $M_{k_i}(\mathbb{C})$ of $\mathcal{A}$. If X denotes complex conjugation on $\mathbb{C}^{n_i}$, it then follows from this that

$$\operatorname{sgn}(\mu_{\mathbf{n}_i\beta})1_{n_i}\otimes 1_{n_\beta} = (X\otimes 1_{n_\beta})(\operatorname{sgn}(\mu_{\mathbf{n}_i\beta})1_{n_i}\otimes 1_{n_\beta})(X\otimes 1_{n_\beta}) = \operatorname{sgn}(\mu_{\overline{\mathbf{n}}_i\beta})1_{n_i}\otimes 1_{n_\beta},$$

so that $\operatorname{sgn}(\mu_{\mathbf{n}_i\beta}) = \operatorname{sgn}(\mu_{\overline{\mathbf{n}}_i\beta})$, or equivalently $\mu_{\mathbf{n}_i\beta}\mu_{\overline{\mathbf{n}}_i\beta} > 0$. One can similarly show that the other two conditions hold.

Now, suppose instead that the three conditions on μ hold. Then for all i, $j \in \{1,\ldots,N\}$, all non-zero entries $\mu_{\alpha\beta}$ for $\alpha \in \widehat{M_{k_i}(\mathbb{K}_i)}$, $\beta \in \widehat{M_{k_j}(\mathbb{K}_j)}$, have the same sign, so set γ_{ij} equal to this common value of non-zero $\operatorname{sgn}(\mu_{\alpha\beta})$ if at least one such $\mu_{\alpha\beta}$ is non-zero, and set $\gamma_{ij} = 0$ otherwise. One can then easily check that $\gamma = \sum_{i,j=1}^{N}\lambda(\gamma_{ij}e_i)\rho(e_j)$, so that $(\mathcal{H},\gamma)$ is indeed orientable. Moreover, using the same three conditions, one can readily check that $\gamma_{ij} = \operatorname{sgn}(\widehat{\mu}_{ij})$, which yields the last part of the claim. □

Let us now turn to intersection forms and Poincaré duality. In particular, we are now able to provide explicit expressions for intersection forms in terms of multiplicity matrices.

Recall that for $\mathbb{K} = \mathbb{R}$, $\mathbb{C}$ or $\mathbb{H}$, $KO_0(M_k(\mathbb{K}))$ is the infinite cyclic group generated by $[p]$ for $p \in M_k(\mathbb{K})$ a minimal projection, so that for $\mathcal{A}$ a real C^*-algebra with Wedderburn decomposition $\oplus_{i=1}^N M_{n_i}(\mathbb{K}_i)$,

$$KO_0(\mathcal{A}) \cong \prod_{i=1}^N KO_0(M_{n_i}(\mathbb{K}_i)) \cong \mathbb{Z}^N,$$

which can be viewed as the infinite abelian group generated by $\{[p_i]\}_{i=1}^N$ for p_i a minimal projection in $M_{n_i}(\mathbb{K}_i)$. Since

$$\tau_i := \operatorname{tr}(p_i) = \begin{cases} 2 & \text{if } \mathbb{K}_i = \mathbb{H}, \\ 1 & \text{otherwise,} \end{cases}$$

it follows that for $\alpha \in \widehat{\mathcal{A}}$,

$$\operatorname{tr}(\lambda_\alpha(p_i)) = \begin{cases} \tau_i & \text{if } \alpha \in \widehat{M_{n_i}(\mathbb{K}_i)}, \\ 0 & \text{otherwise.} \end{cases} \tag{3.9}$$

Now, if $(\mathcal{H}, \gamma)$ is an even $\mathcal{A}$-bimodule with intersection form $\langle \cdot, \cdot \rangle$, we can define a matrix $\cap \in M_N(\mathbb{Z})$ by

$$\cap_{ij} := \langle [p_i], [p_j] \rangle . \tag{3.10}$$

The intersection form $\langle \cdot, \cdot \rangle$ is completely determined by the matrix $\cap$, and in particular, $\langle \cdot, \cdot \rangle$ is non-degenerate (*i.e.* $(\mathcal{H}, \gamma)$ satisfies Poincaré duality) if and only if $\cap$ is non-degenerate.

Proposition 3.14 (Krajewski [18, §3.3], Paschke–Sitarz [20, §2.4]). *Let $(\mathcal{H}, \gamma)$ be an even $\mathcal{A}$-bimodule with pair of multiplicity matrices $(m^{\text{even}}, m^{\text{odd}})$. Then*

$$\cap_{ij} = \tau_i \tau_j \left(\widehat{m^{\text{even}}}_{ij} - \widehat{m^{\text{odd}}}_{ij} \right), \tag{3.11}$$

so that $(\mathcal{H}, \gamma)$ satisfies Poincaré duality if and only if the matrix $\widehat{m^{\text{even}}} - \widehat{m^{\text{odd}}}$ is non-degenerate.

Proof. First, since $\mathcal{H} = \mathcal{H}^{\text{even}} \oplus \mathcal{H}^{\text{odd}}$, we can write

$$\gamma = \bigoplus_{\alpha, \beta \in \widehat{\mathcal{A}}} 1_{n_\alpha} \otimes \gamma_{\alpha\beta} \otimes 1_{n_\beta},$$

where $\gamma_{\alpha\beta} = 1_{m^{\text{even}}_{\alpha\beta}} \oplus (-1_{m^{\text{odd}}_{\alpha\beta}})$. Then,

$$\begin{aligned}
\cap_{ij} &= \langle [p_i], [p_j] \rangle \\
&= \operatorname{tr}(\gamma\lambda(p_i)\rho(p_j)) \\
&= \operatorname{tr}\left(\bigoplus_{\alpha,\beta\in\widehat{\mathcal{A}}} \lambda_\alpha(p_i) \otimes \gamma_{\alpha\beta} \otimes \lambda_\beta(p_j) \right) \\
&= \sum_{\alpha,\beta\in\widehat{\mathcal{A}}} \operatorname{tr}(\lambda_\alpha(p_i)) \operatorname{tr}(\lambda_\beta(p_j))(m^{\text{even}}_{\alpha\beta} - m^{\text{odd}}_{\alpha\beta}) \\
&= \sum_{i,j=1}^{N} \tau_i\tau_j(\widehat{m^{\text{even}}}_{ij} - \widehat{m^{\text{odd}}}_{ij}).
\end{aligned}$$

This calculation implies, in particular, that $\cap$ can be obtained from $\widehat{m^{\text{even}}} - \widehat{m^{\text{odd}}}$ by a finite sequence of elementary row or column operations, so that $\cap$ is indeed non-degenerate if and only if $\widehat{m^{\text{even}}} - \widehat{m^{\text{odd}}}$ is. □

COROLLARY 3.15. *Let $(\mathcal{H}, \gamma)$ be a quasi-orientable $\mathcal{A}$-bimodule with signed multiplicity matrix μ. Then $(\mathcal{H}, \gamma)$ satisfies Poincaré duality if and only if $\widehat{\mu}$ is non-degenerate.*

In particular, if we restrict ourselves to complex C^*-algebras and complex-linear representations, a quasi-orientable bimodule is completely characterised by the K-theoretic datum of its intersection form.

3.3. Real bimodules of odd KO-dimension. Let us now consider real bimodules of odd KO-dimension. Before continuing, recall that

$$\text{Bimod}(\mathcal{A}, 1) = \text{Bimod}(\mathcal{A}, 7), \quad \text{Bimod}(\mathcal{A}, 3) = \text{Bimod}(\mathcal{A}, 5).$$

For $m \in \text{Sym}_S(\mathbb{Z}_{\geq 0})$, we define an antilinear operator X_m on $\mathcal{H}_m$ by defining $(X_m)^{\gamma\delta}_{\alpha\beta} : \mathbb{C}^{n_\alpha} \otimes \mathbb{C}^{m_{\alpha\beta}} \otimes \mathbb{C}^{n_\beta} \to \mathbb{C}^{n_\gamma} \otimes \mathbb{C}^{m_{\gamma\delta}} \otimes \mathbb{C}^{n_\delta}$ by

$$(X_m)^{\beta\alpha}_{\alpha\beta} : \xi_1 \otimes \xi_2 \otimes \xi_3 \mapsto \overline{\xi_3} \otimes \overline{\xi_2} \otimes \overline{\xi_1}, \tag{3.12}$$

and by setting $(X_m)^{\gamma\delta}_{\alpha\beta} = 0$ whenever $(\gamma, \delta) \neq (\beta, \alpha)$.

3.3.1. *KO-dimension 1 or 7 mod 8.* We begin by determining the form of the multiplicity matrix for a real bimodule of KO-dimension 1 or 7 mod 8.

LEMMA 3.16 (Krajewski [18, §3.2], Paschke–Sitarz [20, Lemma 4]). *Let $(\mathcal{H}, J)$ be a real $\mathcal{A}$-bimodule of KO-dimension 1 or 7 mod 8 with multiplicity matrix m. Then m is symmetric, and the only non-zero components of J are of the form $J^{\beta\alpha}_{\alpha\beta}$ for α, $\beta \in \widehat{\mathcal{A}}$, which are anti-unitaries $\mathcal{H}_{\alpha\beta} \to \mathcal{H}_{\beta\alpha}$ satisfing the relations $J^{\alpha\beta}_{\beta\alpha} = (J^{\beta\alpha}_{\alpha\beta})^*$.*

PROOF. Let the projections P^L_α, P^R_β and $P_{\alpha\beta}$ be defined as in the proof of Proposition 3.1, and recall that $P_{\alpha\beta} = P^L_\alpha P^R_\beta$. By Equation 2.4, it follows that for all $\alpha \in \widehat{\mathcal{A}}, JP^L_\alpha = P^R_\alpha J$ and $JP^R_\alpha = P^L_\alpha J$, and hence that for all α, $\beta \in \widehat{\mathcal{A}}$, $JP_{\alpha\beta} = JP^L_\alpha P^R_\beta = P^R_\alpha P^L_\beta J = P_{\beta\alpha} J$. Thus, the only non-zero components of J are the anti-unitaries $J^{\beta\alpha}_{\alpha\beta} : \mathcal{H}_{\alpha\beta} \to \mathcal{H}_{\beta\alpha}$ which satisfy $J^{\alpha\beta}_{\beta\alpha} = (J^{\beta\alpha}_{\alpha\beta})^*$; this, in turn, implies that m is indeed symmetric. □

Next, we show that for every $m \in \mathrm{Sym}_S(\mathbb{Z}_{\geq 0})$, not only does $\mathcal{H}_m$ admit a real structure of KO-dimension 1 or 7 mod 8, but it is also unique up to unitary equivalence.

LEMMA 3.17 (Krajewski [18, §3.2], Paschke–Sitarz [20, Lemma 5]). *Let $m \in \mathrm{Sym}_S(\mathbb{Z}_{\geq 0})$. Then, up to unitary equivalence, $J_m := X_m$ is the unique real structure on $\mathcal{H}_m$ of KO-dimension* 1 *or* 7 mod 8.

PROOF. First, X_m is indeed by construction a real structure on $\mathcal{H}_m$ of KO-dimension 1 or 7 mod 8.

Now, let J be another real structure on $\mathcal{H}_m$ of KO-dimension 1 or 7 mod 8. Define a unitary K on $\mathcal{H}$ by $K = JX_m$; thus, $J = KX_m$. Since the intertwining condition of Equation 2.4 applies to both J and X_m, we have, in fact, that $K \in \mathrm{U}^{\mathrm{LR}}_{\mathcal{A}}(\mathcal{H}_m)$, and hence

$$K = \bigoplus_{\alpha,\beta \in \widehat{\mathcal{A}}} 1_{n_\alpha} \otimes K_{\alpha\beta} \otimes 1_{n_\beta},$$

for $K_{\alpha\beta} \in \mathrm{U}(m_{\alpha\beta})$. In particular, since $K^* = X_m J = X_m K X_m$, we have that $K_{\beta\alpha} = K^T_{\alpha\beta}$.

Let $(\alpha,\beta) \in \mathrm{supp}(m)$, and suppose that $\alpha < \beta$. Let $K_{\alpha\beta} = V_{\alpha\beta}\tilde{K}_{\alpha\beta}V^*_{\alpha\beta}$ be a unitary diagonalisation of $K_{\alpha\beta}$, and let $L_{\alpha\beta}$ be a diagonal square root of $\tilde{K}_{\alpha\beta}$. Then$K_{\alpha\beta} = V_{\alpha\beta}L_{\alpha\beta}L_{\alpha\beta}V^*_{\alpha\beta} = (V_{\alpha\beta}L_{\alpha\beta})(\overline{V_{\alpha\beta}}L_{\alpha\beta})^T$, and hence $K_{\beta\alpha} = (\overline{V_{\alpha\beta}}L_{\alpha\beta})(V_{\alpha\beta}L_{\alpha\beta})^T$. If, instead, $\alpha = \beta$, then $K_{\alpha\alpha}$ is unitary and complex symmetric, so that there exists a unitary $W_{\alpha\alpha}$ such that $K_{\alpha\alpha} = W_{\alpha\alpha}W^T_{\alpha\alpha}$. We can now define a unitary $U \in \mathrm{U}^{\mathrm{LR}}_{\mathcal{A}}(\mathcal{H}_m)$ by

$$U = \bigoplus_{\alpha,\beta \in \widehat{\mathcal{A}}} 1_{n_\alpha} \otimes U_{\alpha\beta} \otimes 1_{n_\beta},$$

where $U_{\alpha\beta} = 0$ if $m_{\alpha\beta} = 0$, and for $(\alpha,\beta) \in \mathrm{supp}(m)$,

$$U_{\alpha\beta} = \begin{cases} V_{\alpha\beta}L_{\alpha\beta}, & \text{if } \alpha < \beta, \\ \overline{V_{\beta\alpha}}L_{\beta\alpha}, & \text{if } \alpha > \beta, \\ W_{\alpha\alpha}, & \text{if } \alpha = \beta. \end{cases}$$

Then, by construction, $K = UX_mU^*X_m$, and hence, $J = UX_mU^*$, so that U is the required unitary equivalence between $(\mathcal{H}_m, X_m)$ and $(\mathcal{H}_m, J)$. □

We can now give our characterisation of real bimodules of KO-dimension 1 or 7 mod 8:

PROPOSITION 3.18 (Krajewski [18, §3.2]). *Let $n = 1$ or $7 \bmod 8$. Then the map $\iota_n : \mathrm{Bimod}(\mathcal{A}, n) \to \mathrm{Bimod}(\mathcal{A})$ defined by $\iota_n : [(\mathcal{H}, J)] \mapsto [\mathcal{H}]$ is injective, and*

$$(\mathrm{mult} \circ \iota_n)(\mathrm{Bimod}(\mathcal{A}, n)) = \mathrm{Sym}_S(\mathbb{Z}_{\geq 0}). \tag{3.13}$$

PROOF. First, since a unitary equivalence of real $\mathcal{A}$-bimodules of KO-dimension n mod 8 is, in particular, a unitary equivalence of odd $\mathcal{A}$-bimodules, the map ι_n is well defined.

Next, let $(\mathcal{H}, J)$ and $(\mathcal{H}', J')$ be real $\mathcal{A}$-bimodules of KO-dimension n mod 8, and suppose that $\mathcal{H}$ and $\mathcal{H}'$ are unitarily equivalent as bimodules; let $U \in \mathrm{U}^{\mathrm{LR}}_{\mathcal{A}}(\mathcal{H}', \mathcal{H})$. Now, if m is the multiplicity matrix of $\mathcal{H}$, then $\mathcal{H}$ and $\mathcal{H}_m$ are unitarily equivalent, so let $V \in \mathrm{U}^{\mathrm{LR}}_{\mathcal{A}}(\mathcal{H}, \mathcal{H}_m)$. Then VJV^* and $VUJ'U^*V^*$ are both real structures of KO-dimension $n \bmod 8$, so by Lemma 3.17, they are both unitarily equivalent to J_m. This implies that J and $UJ'U^*$ are unitarily equivalent as real structures on $\mathcal{H}$, and hence that $(\mathcal{H}, J)$ and $(\mathcal{H}', J')$ are unitarily equivalent. Thus, ι_n is injective.

Finally, Lemma 3.16 implies that $(\mathrm{mult} \circ \iota_n)(\mathrm{Bimod}(\mathcal{A}, n)) \subseteq \mathrm{Sym}_S(\mathbb{Z}_{\geq 0})$, while Lemma 3.17 implies the reverse inclusion. □

Thus, without any loss of generality, a real bimodule $\mathcal{H}$ of KO-dimension 1 or 7 mod 8 with multiplicity matrix m can be assumed to be simply $(\mathcal{H}_m, J_m)$.

One following characterisation of $\mathrm{U}^{\mathrm{LR}}_{\mathcal{A}}(\mathcal{H}, J)$ now follows by direct calculation:

PROPOSITION 3.19. *Let $(\mathcal{H}, J)$ be a real $\mathcal{A}$-bimodule of KO-dimension* 1 *or* 7 mod 8 *with multiplicity matrix m. Then*

(3.14)

$$\mathrm{comp}(\mathrm{U}^{\mathrm{LR}}_{\mathcal{A}}(\mathcal{H}, J)) = \{(1_{n_\alpha} \otimes U_{\alpha\beta} \otimes 1_{n_\beta})_{\alpha,\beta\in\widehat{\mathcal{A}}} \in \mathrm{comp}(\mathrm{U}^{\mathrm{LR}}_{\mathcal{A}}(\mathcal{H})) \mid U_{\beta\alpha} = \overline{U_{\alpha\beta}}\}$$
$$\cong \prod_{\alpha\in\widehat{\mathcal{A}}} \Big(\mathrm{O}(m_{\alpha\alpha}) \times \prod_{\substack{\beta\in\widehat{\mathcal{A}}\\ \beta>\alpha}} \mathrm{U}(m_{\alpha\beta})\Big).$$

3.3.2. *KO-dimension* 3 *or* 5 mod 8. Let us now turn to real bimodules of KO-dimension 3 or 5 mod 8. We begin with the relevant analogue of Lemma 3.16.

LEMMA 3.20. *Let $(\mathcal{H}, J)$ be a real $\mathcal{A}$-bimodule of KO-dimension* 3 *or* 5 mod 8 *with multiplicity matrix m. Then m is symmetric with even diagonal entries, and the only non-zero components of J are of the form $J^{\beta\alpha}_{\alpha\beta}$ for α, $\beta \in \widehat{\mathcal{A}}$, which are anti-unitaries $\mathcal{H}_{\alpha\beta} \to \mathcal{H}_{\beta\alpha}$ satisfying the relations $J^{\alpha\beta}_{\beta\alpha} = -(J^{\beta\alpha}_{\alpha\beta})^*$.*

PROOF. The proof follows just as for Lemma 3.16, except that the equation $J^2 = -1$ forces the relations $J^{\alpha\beta}_{\beta\alpha} = -(J^{\beta\alpha}_{\alpha\beta})^*$, which imply, in particular, that for each $\alpha \in \widehat{\mathcal{A}}$, $(J^{\alpha\alpha}_{\alpha\alpha})^2 = -1$, so that $m_{\alpha\alpha}$ must be even. □

Let us denote by $\mathrm{Sym}^0_S(\mathbb{Z}_{\geq 0})$ the set of all matrices in $\mathrm{Sym}_S(\mathbb{Z}_{\geq 0})$ with even diagonal entries. For $n = 2k$, let

$$\Omega_n = \begin{pmatrix} 0 & -1_k \\ 1_k & 0 \end{pmatrix}.$$

LEMMA 3.21. *Let $m \in \mathrm{Sym}^0_S(\mathbb{Z}_{\geq 0})$. Define an antiunitary J_m on $\mathcal{H}_m$ by*

$$(J_m)^{\gamma\delta}_{\alpha\beta} = \begin{cases} (X_m)^{\beta\alpha}_{\alpha\beta} & \text{if } (\gamma, \delta) = (\beta, \alpha) \text{ and } \alpha < \beta, \\ -(X_m)^{\beta\alpha}_{\alpha\beta} & \text{if } (\gamma, \delta) = (\beta, \alpha) \text{ and } \alpha > \beta, \\ \Omega_{m_{\alpha\alpha}}(X_m)^{\alpha\alpha}_{\alpha\alpha} & \text{if } \alpha = \beta = \gamma = \delta, \\ 0 & \text{otherwise.} \end{cases}$$

Then, up to unitary equivalence, J_m is the unique real structure on $\mathcal{H}_m$ of KO-dimension 3 *or* 5 mod 8.

PROOF. The proof follows that of Lemma 3.17, except we now have that $K^T_{\alpha\alpha} = \Omega_{m_{\alpha\alpha}} K_{\alpha\alpha} \Omega^T_{m_{\alpha\alpha}}$ instead of $K^T_{\alpha\alpha} = K_{\alpha\alpha}$; each $K_{\alpha\alpha}\Omega_{m_{\alpha\alpha}}$ is therefore unitary and complex skew-symmetric, so that we choose $W_{\alpha\alpha}$ unitary such that

$$K_{\alpha\alpha}\Omega_{m_{\alpha\alpha}} = W_{\alpha\alpha}\Omega_{m_{\alpha\alpha}} W^T_{\alpha\alpha},$$

or equivalently, $K_{\alpha\alpha} = W_{\alpha\alpha}\Omega_{m_{\alpha\alpha}}W_{\alpha\alpha}^T\Omega_{m_{\alpha\alpha}}^T$. One can then construct the unitary equivalence U between $(\mathcal{H}_m, J)$ and $(\mathcal{H}, J_m)$ as before. □

Much as in the analogous case of KO-dimension 1 or 7 mod 8, Lemmas 3.20 and 3.21 together imply the following characterisation of real bimodules of KO-dimension 3 or 5 mod 8:

PROPOSITION 3.22. *Let* $n = 3$ *or* 5 mod 8. *Then the map* $\iota_n : \mathrm{Bimod}(\mathcal{A}, n) \to \mathrm{Bimod}(\mathcal{A})$ *defined by* $\iota_n : [(\mathcal{H}, J)] \mapsto [\mathcal{H}]$ *is injective, and*

$$(\mathrm{mult} \circ \iota_n)(\mathrm{Bimod}(\mathcal{A}, n)) = \mathrm{Sym}_S^0(\mathbb{Z}_{\geq 0}). \tag{3.15}$$

Finally, these results immediately imply the following description of $\mathrm{U}_{\mathcal{A}}^{\mathrm{LR}}(\mathcal{H}, J)$:

PROPOSITION 3.23. *Let* $(\mathcal{H}, J)$ *be a real* $\mathcal{A}$*-bimodule of* KO*-dimension* 3 *or* 5 mod 8 *with multiplicity matrix* m. *Then*

(3.16)

$$\mathrm{comp}(\mathrm{U}_{\mathcal{A}}^{\mathrm{LR}}(\mathcal{H}, J)) = \left\{ (1_{n_\alpha} \otimes U_{\alpha\beta} \otimes 1_{n_\beta})_{\alpha,\beta\in\widehat{\mathcal{A}}} \in \mathrm{comp}(\mathrm{U}_{\mathcal{A}}^{\mathrm{LR}}(\mathcal{H})) \mid {}^{U_{\alpha\alpha}\in\mathrm{Sp}(m_{\alpha\alpha}),}_{U_{\beta\alpha}=\overline{U_{\alpha\beta}},\ \alpha\neq\beta} \right\}$$
$$\cong \prod_{\alpha\in\widehat{\mathcal{A}}} \left(\mathrm{Sp}(m_{\alpha\alpha}) \times \prod_{\substack{\beta\in\widehat{\mathcal{A}} \\ \beta>\alpha}} \mathrm{U}(m_{\alpha\beta}) \right).$$

3.4. Real bimodules of even KO**-dimension.** We now come to the case of even KO-dimension. Before continuing, note that for $(\mathcal{H}, \gamma, J)$ a real bimodule of even KO-dimension,

$$\forall p, q \in KO_0(\mathcal{A}),\ \langle q, p\rangle = \varepsilon'' \langle p, q\rangle,$$

as a direct result of the relation $J\gamma = \varepsilon''\gamma J$; this is then equivalent to the condition

$$\cap = \varepsilon'' \cap^T, \tag{3.17}$$

where $\cap$ is the matrix of the intersection form. Thus, for KO-dimension 0 or 4 mod 8, the intersection form is symmetric, whilst for KO-dimension 2 or 6 mod 8, it is anti-symmetric. It then follows, in particular, that a real $\mathcal{A}$-bimodule of KO-dimension 2 or 6 mod 8 satisfies Poincaré duality only if $\mathcal{A}$ has an even number of direct summands in its Wedderburn decomposition, as an anti-symmetric $k \times k$ matrix for k odd is necessarily degenerate.

3.4.1. *KO-dimension* 0 *or* 4 mod 8. We begin with the case where $\varepsilon'' = 1$ and hence $[\gamma, J] = 0$, *i.e.* of KO-dimension 0 or 4 mod 8.

Let $(\mathcal{H}, \gamma, J)$ be a real $\mathcal{A}$-bimodule of KO-dimension n mod 8, for $n = 0$ or 4; let the mutually orthogonal projections P^{even} and P^{odd} on $\mathcal{H}$ be defined as before. Then, since $[J, \gamma] = 0$, we have that $J = J^{\mathrm{even}} \oplus J^{\mathrm{odd}}$, where $J^{\mathrm{even}} = P^{\mathrm{even}}JP^{\mathrm{even}}$ and $J^{\mathrm{odd}} = P^{\mathrm{odd}}JP^{\mathrm{odd}}$. One can then check that $(\mathcal{H}^{\mathrm{even}}, J^{\mathrm{even}})$ and $(\mathcal{H}^{\mathrm{odd}}, J^{\mathrm{odd}})$ are real $\mathcal{A}$-bimodules of KO-dimension 1 or 7 mod 8 if $n = 0$, and 3 or 5 mod 8 if $n = 4$. On the other hand, given $(\mathcal{H}^{\mathrm{even}}, J^{\mathrm{even}})$ and $(\mathcal{H}^{\mathrm{odd}}, J^{\mathrm{odd}})$, one can immediately reconstruct $(\mathcal{H}, \gamma, J)$ by setting $\gamma = 1_{\mathcal{H}^{\mathrm{even}}} \oplus (-1_{\mathcal{H}^{\mathrm{odd}}})$ and $J = J^{\mathrm{even}} \oplus J^{\mathrm{odd}}$. Thus we have proved the following analogue of Proposition 3.6:

PROPOSITION 3.24. *Let* $\mathcal{A}$ *be a real* C^**-algebra. Let* k_0 *denote* 1 *or* 7 mod 8, *and let* k_4 *denote* 3 *or* 5 mod 8. *Then for* $n = 0, 4$ mod 8, *the map*

$$C_n : \mathrm{Bimod}(\mathcal{A}, n) \to \mathrm{Bimod}(\mathcal{A}, k_n) \times \mathrm{Bimod}(\mathcal{A}, k_n)$$

given by $C_n([(\mathcal{H}, \gamma, J)]) := ([(\mathcal{H}^{\mathrm{even}}, J^{\mathrm{even}})], [(\mathcal{H}^{\mathrm{odd}}, J^{\mathrm{odd}})])$ *is an isomorphism of monoids.*

One can then apply this decomposition to the group $\mathrm{U}^{\mathrm{LR}}_{\mathcal{A}}(\mathcal{H},\gamma,J)$ to find:

COROLLARY 3.25. *Let $(\mathcal{H},\gamma,J)$ be a real $\mathcal{A}$-bimodule of KO-dimension* 0 *or* 4 mod 8. *Then*

$$\mathrm{U}^{\mathrm{LR}}_{\mathcal{A}}(\mathcal{H},\gamma,J) = \mathrm{U}^{\mathrm{LR}}_{\mathcal{A}}(\mathcal{H}^{\mathrm{even}},J^{\mathrm{even}}) \oplus \mathrm{U}^{\mathrm{LR}}_{\mathcal{A}}(\mathcal{H}^{\mathrm{odd}},J^{\mathrm{odd}}). \tag{3.18}$$

Combining Proposition 3.24 with our earlier characterisations of real bimodules of odd KO-dimension, we immediately obtain the following:

PROPOSITION 3.26. *Let $n = 0$ or $4 \bmod 8$. Then the map $\iota_n : \mathrm{Bimod}(\mathcal{A},n) \to \mathrm{Bimod}^{\mathrm{even}}(\mathcal{A})$ defined by $[(\mathcal{H},\gamma,J)] \mapsto ([(\mathcal{H},\gamma)])$ is injective, and*

$$(\mathrm{mult}^{\mathrm{even}} \circ \iota_n)(\mathrm{Bimod}(\mathcal{A},n)) = \begin{cases} \mathrm{Sym}_S(\mathbb{Z}_{\geq 0}) \times \mathrm{Sym}_S(\mathbb{Z}_{\geq 0}) & \text{if } n = 0 \bmod 8, \\ \mathrm{Sym}^0_S(\mathbb{Z}_{\geq 0}) \times \mathrm{Sym}^0_S(\mathbb{Z}_{\geq 0}) & \text{if } n = 4 \bmod 8. \end{cases}$$

In particular,

$$\mathrm{Bimod}_q(\mathcal{A},n) := \iota_n^{-1}(\mathrm{Bimod}^{\mathrm{even}}_q(\mathcal{A}))$$

is thus the set of all equivalence classes of quasi-orientable real $\mathcal{A}$-bimodules of KO-dimension n mod 8; the last Proposition then implies the following:

COROLLARY 3.27. *Let $n = 0$ or $4 \bmod 8$. Then*

$$(\mathrm{mult}_q \circ \iota_n)(\mathrm{Bimod}_q(\mathcal{A},n)) = \mathrm{Sym}_S(\mathbb{Z}). \tag{3.19}$$

3.4.2. *KO-dimension* 2 *or* 6 mod 8. Finally, let us consider the remaining case where $\varepsilon'' = -1$ and hence $\{\gamma, J\} = 0$, *i.e.* of KO-dimensions 2 and 6 mod 8.

Let $(\mathcal{H},\gamma,J)$ be a real $\mathcal{A}$-bimodule of KO-dimension n mod 8 for $n = 2$ or 6. Since $\{J,\gamma\}$, we have that

$$J = \begin{pmatrix} 0 & \varepsilon\tilde{J}^* \\ \tilde{J} & 0 \end{pmatrix},$$

where $\tilde{J} := P^{\mathrm{odd}} J P^{\mathrm{even}}$ is an antiunitary $\mathcal{H}^{\mathrm{even}} \to \mathcal{H}^{\mathrm{odd}}$, so that for $a \in \mathcal{A}$,

$$\lambda^{\mathrm{odd}}(a) = \tilde{J}\rho^{\mathrm{even}}(a^*)\tilde{J}^*, \quad \rho^{\mathrm{odd}}(a) = \tilde{J}\lambda^{\mathrm{even}}(a^*)\tilde{J}^*.$$

It then follows, in particular, that $\mathrm{mult}[\mathcal{H}^{\mathrm{odd}}] = \mathrm{mult}[\mathcal{H}^{\mathrm{even}}]^T$.

Now, let J' be another real structure on $(\mathcal{H},\gamma)$ of KO-dimension n mod 8, and let $\tilde{J}' = P^{\mathrm{odd}} J' P^{\mathrm{even}}$. Define $K \in \mathrm{U}^{\mathrm{LR}}_{\mathcal{A}}(\mathcal{H},\gamma)$ by $K = 1_{\mathcal{H}^{\mathrm{even}}} \oplus (\tilde{J}'\tilde{J}^*)$. Then, by construction, $J' = KJK^*$, *i.e.* K is a unitary equivalence of real structures between J and J'. Thus, real structures of KO-dimension 2 or 6 mod 8 are unique. As a result, we have proved the following analogue of Proposition 2.17:

PROPOSITION 3.28. *Let $\mathcal{A}$ be a real C^*-algebra, and let $n = 2$ or $6 \bmod 8$. Then the map*

$$C_n : \mathrm{Bimod}(\mathcal{A},n) \to \mathrm{Bimod}(\mathcal{A})$$

given by $C_n([(\mathcal{H},\gamma,J)]) := ([\mathcal{H}^{\mathrm{even}}])$ is an isomorphism of monoids.

Again, as an immediate consequence, we obtain the following characterisation of $\mathrm{U}^{\mathrm{LR}}_{\mathcal{A}}(\mathcal{H},\gamma,J)$:

COROLLARY 3.29. *Let $(\mathcal{H},\gamma,J)$ be a real $\mathcal{A}$-bimodule of KO-dimension* 2 *or* 6 mod 8. *Then*

(3.20)

$$\begin{aligned}\mathrm{U}^{\mathrm{LR}}_{\mathcal{A}}(\mathcal{H},\gamma,J) &= \{U^{\mathrm{even}} \oplus U^{\mathrm{odd}} \in \mathrm{U}^{\mathrm{LR}}_{\mathcal{A}}(\mathcal{H}^{\mathrm{even}}) \oplus \mathrm{U}^{\mathrm{LR}}_{\mathcal{A}}(\mathcal{H}^{\mathrm{odd}}) \mid U^{\mathrm{odd}} = \tilde{J}U^{\mathrm{even}}\tilde{J}^*\} \\ &\cong \mathrm{U}^{\mathrm{LR}}_{\mathcal{A}}(\mathcal{H}^{\mathrm{even}}).\end{aligned}$$

Finally, one can combine Proposition 3.28 with our observation concerning the uniqueness up to unitary equivalence of real structures of KO-dimension 2 or 6 mod 8 and earlier results on multiplicity matrices to obtain the following characterisation:

PROPOSITION 3.30. *Let* $n = 2$ *or* 6 mod 8. *Then the map* $\iota_n : \mathrm{Bimod}(\mathcal{A}, n) \to \mathrm{Bimod}^{\mathrm{even}}(\mathcal{A})$ *defined by* $[(\mathcal{H}, \gamma, J)] \mapsto ([\mathcal{H}, \gamma])$ *is injective, and*

$$(\mathrm{mult}^{\mathrm{even}} \circ \iota_n)(\mathrm{Bimod}(\mathcal{A}, n)) = \{(m^{\mathrm{even}}, m^{\mathrm{odd}}) \in M_S(\mathbb{Z}_{\geq 0})^2 \mid m^{\mathrm{odd}} = (m^{\mathrm{even}})^T\} \cong M_S(\mathbb{Z}_{\geq 0}). \tag{3.21}$$

Once more, it follows that

$$\mathrm{Bimod}_q(\mathcal{A}, n) := \iota_n^{-1}(\mathrm{Bimod}_q^{\mathrm{even}}(\mathcal{A})),$$

is the set of all equivalence classes of quasi-orientable real $\mathcal{A}$-bimodules of KO-dimension n mod 8, for which we can again obtain a characterisation in terms of signed multiplicity matrices:

COROLLARY 3.31. *Let* $n = 2$ *or* 6 mod 8. *Then*

$$(\mathrm{mult}_q \circ \iota_n)(\mathrm{Bimod}_q(\mathcal{A}, n)) = \{m \in M_S(\mathbb{Z}) \mid m^T = -m\}. \tag{3.22}$$

3.4.3. *S^0-real bimodules of even KO-dimension.* Let us now characterise quasi-orientability, orientability and Poincaré duality for an even KO-dimensional S^0-real $\mathcal{A}$-bimodule $(\mathcal{H}, \gamma, J, \epsilon)$ by means of suitable conditions on $(\mathcal{H}_i, \gamma_i)$ expressible entirely in terms of the pair of multiplicity matrices of $(\mathcal{H}_i, \gamma_i)$

We begin by considering quasi-orientability:

PROPOSITION 3.32. *Let* $(\mathcal{H}, \gamma, J, \epsilon)$ *be an S^0-real $\mathcal{A}$-bimodule of even KO-dimension* n mod 8. *Then* $(\mathcal{H}, \gamma)$ *is quasi-orientable if and only if* $(\mathcal{H}_i, \gamma_i)$ *is quasi-orientable and*

$$\begin{cases} \mathrm{supp}(m_i^{\mathrm{even}}) \cap \mathrm{supp}((m_i^{\mathrm{odd}})^T) = \emptyset & \text{if } n = 0,\ 4, \\ \mathrm{supp}(m_i^{\mathrm{even}}) \cap \mathrm{supp}((m_i^{\mathrm{even}})^T) = \mathrm{supp}(m_i^{\mathrm{odd}}) \cap \mathrm{supp}((m_i^{\mathrm{odd}})^T) = \emptyset & \text{if } n = 2,\ 6, \end{cases}$$

for $(m_i^{\mathrm{even}}, m_i^{\mathrm{odd}})$ *the multiplicity matrices of* $(\mathcal{H}_i, \gamma_i)$, *in which case, if* μ *and* $\mu_i = m_i^{\mathrm{even}} - m_i^{\mathrm{odd}}$ *are the signed multiplicity matrices of* $(\mathcal{H}, \gamma)$ *and* $(\mathcal{H}_i, \gamma_i)$, *respectively, then*

$$\mu = \mu_i + \varepsilon'' \mu_i^T. \tag{3.23}$$

PROOF. First, let $(m^{\mathrm{even}}, m^{\mathrm{odd}})$ and $(m_i^{\mathrm{even}}, m_i^{\mathrm{odd}})$ denote the pairs of multipicity matrices of $(\mathcal{H}, \gamma)$ and $(\mathcal{H}_i, \gamma_i)$, respectively. It then follows that

$$m^{\mathrm{even}} = \begin{cases} m_i^{\mathrm{even}} + (m_i^{\mathrm{even}})^T & \text{if } n = 0,\ 4, \\ m_i^{\mathrm{even}} + (m_i^{\mathrm{odd}})^T & \text{if } n = 2,\ 6; \end{cases}$$

$$m^{\mathrm{odd}} = \begin{cases} m_i^{\mathrm{odd}} + (m_i^{\mathrm{odd}})^T & \text{if } n = 0,\ 4, \\ m_i^{\mathrm{odd}} + (m_i^{\mathrm{even}})^T & \text{if } n = 2,\ 6. \end{cases}$$

Thus $\mathrm{supp}(m^{\mathrm{even}}) = \mathrm{supp}(m_i^{\mathrm{even}}) \cup S^{\mathrm{even}}$, $\mathrm{supp}(m^{\mathrm{odd}}) = \mathrm{supp}(m_i^{\mathrm{odd}}) \cup S^{\mathrm{odd}}$, where

$$S^{\mathrm{even}} = \begin{cases} \mathrm{supp}((m_i^{\mathrm{even}})^T) & \text{if } n = 0,\ 4, \\ \mathrm{supp}((m_i^{\mathrm{odd}})^T) & \text{if } n = 2,\ 6; \end{cases} \qquad S^{\mathrm{odd}} = \begin{cases} \mathrm{supp}((m_i^{\mathrm{odd}})^T) & \text{if } n = 0,\ 4, \\ \mathrm{supp}((m_i^{\mathrm{even}})^T) & \text{if } n = 2,\ 6. \end{cases}$$

Then,

$$\operatorname{supp}(m^{\text{even}}) \cap \operatorname{supp}(m^{\text{odd}}) = (\operatorname{supp}(m_i^{\text{even}}) \cap \operatorname{supp}(m_i^{\text{odd}})) \cup (S^{\text{even}} \cap \operatorname{supp}(m_i^{\text{odd}})) \\ \cup (\operatorname{supp}(m_i^{\text{even}}) \cap S^{\text{odd}}) \cup (S^{\text{even}} \cap S^{\text{odd}}),$$

so that $(\mathcal{H}, \gamma)$ is quasi-orientable if and only if $(\mathcal{H}_i, \gamma_i)$ is quasi-orientable and

$$(S^{\text{even}} \cap \operatorname{supp}(m_i^{\text{odd}})) \cup (\operatorname{supp}(m_i^{\text{even}}) \cap S^{\text{odd}}) = \emptyset,$$

as required.

Finally, if $\mu = m^{\text{even}} - m^{\text{odd}}$ and $\mu_i = m_i^{\text{even}} - m_i^{\text{odd}}$ are the signed multiplicity matrices of $(\mathcal{H}, \gamma)$ and $(\mathcal{H}_i, \gamma_i)$, respectively, then the relations amongst m^{even}, m^{odd}, m_i^{even}, and m_i^{odd} given at the beginning immediately yield the equation $\mu = \mu_i + \varepsilon'' \mu_i^T$. □

Let us now turn to orientability:

PROPOSITION 3.33. *Let $(\mathcal{H}, \gamma, J, \epsilon)$ be a quasi-orientable S^0-real $\mathcal{A}$-bimodule of even KO-dimension $n \bmod 8$. Then $(\mathcal{H}, \gamma)$ is orientable if and only if $(\mathcal{H}_i, \gamma_i)$ is orientable and, if $n = 2$ or $6 \bmod 8$, for all $j \in \{1, \dots, N\}$ such that $\mathbb{K}_j = \mathbb{C}$,*

$$(\mu_i)_{\mathbf{n}_j \overline{\mathbf{n}}_j} = (\mu_i)_{\overline{\mathbf{n}}_j \mathbf{n}_j}, \tag{3.24}$$

where μ_i is the signed multiplicity matrix of $(\mathcal{H}_i, \gamma_i)$.

PROOF. Let μ be the signed multiplicity matrix of $(\mathcal{H}, \gamma)$. Propositions 2.19 and 3.13 together imply that $(\mathcal{H}, \gamma, J, \epsilon)$ is orientable if and only if

$$\gamma_i = \sum_{k,l=1}^{N} \lambda_i(\operatorname{sgn}(\widehat{\mu}_{kl}) e_k) \rho_i(e_l) = \varepsilon'' \sum_{k,l=1}^{N} \lambda_i(e_l) \rho_i(\operatorname{sgn}(\widehat{\mu}_{kl}) e_k),$$

and by considering individual components $(\gamma_i)_{\alpha\beta}$, one can easily check that this in turn holds if and only if $(\mathcal{H}_i, \gamma_i)$ is orientable and for all $k \in \{1, \dots, N\}$,

$$\operatorname{sgn}(\widehat{\mu}_{kk}) = \varepsilon'' \operatorname{sgn}(\widehat{\mu}_{kk}).$$

This last condition is trivial when $\varepsilon'' = 1$, *i.e.* when $n = 0$ or $4 \bmod 8$, so let us suppose instead that $n = 2$ or $6 \bmod 8$, so that $\varepsilon'' = -1$. If $(\mathcal{H}, \gamma)$ is orientable, then, by the above discussion, $(\mathcal{H}_i, \gamma_i)$ is orientable and the diagonal entries of $\widehat{\mu}$ vanish, which in turn implies by Proposition 3.13 that for each $l \in \{1, \dots, N\}$ and all α, $\beta \in \widehat{M_{k_l}(\mathbb{K}_l)}$, $\mu_{\alpha\beta} = 0$. By antisymmetry of μ, this is equivalent to having, for all $l \in \{1, \dots, N\}$ such that $\mathbb{K}_l = \mathbb{C}$, $\mu_{\mathbf{n}_l \overline{\mathbf{n}}_l} = 0$, or equivalently,

$$(\mu_i)_{\mathbf{n}_j \overline{\mathbf{n}}_j} = (\mu_i)_{\overline{\mathbf{n}}_j \mathbf{n}_j},$$

where μ_i is the signed multiplicity matrix of $(\mathcal{H}_i, \gamma_i)$. On the other hand, if $(\mathcal{H}_i, \gamma_i)$ is orientable and this condition on μ_i holds, then μ certainly satisfies the above condition, so that $(\mathcal{H}, \gamma)$ is indeed orientable. □

Finally, let us consider Poincaré duality.

PROPOSITION 3.34. *Let $(\mathcal{H}, \gamma, J, \epsilon)$ be an S^0-real $\mathcal{A}$-bimodule of even KO-dimension $n \bmod 8$, let $(m_i^{\text{even}}, m_i^{\text{odd}})$ denote the multiplicity matrices of $(\mathcal{H}_i, \gamma_i)$, and let $\cap$ denote the matrix of the intersection form of $(\mathcal{H}, \gamma)$. Finally, let $\mu_i = m_i^{\text{even}} - m_i^{\text{odd}}$. Then*

$$\cap_{kl} = \tau_k \tau_l (\widehat{\mu_i} + \varepsilon'' \widehat{\mu_i}^T)_{kl}, \tag{3.25}$$

so that $(\mathcal{H}, \gamma)$ satisfies Poincaré duality if and only if $\widehat{\mu_i} + \varepsilon'' \widehat{\mu_i}^T$ is non-degenerate.

PROOF. By Proposition 2.20, $\cap = \cap_i + \varepsilon'' \cap_i^T$ for $\cap_i$ the matrix of the intersection form of $(\mathcal{H}_i, \gamma_i)$, which, together with Proposition 3.14, yields the desired result. □

3.5. Bimodules in the Chamseddine–Connes–Marcolli model. To illustrate the structure theory outlined thus far, let us apply it to the construction of the finite spectral triple of the NCG Standard Model given by Chamseddine, Connes and Marcolli [4, §§2.1, 2.2, 2.4] (cf. also [8, §1.13]).

Let $\mathcal{A}_{LR} = \mathbb{C} \oplus \mathbb{H}_L \oplus \mathbb{H}_R \oplus M_3(\mathbb{C})$, where the labels L and R serve to distinguish the two copies of $\mathcal{H}$; we can therefore write $\widehat{\mathcal{A}_{LR}} = \{\mathbf{1}, \overline{\mathbf{1}}, \mathbf{2}_L, \mathbf{2}_R, \mathbf{3}, \overline{\mathbf{3}}\}$ without ambiguity. Now, let $(\mathcal{M}_F, \gamma_F, J_F)$ be the orientable real $\mathcal{A}_{LR}$-bimodule of KO-dimension 6 mod 8 with signed multiplicity matrix

$$\mu = \begin{pmatrix} 0 & 0 & -1 & 1 & 0 & 0 \\ 0 & 0 & 0 & 0 & 0 & 0 \\ 1 & 0 & 0 & 0 & 1 & 0 \\ -1 & 0 & 0 & 0 & -1 & 0 \\ 0 & 0 & -1 & 1 & 0 & 0 \\ 0 & 0 & 0 & 0 & 0 & 0 \end{pmatrix}.$$

This bimodule is, in fact, an S^0-real bimodule for $\epsilon_F = \lambda(-1,1,1,-1)$; $\mathcal{E} = (\mathcal{M}_F)_i$ is then the orientable even $\mathcal{A}_{LR}$-bimodule with signed multiplicity matrix

$$\mu_{\mathcal{E}} = \begin{pmatrix} 0 & 0 & 0 & 0 & 0 & 0 \\ 0 & 0 & 0 & 0 & 0 & 0 \\ 1 & 0 & 0 & 0 & 1 & 0 \\ -1 & 0 & 0 & 0 & -1 & 0 \\ 0 & 0 & 0 & 0 & 0 & 0 \\ 0 & 0 & 0 & 0 & 0 & 0 \end{pmatrix}.$$

Note, however, that neither $\mathcal{M}_F$ nor $\mathcal{E}$ satisfies Poincaré duality, as

$$\hat{\mu} = \begin{pmatrix} 0 & -1 & 1 & 0 \\ 1 & 0 & 0 & 1 \\ -1 & 0 & 0 & -1 \\ 0 & -1 & 1 & 0 \end{pmatrix}, \quad \widehat{\mu_{\mathcal{E}}} = \begin{pmatrix} 0 & 0 & 0 & 0 \\ 1 & 0 & 0 & 1 \\ -1 & 0 & 0 & -1 \\ 0 & 0 & 0 & 0 \end{pmatrix}$$

are both clearly degenerate; the intersection forms of $\mathcal{M}_F$ and $\mathcal{E}$ are given by the matrices $\cap = 2\hat{\mu}$ and $\cap_{\mathcal{E}} = 2\widehat{\mu_{\mathcal{E}}}$, respectively.

In order to introduce N generations of fermions and anti-fermions, one now considers the real $\mathcal{A}$-bimodule $\mathcal{H}_F := (\mathcal{M}_F)^{\oplus N}$; by abuse of notation, γ_F, J_F and ϵ_F now also denote the relevant structure operators on $\mathcal{H}_F$. In terms of multiplicity matrices and intersection forms, the sole difference from our discussion of $\mathcal{M}_F$ is that all matrices are now multiplied by N.

Now, let $\mathcal{A}_F = \mathbb{C} \oplus \mathcal{H} \oplus M_3(\mathbb{C})$, which we consider as a subalgebra of $\mathcal{A}_{LR}$ by means of the embedding

$$(\zeta, q, m) \mapsto \left(\zeta, q, \begin{pmatrix} \lambda & 0 \\ 0 & \overline{\lambda} \end{pmatrix}, m\right);$$

just as we could for $\mathcal{A}_{LR}$, we can write $\widehat{\mathcal{A}_F} = \{\mathbf{1}, \overline{\mathbf{1}}, \mathbf{2}, \mathbf{3}, \overline{\mathbf{3}}\}$ without ambiguity. We can therefore view $\mathcal{H}_F$ as a real $\mathcal{A}_F$-bimodule of KO-dimension 6 mod 8, whose pair of multiplicity matrices $(m^{\text{even}}, m^{\text{odd}})$ is then given by

$$m^{\text{even}} = N \begin{pmatrix} 1 & 1 & 0 & 0 & 0 \\ 0 & 0 & 0 & 0 & 0 \\ 1 & 0 & 0 & 1 & 0 \\ 1 & 1 & 0 & 0 & 0 \\ 0 & 0 & 0 & 0 & 0 \end{pmatrix}, \quad m^{\text{odd}} = N \begin{pmatrix} 1 & 0 & 1 & 1 & 0 \\ 1 & 0 & 0 & 1 & 0 \\ 0 & 0 & 0 & 0 & 0 \\ 0 & 0 & 1 & 0 & 0 \\ 0 & 0 & 0 & 0 & 0 \end{pmatrix};$$

the essential observation is that the irreducible representation $\mathbf{2}_R$ of $\mathcal{A}_{LR}$ corresponds to the representation $\mathbf{1} \oplus \overline{\mathbf{1}}$ of $\mathcal{A}_F$, whilst $\mathbf{2}_L$, $\mathbf{3}$ and $\overline{\mathbf{3}}$ correspond to $\mathbf{2}$, $\mathbf{3}$ and $\overline{\mathbf{3}}$, respectively.

Note that $\mathcal{H}_F$ now fails even to be quasi-orientable let alone orientable, with the sub-bimodule $(\mathcal{H}_F)_{11}$ providing the obstruction, and even if we were to restore quasi-orientability by setting $(\mathcal{H}_F)_{11} = 0$, $(\mathcal{H}_F)_{1\overline{1}}$ and $(\mathcal{H}_F)_{\overline{1}1}$ would still present an obstruction to orientability by Proposition 3.13. Note also that $\mathcal{H}_F$ must necessarily fail to satisfy Poincaré duality, as the matrix $\cap_F$ of its intersection form is a 3×3 anti-symmetric matrix, and thus *a priori* degenerate. Let us nonetheless compute $\cap_F$:

$$\widehat{m^{\text{even}}} - \widehat{m^{\text{odd}}} = N \begin{pmatrix} 2 & 0 & 0 \\ 1 & 0 & 1 \\ 2 & 0 & 0 \end{pmatrix} - N \begin{pmatrix} 2 & 1 & 2 \\ 0 & 0 & 0 \\ 0 & 1 & 0 \end{pmatrix} = N \begin{pmatrix} 0 & -1 & -2 \\ 1 & 0 & 1 \\ 2 & -1 & 0 \end{pmatrix},$$

and hence, by Proposition 3.14,

$$\cap_F = 2N \begin{pmatrix} 0 & -1 & -1 \\ 1 & 0 & 1 \\ 1 & -1 & 0 \end{pmatrix}.$$

Finally, let us consider the S^0-real structure on $\mathcal{H}_F$ the $\mathcal{A}_F$-bimodule, inherited from $\mathcal{H}_F$ as an $\mathcal{A}_{LR}$-bimodule; we now denote $(\mathcal{H}_F)_i$ by $\mathcal{H}_f$. One still has that $\mathcal{H}_f = \mathcal{E}^{\oplus N}$, which is still orientable and thus specified by the signed multiplicity matrix

$$\mu_f = N \begin{pmatrix} -1 & 0 & 0 & -1 & 0 \\ -1 & 0 & 0 & -1 & 0 \\ 1 & 0 & 0 & 1 & 0 \\ 0 & 0 & 0 & 0 & 0 \\ 0 & 0 & 0 & 0 & 0 \end{pmatrix};$$

the intersection form is then given by the matrix

$$\cap_f = 2N \begin{pmatrix} -1 & 0 & -1 \\ 1 & 0 & 1 \\ 0 & 0 & 0 \end{pmatrix},$$

so that $\mathcal{H}_f$ fails to satisfy Poincaré duality as an $\mathcal{A}_F$-bimodule.

4. Dirac Operators and their Structure

4.1. The order one condition. We now examine the structure of Dirac operators in detail. We will find it useful to begin with the study of operators between $\mathcal{A}$-bimodules (for fixed $\mathcal{A}$) satisfying a further generalisation of the order one condition. Thus, let $\mathcal{A}$ be a fixed real C^*-algebra, and let $\mathcal{H}_1$ and $\mathcal{H}_2$ be fixed $\mathcal{A}$-bimodules with multiplicity matrices m_1 and m_2, respectively.

DEFINITION 4.1. We shall say that a map $T \in \mathcal{L}(\mathcal{H}_1, \mathcal{H}_2)$ satisfies the *generalised order one condition* if

$$\forall a, b \in \mathcal{A},\ (\lambda_2(a)T - T\lambda_1(a))\rho_1(b) = \rho_2(b)(\lambda_2(a)T - T\lambda_1(a)). \tag{4.1}$$

Note that if $\mathcal{H}_1 = \mathcal{H}_2$, then the generalised order one condition reduces to the usual order one condition on Dirac operators.

It is easy to check that the generalised order one condition is, in fact, equivalent to the following alternative condition:

$$\forall a, b \in \mathcal{A},\ (\rho_2(a)T - T\rho_1(a))\lambda_1(b) = \lambda_2(b)(\rho_2(a)T - T\rho_1(a)). \tag{4.2}$$

Thus, the following are equivalent for $T \in \mathcal{L}(\mathcal{H}_1, \mathcal{H}_2)$:

(1) T satisfies the generalised order one condition;
(2) For all $a \in \mathcal{A}$, $\lambda_2(a)T - T\lambda_1(a)$ is right $\mathcal{A}$-linear;
(3) For all $a \in \mathcal{A}$, $\rho_2(a)T - T\rho_1(a)$ is left $\mathcal{A}$-linear.

Now, since the unitary group $\mathrm{U}(\mathcal{A})$ of $\mathcal{A}$ is a compact Lie group, let μ be the normalised bi-invariant Haar measure on $\mathrm{U}(\mathcal{A})$.

LEMMA 4.2. *Let $\mathcal{H}_1$ and $\mathcal{H}_2$ be $\mathcal{A}$-bimodules. Define operators E_λ and E_ρ on $\mathcal{L}^1_{\mathcal{A}}(\mathcal{H}_1, \mathcal{H}_2)$ by*

$$E_\lambda(T) := \int_{\mathrm{U}(\mathcal{A})} \mathrm{d}\mu(u)\lambda_2(u)T\lambda_1(u^{-1}), \quad E_\rho(T) := \int_{\mathrm{U}(\mathcal{A})} \mathrm{d}\mu(u)\rho_2(u^{-1})T\rho_1(u). \tag{4.3}$$

Then E_λ and E_ρ are commuting idempotents such that

$$\mathrm{im}(E_\lambda) = \mathcal{L}^{\mathrm{L}}_{\mathcal{A}}(\mathcal{H}_1, \mathcal{H}_2), \quad \mathrm{im}(E_\rho) = \mathcal{L}^{\mathrm{R}}_{\mathcal{A}}(\mathcal{H}_1, \mathcal{H}_2),$$

and

$$\ker(E_\lambda) = \mathrm{im}(\mathrm{id} - E_\lambda) \subseteq \mathcal{L}^{\mathrm{R}}_{\mathcal{A}}(\mathcal{H}_1, \mathcal{H}_2), \quad \ker(E_\rho) = \mathrm{im}(\mathrm{id} - E_\rho) \subseteq \mathcal{L}^{\mathrm{L}}_{\mathcal{A}}(\mathcal{H}_1, \mathcal{H}_2),$$

while

$$\mathrm{im}(E_\lambda E_\rho) = \mathcal{L}^{\mathrm{LR}}_{\mathcal{A}}(\mathcal{H}_1, \mathcal{H}_2).$$

PROOF. First, the fact that E_λ and E_ρ are idempotents follows immediately from the Fubini-Tonelli theorem together with translation invariance of the Haar measure μ, whilst commutation of E_λ and E_ρ follows from the Fubini-Tonelli theorem together with the commutation of left and right actions on $\mathcal{H}_1$ and on $\mathcal{H}_2$. Moreover, by construction, E_λ and E_ρ act as the identity on $\mathcal{L}^{\mathrm{L}}_{\mathcal{A}}(\mathcal{H}_1, \mathcal{H}_2)$ and $\mathcal{L}^{\mathrm{R}}_{\mathcal{A}}(\mathcal{H}_1, \mathcal{H}_2)$, respectively, so that

$$\mathrm{im}(E_\lambda) \supseteq \mathcal{L}^{\mathrm{L}}_{\mathcal{A}}(\mathcal{H}_1, \mathcal{H}_2), \quad \mathrm{im}(E_\rho) \supseteq \mathcal{L}^{\mathrm{R}}_{\mathcal{A}}(\mathcal{H}_1, \mathcal{H}_2).$$

Now, let $T \in \mathcal{L}^1_{\mathcal{A}}(\mathcal{H}_1, \mathcal{H}_2)$. Then, by translation invariance of the Haar measure, it follows that for any $u \in \mathrm{U}(\mathcal{A})$,

$$E_\lambda(T) = \lambda_2(u)E_\lambda(T)\lambda_1(u)^*, \quad E_\rho(T) = \rho_2(u)E_\rho(T)\rho_1(u)^*,$$

or equivalently,

$$\lambda_2(u)E_\lambda(T) = E_\lambda(T)\lambda_1(u), \quad \rho_2(u)E_\rho(T) = E_\rho(T)\rho_1(u).$$

By the real analogue of the Russo-Dye theorem [19, Lemma 2.15.16], the convex hull of $\mathrm{U}(\mathcal{A})$ is weakly dense in the unit ball of $\mathcal{A}$, so that

$$\lambda_2(a)E_\lambda(T) = E_\lambda(T)\lambda_1(a), \quad \rho_2(a)E_\rho(T) = E_\rho(T)\rho_1(a)$$

for all $a \in \mathcal{A}$, *i.e.* $E_\lambda(T) \in \mathcal{L}^{\mathrm{L}}_{\mathcal{A}}(\mathcal{H}_1, \mathcal{H}_2)$ and $E_\rho(T) \in \mathcal{L}^{\mathrm{R}}_{\mathcal{A}}(\mathcal{H}_1, \mathcal{H}_2)$.

On the other hand,

$$(\mathrm{id} - E_\lambda)(T) = \int_{\mathrm{U}(\mathcal{A})} \mathrm{d}\mu(u)(T\lambda_1(u) - \lambda_2(u)T)\lambda_1(u^{-1}),$$

$$(\mathrm{id} - E_\rho)(T) = \int_{\mathrm{U}(\mathcal{A})} \mathrm{d}\mu(u)(T\rho_1(u^{-1}) - \rho_2(u^{-1})T)\rho_1(u),$$

so that by the generalised order one condition, $(\mathrm{id} - E_\lambda)(T) \in \mathcal{L}^{\mathrm{R}}_{\mathcal{A}}(\mathcal{H}_1, \mathcal{H}_2)$ and $(\mathrm{id} - E_\rho)(T) \in \mathcal{L}^{\mathrm{L}}_{\mathcal{A}}(\mathcal{H}_1, \mathcal{H}_2)$.

Finally, the commutation of E_λ and E_ρ together with our identification of $\mathrm{im}(E_\lambda)$ and of $\mathrm{im}(E_\rho)$ imply the desired result about $\mathrm{im}(E_\lambda E_\rho)$. □

Now, since

$$\mathrm{im}(\mathrm{id} - E_\lambda) \subseteq \mathrm{im}(E_\rho), \quad \mathrm{im}(\mathrm{id} - E_\rho) \subseteq \mathrm{im}(E_\lambda),$$

one has that

$$(\mathrm{id} - E_\lambda)E_\rho = \mathrm{id} - E_\lambda, \quad (\mathrm{id} - E_\rho)E_\lambda = \mathrm{id} - E_\rho,$$

which implies in turn that $\mathrm{id} - E_\rho$, $E_\lambda E_\rho$ and $\mathrm{id} - E_\lambda$ are mutually orthogonal idempotents such that

$$(\mathrm{id} - E_\rho) + E_\lambda E_\rho + (\mathrm{id} - E_\lambda) = \mathrm{id}.$$

We have therefore proved the following:

PROPOSITION 4.3 (Krajewski [18, §3.4]). *Let* $\mathcal{L}^{\mathrm{R}}_{\mathcal{A}}(\mathcal{H}_1, \mathcal{H}_2)^0$ *denote* $\ker(E_\lambda)$, *and let* $\mathcal{L}^{\mathrm{L}}_{\mathcal{A}}(\mathcal{H}_1, \mathcal{H}_2)^0$ *denote* $\ker(E_\rho)$. *Then*

$$\mathcal{L}^{1}_{\mathcal{A}}(\mathcal{H}_1, \mathcal{H}_2) = \mathcal{L}^{\mathrm{L}}_{\mathcal{A}}(\mathcal{H}_1, \mathcal{H}_2)^0 \oplus \mathcal{L}^{\mathrm{LR}}_{\mathcal{A}}(\mathcal{H}_1, \mathcal{H}_2) \oplus \mathcal{L}^{\mathrm{R}}_{\mathcal{A}}(\mathcal{H}_1, \mathcal{H}_2)^0, \tag{4.4}$$

where

$$\mathcal{L}^{\mathrm{L}}_{\mathcal{A}}(\mathcal{H}_1, \mathcal{H}_2)^0 \oplus \mathcal{L}^{\mathrm{LR}}_{\mathcal{A}}(\mathcal{H}_1, \mathcal{H}_2) = \mathcal{L}^{\mathrm{L}}_{\mathcal{A}}(\mathcal{H}_1, \mathcal{H}_2) \tag{4.5}$$

and

$$\mathcal{L}^{\mathrm{LR}}_{\mathcal{A}}(\mathcal{H}_1, \mathcal{H}_2) \oplus \mathcal{L}^{\mathrm{R}}_{\mathcal{A}}(\mathcal{H}_1, \mathcal{H}_2)^0 = \mathcal{L}^{\mathrm{R}}_{\mathcal{A}}(\mathcal{H}_1, \mathcal{H}_2). \tag{4.6}$$

Thus, elements of $\mathcal{L}^{\mathrm{L}}_{\mathcal{A}}(\mathcal{H}_1, \mathcal{H}_2)^0$ can be interpreted as the "purely" left $\mathcal{A}$-linear maps $\mathcal{H}_1 \to \mathcal{H}_2$, whilst elements of $\mathcal{L}^{\mathrm{R}}_{\mathcal{A}}(\mathcal{H}_1, \mathcal{H}_2)^0$ can be interpreted as the "purely" right $\mathcal{A}$-linear maps $\mathcal{H}_1 \to \mathcal{H}_2$.

One can readily check that the decomposition of Proposition 4.3 is respected by left multiplication by elements of $\mathcal{L}^{\mathrm{LR}}_{\mathcal{A}}(\mathcal{H}_2)$ and right multiplication by elements of $\mathcal{L}^{\mathrm{LR}}_{\mathcal{A}}(\mathcal{H}_1)$:

PROPOSITION 4.4. *For any* $T \in \mathcal{L}^{1}_{\mathcal{A}}(\mathcal{H}_1, \mathcal{H}_2)$, $A \in \mathcal{L}^{\mathrm{LR}}_{\mathcal{A}}(\mathcal{H}_1)$, $B \in \mathcal{L}^{\mathrm{LR}}_{\mathcal{A}}(\mathcal{H}_2)$,

$$E_\lambda(AT) = AE_\lambda(T), \quad E_\rho(TB) = E_\rho(T)B.$$

Now, if $T \in \mathcal{L}(\mathcal{H}_1, \mathcal{H}_2)$, it is easy to see that T satisfies the generalised order one condition if and only if each $T^{\gamma\delta}_{\alpha\beta}$ satisfies the generalised order one condition within $\mathcal{L}((\mathcal{H}_1)_{\alpha\beta}, (\mathcal{H}_2)_{\gamma\delta})$; by abuse of notation, we will also denote by E_λ and E_ρ the appropriate idempotents on each $\mathcal{L}((\mathcal{H}_1)_{\alpha\beta}, (\mathcal{H}_2)_{\gamma\delta})$. It then follows that

$$E_\lambda(T)^{\gamma\delta}_{\alpha\beta} = E_\lambda(T^{\gamma\delta}_{\alpha\beta}), \quad E_\rho(T)^{\gamma\delta}_{\alpha\beta} = E_\rho(T^{\gamma\delta}_{\alpha\beta}).$$

Finally, let us turn to characterising $\ker(E_\lambda)$ and $\ker(E_\rho)$; before proceeding, we first need a technical lemma:

LEMMA 4.5. *Let G be a compact Lie group, and let μ be the bi-invariant Haar measure on G. Let $(\mathcal{H}, \pi)$ and $(\mathcal{H}', \pi')$ be finite-dimensional irreducible unitary matrix representations of G. Then for any $T \in \mathcal{L}(\mathcal{H}', \mathcal{H})$, if $\pi \ncong \pi'$ then*

$$\int_G d\mu(g)\pi(g)T\pi'(g^{-1}) = 0, \tag{4.7}$$

and if $\pi \cong \pi'$, then for any unitary G-isomorphism $U : \mathcal{H}' \to \mathcal{H}$,

$$\int_G d\mu(g)\pi(g)T\pi'(g^{-1}) = \frac{1}{\dim \mathcal{H}} \operatorname{tr}(TU^*)U. \tag{4.8}$$

PROOF. Let

$$\tilde{T} = \int_G d\mu(g)\pi(g)T\pi'(g^{-1}).$$

which, by translation invariance of the Haar measure μ, is a G-invariant map. If $\pi \ncong \pi'$, then Schur's Lemma forces $\tilde{T}$ to vanish. If instead $\pi \cong \pi'$, let $U : \mathcal{H} \to \mathcal{H}'$ be a unitary G-isomorphism. Then by Schur's Lemma there exists some $\alpha \in \mathbb{C}$ such that $\tilde{T} = \alpha U$; in fact,

$$\alpha = \alpha \frac{1}{\dim \mathcal{H}} \operatorname{tr}(UU^*) = \frac{1}{\dim \mathcal{H}} \operatorname{tr}(\tilde{T}U^*).$$

One can then show that $\operatorname{tr}(\tilde{T}U^*) = \operatorname{tr}(TU^*)$ by introducing an orthonormal basis of $\mathcal{H}$ and then calculating directly. □

We now arrive at the desired characterisation:

PROPOSITION 4.6. *If $T \in \mathcal{L}^{\mathrm{R}}_{\mathcal{A}}(\mathcal{H}_1, \mathcal{H}_2)$, then $E_\lambda(T) = 0$ if and only if for all α, $\beta \in \operatorname{supp}(m_1) \cap \operatorname{supp}(m_2)$,*

$$T^{\alpha\beta}_{\alpha\beta} \in \mathfrak{sl}(n_\alpha) \otimes M_{(m_2)_{\alpha\beta} \times (m_1)_{\alpha\beta}}(\mathbb{C}) \otimes 1_{n_\beta},$$

and if $T \in \mathcal{L}^{\mathrm{L}}_{\mathcal{A}}(\mathcal{H}_1, \mathcal{H}_2)$, then $E_\rho(T) = 0$ if and only if for all α, $\beta \in \operatorname{supp}(m_1) \cap \operatorname{supp}(m_2)$,

$$T^{\alpha\beta}_{\alpha\beta} \in 1_{n_\alpha} \otimes M_{(m_2)_{\alpha\beta} \times (m_1)_{\alpha\beta}}(\mathbb{C}) \otimes \mathfrak{sl}(n_\beta).$$

PROOF. Let $T \in \mathcal{L}^{\mathrm{R}}_{\mathcal{A}}(\mathcal{H}_1, \mathcal{H}_2)$. Then, by Proposition 3.4, it suffices to consider components $T^{\gamma\beta}_{\alpha\beta}$ for α, β, $\gamma \in \widehat{\mathcal{A}}$, which take the form

$$T^{\gamma\beta}_{\alpha\beta} = M^{\gamma}_{\alpha\beta} \otimes 1_{n_\beta}$$

for $M^{\gamma}_{\alpha\beta} \in M_{n_\gamma \times n_\alpha}(\mathbb{C}) \otimes M_{(m_2)_{\gamma\beta} \times (m_1)_{\alpha\beta}}(\mathbb{C})$.

Now fix α, β, $\gamma \in \widehat{\mathcal{A}}$, and write

$$M^{\gamma}_{\alpha\beta} = \sum_{i=1}^{k} A_i \otimes B_i$$

for $A_i \in M_{n_\gamma \times n_\alpha}(\mathbb{C})$ and for $B_i \in M_{(m_2)_{\gamma\beta} \times (m_1)_{\alpha\beta}}(\mathbb{C})$ linearly independent. It then follows by direct computation together with Lemma 4.5 that

$$E_\lambda(T^{\gamma\beta}_{\alpha\beta}) = \begin{cases} \frac{1}{n_\alpha}\left(\sum_{i=1}^{k} \operatorname{tr}(A_i) 1_{n_\alpha} \otimes B_i\right) \otimes 1_{n_\beta} & \text{if } \alpha = \gamma, \\ 0 & \text{otherwise,} \end{cases}$$

so that by linear independence of the B_i, $E_\lambda(T^{\gamma\beta}_{\alpha\beta})$ vanishes if and only if either $\alpha \neq \gamma$ or, $\alpha = \beta$ and each A_i is traceless, and hence, if and only $\alpha \neq \gamma$ or, $\alpha = \beta$ and $M^{\alpha}_{\alpha\beta} \in \mathfrak{sl}(n_\alpha) \otimes M_{(m_2)_{\gamma\beta} \times (m_1)_{\alpha\beta}}(\mathbb{C})$, as required.

Mutatis mutandis, this argument also establishes the desired characterisation of $\ker(E_\rho)$. □

4.2. Odd bilateral spectral triples. Let us now take $\mathcal{H}_1 = \mathcal{H}_2 = \mathcal{H}$. By construction of E_λ and E_ρ, the following conditions are readily seen to be equivalent for $T \in \mathcal{L}^1_{\mathcal{A}}(\mathcal{H})$:

(1) T is self-adjoint;
(2) $E_\lambda(T)$ and $(\mathrm{id} - E_\lambda)(T)$ are self-adjoint;
(3) $(\mathrm{id} - E_\rho)(T)$ and $E_\rho(T)$ are self-adjoint;
(4) $(\mathrm{id} - E_\rho)(T)$, $(E_\lambda E_\rho)(T)$ and $(\mathrm{id} - E_\lambda)(T)$ are self-adjoint.

Thus, in particular,

$$\mathcal{D}_0(\mathcal{A},\mathcal{H}) = \mathcal{L}^{\mathrm{L}}_{\mathcal{A}}(\mathcal{H})^0_{\mathrm{sa}} \oplus \mathcal{L}^{\mathrm{LR}}_{\mathcal{A}}(\mathcal{H})_{\mathrm{sa}} \oplus \mathcal{L}^{\mathrm{R}}_{\mathcal{A}}(\mathcal{H})^0_{\mathrm{sa}}. \tag{4.9}$$

In light of Proposition 4.4, we therefore have the following description of $\mathcal{D}(\mathcal{A},\mathcal{H})$:

PROPOSITION 4.7. *Let $\mathcal{H}$ be an $\mathcal{A}$-bimodule. Then*

$$\mathcal{D}(\mathcal{A},\mathcal{H}) = \left(\mathcal{L}^{\mathrm{L}}_{\mathcal{A}}(\mathcal{H})^0_{\mathrm{sa}} \times \mathcal{L}^{\mathrm{LR}}_{\mathcal{A}}(\mathcal{H})_{\mathrm{sa}} \times \mathcal{L}^{\mathrm{R}}_{\mathcal{A}}(\mathcal{H})^0_{\mathrm{sa}}\right) / \mathrm{U}^{\mathrm{LR}}_{\mathcal{A}}(\mathcal{H}), \tag{4.10}$$

where $\mathrm{U}^{\mathrm{LR}}_{\mathcal{A}}(\mathcal{H})$ acts diagonally by conjugation.

Now, in light of Propositions 3.4, 4.3 and 4.6, we can describe how to construct an arbitrary Dirac operator on an odd $\mathcal{A}$-bimodule $\mathcal{H}$ with multiplicity matrix m:

(1) For α, β, $\gamma \in \widehat{\mathcal{A}}$ such that $\alpha < \gamma$, choose $M^\gamma_{\alpha\beta} \in M_{n_\gamma m_{\gamma\beta} \times n_\alpha m_{\alpha\beta}}(\mathbb{C})$;
(2) For α, β, $\delta \in \widehat{\mathcal{A}}$ such that $\beta < \delta$, choose $N^\delta_{\alpha\beta} \in M_{m_{\alpha\delta} n_\delta \times m_{\alpha\beta} n_\beta}(\mathbb{C})$;
(3) For α, $\beta \in \widehat{\mathcal{A}}$, choose $M^\alpha_{\alpha\beta} \in M_{n_\alpha m_{\alpha\beta}}(\mathbb{C})_{\mathrm{sa}}$ and $N^\beta_{\alpha\beta} \in M_{m_{\alpha\beta} n_\beta}(\mathbb{C})_{\mathrm{sa}}$;
(4) Finally, for α, β, γ, $\delta \in \widehat{\mathcal{A}}$, set

$$D^{\gamma\delta}_{\alpha\beta} = \begin{cases} M^\gamma_{\alpha\beta} \otimes 1_{n_\beta} & \text{if } \alpha < \gamma \text{ and } \beta = \delta, \\ (M^\alpha_{\gamma\beta})^* \otimes 1_{n_\beta} & \text{if } \alpha > \gamma \text{ and } \beta = \delta, \\ 1_{n_\alpha} \otimes N^\delta_{\alpha\beta} & \text{if } \alpha = \gamma \text{ and } \beta < \delta, \\ 1_{n_\alpha} \otimes (N^\beta_{\alpha\delta})^* & \text{if } \alpha = \gamma \text{ and } \beta > \delta, \\ M^\alpha_{\alpha\beta} \otimes 1_{n_\beta} + 1_{n_\alpha} \otimes N^\beta_{\alpha\beta} & \text{if } (\alpha,\beta) = (\gamma,\delta), \\ 0 & \text{otherwise.} \end{cases} \tag{4.11}$$

Note that for any $K = (1_{n_\alpha} \otimes K_{\alpha\beta} \otimes 1_{n_\beta})_{\alpha,\beta\in\widehat{\mathcal{A}}} \in \mathcal{L}^{\mathrm{LR}}_{\mathcal{A}}(\mathcal{H})_{\mathrm{sa}}$ (so that each $K_{\alpha\beta}$ is self-adjoint), we can make the replacements

$$M^\alpha_{\alpha\beta} \mapsto M^\alpha_{\alpha\beta} + 1_{n_\alpha} \otimes K_{\alpha\beta}, \quad N^\beta_{\alpha\beta} \mapsto N^\beta_{\alpha\beta} - K_{\alpha\beta} \otimes 1_{n_\beta},$$

and still obtain the same Dirac operator D; by Proposition 4.3, this freedom is removed by requiring either that $M^\alpha_{\alpha\beta} \in \mathfrak{sl}(n_\alpha) \otimes M_{m_{\alpha\beta}}(\mathbb{C})$ or that $N^\beta_{\alpha\beta} \in M_{m_{\alpha\beta}}(\mathbb{C}) \otimes \mathfrak{sl}(n_\beta)$.

We now turn to the moduli space $\mathcal{D}(\mathcal{A},\mathcal{H})$ itself. By the above discussion and Corollary 3.5, we can identify the space $\mathcal{D}_0(\mathcal{A},\mathcal{H})$ with

$$\mathcal{D}_0(\mathcal{A},m) := \prod_{\alpha,\beta\in\widehat{\mathcal{A}}} \prod_{\substack{\gamma\in\widehat{\mathcal{A}} \\ \gamma>\alpha}} M_{n_\gamma m_{\gamma\beta} \times n_\alpha m_{\alpha\beta}}(\mathbb{C}) \times \left(\mathfrak{sl}(n_\alpha) \otimes M_{m_{\alpha\beta}}(\mathbb{C})\right)_{\mathrm{sa}} \times \prod_{\substack{\delta\in\widehat{\mathcal{A}} \\ \delta\geq\alpha}} M_{m_{\alpha\delta} n_\delta \times m_{\alpha\beta} n_\beta}(\mathbb{C}) \times M_{m_{\alpha\beta} n_\beta}(\mathbb{C})_{\mathrm{sa}}, \tag{4.12}$$

and identify $\mathrm{U}^{\mathrm{LR}}_{\mathcal{A}}(\mathcal{H})$ with

$$\mathrm{U}(\mathcal{A},m) := \prod_{\alpha,\beta\in\widehat{\mathcal{A}}} \mathrm{U}(m_{\alpha\beta}). \tag{4.13}$$

By checking at the level of components, one sees that the action of $\mathrm{U}^{\mathrm{LR}}_{\mathcal{A}}(\mathcal{H})$ on the space $\mathcal{D}_0(\mathcal{A},\mathcal{H})$ corresponds under these identifications to the action of $\mathrm{U}(\mathcal{A},m)$ on $\mathcal{D}_0(\mathcal{A},m)$ defined by having $(U_{\alpha\beta}) \in \mathrm{U}(\mathcal{A},m)$ act on

$$(M^{\gamma}_{\alpha\beta}; M^{\alpha}_{\alpha\beta}; N^{\delta}_{\alpha\beta}; N^{\beta}_{\alpha\beta}) \in \mathcal{D}_0(\mathcal{A},m)$$

by

$$M^{\gamma}_{\alpha\beta} \mapsto (1_{n_\gamma} \otimes U_{\gamma\beta}) M^{\gamma}_{\alpha\beta} (1_{n_{\alpha\beta}} \otimes U^*_{\alpha\beta}), \quad N^{\delta}_{\alpha\beta} \mapsto (U_{\alpha\delta} \otimes 1_{n_\delta}) N^{\delta}_{\alpha\beta} (U^*_{\alpha\beta} \otimes 1_{n_\beta}).$$

We have therefore proved the following:

PROPOSITION 4.8. *Let $\mathcal{H}$ be an odd $\mathcal{A}$-bimodule with multiplicity matrix m. Then*

$$\mathcal{D}(\mathcal{A},\mathcal{H}) \cong \mathcal{D}_0(\mathcal{A},m)/\,\mathrm{U}(\mathcal{A},m). \tag{4.14}$$

4.3. Even bilateral spectral triples. For this section, let $(\mathcal{H},\gamma)$ be a fixed even $\mathcal{A}$-bimodule with pair of multiplicity matrices $(m^{\mathrm{even}}, m^{\mathrm{odd}})$.

Now, let D be a self-adjoint operator on $\mathcal{H}$ anticommuting with γ. Then, with respect to the decomposition $\mathcal{H} = \mathcal{H}^{\mathrm{even}} \oplus \mathcal{H}^{\mathrm{odd}}$ we can write

$$D = \begin{pmatrix} 0 & \Delta^* \\ \Delta & 0 \end{pmatrix},$$

where $\Delta = P^{\mathrm{odd}} D P^{\mathrm{even}}$, viewed as a map $\mathcal{H}^{\mathrm{even}} \to \mathcal{H}^{\mathrm{odd}}$. Thus, D is uniquely determined by Δ and *vice versa*. Moreover, one can check that D satisfies the order one condition if and only if Δ satisfies the generalised order one condition as a map $\mathcal{H}^{\mathrm{even}} \to \mathcal{H}^{\mathrm{odd}}$. We therefore have the following:

LEMMA 4.9. *Let $(\mathcal{H},\gamma)$ be an even $\mathcal{A}$-bimodule. Then the map $\mathcal{D}_0(\mathcal{A},\mathcal{H},\gamma) \to \mathcal{L}^1_{\mathcal{A}}(\mathcal{H}^{\mathrm{even}},\mathcal{H}^{\mathrm{odd}})$ defined by $D \mapsto P^{\mathrm{odd}} D P^{\mathrm{even}}$ is an isomorphism.*

We now apply this Lemma to obtain our first result regarding the form of $\mathcal{D}(\mathcal{A},\mathcal{H},\gamma)$:

PROPOSITION 4.10. *The map*

$$\mathcal{D}(\mathcal{A},\mathcal{H},\gamma) \to \mathrm{U}^{\mathrm{LR}}_{\mathcal{A}}(\mathcal{H}^{\mathrm{odd}}) \backslash \mathcal{L}^1_{\mathcal{A}}(\mathcal{H}^{\mathrm{even}},\mathcal{H}^{\mathrm{odd}}) /\, \mathrm{U}^{\mathrm{LR}}_{\mathcal{A}}(\mathcal{H}^{\mathrm{even}})$$

defined by $[D] \mapsto [P^{\mathrm{odd}} D P^{\mathrm{even}}]$ is a homeomorphism.

PROOF. Recall that $\mathrm{U}^{\mathrm{LR}}_{\mathcal{A}}(\mathcal{H},\gamma) = \mathrm{U}^{\mathrm{LR}}_{\mathcal{A}}(\mathcal{H}^{\mathrm{even}},\mathcal{H}^{\mathrm{odd}})$. We therefore have for $D \in \mathcal{D}_0(\mathcal{A},\mathcal{H},\gamma)$ and $U = U^{\mathrm{even}} \oplus U^{\mathrm{odd}} \in \mathrm{U}^{\mathrm{LR}}_{\mathcal{A}}(\mathcal{H}^{\mathrm{even}},\mathcal{H}^{\mathrm{odd}})$ that

$$P^{\mathrm{odd}} U D U^* P^{\mathrm{even}} = U^{\mathrm{odd}} P^{\mathrm{odd}} D P^{\mathrm{even}} (U^{\mathrm{even}})^*.$$

Thus, under the correspondence $\mathcal{D}_0(\mathcal{A},\mathcal{H},\gamma) \cong \mathcal{L}^1_{\mathcal{A}}(\mathcal{H}^{\mathrm{even}},\mathcal{H}^{\mathrm{odd}})$, the action of $\mathrm{U}^{\mathrm{LR}}_{\mathcal{A}}(\mathcal{H},\gamma)$ decouples into an action of $\mathrm{U}^{\mathrm{LR}}_{\mathcal{A}}(\mathcal{H}^{\mathrm{odd}})$ by multiplication on the left and an action of $\mathrm{U}^{\mathrm{LR}}_{\mathcal{A}}(\mathcal{H}^{\mathrm{even}})$ by multiplication by the inverse on the right. Thus, the map $[D] \to [P^{\mathrm{odd}} D P^{\mathrm{even}}]$ is not only well-defined but manifestly homeomorphic. □

Combining this last Proposition with Proposition 4.3, we immediately obtain the following:

COROLLARY 4.11. *Let $(\mathcal{H},\gamma)$ be an even $\mathcal{A}$-bimodule. Then*

$$(4.15)\quad \mathcal{D}(\mathcal{A},\mathcal{H},\gamma) \cong \mathrm{U}^{\mathrm{LR}}_{\mathcal{A}}(\mathcal{H}^{\mathrm{odd}})\backslash(\mathcal{L}^{\mathrm{L}}_{\mathcal{A}}(\mathcal{H}^{\mathrm{even}},\mathcal{H}^{\mathrm{odd}})^0 \times \mathcal{L}^{\mathrm{LR}}_{\mathcal{A}}(\mathcal{H}^{\mathrm{even}},\mathcal{H}^{\mathrm{odd}}) \times \mathcal{L}^{\mathrm{R}}_{\mathcal{A}}(\mathcal{H}^{\mathrm{even}},\mathcal{H}^{\mathrm{odd}})^0)/\,\mathrm{U}^{\mathrm{LR}}_{\mathcal{A}}(\mathcal{H}^{\mathrm{even}}),$$

where $\mathrm{U}^{\mathrm{LR}}_{\mathcal{A}}(\mathcal{H}^{\mathrm{odd}})$ acts diagonally by multiplication on the left, and $\mathrm{U}^{\mathrm{LR}}_{\mathcal{A}}(\mathcal{H}^{\mathrm{even}})$ acts diagonally by multiplication on the right by the inverse.

Now, just as we did in the odd case, let us describe the construction of an arbitary Dirac operator D on $(\mathcal{H},\gamma)$:

(1) For α, β, $\gamma \in \widehat{\mathcal{A}}$, choose $M^{\gamma}_{\alpha\beta} \in M_{n_\gamma m^{\mathrm{odd}}_{\gamma\beta} \times n_\alpha m^{\mathrm{even}}_{\alpha\beta}}(\mathbb{C})$;

(2) For α, β, $\delta \in \widehat{\mathcal{A}}$, choose $N^{\delta}_{\alpha\beta} \in M_{m^{\mathrm{odd}}_{\alpha\delta} n_\delta \times m^{\mathrm{even}}_{\alpha\beta} n_\beta}(\mathbb{C})$;

(3) Construct $\Delta \in \mathcal{L}^1_{\mathcal{A}}(\mathcal{H}^{\mathrm{even}},\mathcal{H}^{\mathrm{odd}})$ by setting, for α, β, γ, $\delta \in \widehat{\mathcal{A}}$,

$$(4.16)\quad \Delta^{\gamma\delta}_{\alpha\beta} = \begin{cases} M^{\gamma}_{\alpha\beta} \otimes 1_{n_\beta} & \text{if } \alpha \neq \gamma \text{ and } \beta = \delta, \\ 1_{n_\alpha} \otimes N^{\delta}_{\alpha\beta} & \text{if } \alpha = \gamma \text{ and } \beta \neq \delta, \\ M^{\alpha}_{\alpha\beta} \otimes 1_{n_\beta} + 1_{n_\alpha} \otimes N^{\beta}_{\alpha\beta} & \text{if } (\alpha,\beta) = (\gamma,\delta), \\ 0 & \text{otherwise;} \end{cases}$$

(4) Finally, set $D = \left(\begin{smallmatrix} 0 & \Delta^* \\ \Delta & 0 \end{smallmatrix}\right)$.

Again, note that for any $K = (1_{n_\alpha} \otimes K_{\alpha\beta} \otimes 1_{n_\beta})_{\alpha,\beta\in\widehat{\mathcal{A}}} \in \mathcal{L}^{\mathrm{LR}}_{\mathcal{A}}(\mathcal{H}^{\mathrm{even}},\mathcal{H}^{\mathrm{odd}})$, we can make the replacements

$$M^{\alpha}_{\alpha\beta} \mapsto M^{\alpha}_{\alpha\beta} + 1_{n_\alpha} \otimes K_{\alpha\beta}, \quad N^{\beta}_{\alpha\beta} \mapsto N^{\beta}_{\alpha\beta} - K_{\alpha\beta} \otimes 1_{n_\beta},$$

and still obtain the same Dirac operator D; by Proposition 4.3, this freedom is removed by requiring either that

$$M^{\alpha}_{\alpha\beta} \in \mathfrak{sl}(n_\alpha) \otimes M_{m^{\mathrm{odd}}_{\alpha\beta} \times m^{\mathrm{even}}_{\alpha\beta}}(\mathbb{C})$$

or that

$$N^{\beta}_{\alpha\beta} \in M_{m^{\mathrm{odd}}_{\alpha\beta} \times m^{\mathrm{even}}_{\alpha\beta}}(\mathbb{C}) \otimes \mathfrak{sl}(n_\beta).$$

Just as in the odd case, the above discussion and Corollary 3.5 imply that we can identify $\mathcal{D}_0(\mathcal{A},\mathcal{H},\gamma)$ with

$$(4.17)\quad \mathcal{D}_0(\mathcal{A},m^{\mathrm{even}},m^{\mathrm{odd}}) := \prod_{\alpha,\beta\in\widehat{\mathcal{A}}} \prod_{\substack{\gamma\in\widehat{\mathcal{A}} \\ \gamma\neq\alpha}} M_{n_\gamma m^{\mathrm{odd}}_{\gamma\beta} \times n_\alpha m^{\mathrm{even}}_{\alpha\beta}}(\mathbb{C}) \times \left(\mathfrak{sl}(n_\alpha) \otimes M_{m^{\mathrm{odd}}_{\alpha\beta} \times m^{\mathrm{even}}_{\alpha\beta}}(\mathbb{C})\right) \times \prod_{\delta\in\widehat{\mathcal{A}}} M_{m^{\mathrm{odd}}_{\alpha\delta} n_\delta \times m^{\mathrm{even}}_{\alpha\beta} n_\beta}(\mathbb{C}),$$

and identify $\mathrm{U}^{\mathrm{LR}}_{\mathcal{A}}(\mathcal{H}^{\mathrm{even}})$ and $\mathrm{U}^{\mathrm{LR}}_{\mathcal{A}}(\mathcal{H}^{\mathrm{odd}})$ with $\mathrm{U}(\mathcal{A},m^{\mathrm{even}})$ and $\mathrm{U}(\mathcal{A},m^{\mathrm{odd}})$, respectively, which are defined according to Equation 4.13. The actions of $\mathrm{U}^{\mathrm{LR}}_{\mathcal{A}}(\mathcal{H}^{\mathrm{even}})$ and $\mathrm{U}^{\mathrm{LR}}_{\mathcal{A}}(\mathcal{H}^{\mathrm{odd}})$ on $\mathcal{L}^1_{\mathcal{A}}(\mathcal{H}^{\mathrm{even}},\mathcal{H}^{\mathrm{odd}})$ therefore correspond under these identifications to the actions of $\mathrm{U}(\mathcal{A},m^{\mathrm{even}})$ and $\mathrm{U}(\mathcal{A},m^{\mathrm{odd}})$, respectively, on $\mathcal{D}_0(\mathcal{A},m^{\mathrm{even}},m^{\mathrm{odd}})$ defined by having $(U^{\mathrm{odd}}_{\alpha\beta}) \in \mathrm{U}(\mathcal{A},m^{\mathrm{odd}})$ and $(U^{\mathrm{even}}_{\alpha\beta}) \in \mathrm{U}(\mathcal{A},m^{\mathrm{even}})$ act on

$$(M^{\gamma}_{\alpha\beta}; M^{\alpha}_{\alpha\beta}; N^{\delta}_{\alpha\beta}) \in \mathcal{D}_0(\mathcal{A},m^{\mathrm{even}},m^{\mathrm{odd}})$$

by

$$M^{\gamma}_{\alpha\beta} \mapsto (1_{n_\gamma} \otimes U^{\mathrm{odd}}_{\gamma\beta})M^{\gamma}_{\alpha\beta}, \quad N^{\delta}_{\alpha\beta} \mapsto (U^{\mathrm{odd}}_{\alpha\delta} \otimes 1_{n_\delta})N^{\delta}_{\alpha\beta},$$

and

$$M_{\alpha\beta}^{\gamma} \mapsto M_{\alpha\beta}^{\gamma}(1_{n_\alpha} \otimes (U_{\alpha\beta}^{\text{even}})^*), \quad N_{\alpha\beta}^{\delta} \mapsto N_{\alpha\beta}^{\delta}((U_{\alpha\beta}^{\text{even}})^* \otimes 1_{n_\beta}),$$

respectively. Thus we have proved the following:

PROPOSITION 4.12. *Let $(\mathcal{H}, \gamma)$ be an even $\mathcal{A}$-bimodule with multiplicity matrices $(m^{\text{even}}, m^{\text{odd}})$. Then*

$$\mathcal{D}(\mathcal{A}, \mathcal{H}, \gamma) \cong \mathrm{U}(\mathcal{A}, m^{\text{odd}}) \backslash \mathcal{D}_0(\mathcal{A}, m^{\text{even}}, m^{\text{odd}}) / \mathrm{U}(\mathcal{A}, m^{\text{even}}). \tag{4.18}$$

In the quasi-orientable case, the picture simplifies considerably, as all components $\Delta_{\alpha\beta}^{\alpha\beta}$ necessarily vanish. One is then left, essentially, with the situation described by Krajewski [18, §3.4] and Paschke–Sitarz [20, §2.II] ; as mentioned before, one can find in the former the original definition of what are now called *Krajewski diagrams*. These diagrams, used extensively by Iochum, Jureit, Schücker and Stephan [12–16, 22], offer a concise, diagrammatic approach to the study of quasi-orientable even bilateral spectral triples that strongly emphasizes the underlying combinatorics. Though they do admit ready generalisation to the non-quasi-orientable case, we will not discuss them here.

We conclude our discussion of even bilateral spectral triples by recalling a result of Paschke and Sitarz of particular interest in relation to the NCG Standard Model.

PROPOSITION 4.13 (Paschke–Sitarz [20, Lemma 7]). *Let $(\mathcal{H}, \gamma)$ be an orientable $\mathcal{A}$-bimodule. Then for all $D \in \mathcal{D}_0(\mathcal{A}, \mathcal{H}, \gamma)$,*

$$D = \sum_{\substack{i,j=1 \\ i \neq j}}^{N} \lambda(e_i)[D, \lambda(e_j)] + \sum_{\substack{k,l=1 \\ k \neq l}}^{N} \rho(e_k)[D, \rho(e_l)]. \tag{4.19}$$

PROOF. Fix $D \in \mathcal{D}_0(\mathcal{A}, \mathcal{H}, \gamma)$, and let

$$\begin{aligned} T &:= D - \sum_{\substack{i,j=1 \\ i \neq j}}^{N} \lambda(e_i)[D, \lambda(e_j)] - \sum_{\substack{k,l=1 \\ k \neq l}}^{N} \rho(e_k)[D, \rho(e_l)] \\ &= D - \sum_{\substack{i,j=1 \\ i \neq j}}^{N} \lambda(e_i) D \lambda(e_j) - \sum_{\substack{k,l=1 \\ k \neq l}}^{N} \rho(e_k) D \rho(e_l). \end{aligned}$$

Then for all α, β, γ, $\delta \in \widehat{\mathcal{A}}$,

$$T_{\alpha\beta}^{\gamma\delta} = \begin{cases} D_{\alpha\beta}^{\gamma\delta} & \text{if } r(\alpha) = r(\gamma),\ r(\beta) = r(\delta), \\ -D_{\alpha\beta}^{\gamma\delta} & \text{if } r(\alpha) \neq r(\gamma),\ r(\beta) \neq r(\delta), \\ 0 & \text{otherwise,} \end{cases}$$

where for $\alpha \in \widehat{\mathcal{A}}$, $r(\alpha)$ is the value of $j \in \{1, \dots, N\}$ such that $\alpha \in \widehat{M_{k_j}(\mathbb{K}_j)}$. However, by Proposition 4.3, $D_{\alpha\beta}^{\gamma\delta}$ must vanish in the second case, whilst by Proposition 3.13, $D_{\alpha\beta}^{\gamma\delta}$ must vanish in the first, so that $T = 0$. □

Now, let $(\mathcal{A}, \mathcal{H}, D, J, \gamma)$ be a real spectral triple of even KO-dimension. A *gauge potential* for the triple is then a self-adjoint operator on $\mathcal{H}$ of the form

$$\sum_{k=1}^{n} \lambda(a_k)[D, \lambda(b_k)],$$

where $a_1, \ldots, a_n, b_1, \ldots, b_n \in \mathcal{A}$, and an *inner fluctuation of the metric* is a Dirac operator $D_A \in \mathcal{D}_0(\mathcal{A}, \mathcal{H}, J, \gamma)$ of the form

$$D_A := D + A + \varepsilon' JAJ^* = D + A + JAJ^*,$$

where A is a gauge potential. One then has that for any gauge potential A, $(\mathcal{A}, \mathcal{H}, D, J, \gamma)$ and $(\mathcal{A}, \mathcal{H}, D_A, J, \gamma)$ are Morita equivalent. In this light, the last Proposition admits the following interpretation:

COROLLARY 4.14. *Let $(\mathcal{H}, J, \gamma)$ be an orientable real $\mathcal{A}$-bimodule of even KO-dimension. Then for all $D \in \mathcal{D}_0(\mathcal{A}, \mathcal{H}, \gamma, J)$,*

$$A = -\sum_{\substack{i,j=1 \\ i \neq j}}^{N} \lambda(e_i)[D, \lambda(e_j)] \tag{4.20}$$

is a gauge potential for the real spectral triple $(\mathcal{A}, \mathcal{H}, D, J, \gamma)$ such that $D_A = 0$.

Thus, every finite orientable real spectral triple $(\mathcal{A}, \mathcal{H}, D, J, \gamma)$ of even KO-dimension is Morita equivalent to the dynamically trivial triple $(\mathcal{A}, \mathcal{H}, 0, J, \gamma)$.

4.4. Real spectral triples of odd *KO*-dimension. For this section, let $(\mathcal{H}, J)$ be a real $\mathcal{A}$-bimodule of odd KO-dimension $n \bmod 8$ with multiplicity matrix m. We begin by reducing the study of Dirac operators on $(\mathcal{H}, J)$ to that of self-adjoint right $\mathcal{A}$-linear operators on $\mathcal{H}$.

PROPOSITION 4.15 (Krajewski [18, §3.4]). *Let $(\mathcal{H}, J)$ be a real $\mathcal{A}$-bimodule of odd KO-dimension $n \bmod 8$. Then the map $R_n : \mathcal{L}^{\mathrm{R}}_{\mathcal{A}}(\mathcal{H})_{\mathrm{sa}} \to \mathcal{D}_0(\mathcal{A}, \mathcal{H}, J)$ defined by $R_n(M) := M + \varepsilon' JMJ^*$ is a surjection interwining the action of $\mathrm{U}^{\mathrm{LR}}_{\mathcal{A}}(\mathcal{H}, J)$ on $\mathcal{L}^{\mathrm{R}}_{\mathcal{A}}(\mathcal{H})_{\mathrm{sa}}$ by conjugation with the action on $\mathcal{D}_0(\mathcal{A}, \mathcal{H}, J)$ by conjugation, and $\ker(R_n) \subseteq \mathcal{L}^{\mathrm{LR}}_{\mathcal{A}}(\mathcal{H})_{\mathrm{sa}}$.*

PROOF. First, note that R_n is indeed well-defined, since by Equation 2.4, for any $M \in \mathcal{L}^{\mathrm{R}}_{\mathcal{A}}(\mathcal{H})_{\mathrm{sa}}$, $JMJ^* \in \mathcal{L}^{\mathrm{L}}_{\mathcal{A}}(\mathcal{H})_{\mathrm{sa}}$, and hence $R_n(M) \in \mathcal{D}_0(\mathcal{A}, \mathcal{H}, J)$.

Now, let E_λ and E_ρ be defined as in Lemma 4.2, and let $E'_\lambda = \mathrm{id} - E_\lambda$, $E'_\rho = \mathrm{id} - E_\rho$. Then, by construction of E_λ and E_ρ and Equation 2.4, for any $T \in \mathcal{L}^1_{\mathcal{A}}(\mathcal{H})$,

$$E_\lambda(JTJ^*) = JE_\rho(T)J^*, \quad E_\rho(JTJ^*) = JE_\lambda(T)J^*.$$

Hence, in particular, for $D \in \mathcal{D}_0(\mathcal{A}, \mathcal{H}, J)$, since $JDJ^* = \varepsilon' D$,

$$\begin{aligned} D &= \frac{1}{2}(E'_\lambda + E_\rho)(T) + \frac{1}{2}(E_\lambda + E'_\rho)(T) \\ &= \frac{1}{2}(E'_\lambda + E_\rho)(T) + \varepsilon' J \frac{1}{2}(E'_\lambda + E_\rho)(T)J^* \\ &= R_n\left(\frac{1}{2}(E'_\lambda + E_\rho)(T)\right), \end{aligned}$$

where $\frac{1}{2}(E'_\lambda + E_\rho)(T) \in \mathcal{L}^{\mathrm{R}}_{\mathcal{A}}(\mathcal{H})_{\mathrm{sa}}$.

Finally, that R_n interwtines the actions of $\mathrm{U}^{\mathrm{LR}}_{\mathcal{A}}(\mathcal{H}, J)$ follows from Proposition 4.4 together with the fact that elements of $\mathrm{U}^{\mathrm{LR}}_{\mathcal{A}}(\mathcal{H}, J)$, by definition, commute with J, whilst the fact that $R_n(M) = 0$ if and only if $M = -\varepsilon' JMJ^*$ implies that $\ker(R_n) \subseteq \mathcal{L}^{\mathrm{LR}}_{\mathcal{A}}(\mathcal{H})_{\mathrm{sa}}$. □

It follows, in particular, that $\ker(R_n)$ is invariant under the action of $\mathrm{U}^{\mathrm{LR}}_{\mathcal{A}}(\mathcal{H},J)$ by conjugation, so that the action of $\mathrm{U}^{\mathrm{LR}}_{\mathcal{A}}(\mathcal{H},J)$ on $\mathcal{L}^{\mathrm{R}}_{\mathcal{A}}(\mathcal{H})_{\mathrm{sa}}$ induces an action on the quotient $\mathcal{L}^{\mathrm{R}}_{\mathcal{A}}(\mathcal{H})_{\mathrm{sa}}/\ker(R_n)$, and hence R_n induces an isomorphism

$$\mathcal{D}_0(\mathcal{A},\mathcal{H},J) \cong \mathcal{L}^{\mathrm{R}}_{\mathcal{A}}(\mathcal{H})_{\mathrm{sa}}/\ker(R_n) \tag{4.21}$$

of $\mathrm{U}^{\mathrm{LR}}_{\mathcal{A}}(\mathcal{H},J)$-representations. Thus we have proved the following:

COROLLARY 4.16. *Let $(\mathcal{H},J)$ be a real $\mathcal{A}$-bimodule of odd KO-dimension n mod 8. Then*

$$\mathcal{D}(\mathcal{A},\mathcal{H},J) \cong \left(\mathcal{L}^{\mathrm{R}}_{\mathcal{A}}(\mathcal{H})_{\mathrm{sa}}/\ker(R_n)\right)/\,\mathrm{U}^{\mathrm{LR}}_{\mathcal{A}}(\mathcal{H},J). \tag{4.22}$$

Discussion of $\mathcal{D}(\mathcal{A},\mathcal{H},J)$ thus requires discussion first of $\ker(R_n)$:

LEMMA 4.17. *If $K = (1_{n_\alpha}\otimes K_{\alpha\beta}\otimes 1_{n_\beta})_{\alpha,\beta\in\widehat{\mathcal{A}}} \in \mathcal{L}^{\mathrm{LR}}_{\mathcal{A}}(\mathcal{H})_{\mathrm{sa}}$, then $K\in\ker(R_n)$ if and only if for each α, $\beta\in\widehat{\mathcal{A}}$ such that $\alpha\neq\beta$,*

$$K_{\beta\alpha} = -\varepsilon' K^T_{\alpha\beta}, \tag{4.23}$$

and for each $\alpha\in\widehat{\mathcal{A}}$,

$$K_{\alpha\alpha}\in\mathcal{R}_\alpha(n) = \begin{cases} \mathrm{Sym}_{m_{\alpha\alpha}}(\mathbb{R}) & \text{if } n=1,\\ i\mathfrak{sp}(m_{\alpha\alpha}) & \text{if } n=3,\\ M_{m_{\alpha\alpha}/2}(\mathbb{H})_{\mathrm{sa}} & \text{if } n=5,\\ i\mathfrak{so}(m_{\alpha\alpha}) & \text{if } n=7. \end{cases} \tag{4.24}$$

PROOF. By definition of R_n, $K\in\ker(R_n)$ if and only if $K=-\varepsilon' JKJ^* = -\varepsilon\varepsilon' JKJ$, and this in turn holds if and only if, for α, $\beta\in\widehat{\mathcal{A}}$ such that $\alpha\neq\beta$,

$$K_{\alpha\beta} = -\varepsilon' K^T_{\beta\alpha},$$

while for $\alpha\in\widehat{\mathcal{A}}$,

$$K_{\alpha\alpha} = \begin{cases} -\varepsilon'\overline{K_{\alpha\alpha}}, & \text{if } n = 1 \text{ or } 7,\\ \varepsilon' I_\alpha K_{\alpha\alpha} I^*_\alpha & \text{if } n=3 \text{ or } 5, \end{cases}$$

where $I_\alpha = \Omega_{m_{\alpha\alpha}}\circ$ complex conjugation. In the case that $n=3$ or 5, however, by construction, $M_{m_{\alpha\alpha}/2}(\mathbb{H})$, viewed in the usual way as a real form of $M_{m_{\alpha\alpha}}(\mathbb{C})$, is precisely the set of matrices in $M_{m_{\alpha\alpha}}(\mathbb{C})$ commuting with I_α. This, together with the hypothesis that K is self-adjoint, so that each $K_{\alpha\beta}$ is self-adjoint, yields the desired result. □

We can now describe the the construction of an arbitrary Dirac operator D on $(\mathcal{H},J)$:

(1) For α, β, $\gamma\in\widehat{\mathcal{A}}$ such that $\alpha<\gamma$, choose $M^\gamma_{\alpha\beta}\in M_{n_\gamma m_{\gamma\beta}\times n_\alpha m_{\alpha\beta}}(\mathbb{C})$;
(2) For α, $\beta\in\widehat{\mathcal{A}}$, choose $M^\alpha_{\alpha\beta}\in M_{n_\alpha m_{\alpha\beta}}(\mathbb{C})_{\mathrm{sa}}$;
(3) For α, β, γ, $\delta\in\widehat{\mathcal{A}}$, set

$$M^{\gamma\delta}_{\alpha\beta} = \begin{cases} M^\gamma_{\alpha\beta}\otimes 1_{n_\beta} & \text{if } \alpha<\gamma \text{ and } \beta=\delta,\\ (M^\alpha_{\gamma\beta})^*\otimes 1_{n_\beta} & \text{if } \alpha>\gamma \text{ and } \beta=\delta,\\ M^\alpha_{\alpha\beta}\otimes 1_{n_\beta} & \text{if } (\alpha,\beta)=(\gamma,\delta),\\ 0 & \text{otherwise.} \end{cases} \tag{4.25}$$

(4) Finally, set $D = R_n(M)$.

Now, let $K = (1_{n_\alpha} \otimes K_{\alpha\beta} \otimes 1_{n_\beta})_{\alpha,\beta\in\widehat{\mathcal{A}}} \in \ker(R_n)$, so that each $K_{\alpha\beta}$ is self-adjoint, and for α, $\beta \in \widehat{\mathcal{A}}$ such that $\alpha \neq \beta$, $K_{\beta\alpha} = -\varepsilon' K_{\alpha\beta}^T$ and $K_{\alpha\alpha} \in \mathcal{R}_\alpha(n)$. Thus, K is uniquely specified by the matrices $K_{\alpha\beta} \in M_{m_{\alpha\beta}}(\mathbb{C})_{\mathrm{sa}}$ for $\alpha < \beta$ and by the $K_{\alpha\alpha} \in \mathcal{R}_\alpha(n)$. Then, we can replace M by $M + K$, *i.e.* make the replacements, for α, $\beta \in \widehat{\mathcal{A}}$ such that $\alpha < \beta$,

$$M_{\alpha\beta}^\alpha \mapsto M_{\alpha\beta}^\alpha + 1_{n_\alpha} \otimes K_{\alpha\beta}, \quad M_{\beta\alpha}^\beta \mapsto M_{\beta\alpha}^\beta + 1_{n_\beta} \otimes (-\varepsilon' K_{\alpha\beta}^T),$$
$$M_{\alpha\alpha}^\alpha \mapsto M_{\alpha\alpha}^\alpha + 1_{n_\alpha} \otimes K_{\alpha\alpha}$$

and obtain the same Dirac operator D. However, this is a freedom cannot generally be removed as we did in earlier cases, as it reflects precisely the non-injectivity of R_n.

By the above discussion and Propositions 3.19 and 3.23, we can identify the space $\mathcal{D}_0(\mathcal{A}, \mathcal{H}, J)$ with

$$(4.26) \quad \mathcal{D}_0(\mathcal{A}, m, n) := \prod_{\alpha\in\widehat{\mathcal{A}}} \Bigg[M_{n_\alpha m_{\alpha\alpha}}(\mathbb{C})_{\mathrm{sa}} / (1_{n_\alpha} \otimes \mathcal{R}_\alpha(n))$$
$$\times \prod_{\substack{\beta\in\widehat{\mathcal{A}} \\ \beta>\alpha}} (M_{n_\alpha m_{\alpha\beta}}(\mathbb{C})_{\mathrm{sa}} \oplus M_{n_\beta m_{\alpha\beta}}(\mathbb{C})_{\mathrm{sa}}) / M_{m_\alpha\beta}(\mathbb{C})_{\mathrm{sa}} \times \prod_{\substack{\beta,\gamma\in\widehat{\mathcal{A}} \\ \gamma>\alpha}} M_{n_\gamma m_{\gamma\beta} \times n_\alpha m_{\alpha\beta}}(\mathbb{C}) \Bigg],$$

where $M_{m_{\alpha\beta}}(\mathbb{C})_{\mathrm{sa}}$ is viewed as embedded in $M_{n_\alpha m_{\alpha\beta}}(\mathbb{C})_{\mathrm{sa}} \oplus M_{n_\beta m_{\alpha\beta}}(\mathbb{C})_{\mathrm{sa}}$ via the map

$$K \mapsto (1_{n_\alpha} \otimes K) \oplus (-\varepsilon' 1_{n_\beta} \otimes K^T),$$

and $\mathrm{U}_{\mathcal{A}}^{\mathrm{LR}}(\mathcal{H}, J)$ with

$$(4.27) \qquad \mathrm{U}(\mathcal{A}, m, n) := \prod_{\alpha\in\widehat{\mathcal{A}}} \Bigg(\mathcal{U}_\alpha(n) \times \prod_{\substack{\beta\in\widehat{\mathcal{A}} \\ \beta>\alpha}} \mathrm{U}(m_{\alpha\beta}) \Bigg),$$

where

$$\mathcal{U}_\alpha(n) := \begin{cases} \mathrm{O}(m_{\alpha\alpha}) & \text{if } n = 1 \text{ or } 7, \\ \mathrm{Sp}(m_{\alpha\alpha}) & \text{if } n = 3 \text{ or } 5. \end{cases}$$

Then the action of $\mathrm{U}_{\mathcal{A}}^{\mathrm{LR}}(\mathcal{H}; J)$ on $\mathcal{D}_0(\mathcal{A}, \mathcal{H}; J)$ corresponds under these identifications to the action of $\mathrm{U}(\mathcal{A}, m, n)$ on $\mathcal{D}_0(\mathcal{A}, m, n)$ defined by having the element $(U_{\alpha\alpha}; U_{\alpha\beta}) \in \mathrm{U}(\mathcal{A}, m, n)$ act on $\big([M_{\alpha\alpha}^\alpha]; [(M_{\alpha\beta}^\alpha, M_{\beta\alpha}^\beta)]; M_{\alpha\beta}^\gamma\big) \in \mathcal{D}_0(\mathcal{A}, m, n)$ by

$$[M_{\alpha\alpha}^\alpha] \mapsto \big[(1_{n_\alpha} \otimes U_{\alpha\alpha}) M_{\alpha\alpha} (1_{n_\alpha} \otimes U_{\alpha\alpha}^*)\big];$$
$$[(M_{\alpha\beta}^\alpha, M_{\beta\alpha}^\beta)] \mapsto \Big[\big((1_{n_\alpha} \otimes U_{\alpha\beta}) M_{\alpha\beta}^\alpha (1_{n_\alpha} \otimes U_{\alpha\beta}^*), (1_{n_\beta} \otimes \overline{U_{\alpha\beta}}) M_{\beta\alpha}^\beta (1_{n_\beta} \otimes U_{\alpha\beta}^T)\big)\Big];$$
$$M_{\alpha\beta}^\gamma \mapsto \begin{cases} (1_{n_\gamma} \otimes U_{\gamma\beta}) M_{\alpha\beta}^\gamma (1_{n_\alpha} \otimes U_{\alpha\beta}^*) & \text{if } \alpha < \beta, \gamma < \delta, \\ (1_{n_\gamma} \otimes U_{\gamma\beta}) M_{\alpha\beta}^\gamma (1_{n_\alpha} \otimes U_{\beta\alpha}^T) & \text{if } \alpha > \beta, \gamma < \delta, \\ (1_{n_\gamma} \otimes \overline{U_{\beta\gamma}}) M_{\alpha\beta}^\gamma (1_{n_\alpha} \otimes U_{\alpha\beta}^*) & \text{if } \alpha < \beta, \gamma > \delta, \\ (1_{n_\gamma} \otimes \overline{U_{\beta\gamma}}) M_{\alpha\beta}^\gamma (1_{n_\alpha} \otimes U_{\beta\alpha}^T) & \text{if } \alpha > \beta, \gamma > \delta. \end{cases}$$

We have therefore proved the following:

Proposition 4.18. *Let* $(\mathcal{H}, J)$ *be a real* $\mathcal{A}$*-bimodule of odd* KO*-dimension* $n \bmod 8$ *with multiplicity matrix* m. *Then*

$$\mathcal{D}(\mathcal{A}, \mathcal{H}, J) \cong \mathcal{D}_0(\mathcal{A}, m, n) / \operatorname{U}(\mathcal{A}, m, n). \tag{4.28}$$

4.5. Real spectral triples of even KO**-dimension.** We now turn to real spectral triples of even KO-dimension. Because of the considerable qualitative differences between the two cases, we consider separately the case of KO-dimension 0 or 4 mod 8 and KO-dimension 2 or 6 mod 8.

In what follows, $(\mathcal{H}, \gamma, J)$ is a fixed real $\mathcal{A}$-bimodule of even KO-dimension $n \bmod 8$ with multiplicity matrices $(m^{\text{even}}, m^{\text{odd}})$; we denote by $\mathcal{L}^1_{\mathcal{A}}(\mathcal{H}^{\text{even}}, \mathcal{H}^{\text{odd}}; J)$ the subspace of $\mathcal{L}^1_{\mathcal{A}}(\mathcal{H}^{\text{even}}, \mathcal{H}^{\text{odd}})$ consisting of δ such that

$$\begin{pmatrix} 0 & \Delta^* \\ \Delta & 0 \end{pmatrix} \in \mathcal{D}_0(\mathcal{A}, \mathcal{H}; \gamma, J).$$

It then follows that

$$\mathcal{D}_0(\mathcal{A}, \mathcal{H}, \gamma, J) \cong \mathcal{L}^1_{\mathcal{A}}(\mathcal{H}^{\text{even}}, \mathcal{H}^{\text{odd}}; J) \tag{4.29}$$

via the map $D \mapsto P^{\text{odd}} D P^{\text{even}}$.

4.5.1. *KO-dimension* 0 *or* 4 mod 8. Let us first consider the case where $n = 0$ or 4 mod 8, *i.e.* where $\varepsilon' = 1$. Then $J = J^{\text{even}} \oplus J^{\text{odd}}$ for anti-unitaries J^{even} and J^{odd} on $\mathcal{H}^{\text{even}}$ and $\mathcal{H}^{\text{odd}}$, respectively, such that $(\mathcal{H}^{\text{even}}, J^{\text{even}})$ and $(\mathcal{H}^{\text{odd}}, J^{\text{odd}})$ are real $\mathcal{A}$-bimodules of KO-dimension n' mod 8, where $n' = 1$ or 7 if $n = 0$, 3 or 5 if $n = 4$. In light of Corollary 3.25, one can readily check the following analogue of Proposition 4.10:

Proposition 4.19. *The map*

$$\mathcal{D}(\mathcal{A}, \mathcal{H}, \gamma, J) \to \operatorname{U}^{\text{LR}}_{\mathcal{A}}(\mathcal{H}^{\text{odd}}, J^{\text{odd}}) \backslash \mathcal{L}^1_{\mathcal{A}}(\mathcal{H}^{\text{even}}, \mathcal{H}^{\text{odd}}; J) / \operatorname{U}^{\text{LR}}_{\mathcal{A}}(\mathcal{H}^{\text{even}}, J^{\text{even}})$$

defined by $[D] \mapsto [P^{\text{odd}} D P^{\text{even}}]$ *is a homeomorphism.*

Here, as before, $\operatorname{U}^{\text{LR}}_{\mathcal{A}}(\mathcal{H}^{\text{odd}}, J^{\text{odd}})$ acts by multiplication on the left, whilst the group $\operatorname{U}^{\text{LR}}_{\mathcal{A}}(\mathcal{H}^{\text{even}}, J^{\text{even}})$ acts by multiplication on the right by the inverse.

We now prove the relevant analogue of Proposition 4.15:

Proposition 4.20. *The map* $R_n : \mathcal{L}^{\text{R}}_{\mathcal{A}}(\mathcal{H}^{\text{even}}, \mathcal{H}^{\text{odd}}) \to \mathcal{L}^1_{\mathcal{A}}(\mathcal{H}^{\text{even}}, \mathcal{H}^{\text{odd}}, J)$ *defined by* $R_n(M) := M + J^{\text{odd}} M (J^{\text{even}})^*$ *is a surjection interwining the actions of* $\operatorname{U}^{\text{LR}}_{\mathcal{A}}(\mathcal{H}^{\text{odd}}, J^{\text{odd}})$ *by multiplication on the left and of* $\operatorname{U}^{\text{LR}}_{\mathcal{A}}(\mathcal{H}^{\text{even}}, J^{\text{even}})$ *by multiplication on the right by the inverse on* $\mathcal{L}^{\text{R}}_{\mathcal{A}}(\mathcal{H}^{\text{even}}, \mathcal{H}^{\text{odd}})$ *and* $\mathcal{L}^1_{\mathcal{A}}(\mathcal{A}, \mathcal{H}^{\text{even}}, \mathcal{H}^{\text{odd}}, J)$, *and* $\ker(R_n) \subseteq \mathcal{L}^{\text{LR}}_{\mathcal{A}}(\mathcal{H}^{\text{even}}, \mathcal{H}^{\text{odd}})$.

Proof. First note that

$$\mathcal{L}^1_{\mathcal{A}}(\mathcal{H}^{\text{even}}, \mathcal{H}^{\text{odd}}, J) = \{\Delta \in \mathcal{L}^1_{\mathcal{A}}(\mathcal{H}^{\text{even}}, \mathcal{H}^{\text{odd}}) \mid \Delta = J^{\text{odd}} \Delta (J^{\text{even}})^*\},$$

so that R_n is indeed well-defined by construction. Moreover, since $\operatorname{U}^{\text{LR}}_{\mathcal{A}}(\mathcal{H}^{\text{even}}, J^{\text{even}})$ and $\operatorname{U}^{\text{LR}}_{\mathcal{A}}(\mathcal{H}^{\text{odd}}, J^{\text{odd}})$ commute by definition with J^{even} and J^{odd}, respectively, it then follows by construction of R_n that R_n does indeed have the desired intertwining properties.

Next, for $M \in \mathcal{L}^{\text{R}}_{\mathcal{A}}(\mathcal{H}^{\text{even}}, \mathcal{H}^{\text{odd}})$, we have that $R_n(M) = 0$ if and only if $M = -J^{\text{odd}} M (J^{\text{even}})^*$, but M is right $\mathcal{A}$-linear if and only if $J^{\text{odd}} M (J^{\text{even}})^* = \varepsilon J^{\text{odd}} M J^{\text{even}}$ is left $\mathcal{A}$-linear, so that $M \in \mathcal{L}^{\text{LR}}_{\mathcal{A}}(\mathcal{H}^{\text{even}}, \mathcal{H}^{\text{odd}})$ as claimed.

Finally, it is easy to check, just as in the proof of Proposition 4.15, that for $\Delta \in \mathcal{L}^1_{\mathcal{A}}(\mathcal{H}^{\text{even}}, \mathcal{H}^{\text{odd}}, J)$,

$$\Delta = R_n\left(\frac{1}{2}(E'_\lambda + E_\rho)(\Delta)\right),$$

where $\frac{1}{2}(E'_\lambda + E_\rho)(\Delta) \in \mathcal{L}^{\mathrm{R}}_{\mathcal{A}}(\mathcal{H}^{\text{even}}, \mathcal{H}^{\text{odd}})$. □

Again, just as in the case of odd KO-dimension, this last result not only implies that the actions of $\mathrm{U}^{\mathrm{LR}}_{\mathcal{A}}(\mathcal{H}^{\text{even}}, J^{\text{even}})$ and $\mathrm{U}^{\mathrm{LR}}_{\mathcal{A}}(\mathcal{H}^{\text{odd}}, J^{\text{odd}})$ on $\mathcal{L}^{\mathrm{R}}_{\mathcal{A}}(\mathcal{H}^{\text{even}}, \mathcal{H}^{\text{odd}})$ descend to actions on $\mathcal{L}^{\mathrm{R}}_{\mathcal{A}}(\mathcal{H}^{\text{even}}, \mathcal{H}^{\text{odd}})/\ker(R_n)$, but that R_n descends to an isomorphism $\mathcal{L}^{\mathrm{R}}_{\mathcal{A}}(\mathcal{H}^{\text{even}}, \mathcal{H}^{\text{odd}})/\ker(R_n) \cong \mathcal{L}^1_{\mathcal{A}}(\mathcal{H}^{\text{even}}, \mathcal{H}^{\text{odd}}; J)$ intertwining the actions of $\mathrm{U}^{\mathrm{LR}}_{\mathcal{A}}(\mathcal{H}^{\text{even}}, J^{\text{even}})$ and $\mathrm{U}^{\mathrm{LR}}_{\mathcal{A}}(\mathcal{H}^{\text{odd}}, J^{\text{odd}})$, thereby yielding the following

COROLLARY 4.21. *Let $(\mathcal{H}, \gamma, J)$ be a real $\mathcal{A}$-bimodule of KO-dimension n* mod 8 *for $n = 0$ or 4. Then*

(4.30)

$$\mathcal{D}(\mathcal{A}, \mathcal{H}, \gamma, J) \cong \mathrm{U}^{\mathrm{LR}}_{\mathcal{A}}(\mathcal{H}^{\text{odd}}, J^{\text{odd}}) \backslash \left(\mathcal{L}^{\mathrm{R}}_{\mathcal{A}}(\mathcal{H}^{\text{even}}, \mathcal{H}^{\text{odd}})/\ker(R_n)\right) / \mathrm{U}^{\mathrm{LR}}_{\mathcal{A}}(\mathcal{H}^{\text{even}}, J^{\text{even}})$$

Mutatis mutandis, the proof of Lemma 4.17 yields the following characterisation of $\ker(R_n)$:

LEMMA 4.22. *If $K = (1_{n_\alpha} \otimes K_{\alpha\beta} \otimes 1_{n_\beta})_{\alpha,\beta \in \widehat{\mathcal{A}}} \in \mathcal{L}^{\mathrm{LR}}_{\mathcal{A}}(\mathcal{H}^{\text{even}}, \mathcal{H}^{\text{odd}})$, then $K \in \ker(R_n)$ if and only if for each α, $\beta \in \widehat{\mathcal{A}}$ such that $\alpha \neq \beta$,*

$$K_{\beta\alpha} = -\overline{K_{\alpha\beta}}, \tag{4.31}$$

and for each $\alpha \in \widehat{\mathcal{A}}$,

$$K_{\alpha\alpha} \in \mathcal{R}_\alpha(n) = \begin{cases} iM_{m^{\text{odd}}_{\alpha\alpha} \times m^{\text{even}}_{\alpha\alpha}}(\mathbb{R}) & \text{if } n = 0, \\ iM_{m^{\text{odd}}_{\alpha\alpha}/2 \times m^{\text{even}}_{\alpha\alpha}/2}(\mathbb{H}) & \text{if } n = 4. \end{cases} \tag{4.32}$$

Note that such a map $K \in \mathcal{L}^{\mathrm{LR}}_{\mathcal{A}}(\mathcal{H}^{\text{even}}, \mathcal{H}^{\text{odd}})$ is therefore entirely specified by the $K_{\alpha\beta} \in M_{m^{\text{odd}}_{\alpha\beta} \times m^{\text{even}}_{\alpha\beta}}(\mathbb{C})$ for $\alpha < \beta$ and by the $K_{\alpha\alpha} \in \mathcal{R}_\alpha(n)$.

Let us now describe the construction of an arbitrary Dirac operator D on the real $\mathcal{A}$-bimodule $(\mathcal{H}, \gamma, J)$ of KO-dimension $n = 0$ or 4 mod 8:

(1) For α, β, $\gamma \in \widehat{\mathcal{A}}$, choose $M^\gamma_{\alpha\beta} \in M_{n_\gamma m^{\text{odd}}_{\gamma\beta} \times n_\alpha m^{\text{even}}_{\alpha\beta}}(\mathbb{C})$;

(2) Construct $M \in \mathcal{L}^{\mathrm{R}}_{\mathcal{A}}(\mathcal{H}^{\text{even}}, \mathcal{H}^{\text{odd}})$ by setting for α, β, γ, $\delta \in \widehat{\mathcal{A}}$,

$$M^{\gamma\delta}_{\alpha\beta} = \begin{cases} M^\gamma_{\alpha\beta} \otimes 1_{n_\beta} & \text{if } \beta = \delta, \\ 0 & \text{otherwise;} \end{cases} \tag{4.33}$$

(3) Finally, set

$$D = \begin{pmatrix} 0 & R_n(M)^* \\ R_n(M) & 0 \end{pmatrix}. \tag{4.34}$$

Just as before, if R_n is non-injective, we can make the substitution $M \mapsto M + K$ for any $K \in \ker(R_n)$ and obtain the same Dirac operator D; at the level of components, we have for α, $\beta \in \widehat{\mathcal{A}}$ such that $\alpha < \beta$,

$$M^\alpha_{\alpha\beta} \mapsto M^\alpha_{\alpha\beta} + 1_{n_\alpha} \otimes K_{\alpha\beta}, \quad M^\beta_{\beta\alpha} \mapsto M^\beta_{\beta\alpha} + 1_{n_\alpha} \otimes (-\overline{K_{\alpha\beta}})$$
$$M^\alpha_{\alpha\alpha} \mapsto M^\alpha_{\alpha\alpha} + 1_{n_\alpha} \otimes K_{\alpha\alpha}.$$

With these observations in hand, we can revisit the moduli space $\mathcal{D}(\mathcal{A}, \mathcal{H}, \gamma, J)$.

By the discussion above and Corollaries 3.19 and 3.23, we can identify the space $\mathcal{D}_0(\mathcal{A},\mathcal{H},\gamma,J)$ with

$$(4.35)\quad \mathcal{D}_0(\mathcal{A},m^{\text{even}},m^{\text{odd}},n) := \prod_{\alpha\in\widehat{\mathcal{A}}}\Bigg[M_{n_\alpha m^{\text{odd}}_{\alpha\alpha}\times n_\alpha m^{\text{even}}_{\alpha\alpha}}(\mathbb{C})/(1_{n_\alpha}\otimes\mathcal{R}_\alpha(n))$$
$$\times\prod_{\substack{\beta\in\widehat{\mathcal{A}}\\ \beta>\alpha}}(M_{n_\alpha m^{\text{odd}}_{\alpha\beta}\times n_\alpha m^{\text{even}}_{\alpha\beta}}(\mathbb{C})\oplus M_{n_\beta m^{\text{odd}}_{\alpha\beta}\times n_\beta m^{\text{even}}_{\alpha\beta}}(\mathbb{C}))/M_{m^{\text{odd}}_{\alpha\beta}\times m^{\text{even}}_{\alpha\beta}}(\mathbb{C})$$
$$\times\prod_{\substack{\beta,\gamma\in\widehat{\mathcal{A}}\\ \gamma\neq\alpha}} M_{n_\gamma m^{\text{odd}}_{\gamma\beta}\times n_\alpha m^{\text{even}}_{\alpha\beta}}(\mathbb{C})\Bigg],$$

where $M_{m^{\text{odd}}_{\alpha\beta}\times m^{\text{even}}_{\alpha\beta}}(\mathbb{C})$ is viewed as embedded in the space

$$M_{n_\alpha m^{\text{odd}}_{\alpha\beta}\times n_\alpha m^{\text{even}}_{\alpha\beta}}(\mathbb{C})\oplus M_{n_\beta m^{\text{odd}}_{\alpha\beta}\times n_\beta m^{\text{even}}_{\alpha\beta}}(\mathbb{C})$$

via the map $K\mapsto(1_{n_\alpha}\otimes K)\oplus(-1_{n_\beta}\otimes\overline{K})$, and identify the groups $\mathrm{U}^{\mathrm{LR}}_{\mathcal{A}}(\mathcal{H}^{\text{even}};J^{\text{even}})$ and $\mathrm{U}^{\mathrm{LR}}_{\mathcal{A}}(\mathcal{H}^{\text{odd}};J^{\text{odd}})$ with $\mathrm{U}(\mathcal{A},m^{\text{even}},n')$ and $\mathrm{U}(\mathcal{A},m^{\text{odd}},n')$, respectively. Then the actions of $\mathrm{U}^{\mathrm{LR}}_{\mathcal{A}}(\mathcal{H}^{\text{even}};J^{\text{even}})$ and $\mathrm{U}^{\mathrm{LR}}_{\mathcal{A}}(\mathcal{H}^{\text{odd}};J^{\text{odd}})$ on $\mathcal{L}^1_{\mathcal{A}}(\mathcal{H}^{\text{even}},\mathcal{H}^{\text{odd}};J)$ corresponds under these identifications to the actions of the groups $\mathrm{U}(\mathcal{A},m^{\text{even}},n')$ and $\mathrm{U}(\mathcal{A},m^{\text{odd}},n')$, respectively, on $\mathcal{D}_0(\mathcal{A},m^{\text{even}},m^{\text{odd}},n)$ defined by having

$$(U^{\text{odd}}_{\alpha\alpha};U^{\text{odd}}_{\alpha\beta})\in\mathrm{U}(\mathcal{A},m^{\text{odd}};n'),\qquad(U^{\text{even}}_{\alpha\alpha};U^{\text{odd}}_{\alpha\beta})\in\mathrm{U}(\mathcal{A},m^{\text{even}};n')$$

act on $\left([M^\alpha_{\alpha\alpha}];[(M^\alpha_{\alpha\beta},M^\beta_{\beta\alpha})];M^\gamma_{\alpha\beta}\right)\in\mathcal{D}_0(\mathcal{A},m,n)$ by

$$[M^\alpha_{\alpha\alpha}]\mapsto\left[(1_{n_\alpha}\otimes U^{\text{odd}}_{\alpha\alpha})M_{\alpha\alpha}\right];$$
$$[(M^\alpha_{\alpha\beta},M^\beta_{\beta\alpha})]\mapsto\left[\left((1_{n_\alpha}\otimes U^{\text{odd}}_{\alpha\beta})M^\alpha_{\alpha\beta},(1_{n_\beta}\otimes\overline{U^{\text{odd}}_{\alpha\beta}})M^\beta_{\beta\alpha}\right)\right];$$
$$M^\gamma_{\alpha\beta}\mapsto\begin{cases}(1_{n_\gamma}\otimes U^{\text{odd}}_{\gamma\beta})M^\gamma_{\alpha\beta} & \text{if }\gamma<\delta,\\ (1_{n_\gamma}\otimes\overline{U^{\text{odd}}_{\beta\gamma}})M^\gamma_{\alpha\beta} & \text{if }\gamma>\delta;\end{cases}$$

and

$$[M^\alpha_{\alpha\alpha}]\mapsto\left[M_{\alpha\alpha}(1_{n_\alpha}\otimes(U^{\text{even}}_{\alpha\alpha})^*)\right];$$
$$[(M^\alpha_{\alpha\beta},M^\beta_{\beta\alpha})]\mapsto\left[\left(M^\alpha_{\alpha\beta}(1_{n_\alpha}\otimes(U^{\text{even}}_{\alpha\beta})^*),M^\beta_{\beta\alpha}(1_{n_\beta}\otimes(U^{\text{even}}_{\alpha\beta})^T)\right)\right];$$
$$M^\gamma_{\alpha\beta}\mapsto\begin{cases}M^\gamma_{\alpha\beta}(1_{n_\alpha}\otimes(U^{\text{even}}_{\alpha\beta})^*) & \text{if }\alpha<\beta,\\ M^\gamma_{\alpha\beta}(1_{n_\alpha}\otimes(U^{\text{even}}_{\beta\alpha})^T) & \text{if }\alpha>\beta;\end{cases}$$

respectively. We have therefore proved the following:

PROPOSITION 4.23. *Let $(\mathcal{H},\gamma,J)$ be a real $\mathcal{A}$-bimodule of even KO-dimension $n \bmod 8$ for $n=0$ or 4, with multiplicity matrices $(m^{\text{even}},m^{\text{odd}})$. Then*

$$(4.36)\quad \mathcal{D}(\mathcal{A},\mathcal{H},\gamma,J)\cong\mathrm{U}(\mathcal{A},m^{\text{odd}},n')\backslash\mathcal{D}_0(\mathcal{A},m^{\text{even}},m^{\text{odd}},n)/\mathrm{U}(\mathcal{A},m^{\text{even}},n').$$

It is worth noting that considerable simplifications are obtained in the quasi-orientable case, as all components of the form $M^\alpha_{\alpha\beta}\otimes 1_{n_\beta}$ of $M\in\mathcal{L}^{\mathrm{R}}_{\mathcal{A}}(\mathcal{H}^{\text{even}},\mathcal{H}^{\text{odd}})$

must necessarily vanish, as must $\ker(R_n)$ itself. In particular, then, one is left with

$$\mathcal{D}_0(\mathcal{A}, m^{\text{even}}, m^{\text{odd}}, n) = \prod_{\alpha,\beta,\gamma\in\widehat{\mathcal{A}}} M_{n_\gamma m^{\text{odd}}_{\gamma\beta} \times n_\alpha m^{\text{even}}_{\alpha\beta}}(\mathbb{C}).$$

4.5.2. *KO-dimension* 2 *or* 6 mod 8. Let us now consider the case where $n = 2$ or $n = 6 \bmod 8$, *i.e.* where $\varepsilon' = -1$. Then

$$J = \begin{pmatrix} 0 & \varepsilon\tilde{J}^* \\ \tilde{J} & 0 \end{pmatrix}$$

for $\tilde{J} : \mathcal{H}^{\text{even}} \to \mathcal{H}^{\text{odd}}$ anti-unitary, and $m^{\text{odd}} = (m^{\text{even}})^T$. In light of Corollary 3.29, one can easily establish, along the lines of Propositions 4.10 and 4.20, the following result:

PROPOSITION 4.24. *Let* $\mathrm{U}^{\mathrm{LR}}_{\mathcal{A}}(\mathcal{H}^{\text{even}})$ *act on* $\mathcal{L}^1_{\mathcal{A}}(\mathcal{H}^{\text{even}}, \mathcal{H}^{\text{odd}}; J)$ *by*

$$(U, \Delta) \mapsto \tilde{J}U\tilde{J}^*\Delta U^*$$

for $U \in \mathrm{U}^{\mathrm{LR}}_{\mathcal{A}}(\mathcal{H}^{\text{even}})$ *and* $\Delta \in \mathcal{L}^1_{\mathcal{A}}(\mathcal{H}^{\text{even}}, \mathcal{H}^{\text{odd}}; J)$. *Then the map*

$$\mathcal{D}(\mathcal{A}, \mathcal{H}, \gamma, J) \to \mathcal{L}^1_{\mathcal{A}}(\mathcal{H}^{\text{even}}, \mathcal{H}^{\text{odd}}; J)/\,\mathrm{U}^{\mathrm{LR}}_{\mathcal{A}}(\mathcal{H}^{\text{even}})$$

defined by $[D] \mapsto [P^{\text{odd}} D P^{\text{even}}]$ *is a homeomorphism.*

In the same way, we can define an action of $\mathrm{U}^{\mathrm{LR}}_{\mathcal{A}}(\mathcal{H}^{\text{even}})$ on $\mathcal{L}^{\mathrm{R}}_{\mathcal{A}}(\mathcal{H}^{\text{even}}, \mathcal{H}^{\text{odd}})$. We now give the relevant analogue of Propositions 4.15 and 4.20:

PROPOSITION 4.25. *The map* $R_n : \mathcal{L}^{\mathrm{R}}_{\mathcal{A}}(\mathcal{H}^{\text{even}}, \mathcal{H}^{\text{odd}}) \to \mathcal{L}^1_{\mathcal{A}}(\mathcal{H}^{\text{even}}, \mathcal{H}^{\text{odd}}; J)$ *defined by* $R_n(M) := M + \varepsilon\tilde{J}M^*\tilde{J}$ *is a surjection intertwining the actions of the group* $\mathrm{U}^{\mathrm{LR}}_{\mathcal{A}}(\mathcal{H}^{\text{even}})$ *on* $\mathcal{L}^{\mathrm{R}}_{\mathcal{A}}(\mathcal{H}^{\text{even}}, \mathcal{H}^{\text{odd}})$ *and* $\mathcal{L}^1_{\mathcal{A}}(\mathcal{H}^{\text{even}}, \mathcal{H}^{\text{odd}}; J)$, *and* $\ker(R_n) \subset \mathcal{L}^{\mathrm{LR}}_{\mathcal{A}}(\mathcal{H}^{\text{even}}, \mathcal{H}^{\text{odd}})$.

PROOF. First note that

$$\mathcal{L}^1_{\mathcal{A}}(\mathcal{H}^{\text{even}}, \mathcal{H}^{\text{odd}}; J) = \{\Delta \in \mathcal{L}^1_{\mathcal{A}}(\mathcal{H}^{\text{even}}, \mathcal{H}^{\text{odd}}) \mid \Delta = \varepsilon\tilde{J}\Delta^*\tilde{J}\},$$

as can be checked by direct calculation, so that R_n is indeed well-defined. It also readily follows by construction of R_n and the definition of the actions of $\mathrm{U}^{\mathrm{LR}}_{\mathcal{A}}(\mathcal{H}^{\text{even}})$ that R_n has the desired intertwining properties.

Now, for $M \in \mathcal{L}^{\mathrm{R}}_{\mathcal{A}}(\mathcal{H}^{\text{even}}, \mathcal{H}^{\text{odd}})$, one has that $R_n(M) = 0$ if and only if $M = -\varepsilon\tilde{J}M^*\tilde{J}$, but $\tilde{J}M^*\tilde{J}$ is manifestly left $\mathcal{A}$-linear, so that $M \in \mathcal{L}^{\mathrm{LR}}_{\mathcal{A}}(\mathcal{H}^{\text{even}}, \mathcal{H}^{\text{odd}})$, as claimed.

Finally, just as in the proof of Propositions 4.15 and 4.20, one can easily check that for $\Delta \in \mathcal{L}^1_{\mathcal{A}}(\mathcal{H}^{\text{even}}, \mathcal{H}^{\text{odd}}; J)$,

$$\Delta = R_n\Big(\frac{1}{2}(E'_\lambda + E_\rho)(\Delta)\Big),$$

where $\frac{1}{2}(E'_\lambda + E_\rho)(\Delta)$ is right $\mathcal{A}$-linear. □

Just as in the earlier cases, the action of $\mathrm{U}^{\mathrm{LR}}_{\mathcal{A}}(\mathcal{H}^{\text{even}})$ on $\mathcal{L}^{\mathrm{R}}_{\mathcal{A}}(\mathcal{H}^{\text{even}}, \mathcal{H}^{\text{odd}})$ descends to an action on the quotient $\mathcal{L}^{\mathrm{R}}_{\mathcal{A}}(\mathcal{H}^{\text{even}}, \mathcal{H}^{\text{odd}})/\ker(R_n)$, so that R_n descends to an $\mathrm{U}^{\mathrm{LR}}_{\mathcal{A}}(\mathcal{H}^{\text{even}}, \mathcal{H}^{\text{odd}})$-isomorphism

$$\mathcal{L}^{\mathrm{R}}_{\mathcal{A}}(\mathcal{H}^{\text{even}}, \mathcal{H}^{\text{odd}})/\ker(R_n) \cong \mathcal{L}^1_{\mathcal{A}}(\mathcal{H}^{\text{even}}, \mathcal{H}^{\text{odd}}; J), \tag{4.37}$$

thereby yielding the following:

COROLLARY 4.26. *Let $(\mathcal{H},\gamma,J)$ be a real $\mathcal{A}$-bimodule of KO-dimension $n \bmod 8$ for $n = 2$ or 6. Then*

$$\mathcal{D}(\mathcal{A},\mathcal{H},\gamma,J) \cong (\mathcal{L}^{\mathrm{R}}_{\mathcal{A}}(\mathcal{H}^{\mathrm{even}},\mathcal{H}^{\mathrm{odd}})/\ker(R_n))/\,\mathrm{U}^{\mathrm{LR}}_{\mathcal{A}}(\mathcal{H}^{\mathrm{even}}). \tag{4.38}$$

Again, *mutatis mutandis*, the proof of Lemma 4.17 yields the following characterisation of $\ker(R_n)$:

LEMMA 4.27. *If $K = (1_{n_\alpha} \otimes K_{\alpha\beta} \otimes 1_{n_\beta})_{\alpha,\beta\in\widehat{\mathcal{A}}} \in \mathcal{L}^{\mathrm{LR}}_{\mathcal{A}}(\mathcal{H}^{\mathrm{even}},\mathcal{H}^{\mathrm{odd}})$, then $K \in \ker(R_n)$ if and only if for each α, $\beta \in \widehat{\mathcal{A}}$ such that $\alpha \neq \beta$,*

$$K_{\beta\alpha} = -\varepsilon K^T_{\alpha\beta}, \tag{4.39}$$

and for each $\alpha \in \widehat{\mathcal{A}}$,

$$K_{\alpha\alpha} \in \mathcal{R}_\alpha(n) = \begin{cases} \mathrm{Sym}_{m^{\mathrm{even}}_{\alpha\alpha}}(\mathbb{C}) & \text{if } n = 2, \\ \mathfrak{so}(m^{\mathrm{even}}_{\alpha\alpha},\mathbb{C}) & \text{if } n = 6. \end{cases} \tag{4.40}$$

Thus, such a map $K \in \ker(R_n)$ is entirely specified by the components $K_{\alpha\beta} \in M_{m^{\mathrm{even}}_{\beta\alpha}\times m^{\mathrm{even}}_{\alpha\beta}}(\mathbb{C})$ for $\alpha < \beta$ and by the $K_{\alpha\alpha} \in \mathcal{R}_\alpha(n)$.

Note that the discussion of the construction of Dirac operators and of the freedom in the construction provided by $\ker(R_n)$ in the case of KO-dimension 0 or 4 mod 8 holds also in this case. Thus we can identify $\mathcal{D}(\mathcal{A},\mathcal{H},\gamma,J)$ with

$$\begin{aligned} \mathcal{D}_0(\mathcal{A},m^{\mathrm{even}},n) := \prod_{\alpha\in\widehat{\mathcal{A}}} \Bigg[& M_{n_\alpha m^{\mathrm{even}}_{\alpha\alpha}}(\mathbb{C})/(1_{n_\alpha}\otimes\mathcal{R}_\alpha(n)) \\ & \times \prod_{\substack{\beta\in\widehat{\mathcal{A}}\\ \beta>\alpha}} \left(M_{n_\alpha m^{\mathrm{even}}_{\beta\alpha}\times n_\alpha m^{\mathrm{even}}_{\alpha\beta}}(\mathbb{C}) \oplus M_{n_\beta m^{\mathrm{even}}_{\alpha\beta}\times n_\beta m^{\mathrm{even}}_{\beta\alpha}}(\mathbb{C})\right)/M_{m^{\mathrm{even}}_{\beta\alpha}\times m^{\mathrm{even}}_{\alpha\beta}}(\mathbb{C}) \\ & \times \prod_{\substack{\beta,\gamma\in\widehat{\mathcal{A}}\\ \gamma\neq\alpha}} M_{n_\gamma m^{\mathrm{even}}_{\beta\gamma}\times n_\alpha m^{\mathrm{even}}_{\alpha\beta}}(\mathbb{C})\Bigg], \end{aligned} \tag{4.41}$$

where $M_{m^{\mathrm{even}}_{\beta\alpha}\times m^{\mathrm{even}}_{\alpha\beta}}(\mathbb{C})$ is viewed as embedded in the space

$$M_{n_\alpha m^{\mathrm{even}}_{\beta\alpha}\times n_\alpha m^{\mathrm{even}}_{\alpha\beta}}(\mathbb{C}) \oplus M_{n_\beta m^{\mathrm{even}}_{\alpha\beta}\times n_\beta m^{\mathrm{even}}_{\beta\alpha}}(\mathbb{C})$$

via the map $K \mapsto (1_{n_\alpha}\otimes K) \oplus (-\varepsilon 1_{n_\beta}\otimes K^T)$, and identify $\mathrm{U}^{\mathrm{LR}}_{\mathcal{A}}(\mathcal{H}^{\mathrm{even}})$ with $\mathrm{U}(\mathcal{A},m^{\mathrm{even}})$. Then the action of $\mathrm{U}^{\mathrm{LR}}_{\mathcal{A}}(\mathcal{H}^{\mathrm{even}})$ on $\mathcal{L}^1_{\mathcal{A}}(\mathcal{H}^{\mathrm{even}},\mathcal{H}^{\mathrm{odd}};J)$ corresponds under these identifications with the action of $\mathrm{U}(\mathcal{A},m^{\mathrm{even}})$ on $\mathcal{D}_0(\mathcal{A},m^{\mathrm{even}},n)$ defined by having $(U_{\alpha\beta}) \in \mathrm{U}(\mathcal{A},m^{\mathrm{even}})$ act on $\left([M^\alpha_{\alpha\alpha}];[(M^\alpha_{\alpha\beta},M^\beta_{\beta\alpha})];M^\gamma_{\alpha\beta}\right) \in \mathcal{D}_0(\mathcal{A},m,n)$ by

$$\begin{aligned} [M^\alpha_{\alpha\alpha}] &\mapsto \left[(1_{n_\alpha}\otimes\overline{U_{\alpha\alpha}})M^\alpha_{\alpha\alpha}(1_{n_\alpha}\otimes U^*_{\alpha\alpha})\right]; \\ [(M^\alpha_{\alpha\beta},M^\beta_{\beta\alpha})] &\mapsto \left[((1_{n_\alpha}\otimes\overline{U_{\beta\alpha}})M^\alpha_{\alpha\beta}(1_{n_\alpha}\otimes U^*_{\alpha\beta}),(1_{n_\beta}\otimes\overline{U_{\alpha\beta}})M^\alpha_{\beta\alpha}(1_{n_\beta}\otimes U^*_{\beta\alpha}))\right]; \\ M^\gamma_{\alpha\beta} &\mapsto (1_{n_\gamma}\otimes\overline{U_{\beta\gamma}})M^\gamma_{\alpha\beta}(1_{n_\alpha}\otimes U^*_{\alpha\beta}). \end{aligned}$$

This, then, proves the following:

PROPOSITION 4.28. *Let $(\mathcal{H},\gamma,J)$ be a real $\mathcal{A}$-bimdoule of even KO-dimension $n \bmod 8$ for $n = 2$ or 6, with multiplicity matrices $(m^{\mathrm{even}},(m^{\mathrm{even}})^T)$. Then*

$$\mathcal{D}(\mathcal{A},\mathcal{H},\gamma,J) \cong \mathcal{D}_0(\mathcal{A},m^{\mathrm{even}},n)/\,\mathrm{U}(\mathcal{A},m^{\mathrm{even}}). \tag{4.42}$$

Again, considerable simplifications are obtained in the quasi-orientable case, just as for KO-dimension 0 or 4 mod 8.

4.6. Dirac operators in the Chamseddine–Connes–Marcolli model. Let us now apply the above results on Dirac operators and moduli spaces thereof to the bimodules appearing in the Chamseddine–Connes–Marcolli model.

We begin with $(\mathcal{H}_F, \gamma_F, J_F, \epsilon_F)$ as an S^0-real $\mathcal{A}_{LR}$-bimodule of KO-dimension 6 mod 8, which, as we shall now see, is essentially S^0-real in structure:

PROPOSITION 4.29. *For the S^0-real $\mathcal{A}_{LR}$-bimodule $(\mathcal{H}_F, \gamma_F, J_F, \epsilon_F)$ of KO-dimension* 6 mod 8,

$$\mathcal{D}_0(\mathcal{A}_{LR}, \mathcal{H}_F, \gamma_F, J_F) = \mathcal{D}_0(\mathcal{A}_{LR}, \mathcal{H}_F, \gamma_F, J_F, \epsilon_F),$$

and

$$\mathrm{U}^{\mathrm{LR}}_{\mathcal{A}_{LR}}(\mathcal{H}, \gamma_F, J_F) = \mathrm{U}^{\mathrm{LR}}_{\mathcal{A}_{LR}}(\mathcal{H}, \gamma_F, J_F, \epsilon_F),$$

so that

$$\mathcal{D}(\mathcal{A}_{LR}, \mathcal{H}_F, \gamma_F, J_F) = \mathcal{D}(\mathcal{A}_{LR}, \mathcal{H}_F, \gamma_F, J_F, \epsilon_F).$$

PROOF. To prove the first part of the claim, by Proposition 4.25, it suffices to show that any right $\mathcal{A}_{LR}$-linear operator $\mathcal{H}_F^{\mathrm{even}} \to \mathcal{H}_F^{\mathrm{odd}}$ commutes with ϵ_F. Thus, let $T \in \mathcal{L}^{\mathrm{R}}_{\mathcal{A}_{LR}}(\mathcal{H}_F^{\mathrm{even}}, \mathcal{H}_F^{\mathrm{odd}})$. Then, since the signed multiplicity matrix μ of $(\mathcal{H}_F, \gamma_F)$ as an orientable even $\mathcal{A}_{LR}$-bimodule is given by

$$\mu = N \begin{pmatrix} 0 & 0 & -1 & +1 & 0 & 0 \\ 0 & 0 & 0 & 0 & 0 & 0 \\ +1 & 0 & 0 & 0 & +1 & 0 \\ -1 & 0 & 0 & 0 & -1 & 0 \\ 0 & 0 & -1 & +1 & 0 & 0 \\ 0 & 0 & 0 & 0 & 0 & 0 \end{pmatrix},$$

it follows from Proposition 3.4 that the only non-zero components of T are $T^{\mathbf{2}_L\mathbf{1}}_{\mathbf{2}_R\mathbf{1}}$ and $T^{\mathbf{2}_L\mathbf{3}}_{\mathbf{2}_R\mathbf{3}}$, which both have domain and range within $\mathcal{H}_f = (\mathcal{H}_F)_i$, where ϵ acts as the identity. Thus, T commutes with ϵ_F.

To prove the next part of the claim, it suffices to show that any left and right $\mathcal{A}_{LR}$-linear operator on $\mathcal{H}_F$ commutes with ϵ_F. But again, if $K \in \mathcal{L}^{\mathrm{LR}}_{\mathcal{A}_{LR}}(\mathcal{H}_F)$, then the only non-zero components of K are of the form $K^{\alpha\beta}_{\alpha\beta}$, each of which therefore has both domain and range either within $\mathcal{H}_f$ or $\mathcal{H}_{\overline{f}} = J_F\mathcal{H}_f$, so that K commutes with ϵ_F. The last part of the claim is then an immediate consequence of the first two parts. □

Thus, by Proposition 2.18, we have that

$$\mathcal{D}_0(\mathcal{A}_{LR}, \mathcal{H}_F, \gamma_F, J_F) = \mathcal{D}_0(\mathcal{A}_{LR}, \mathcal{H}_F, \gamma_F, J_F, \epsilon_F) \cong \mathcal{D}_0(\mathcal{A}_{LR}, \mathcal{H}_f, \gamma_f) \tag{4.43}$$

and

$$\mathcal{D}(\mathcal{A}_{LR}, \mathcal{H}_F, \gamma_F, J_F) = \mathcal{D}(\mathcal{A}_{LR}, \mathcal{H}_F, \gamma_F, J_F, \epsilon_F) \cong \mathcal{D}(\mathcal{A}_{LR}, \mathcal{H}_f, \gamma_f), \tag{4.44}$$

where $(\mathcal{H}_f, \gamma_f) = ((\mathcal{H}_F)_i, (\gamma_F)_i)$ is the orientable even $\mathcal{A}_{LR}$-bimodule with signed multiplicity matrix

$$\mu_f = N \begin{pmatrix} 0 & 0 & 0 & 0 & 0 & 0 \\ 0 & 0 & 0 & 0 & 0 & 0 \\ +1 & 0 & 0 & 0 & +1 & 0 \\ -1 & 0 & 0 & 0 & -1 & 0 \\ 0 & 0 & 0 & 0 & 0 & 0 \\ 0 & 0 & 0 & 0 & 0 & 0 \end{pmatrix}.$$

In particular, then, $(\mathcal{H}_F, \gamma_F, J_F)$ as a real $\mathcal{A}_{LR}$-bimodule admits no *off-diagonal* Dirac operators, that is, Dirac operators with non-zero $P_{-i}DP_i : \mathcal{H}_f \to \mathcal{H}_{\overline{f}}$, or equivalently, that have non-vanishing commutator with ϵ_F. Let us now examine $\mathcal{D}_0(\mathcal{A}_{LR}, \mathcal{H}_F, \gamma_F, J_F)$ and $\mathcal{D}(\mathcal{A}_{LR}, \mathcal{H}_F, \gamma_F, J_F)$, or rather, $\mathcal{D}_0(\mathcal{A}_{LR}, \mathcal{H}_f, \gamma_f)$ and $\mathcal{D}(\mathcal{A}_{LR}, \mathcal{H}_f, \gamma_f)$, in more detail.

First, it follows from the form of μ_f and Proposition 3.4 that $\mathcal{L}^{\mathrm{L}}_{\mathcal{A}_{LR}}(\mathcal{H}_f^{\text{even}}, \mathcal{H}_f^{\text{odd}})$ vanishes, whilst

$$\mathcal{L}^{\mathrm{R}}_{\mathcal{A}_{LR}}(\mathcal{H}_f^{\text{even}}, \mathcal{H}_f^{\text{odd}}) = M_{2N}(\mathbb{C}) \oplus (M_{2N}(\mathbb{C}) \otimes 1_3) \cong M_{2N}(\mathbb{C}) \oplus M_{2N}(\mathbb{C}).$$

so that any Dirac operator on $\mathcal{H}_f$ (and hence on $\mathcal{H}_F$) is completely specified by a choice of $M_{\mathbf{2}_L\mathbf{1}}^{\mathbf{2}_R}$, $M_{\mathbf{2}_L\mathbf{3}}^{\mathbf{2}_R} \in M_{2N}(\mathbb{C})$. Indeed, if $(m^{\text{even}}, m^{\text{odd}})$ denotes the pair of multiplicity matrices of $(\mathcal{H}_f, \gamma_f)$, then, in the notation of subsection 4.5.2,

$$\mathcal{D}_0(\mathcal{A}_{LR}, m^{\text{even}}, 6) = M_{2N}(\mathbb{C}) \oplus M_{2N}(\mathbb{C}).$$

At the same time,

$$\mathrm{U}^{\mathrm{LR}}_{\mathcal{A}_{LR}}(\mathcal{H}_f^{\text{even}}) = (1_2 \otimes \mathrm{U}(N)) \oplus (1_2 \otimes \mathrm{U}(N) \otimes 1_3) \cong \mathrm{U}(N) \times \mathrm{U}(N) =: \mathrm{U}(\mathcal{A}_{LR}, m^{\text{even}})$$

and

$$\mathrm{U}^{\mathrm{LR}}_{\mathcal{A}_{LR}}(\mathcal{H}_f^{\text{odd}}) = (1_2 \otimes \mathrm{U}(N)) \oplus (1_2 \otimes \mathrm{U}(N) \otimes 1_3) \cong \mathrm{U}(N) \times \mathrm{U}(N) =: \mathrm{U}(\mathcal{A}_{LR}, m^{\text{odd}}).$$

It then follows that

$$\mathcal{D}(\mathcal{A}_{LR}, \mathcal{H}_f, \gamma_f) \cong \mathrm{U}(\mathcal{A}_{LR}, m^{\text{odd}}) \backslash \mathcal{D}_0(\mathcal{A}_{LR}, m^{\text{even}}, 6) / \mathrm{U}(\mathcal{A}_{LR}, m^{\text{even}}) \tag{4.45}$$

$$= \left(\mathrm{U}(N) \backslash M_{2N}(\mathbb{C}) / \mathrm{U}(N)\right)^2, \tag{4.46}$$

where $\mathrm{U}(N)$ acts on the left by multiplication and on the right by multiplication by the inverse as $1_2 \otimes U(N)$. The two factors of the form $\mathrm{U}(N) \backslash M_{2N}(\mathbb{C}) / \mathrm{U}(N)$ can thus be viewed as the parameter spaces of the components $M_{\mathbf{2}_L\mathbf{1}}^{\mathbf{2}_R}$ and $M_{\mathbf{2}_L\mathbf{3}}^{\mathbf{2}_R}$, respectively.

Let us now consider $(\mathcal{H}_F, \gamma_F, J_F, \epsilon_F)$ as an S^0-real $\mathcal{A}_F$-bimodule, so that the multiplicity matrices $(m^{\text{even}}, m^{\text{odd}})$ of $(\mathcal{H}_F, \gamma_F)$ are given by

$$m^{\text{even}} = N \begin{pmatrix} 1 & 1 & 0 & 0 & 0 \\ 0 & 0 & 0 & 0 & 0 \\ 1 & 0 & 0 & 1 & 0 \\ 1 & 1 & 0 & 0 & 0 \\ 0 & 0 & 0 & 0 & 0 \end{pmatrix}, \quad m^{\text{odd}} = N \begin{pmatrix} 1 & 0 & 1 & 1 & 0 \\ 1 & 0 & 0 & 1 & 0 \\ 0 & 0 & 0 & 0 & 0 \\ 0 & 0 & 1 & 0 & 0 \\ 0 & 0 & 0 & 0 & 0 \end{pmatrix} = (m^{\text{even}})^T.$$

Now it follows from the form of $(m^{\text{even}}, m^{\text{odd}})$ that

$$\mathcal{L}^{\mathrm{R}}_{\mathcal{A}_F}(\mathcal{H}_F^{\text{even}}, \mathcal{H}_F^{\text{odd}}) = M_N(\mathbb{C})^{\oplus 2} \oplus M_{N\times 2N}(\mathbb{C})^{\oplus 2} \oplus M_{N\times 3N}(\mathbb{C})^{\oplus 2} \oplus (M_{N\times 2N}(\mathbb{C}) \otimes 1_3)^{\oplus 2},$$

whilst

$$\ker(R_6) = \mathfrak{sl}(N, \mathbb{C}) \subseteq M_N(\mathbb{C})$$

for the copy of $M_N(\mathbb{C})$ corresponding to $\mathcal{L}^{\mathrm{R}}_{\mathcal{A}}((\mathcal{H}^{\mathrm{even}}_F)_{11}, (\mathcal{H}^{\mathrm{odd}}_F)_{11})$. Since $M_N(\mathbb{C}) = \mathrm{Sym}_N(\mathbb{C}) \oplus \mathfrak{sl}(N, \mathbb{C})$, $M_N(\mathbb{C})/\mathfrak{sl}(N, \mathbb{C})$ can be identified with $\mathrm{Sym}_N(\mathbb{C})$, so that

$$\begin{aligned} &\mathcal{D}_0(\mathcal{A}_F, \mathcal{H}_F, \gamma_F, J_F) \\ &\cong \mathcal{L}^{\mathrm{R}}_{\mathcal{A}_F}(\mathcal{H}^{\mathrm{even}}_F, \mathcal{H}^{\mathrm{odd}}_F)/\ker(R_6) \\ &= \mathrm{Sym}_N(\mathbb{C}) \oplus M_N(\mathbb{C}) \oplus M_{N\times 2N}(\mathbb{C})^{\oplus 2} \oplus M_{N\times 3N}(\mathbb{C})^{\oplus 2} \oplus (M_{N\times 2N}(\mathbb{C}) \otimes 1_3)^{\oplus 2}. \end{aligned}$$

Thus, a Dirac operator D, which is specified by a choice of class

$$[M] \in \mathcal{L}^{\mathrm{R}}_{\mathcal{A}_F}(\mathcal{H}^{\mathrm{even}}_F, \mathcal{H}^{\mathrm{odd}}_F)/\ker(R_6),$$

is therefore specified in turn by the choice of the following matrices:

- $M^{\mathbf{1}}_{\mathbf{11}} \in \mathrm{Sym}_N(\mathbb{C})$, $M^{\overline{\mathbf{1}}}_{\mathbf{11}} \in M_N(\mathbb{C})$;
- $M^{\mathbf{1}}_{\mathbf{21}}$, $M^{\overline{\mathbf{1}}}_{\mathbf{21}} \in M_{N\times 2N}(\mathbb{C})$;
- $M^{\mathbf{1}}_{\mathbf{31}}$, $M^{\overline{\mathbf{1}}}_{\mathbf{31}} \in M_{N\times 3N}(\mathbb{C})$;
- $M^{\mathbf{1}}_{\mathbf{23}}$, $M^{\overline{\mathbf{1}}}_{\mathbf{23}} \in M_{N\times 2N}(\mathbb{C})$.

Indeed, it follows that

$$\begin{aligned} \mathcal{D}_0(\mathcal{A}_F, m^{\mathrm{even}}, 6) = \mathrm{Sym}_N(\mathbb{C}) \oplus M_N(\mathbb{C}) \oplus M_{N\times 2N}(\mathbb{C})^{\oplus 2} \oplus M_{N\times 3N}(\mathbb{C})^{\oplus 2} \\ \oplus M_{N\times 2N}(\mathbb{C})^{\oplus 2}. \end{aligned} \tag{4.47}$$

Next, we have that $\mathrm{U}(\mathcal{A}_F, m^{\mathrm{even}}) = \mathrm{U}(N)^6$, with a copy of $\mathrm{U}(N)$ corresponding to each of $(\mathcal{H}^{\mathrm{even}}_F)_{\mathbf{11}}$, $(\mathcal{H}^{\mathrm{even}}_F)_{\mathbf{1}\overline{\mathbf{1}}}$, $(\mathcal{H}^{\mathrm{even}}_F)_{\mathbf{21}}$, $(\mathcal{H}^{\mathrm{even}}_F)_{\mathbf{23}}$, $(\mathcal{H}^{\mathrm{even}}_F)_{\mathbf{31}}$, and $(\mathcal{H}^{\mathrm{even}}_F)_{\mathbf{3}\overline{\mathbf{1}}}$. Then, by Proposition 4.28,

$$\mathcal{D}(\mathcal{A}_F, \mathcal{H}_F, \gamma_F, J_F) \cong \mathcal{D}_0(\mathcal{A}_F, m^{\mathrm{even}}, 6)/\,\mathrm{U}(\mathcal{A}_F, m^{\mathrm{even}}) \tag{4.48}$$

for the action of $\mathrm{U}(\mathcal{A}_F, m^{\mathrm{even}})$ on $\mathcal{D}_0(\mathcal{A}_F, m^{\mathrm{even}}, 6)$ given by having the element $(U_{\alpha\beta}) \in \mathrm{U}(\mathcal{A}_F, m^{\mathrm{even}})$ act on $(M^{\gamma}_{\alpha\beta}) \in \mathcal{D}_0(\mathcal{A}_F, m^{\mathrm{even}}, 6)$ by

$$M^{\gamma}_{\alpha\beta} \mapsto (1_{n_\gamma} \otimes \overline{U_{\beta\gamma}}) M^{\gamma}_{\alpha\beta} (1_{n_\alpha} \otimes U^*_{\alpha\beta}).$$

Note that in the notation of [8, §§13.4, 13.5], for $(M^{\gamma}_{\alpha\beta}) \in \mathcal{D}_0(\mathcal{A}_F, m^{\mathrm{even}}, 6)$,

$$M^{\mathbf{1}}_{\mathbf{11}} = \frac{1}{2}\Upsilon_R,$$

so that the so-called Majorana mass term is already present in its final form, whilst for $U \in \mathrm{U}(\mathcal{A}_F, m^{\mathrm{even}})$,

$$U = (U_{\mathbf{11}}, U_{\mathbf{1}\overline{\mathbf{1}}}, U_{\mathbf{21}}, U_{\mathbf{23}}, U_{\mathbf{31}}, U_{\mathbf{3}\overline{\mathbf{1}}}) = (\overline{V_2}, \overline{V_1}, V_3, W_3, \overline{W_2}, \overline{W_1}).$$

Finally, let us compute the sub-moduli space $\mathcal{D}(\mathcal{A}_F, \mathcal{H}_F, \gamma_F, J_F; \mathbb{C}_F)$ for

$$\mathbb{C}_F = \{(\zeta, \mathrm{diag}(\zeta, \overline{\zeta}), 0) \in \mathcal{A}_F \mid \lambda \in \mathbb{C}\} \cong \mathbb{C}.$$

It is easy to see that $[M] \in \mathcal{L}^{\mathrm{R}}_{\mathcal{A}_F}(\mathcal{H}^{\mathrm{even}}_F, \mathcal{H}^{\mathrm{odd}}_F)/\ker(R_6)$ yields an element of the subspace $\mathcal{D}_0(\mathcal{A}_F, \mathcal{H}_F, \gamma_F, J_F; \mathbb{C}_F)$ if and only if M commutes with $\lambda(\mathbb{C}_F)$, but this holds if and only if for all $\zeta \in \mathbb{C}$ and $\beta \in \widehat{\mathcal{A}_F}$,

$$\begin{aligned} \zeta M^{\mathbf{1}}_{\mathbf{1}\beta} &= M^{\mathbf{1}}_{\mathbf{1}\beta}\zeta, & \overline{\zeta} M^{\overline{\mathbf{1}}}_{\mathbf{1}\beta} &= M^{\overline{\mathbf{1}}}_{\mathbf{1}\beta}\zeta, \\ \zeta M^{\mathbf{1}}_{\mathbf{2}\beta} &= M^{\mathbf{1}}_{\mathbf{2}\beta}(\mathrm{diag}(\zeta, \overline{\zeta}) \otimes 1_N), & \overline{\zeta} M^{\overline{\mathbf{1}}}_{\mathbf{2}\beta} &= M^{\overline{\mathbf{1}}}_{\mathbf{2}\beta}(\mathrm{diag}(\zeta, \overline{\zeta}) \otimes 1_N), \\ 0 M^{\mathbf{1}}_{\mathbf{3}\beta} &= M^{\mathbf{1}}_{\mathbf{3}\beta}\zeta, & 0 M^{\overline{\mathbf{1}}}_{\mathbf{3}\beta} &= M^{\overline{\mathbf{1}}}_{\mathbf{3}\beta}\zeta, \end{aligned}$$

which is in turn equivalent to having $M_{11}^{\bar{1}}$, M_{31}^{1} and $M_{31}^{\bar{1}}$ all vanish, and

$$M_{21}^{1} = \begin{pmatrix} \Upsilon_\nu & 0 \end{pmatrix}, \quad M_{21}^{\bar{1}} = \begin{pmatrix} 0 & \Upsilon_e \end{pmatrix}, \quad M_{23}^{1} = \begin{pmatrix} \Upsilon_u & 0 \end{pmatrix}, \quad M_{23}^{\bar{1}} = \begin{pmatrix} 0 & \Upsilon_d \end{pmatrix},$$

for Υ_ν, Υ_e, Υ_u, $\Upsilon_d \in M_N(\mathbb{C})$. One can check that our notation is consistent with that of [8, §§13.4, 13.5]. Indeed, if $\mathcal{D}_0(\mathcal{A}_F, m^{\text{even}}, 6; \mathbb{C}_F)$ denotes the subspace of $\mathcal{D}_0(\mathcal{A}_F, m^{\text{even}}, 6)$ corresponding to $\mathcal{D}_0(\mathcal{A}_F, \mathcal{H}_F, \gamma_F, J_F; \mathbb{C}_F)$, then

$$\mathcal{D}(\mathcal{A}_F, \mathcal{H}_F, \gamma_F, J_F; \mathbb{C}_F) \cong \mathcal{D}_0(\mathcal{A}_F, m^{\text{even}}, 6; \mathbb{C}_F)/\operatorname{U}(\mathcal{A}_F, m^{\text{even}}) \cong \mathcal{C}_q \times \mathcal{C}_l \tag{4.49}$$

for

$$\mathcal{C}_q := \left(\operatorname{U}(N) \times \operatorname{U}(N)\right)\backslash\left(M_N(\mathbb{C}) \times M_N(\mathbb{C})\right)/\operatorname{U}(N), \tag{4.50}$$

where $\operatorname{U}(N)$ acts diagonally by multiplication on the right, and

$$\mathcal{C}_l := \left(\operatorname{U}(N) \times \operatorname{U}(N)\right)\backslash\left(M_N(\mathbb{C}) \times M_N(\mathbb{C}) \times \operatorname{Sym}_N(\mathbb{C})\right)/\operatorname{U}(N),$$

where $\operatorname{U}(N) \times \operatorname{U}(N)$ acts trivially on $\operatorname{Sym}_N(\mathbb{C})$ and $\operatorname{U}(N)$ acts on $\operatorname{Sym}_N(\mathbb{C})$ by

$$(V_2, \Upsilon_R) \mapsto V_2 \Upsilon_R V_2^T;$$

note that $\mathcal{C}_q$ is the parameter space for the matrices (Υ_u, Υ_d), whilst $\mathcal{C}_l$ is the parameter space for the matrices $(\Upsilon_\nu, \Upsilon_e, \Upsilon_R)$. Thus we have recovered the sub-moduli space of Dirac operators considered by Chamseddine–Connes–Marcolli [4, §§2.6, 2.7] (*cf.* also [8, §§13.4, 13.5]).

5. Applications to the Recent Work of Chamseddine and Connes

In this section, we reformulate the results of Chamseddine and Connes in [2, 3] and give new proofs thereof using the theory of bimodules and bilateral triples developed above.

Before continuing, recall that, up to automorphisms, the only real forms of $M_n(\mathbb{C})$ are $M_n(\mathbb{C})$, $M_n(\mathbb{R})$, and, if n is even, $M_{n/2}(\mathbb{H})$.

5.1. Admissible real bimodules.

We begin by studying what Chamseddine and Connes call *irreducible triplets*, namely, real $\mathcal{A}$-bimodules satisfying certain representation-theoretic conditions, along the lines of [3, §2]. However, we shall progress by adding Chamseddine and Connes's various requirements for irreducible triplets one by one, bringing us gradually to their classification of irreducible triplets.

In what follows, $\mathcal{A}$ will once more denote a fixed real C^*-algebra, and for $(\mathcal{H}, J)$ a real $\mathcal{A}$-bimodule of odd KO-dimension, $\mathcal{L}_{\mathcal{A}}^{\text{LR}}(\mathcal{H}; J)$ will denote the real $*$-subalgebra of $\mathcal{L}_{\mathcal{A}}^{\text{LR}}(\mathcal{H})$ consisting of elements commuting with J.

Let us now introduce the first explicit requirement for irreducible triplets.

DEFINITION 5.1. Let $(\mathcal{H}, J)$ be a real $\mathcal{A}$-bimodule of odd KO-dimension. We shall say that $(\mathcal{H}, J)$ is *irreducible* if 0 and 1 are the only projections in $\mathcal{L}_{\mathcal{A}}^{\text{LR}}(\mathcal{H}; J)$.

To proceed, we shall need the following:

LEMMA 5.2. *Let $(\mathcal{H}, J)$ be a real $\mathcal{A}$-bimodule of odd KO-dimension n mod 8 with multiplicity matrix m. Then*

$$\mathcal{L}_{\mathcal{A}}^{\text{LR}}(\mathcal{H}; J) \cong \begin{cases} \left(\bigoplus_{\alpha \in \hat{\mathcal{A}}} M_{m_{\alpha\alpha}}(\mathbb{R})\right) \oplus \bigoplus_{\substack{\alpha,\beta \in \hat{\mathcal{A}} \\ \alpha < \beta}} M_{m_{\alpha\beta}}(\mathbb{C}), & \text{if } n = 1 \text{ or } 7 \bmod 8, \\ \left(\bigoplus_{\alpha \in \hat{\mathcal{A}}} M_{m_{\alpha\alpha}/2}(\mathbb{H})\right) \oplus \bigoplus_{\substack{\alpha,\beta \in \hat{\mathcal{A}} \\ \alpha < \beta}} M_{m_{\alpha\beta}}(\mathbb{C}), & \text{if } n = 3 \text{ or } 5 \bmod 8. \end{cases} \tag{5.1}$$

PROOF. Let $T = (1_{n_\alpha} \otimes T_{\alpha\beta} \otimes 1_{n_\beta}) \in \mathcal{L}^{\mathrm{LR}}_{\mathcal{A}}(\mathcal{H})$. Just as for Propositions 3.19 and 3.23, one can show that $[T, J] = 0$ if and only if for all $\alpha, \beta \in \widehat{\mathcal{A}}$, $T_{\beta\alpha} = \overline{T_{\alpha\beta}}$ if $\alpha \neq \beta$ and

$$T_{\alpha\alpha} \in \begin{cases} M_{m_{\alpha\alpha}}(\mathbb{R}), & \text{if } n = 1 \text{ or } 7 \bmod 8, \\ M_{m_{\alpha\alpha}/2}(\mathbb{H}), & \text{if } n = 3 \text{ or } 5 \bmod 8. \end{cases}$$

Thus, $T \in \mathcal{L}^{\mathrm{LR}}_{\mathcal{A}}(\mathcal{H}; J)$ is completely specified by the matrices $T_{\alpha\alpha}$ and $T_{\alpha\beta}$ for $\alpha > \beta$, giving rise to the isomorphisms of the claim. □

We can now formulate the part of the results of [3, §2] that depends only on this notion of irreducibility.

PROPOSITION 5.3. *Let $(\mathcal{H}, J)$ be a real $\mathcal{A}$-bimodule of odd KO-dimension n* mod 8 *with multiplicity matrix m. Then $(\mathcal{H}, J)$ is irreducible if and only if one of the following holds:*

(1) *There exists $\alpha \in \widehat{\mathcal{A}}$ such that $m = 2^{(1-\varepsilon)/2} E_{\alpha\alpha}$;*
(2) *There exist $\alpha, \beta \in \widehat{\mathcal{A}}$, $\alpha \neq \beta$, such that $m = E_{\alpha\beta} + E_{\beta\alpha}$.*

PROOF. By definition, $(\mathcal{H}, J)$ is irreducible if and only if the only projections in the real C^*-algebra $\mathcal{L}^{\mathrm{LR}}_{\mathcal{A}}(\mathcal{H}, J)$ are 0 and 1, but by Lemma 5.2, this in turn holds if and only if one of the following holds:

(1) $\mathcal{L}^{\mathrm{LR}}_{\mathcal{A}}(\mathcal{H}; J) \cong \mathbb{R}$, so that $n = 1$ or $7 \bmod 8$, and $m = E_{\alpha\alpha}$ for some $\alpha \in \widehat{\mathcal{A}}$,
(2) $\mathcal{L}^{\mathrm{LR}}_{\mathcal{A}}(\mathcal{H}; J) \cong \mathbb{H}$, so that $n = 3$ or $5 \bmod 8$, and $m = 2E_{\alpha\alpha}$ for some $\alpha \in \widehat{\mathcal{A}}$,
(3) $\mathcal{L}^{\mathrm{LR}}_{\mathcal{A}}(\mathcal{H}; J) \cong \mathbb{C}$, so that $m = E_{\alpha\beta} + E_{\beta\alpha}$ for some $\alpha, \beta \in \widehat{\mathcal{A}}$, $\alpha \neq \beta$,

which yields in turn the desired result. □

We shall call an irreducible odd KO-dimensional real $\mathcal{A}$-bimodule $(\mathcal{H}, J)$ *type A* if the first case holds, and *type B* if the second case holds; Chamseddine and Connes's first and second case for irreducible triplets [3, Lemma 2.2] correspond to the type A and type B case, respectively. We shall also find it convenient to define the *skeleton* $\operatorname{skel}(\mathcal{H}, J)$ of such a bimodule as follows:

(1) if $(\mathcal{H}, J)$ is type A, then $\operatorname{skel}(\mathcal{H}, J) := \{\alpha\}$, where $\alpha \in \widehat{\mathcal{A}}$ is such that $\operatorname{mult}[\mathcal{H}] = 2^{\frac{1-\varepsilon}{2}} E_{\alpha\alpha}$;
(2) if $(\mathcal{H}, J)$ is type B, then $\operatorname{skel}(\mathcal{H}, J) := \{\alpha, \beta\}$, where $\alpha, \beta \in \widehat{\mathcal{A}}$, $\alpha \neq \beta$, are such that $\operatorname{mult}[\mathcal{H}] = E_{\alpha\beta} + E_{\beta\alpha}$.

Let us now introduce the second explicit requirement for irreducible triplets.

DEFINITION 5.4. An $\mathcal{A}$-bimodule $\mathcal{H}$ is *(left) separating* if there exists some $\xi \in \mathcal{H}$ such that $\lambda(\mathcal{A})'\xi = \mathcal{H}$. Such a vector ξ is then called a *separating vector* for $\mathcal{A}$.

Recall that for a representation $\mathcal{X}$ of a complex C^*-algebra $\mathcal{C}$, $\xi \in \mathcal{X}$ is a separating vector if and only if the map $\mathcal{C} \to \mathcal{X}$ given by $c \mapsto c\xi$ is injective.

LEMMA 5.5. *Let $p, q \in \mathbb{N}$. There exists a separating vector ξ for the usual action of $M_p(\mathbb{C})$ on $\mathbb{C}^p \otimes \mathbb{C}^q$ as $M_p(\mathbb{C}) \otimes 1_q$ if and only if $p \leq q$.*

PROOF. Let $\{e_i\}_{i=1}^p$ be a basis for $\mathbb{C}^p$, and let $\{f_j\}_{j=1}^q$ be a basis for $\mathbb{C}^q$.

First suppose that $p \leq q$. Let $\xi \in \mathbb{C}^p \otimes \mathbb{C}^q$ be given by $\xi = \sum_{i=1}^p e_i \otimes f_i$. Then for any a, $b \in M_p(\mathbb{C})$,

$$(a \otimes 1_q)\,\xi - (b \otimes 1_q)\,\xi = \sum_{i=1}^{p} \left(\sum_{l=1}^{p} (a_i^l - b_i^l) e_l \right) \otimes f_i$$

so that by linear independence of the e_i and f_j, the left-hand side vanishes if and only if for each i and l, $a_i^l - b_i^l = 0$, *i.e.* $a = b$. Hence, ξ is indeed a separating vector.

Now suppose that $p > q$. Then $\dim_{\mathbb{C}} M_p(\mathbb{C}) - \dim_{\mathbb{C}} \mathbb{C}^p \otimes \mathbb{C}^q = p(p-q) > 0$, so that for any $\xi \in \mathbb{C}^p \otimes \mathbb{C}^q$, the map $M_p(\mathbb{C}) \mapsto \mathbb{C}^p \otimes \mathbb{C}^q$ given by $a \mapsto (a \otimes 1_q)\,\xi$ cannot possibly be injective, and hence ξ cannot possibly be separating. □

We can now reformulate that part of the results in [3, §2] that depends only on irreducibility and the existence of a separating vector.

PROPOSITION 5.6. *Let $(\mathcal{H}, J)$ be an irreducible real $\mathcal{A}$-bimodule of odd KO-dimension n* mod 8.

(1) *If $(\mathcal{H}, J)$ is type A, then it is separating;*
(2) *If $(\mathcal{H}, J)$ is type B with skeleton (α, β), then $(\mathcal{H}, J)$ is separating if and only if $n_\alpha = n_\beta$.*

PROOF. First suppose that $(\mathcal{H}, J)$ is type A. Let $\{\alpha\} = \operatorname{skel}(\mathcal{H}, J)$, and let $m_n = 2^{(1-\varepsilon)/2}$. Then $\mathcal{H} = \mathbb{C}^{n_\alpha} \otimes \mathbb{C}^{m_n} \otimes \mathbb{C}^{n_\alpha} = \mathbb{C}^{n_\alpha} \otimes \mathbb{C}^{m_n n_\alpha}$, and the left action λ of $\mathcal{A}$ on $\mathcal{H}$ is thus given by $\lambda_\alpha \otimes 1_{m_n n_\alpha}$. Now

$$\lambda(\mathcal{A})' = (\lambda_\alpha(\mathcal{A}) \otimes 1_{m_n n_\alpha})' = (M_{n_\alpha}(\mathbb{C}) \otimes 1_{m_n n_\alpha})',$$

so that the action λ of $\mathcal{A}$ admits a separating vector if and only if the action of $M_{n_\alpha}(\mathbb{C})$ as $M_{n_\alpha}(\mathbb{C}) \otimes 1_{m_n n_\alpha}$ admits a separating vector, but by Lemma 5.5 this is indeed the case, as $n_\alpha \leq m_n n_\alpha$.

Now, suppose that $(\mathcal{H}, J)$ is type B. Let $\{\alpha, \beta\} = \operatorname{skel}(\mathcal{H}, J)$. Then

$$\mathcal{H} = (\mathbb{C}^{n_\alpha} \otimes \mathbb{C}^{n_\beta}) \oplus (\mathbb{C}^{n_\beta} \otimes \mathbb{C}^{n_\alpha}),$$

and the left action λ of $\mathcal{A}$ on $\mathcal{H}$ is given by $\lambda = (\lambda_\alpha \otimes 1_{n_\beta}) \oplus (\lambda_\beta \otimes 1_{n_\alpha})$. Since $\alpha \neq \beta$,

$$\begin{aligned} \lambda(\mathcal{A})' &= \left((\lambda_\alpha(\mathcal{A}) \otimes 1_{n_\beta}) \oplus (\lambda_\beta(\mathcal{A}) \otimes 1_{n_\alpha})\right)' \\ &= \left((M_{n_\alpha}(\mathbb{C}) \otimes 1_{n_\beta}) \oplus (M_{n_\beta}(\mathbb{C}) \otimes 1_{n_\alpha})\right)', \end{aligned}$$

so that the action λ of $\mathcal{A}$ admits a separating vector if and only if the action of $M_{n_\alpha}(\mathbb{C}) \oplus M_{n_\beta}(\mathbb{C})$ as $(M_{n_\alpha}(\mathbb{C}) \otimes 1_{n_\beta}) \oplus (M_{n_\beta}(\mathbb{C}) \otimes 1_{n_\alpha})$ admits a separating vector. Since $\dim_{\mathbb{C}} M_{n_\alpha}(\mathbb{C}) \oplus M_{n_\beta}(\mathbb{C}) - \dim_{\mathbb{C}} \mathcal{H} = (n_\alpha - n_\beta)^2$, if $n_\alpha \neq n_\beta$ then no injective linear maps $M_{n_\alpha}(\mathbb{C}) \oplus M_{n_\beta}(\mathbb{C}) \to \mathcal{H}$ can exist, and in particular, there exist no separating vectors for the action of $M_{n_\alpha}(\mathbb{C}) \oplus M_{n_\beta}(\mathbb{C})$, and hence for λ. Suppose instead that $n_\alpha = n_\beta = n$. Then

$$\mathcal{H} = (\mathbb{C}^n \otimes \mathbb{C}^n) \oplus (\mathbb{C}^n \otimes \mathbb{C}^n)$$

so that, since $\alpha \neq \beta$, $\lambda(\mathcal{A})' = (M_n(\mathbb{C}) \otimes 1_n)' \oplus (M_n(\mathbb{C}) \otimes 1_n)'$. Thus, if ξ is the separating vector for the action of $M_n(\mathbb{C})$ on $\mathbb{C}^n \otimes \mathbb{C}^n$ given by the proof of Lemma 5.5, then $\xi \oplus \xi$ is also a separating vector for the action λ of $\mathcal{A}$, and hence $(\mathcal{H}, J)$ is indeed separating. □

Let us now introduce the final requirement for irreducible triplets; recall that the complex form of a real C^*-algebra $\mathcal{A}$ a real C^*-algebra is denoted by $\mathcal{A}_{\mathbb{C}}$.

DEFINITION 5.7. We shall call an $\mathcal{A}$-bimodule $\mathcal{H}$ *complex-linear* if both left and right actions of $\mathcal{A}$ on $\mathcal{H}$ extend to $\mathbb{C}$-linear actions of $\mathcal{A}_{\mathbb{C}}$, making $\mathcal{H}$ into a complex $\mathcal{A}_{\mathbb{C}}$-bimodule.

It follows immediately that a $\mathcal{A}$-bimodule $\mathcal{H}$ is complex-linear if and only if for $m = \text{mult}[\mathcal{H}]$, $m_{\alpha\beta} = 0$ whenever α or β is conjugate-linear. In particular, by Proposition 3.13, it follows that a complex-linear quasi-orientable graded bimodule is always orientable.

We can now reformulate Chamseddine and Connes's definition for irreducible triplets:

DEFINITION 5.8. An *irreducible triplet* is a triplet $(\mathcal{A}, \mathcal{H}, J)$, where $\mathcal{A}$ is a finite-dimensional real C^*-algebra and $(\mathcal{H}, J)$ is a complex-linear, separating, irreducible real $\mathcal{A}$-bimodule of odd KO-dimension such that the left action of $\mathcal{A}$ on $\mathcal{H}$ is faithful.

Note that for $\mathcal{H}$ a real $\mathcal{A}$-bimodule, the left action of $\mathcal{A}$ is faithful if and only if the right action is faithful.

By combining the above results, we immediately obtain Chamseddine and Connes's classification of irreducible triplets:

PROPOSITION 5.9 (Chamseddine–Connes [3, Propositions 2.5, 2.8]). *Let $\mathcal{A}$ be a finite-dimensional real C^*-algebra, and let $(\mathcal{H}, J)$ be a real $\mathcal{A}$-bimodule of odd KO-dimension $n \bmod 8$. Then $(\mathcal{A}, \mathcal{H}, J)$ is an irreducible triplet if and only if one of the following cases holds:*

(1) *There exists $n \in \mathbb{N}$ such that $\mathcal{A} = M_k(\mathbb{K})$ for a real form $M_k(\mathbb{K})$ of $M_n(\mathbb{C})$, and*

$$\text{mult}[\mathcal{H}] = 2^{(1-\varepsilon)/2} E_{\mathbf{nn}}; \tag{5.2}$$

(2) *There exists $n \in \mathbb{N}$ such that $\mathcal{A} = M_{k_1}(\mathbb{K}_1) \oplus M_{k_2}(\mathbb{K}_2)$ for real forms $M_{k_1}(\mathbb{K}_1)$ and $M_{k_2}(\mathbb{K}_2)$ of $M_n(\mathbb{C})$, and*

$$\text{mult}[\mathcal{H}] = E_{\mathbf{n_1 n_2}} + E_{\mathbf{n_2 n_1}}. \tag{5.3}$$

5.2. Gradings. We now seek a classification of gradings inducing even KO-dimensional real bimodules from irreducible triplets.

DEFINITION 5.10. Let $(\mathcal{A}, \mathcal{H}, J)$ be an irreducible triplet. We shall call a $\mathbb{Z}_2$-grading γ on $\mathcal{H}$ as a Hilbert space *compatible* with $(\mathcal{A}, \mathcal{H}, J)$ if and only if the following conditions all hold:

(1) For every $a \in \mathcal{A}$, $\gamma\lambda(a)\gamma \in \lambda(\mathcal{A})$;
(2) The operator γ either commutes or anticommutes with J.

Given a compatible grading γ for an irreducible triplet $(\mathcal{A}, \mathcal{H}, J)$, one can view $(\mathcal{H}, \gamma, J)$ as a real $\mathcal{A}^{\text{even}}$-bimodule of even KO-dimension, for $\mathcal{A}^{\text{even}} = \{a \in \mathcal{A} \mid [\lambda(a), \gamma] = 0\}$, with KO-dimension specified by the values of ε and ε'' such that $J^2 = \varepsilon$, $\gamma J = \varepsilon'' J\gamma$.

Now, recall that a $\mathbb{Z}_2$-grading on a real C^*-algebra $\mathcal{A}$ is simply an automorphism Γ on $\mathcal{A}$ satisfying $\Gamma^2 = \text{id}$; we call such a grading *admissible* if and only if Γ extends to a $\mathbb{C}$-linear grading on $\mathcal{A}_{\mathbb{C}}$. Thus, if $(\mathcal{A}, \mathcal{H}, J)$ is an irreducible triplet and γ is a grading on $\mathcal{H}$, then γ satisfies the first condition for compatibility if and only if there exists some admissible grading Γ on $\mathcal{A}$ such that $\mathrm{Ad}_\gamma \circ \lambda = \lambda \circ \Gamma$, where Ad_x denotes conjugation by x.

LEMMA 5.11. *Let $M_k(\mathbb{K})$ be a real form of $M_n(\mathbb{C})$, and let $\alpha \in \mathrm{Aut}(M_n(\mathbb{C}))$. Then α is an admissible grading on $M_k(\mathbb{K})$ if and only if there exists a self-adjoint unitary γ in $M_k(\mathbb{K})$ or $iM_k(\mathbb{K})$, such that $\alpha = \mathrm{Ad}_\gamma$.*

PROOF. Suppose that α is an admissible grading. Let $\mathbb{K}_0$ be $\mathbb{C}$ if $\mathbb{K} = \mathbb{C}$, and $\mathbb{R}$ otherwise. Then $M_k(\mathbb{K})$ is central simple over $\mathbb{K}_0$, so that there exists some invertible element S of $M_k(\mathbb{K})$ such that $\alpha = \mathrm{Ad}_S$. Since α respects the involution, for any $A \in M_k(\mathbb{K})$ we must have

$$(S^{-1})^* A^* S^* = (SAS^{-1})^* = \alpha(A)^* = \alpha(A^*) = SA^*S^{-1},$$

i.e. $[A, S^*S] = 0$, so that S^*S is a positive central element of $M_k(\mathbb{K})$, and hence $S^*S = c1$ for some $c > 0$. Thus, $U = c^{-1/2}S$ is a unitary element of $M_k(\mathbb{K})$ such that $\alpha = \mathrm{Ad}_U$. Now, recall that $\alpha^2 = \mathrm{id}$, so that $\mathrm{Ad}_{U^2} = \mathrm{id}$, and hence $U^2 = \zeta 1$ for some $\zeta \in \mathbb{T} \cap \mathbb{K}_0$. If $\mathbb{K} = \mathbb{C}$, then one can simply set $\gamma = \overline{\lambda} U$ for λ is a square root of ζ. Otherwise, $U^2 = \pm 1$, so that if $U^2 = 1$, set $\gamma = U \in M_k(\mathbb{K})$, and if $U^2 = -1$, set $\gamma = iU \in iM_k(\mathbb{K})$.

On the other hand, if γ is a self-adjoint unitary in either $M_k(\mathbb{K})$ or $iM_k(\mathbb{K})$, then Ad_γ is readily seen to be an admissible grading on $M_k(\mathbb{K})$. □

Let us now give the classification of compatible gradings for a type A irreducible triplet; it is essentially a generalisation of [3, Lemma 3.1].

PROPOSITION 5.12. *Let $(\mathcal{A}, \mathcal{H}, J)$ be a type A irreducible triplet of odd KO-dimension $n \bmod 8$, so that $\mathcal{A}$ is a real form $M_k(\mathbb{K})$ of $M_n(\mathbb{C})$ for some n, and let γ be a grading on $\mathcal{H}$ as a Hilbert space. Then γ is compatible if and only if there exists a self-adjoint unitary g in $M_k(\mathbb{K})$ or $iM_k(\mathbb{K})$ such that*

$$\gamma = \pm g \otimes 1_{m_k} \otimes g^T, \tag{5.4}$$

in which case γ necessarily commutes with J.

PROOF. Let $m_n = 2^{(1-\varepsilon)/2}$. Then $\mathcal{H} = \mathbb{C}^n \otimes \mathbb{C}^{m_n} \otimes \mathbb{C}^n$, and for all $a \in \mathcal{A}$,

$$\lambda(a) = \lambda_\alpha(a) \otimes 1_{m_n} \otimes 1_n = a \otimes 1_{m_n k} \otimes 1_n, \quad \rho(a) = 1_n \otimes 1_{m_n} \otimes \lambda_\alpha(a)^T = 1_n \otimes 1_{m_n} \otimes a^T.$$

Suppose that γ is compatible. Then by Lemma 5.11 there exists some self-adjoint unitary g in either $M_k(\mathbb{K})$ or $iM_k(\mathbb{K})$ such that for all $a \in \mathcal{A}$,

$$\gamma(a \otimes 1_{m_n} \otimes 1_n)\gamma = (gag) \otimes 1_{m_k} \otimes 1_n.$$

Now, let $\gamma_0 = g \otimes 1_{m_n} \otimes g^T$. Then, by construction, γ_0 is a compatible grading for $(\mathcal{A}, \mathcal{H}, J)$ that induces the same admissible grading on $\mathcal{A}$ as γ, and moreover commutes with J. Then $\nu := \gamma\gamma_0 \in \mathrm{U}^{\mathrm{LR}}_{\mathcal{A}}(\mathcal{H}; J)$, so that $\nu = 1_n \otimes \nu_{\mathbf{nn}} \otimes 1_n$ for some

$$\nu_{\mathbf{nn}} \in \begin{cases} \{\pm 1\}, & \text{if } n = 1 \text{ or } 7 \bmod 8, \\ \mathrm{SU}(2), & \text{if } k = 3 \text{ or } 5 \bmod 8. \end{cases}$$

Thus $\gamma = g \otimes \nu_{\mathbf{nn}} \otimes g^T$, and hence, since γ is self-adjoint, $\nu_{\mathbf{nn}}$ must also be self-adjoint. Therefore $\nu_{\alpha\alpha} = \pm 1_{m_k}$, or equivalently, $\gamma' = \pm\gamma$.

On the other hand, if g is a self-adjoint unitary in either $M_k(\mathbb{K})$ or $iM_k(\mathbb{K})$, then $\gamma = g \otimes 1_{m_k} \otimes g^T$ is certainly a compatible grading that commutes with J. □

Thus, irreducible triplets can only give rise to real $\mathcal{A}^{\text{even}}$-bimodules of KO-dimension 0 or 4 mod 8.

Let us now turn to the type B case.

PROPOSITION 5.13. *Let $(\mathcal{A}, \mathcal{H}, J)$ be a type B irreducible triplet of odd KO-dimension $n \bmod 8$, so that for some $n \in \mathbb{N}$, $\mathcal{A} = M_{k_1}(\mathbb{K}_1) \oplus M_{k_2}(\mathbb{K}_2)$ for real forms $M_{k_1}(\mathbb{K}_1)$ and $M_{k_2}(\mathbb{K}_2)$ of $M_n(\mathbb{C})$, and let γ be a grading on $\mathcal{H}$ as a Hilbert space. Then γ is compatible if and only if one of the following holds:*

(1) *There exist gradings γ_1 and γ_2 on $\mathbb{C}^n$, with $\gamma_j \in M_{k_j}(\mathbb{K}_j)$ or $iM_{k_j}(\mathbb{K}_j)$, such that*

$$\gamma = \begin{pmatrix} \gamma_1 \otimes \gamma_2^T & 0 \\ 0 & \varepsilon'' \gamma_2 \otimes \gamma_1^T \end{pmatrix}, \tag{5.5}$$

in which case $\gamma J = \varepsilon'' J \gamma$, and if γ' is any other compatible grading, $\mathrm{Ad}_{\gamma'} = \mathrm{Ad}_\gamma$ if and only if $\gamma' = \pm\gamma$.

(2) *One has that $\mathbb{K}_1 = \mathbb{K}_2 = \mathbb{K}$ and $k_1 = k_2 = k$, and there exist a unitary $u \in M_k(\mathbb{K})$ and $\eta \in \mathbb{T}$ such that*

$$\gamma = \begin{pmatrix} 0 & \overline{\eta} u^* \otimes \overline{u} \\ \eta u \otimes u^T & 0 \end{pmatrix}, \tag{5.6}$$

in which case γ necessarily commutes with J, and if γ' is any other compatible grading, $\mathrm{Ad}_{\gamma'} = \mathrm{Ad}_\gamma$ if and only if $\gamma' = (\zeta 1_{n^2} \oplus \overline{\zeta} 1_{n^2})\gamma$ for some $\zeta \in \mathbb{T}$.

PROOF. Let γ be a compatible grading. Then, with respect to the decomposition $\mathcal{H} = (\mathbb{C}^n \otimes \mathbb{C}^n) \oplus (\mathbb{C}^n \oplus \mathbb{C}^n)$, let us write

$$\gamma = \begin{pmatrix} A & B \\ C & D \end{pmatrix}$$

for A, B, C and $D \in M_n(\mathbb{C}) \otimes M_n(\mathbb{C})$. Applying self-adjointness of γ, we find that A and D must be self-adjoint, and that $B = C^*$, and then applying the fact that $\gamma^2 = 1$, we find that

$$A^2 + C^*C = 1, \qquad CA + DC = 0, \qquad CC^* + D^2 = 1.$$

Finally, applying the condition that γ commutes or anticommutes with J, *i.e.* that $\gamma J = \varepsilon'' J \gamma$ for $\varepsilon'' = \pm 1$, we find that

$$D = \varepsilon'' XAX, \qquad C^* = \varepsilon'' XCX,$$

where X is the antiunitary on $\mathbb{C}^n \otimes \mathbb{C}^n$ given by $X : \xi_1 \otimes \xi_2 \mapsto \overline{\xi_2} \otimes \overline{\xi_1}$.

Now, since γ is compatible, and since $(1,0)$ and $(0,1)$ are projections in $\mathcal{A}$ satisfying $(1,0) + (0,1) = 1$, there exist projections P and Q in $\mathcal{A}$ such that

$$\mathrm{Ad}_\gamma \lambda(1,0) = \lambda(P, 1-Q), \qquad \mathrm{Ad}_\gamma \lambda(0,1) = \lambda(1-P, Q),$$

that is,

$$\begin{pmatrix} P \otimes 1_n & 0 \\ 0 & (1-Q) \otimes 1_n \end{pmatrix} = \gamma \begin{pmatrix} 1 & 0 \\ 0 & 0 \end{pmatrix} \gamma = \begin{pmatrix} A^2 & AC^* \\ CA & CC^* \end{pmatrix}$$

and

$$\begin{pmatrix} (1-P) \otimes 1_n & 0 \\ 0 & Q \otimes 1_n \end{pmatrix} = \gamma \begin{pmatrix} 0 & 0 \\ 0 & 1 \end{pmatrix} \gamma = \begin{pmatrix} C^*C & C^*D \\ DC & D^2 \end{pmatrix}.$$

Thus, A is a self-adjoint partial isometry with support and range projection $P \otimes 1_n$, D is a self-adjoint partial isometry with support and range projection $Q \otimes 1_n$, and

C is a partial isometry with support projection $(1-P)\otimes 1_n$ and range projection $(1-Q)\otimes 1_n$.

Now, recalling that $D = \varepsilon'' XAX$, we see that

$$Q \otimes 1_n = D^2 = XA^2X = XP \otimes 1_n X = 1_n \otimes \overline{P}.$$

If $Q = 0$, then certainly $P = 0$. Suppose instead that $Q \neq 0$, and let $\xi \in Q\mathbb{C}^n \otimes \mathbb{C}^n$ be non-zero. Then

$$\mathrm{id}_{\xi\otimes\mathbb{C}^n} = (Q \otimes 1_n)|_{\xi\otimes\mathbb{C}^n} = (1 \otimes \overline{P})|_{\xi\otimes\mathbb{C}^n},$$

so that $P = 1$ and hence $Q = 1$ also. We therefore have two possible cases:

(1) We have

$$\gamma = \begin{pmatrix} A & 0 \\ 0 & \varepsilon'' XAX \end{pmatrix}$$

for A a grading on $\mathbb{C}^n \otimes \mathbb{C}^n$;

(2) We have

$$\gamma = \begin{pmatrix} 0 & C^* \\ C & 0 \end{pmatrix}$$

for C a unitary on $\mathbb{C}^n \otimes \mathbb{C}^n$ such that $C^* = (-1)^m XCX$.

First suppose that the first case holds. Then, on the one hand, $\mathrm{Ad}_A|_{M_n(\mathbb{C})\otimes 1_n}$ induces an admissible grading for $M_{k_1}(\mathbb{K}_1)$, so that there exists a self-adjoint unitary γ_1 in either $M_{k_1}(\mathbb{K}_1)$ or $iM_{k_1}(\mathbb{K}_1)$ such that $\mathrm{Ad}_A|_{M_n(\mathbb{C})\otimes 1_n} = \mathrm{Ad}_{\gamma_1\otimes 1_n}$, and on the other hand, $\mathrm{Ad}_{\varepsilon''XAX}|_{M_n(\mathbb{C})\otimes 1_n}$ induces an admissible grading for $M_{k_2}(\mathbb{K}_2)$, so that there exists a self-adjoint unitary γ_2 in $M_{k_2}(\mathbb{K}_1)$ or $iM_{k_2}(\mathbb{K}_1)$ such that $\mathrm{Ad}_{\varepsilon''XAX}|_{M_n(\mathbb{C})\otimes 1_n} = \mathrm{Ad}_{\gamma_2\otimes 1_n}$. Since for $a \otimes b \in M_n(\mathbb{C}) \otimes M_n(\mathbb{C})$ we can write

$$a \otimes b = (a \otimes 1_n)X(\bar{b} \otimes 1_n)X,$$

it therefore follows that $\mathrm{Ad}_A = \mathrm{Ad}_{\gamma_1\otimes\gamma_2^T}$ on the central simple algebra $M_n(\mathbb{C}) \otimes M_n(\mathbb{C}) \cong M_{n^2}(\mathbb{C})$ over $\mathbb{C}$. Hence, there exists some non-zero $\eta \in \mathbb{C}$ such that $A = \eta\gamma_1 \otimes \gamma_2^T$, and since both A and $\gamma_1 \otimes \gamma_2^T$ are self-adjoint and unitary, it follows that $\eta = \pm 1$. Absorbing ± 1 into γ_1 or γ_2, we therefore find that

$$\gamma = \begin{pmatrix} \gamma_1 \otimes \gamma_2^T & 0 \\ 0 & \varepsilon''\gamma_2 \otimes \gamma_1^T \end{pmatrix}.$$

On the other hand, γ so constructed is readily seen to be a compatible grading satisfying $\gamma J = \varepsilon'' J\gamma$.

Now suppose that the second case holds. Then, since γ is compatible, it is clear that the automorphisms α, β of $M_n(\mathbb{C})$ specified by

$$\alpha(a) \otimes 1_n = C(a \otimes 1_n)C^*, \qquad \beta(a) \otimes 1_n = C^*(a \otimes 1_n)C,$$

are inverses of each other, and that α, in particular, induces an isomorphism $M_{k_1}(\mathbb{K}_1) \to M_{k_2}(\mathbb{K}_2)$, so that $\mathbb{K}_1 = \mathbb{K}_2 = \mathbb{K}$ and $k_1 = k_2 = k$. Next, by the proof of Lemma 5.11, there exists some unitary u in $M_n(\mathbb{C})_I$ such that $\alpha = \mathrm{Ad}_u$, from which it follows that $\beta = \mathrm{Ad}_{u^*}$. By the same trick as above, we then find that $\mathrm{Ad}_C = \mathrm{Ad}_{u\otimes u^T}$ on the central simple algebra $M_n(\mathbb{C}) \otimes M_n(\mathbb{C}) \cong M_{n^2}(\mathbb{C})$ over $\mathbb{C}$. Hence, there exists some non-zero $\eta \in \mathbb{C}$ such that $C = \eta u \otimes u^T$, and since both C and $u \otimes u^T$ are unitary, it follows that $\eta \in \mathbb{T}$. Thus,

$$\gamma = \begin{pmatrix} 0 & \overline{\eta} u^* \otimes \overline{u} \\ \eta u \otimes u^T & 0 \end{pmatrix}.$$

On the other hand, γ so constructed is readily seen to be a compatible grading satisfying $[\gamma, J] = 0$.

Finally, let γ and γ' be two compatible gradings. Suppose that $\mathrm{Ad}_\gamma = \mathrm{Ad}_{\gamma'}$, and set $U = \gamma'\gamma$. Then, by construction, U is a unitary element of $\mathcal{L}^{\mathrm{LR}}_{\mathcal{A}}(\mathcal{H}; J)$, so that there exists some $\zeta \in \mathbb{T}$ such that

$$U = \zeta 1_{n^2} \oplus \overline{\zeta} 1_{n^2}.$$

If the second case holds, then nothing more can be said, but if the first case holds, so that

$$\gamma = \begin{pmatrix} \gamma_1 \otimes \gamma_2^T & 0 \\ 0 & \varepsilon'' \gamma_2 \otimes \gamma_1^T \end{pmatrix}$$

for suitable γ_1 and γ_2, then

$$\gamma' = \begin{pmatrix} \zeta\gamma_1 \otimes \gamma_2^T & 0 \\ 0 & \varepsilon'' \overline{\zeta} \gamma_2 \otimes \gamma_1^T \end{pmatrix},$$

so that by self-adjointness of γ', γ_1 and γ_2, we must have $\zeta = \pm 1$, as required. □

Thus, we can obtain a real bimodule of KO-dimension 6 mod 8 only from a type B irreducible triplet together with a compatible grading satisfying the first case of the last result.

5.3. Even subalgebras and even KO-dimensional bimodules. We now consider real bimodules of KO-dimension 6 mod 8 obtained from irreducible triplets. Thus, let $(\mathcal{A}, \mathcal{H}, J)$ be a fixed type B irreducible triplet of KO-dimension 1 or 7 mod 8, and let γ be a fixed compatible grading for $(\mathcal{A}, \mathcal{H}, J)$ anticommuting with J, so that for some $n \in \mathbb{N}$,

- $\mathcal{A} = M_{k_1}(\mathbb{K}_1) \oplus M_{k_2}(\mathbb{K}_2)$ for real forms $M_{k_j}(\mathbb{K}_j)$ of $M_n(\mathbb{C})$;
- $\mathrm{mult}[\mathcal{H}] = E_{\mathbf{n}_1\mathbf{n}_2} + E_{\mathbf{n}_2\mathbf{n}_1}$;
- There exist self-adjoint unitaries $\gamma_j \in M_{k_j}(\mathbb{K}_j)$ or $iM_{k_j}(\mathbb{K}_j)$ with signature $(r_j, n - r_j)$ such that

$$\gamma = \begin{pmatrix} \gamma_1 \otimes \gamma_2^T & 0 \\ 0 & -\gamma_2 \otimes \gamma_1^T \end{pmatrix}.$$

It is worth noting that $(\mathcal{H}, J)$ admits, up to sign, a unique S^0-real structure, given by $\epsilon = 1_{n^2} \oplus -1_{n^2}$, which certainly commutes with γ. We can exploit the symmetries present to simplify our discussion by taking, without loss of generality, $r_j > 0$, and requiring that $\gamma_1 \in iM_{k_1}(\mathbb{K}_1)$ only if $\gamma_2 \in iM_{k_2}(\mathbb{K}_2)$, and that $\gamma_1 = 1_n$ only if $\gamma_2 = 1_n$.

Our main goal in this section is to give an explicit description of $\mathcal{A}^{\mathrm{even}}$ and of $(\mathcal{H}, \gamma, J)$ as a real $\mathcal{A}^{\mathrm{even}}$-bimodule. To do so, however, we first need the following:

LEMMA 5.14. *Let $M_k(\mathbb{K})$ be a real form of $M_n(\mathbb{C})$, let g be a self-adjoint unitary in $M_k(\mathbb{K})$ or $iM_k(\mathbb{K})$, and let $r = \mathrm{null}(g - 1)$. Set $M_k(\mathbb{K})^g := \{a \in M_k(\mathbb{K}) \mid [a, g] = 0\}$.*

- *If $g \in M_k(\mathbb{K})$, then $M_k(\mathbb{K})^g \cong M_{kr/n}(\mathbb{K}) \oplus M_{k(n-r)/n}(\mathbb{K})$;*
- *If $g \in iM_k(\mathbb{K})$, then $r = n/2$ and*

$$M_k(\mathbb{K})^g \cong \{(a, b) \in M_{k/2}(\mathbb{C})^2 \mid b = \overline{a}\} \cong M_{k/2}(\mathbb{C}).$$

PROOF. Let $P^+ := \frac{1}{2}(1+g)$ and $P^- := \frac{1}{2}(1-g)$, which are thus projections in $M_n(\mathbb{C})$ of rank r and $n-r$, respectively. Define an injection $\phi : M_k(\mathbb{K})^g \mapsto M_r(\mathbb{C}) \oplus M_{n-r}(\mathbb{C})$ by $\phi(A) := (P^{\text{even}} A P^{\text{even}}, P^{\text{odd}} A P^{\text{odd}})$.

First, suppose that $g \in M_k(\mathbb{K})$. Then P^+ and P^- are also in $M_k(\mathbb{K})$, from which it immediately follows that $\phi(M_k(\mathbb{K})^g) = M_{kr/n}(\mathbb{K}) \oplus M_{k(n-r)/n}(\mathbb{K})$.

Suppose instead that $g \in iM_k(\mathbb{K})$ and $\mathbb{K} \neq \mathbb{C}$. Then $M_k(\mathbb{K}) = \{A \in M_n(\mathbb{C}) \mid [A, I] = 0\}$ for a suitable antiunitary I on $\mathbb{C}^n$ satisfying $I^2 = \alpha 1$, where $\alpha = 1$ if $\mathbb{K} = \mathbb{R}$ and $\alpha = -1$ if $\mathbb{K} = \mathbb{H}$. Then $\{g, I\} = 0$, and hence, with respect to the decomposition $\mathbb{C}^n = P^+\mathbb{C}^n \oplus P^-\mathbb{C}^n \cong \mathbb{C}^r \oplus \mathbb{C}^{n-r}$,

$$I = \begin{pmatrix} 0 & \alpha \tilde{I}^* \\ \tilde{I} & 0 \end{pmatrix},$$

where $\tilde{I} = P^{\text{odd}} I P^{\text{even}}$ is an antiunitary $\mathbb{C}^r \mapsto \mathbb{C}^{n-r}$. Thus, n is even and $r = n/2$, and taking $\tilde{I}$, without loss of generality, to be complex conjugation on $\mathbb{C}^r$, for all $A \in M_n(\mathbb{C})$ commuting with g, $[A, I] = 0$ if and only if $P^- A P^- = \overline{P^+ A P^+}$, and hence $\phi(M_k(\mathbb{K})^g) = \{(a, \overline{a}) \mid a \in M_{n/2}(\mathbb{C})\} \cong M_{n/2}(\mathbb{C})$. □

In light of the form of γ, this last Lemma immediately implies the aforementioned explicit description of $\mathcal{A}^{\text{even}}$ and $(\mathcal{H}, \gamma, J)$:

PROPOSITION 5.15. *Let $(m^{\text{even}}, m^{\text{odd}}) = (m^{\text{even}}, (m^{\text{even}})^T)$ be the pair of multiplicity matrices of $(\mathcal{H}, \gamma, J)$ as an even KO-dimensional real $\mathcal{A}^{\text{even}}$-bimodule. Let $r_i' = n - r_i$, and, when n is even, let $c = n/2$. Then:*

(1) *If $\gamma_1 \in iM_{k_1}(\mathbb{K}_1)$, $\gamma_2 \in iM_{k_2}(\mathbb{K})$, then*

$$\mathcal{A}^{\text{even}} = M_c(\mathbb{C}) \oplus M_c(\mathbb{C}), \tag{5.7}$$

and

$$m^{\text{even}} = E_{\mathbf{c}_1\mathbf{c}_2} + E_{\overline{\mathbf{c}}_1\overline{\mathbf{c}}_2} + E_{\mathbf{c}_2\overline{\mathbf{c}}_1} + E_{\overline{\mathbf{c}}_2\mathbf{c}_1}; \tag{5.8}$$

(2) *If $\gamma_1 \in iM_{k_1}(\mathbb{K}_1)$, $\gamma_2 \in M_{k_2}(\mathbb{K}) \setminus \{1_n\}$, then*

$$\mathcal{A}^{\text{even}} = M_c(\mathbb{C}) \oplus M_{k_2 r_2/n}(\mathbb{K}_2) \oplus M_{k_2 r_2'/n}(\mathbb{K}_2). \tag{5.9}$$

and

$$m^{\text{even}} = E_{\mathbf{c}\mathbf{r}_2} + E_{\overline{\mathbf{c}}\mathbf{r}_2'} + E_{\mathbf{r}_2\overline{\mathbf{c}}} + E_{\mathbf{r}_2'\mathbf{c}}; \tag{5.10}$$

(3) *If $\gamma_1 \in iM_{k_1}(\mathbb{K}_1)$, $\gamma_2 = 1$, then*

$$\mathcal{A}^{\text{even}} = M_c(\mathbb{C}) \oplus M_{k_2}(\mathbb{K}_2), \tag{5.11}$$

and

$$m^{\text{even}} = E_{\mathbf{c}\mathbf{n}} + E_{\mathbf{n}\overline{\mathbf{c}}}; \tag{5.12}$$

(4) *If $\gamma_1 \in M_{k_1}(\mathbb{K}_1) \setminus \{1_n\}$, $\gamma_2 \in M_{k_2}(\mathbb{K}_2) \setminus \{1_n\}$, then*

$$\mathcal{A}^{\text{even}} = M_{k_1 r_1/n}(\mathbb{K}_1) \oplus M_{k_1 r_1'/n}(\mathbb{K}_1) \oplus M_{k_2 r_2/n}(\mathbb{K}_2) \oplus M_{k_2 r_2'/n}(\mathbb{K}_2), \tag{5.13}$$

and

$$m^{\text{even}} = E_{\mathbf{r}_1\mathbf{r}_2} + E_{\mathbf{r}_1'\mathbf{r}_2'} + E_{\mathbf{r}_2\mathbf{r}_1'} + E_{\mathbf{r}_2'\mathbf{r}_1}; \tag{5.14}$$

(5) *If* $\gamma_1 \in M_{k_1}(\mathbb{K}_1) \setminus \{1_n\}$, $\gamma_2 = 1_n$, *then*

$$\mathcal{A}^{\mathrm{even}} = M_{k_1 r_1/n}(\mathbb{K}_1) \oplus M_{k_1 r_1'/n}(\mathbb{K}_1) \oplus M_{k_2}(\mathbb{K}_2), \tag{5.15}$$

and

$$m^{\mathrm{even}} = E_{\mathbf{r_1 n}} + E_{\mathbf{n r_1'}}; \tag{5.16}$$

item If $\gamma_1 = \gamma_2 = 1_n$, *then*

$$\mathcal{A}^{\mathrm{even}} = M_{k_1}(\mathbb{K}_1) \oplus M_{k_2}(\mathbb{K}_2), \tag{5.17}$$

and

$$m^{\mathrm{even}} = E_{\mathbf{n_1 n_2}}. \tag{5.18}$$

One can check in each case that $(\mathcal{H}, \gamma)$ is quasi-orientable as an even $\mathcal{A}^{\mathrm{even}}$-bimodule. However, Propositions 3.13 and 3.14 immediately imply the following:

COROLLARY 5.16. *The following are equivalent for* $(\mathcal{H}, \gamma)$ *as an even* $\mathcal{A}^{\mathrm{even}}$-*bimodule:*

(1) $\gamma_1 \in M_{k_1}(\mathbb{K}_1)$ *and* $\gamma_2 \in M_{k_2}(\mathbb{K}_2)$;
(2) $(\mathcal{H}, \gamma)$ *is orientable;*
(3) $(\mathcal{H}, \gamma)$ *has non-vanishing intersection form;*
(4) $(\mathcal{H}, \gamma)$ *is complex-linear.*

This then motivates us to restrict ourselves to the case where $\gamma_1 \in M_{k_1}(\mathbb{K}_1)$ and $\gamma_2 \in M_{k_2}(\mathbb{K}_2)$. Note, however, that in no case is Poincaré duality possible.

5.4. Off-diagonal Dirac operators. Let us now consider the slightly more general S^0-real $\mathcal{A}^{\mathrm{even}}$-bimodule $(\mathcal{H}_F, \gamma_F, J_F, \epsilon_F)$ of KO-dimension 6 mod 8 given by taking the direct sum of N copies of $(\mathcal{H}, \gamma, J, \epsilon)$, where $N \in \mathbb{N}$. If we modify our earlier conventions slightly to allow for the summand 0 in Wedderburn decompositions, we can therefore write

$$\mathcal{A}^{\mathrm{even}} = M_{k_1 r_1/n}(\mathbb{K}_1) \oplus M_{k_1 r_1'/n}(\mathbb{K}_1) \oplus M_{k_2 r_2/n}(\mathbb{K}_2) \oplus M_{k_2 r_2'}(\mathbb{K}_2), \tag{5.19}$$

so that $(\mathcal{H}_F, \gamma_F, J_F)$ is the real $\mathcal{A}^{\mathrm{even}}$-bimodule of KO-dimension 6 mod 8 with signed multiplicity matrix

$$\mu_F = N(E_{\mathbf{r_1 r_2}} - E_{\mathbf{r_1 r_2'}} - E_{\mathbf{r_1' r_2}} + E_{\mathbf{r_1' r_2'}} - E_{\mathbf{r_2 r_1}} + E_{\mathbf{r_2 r_1'}} + E_{\mathbf{r_2' r_1}} - E_{\mathbf{r_2' r_1'}}), \tag{5.20}$$

whilst $(\mathcal{H}_f, \gamma_f) := ((\mathcal{H}_F)_i, (\gamma_F)_i)$ is the even $\mathcal{A}^{\mathrm{even}}$-bimodule with signed multiplicity matrix

$$\mu_f = N(E_{\mathbf{r_1 r_2}} - E_{\mathbf{r_1 r_2'}} - E_{\mathbf{r_1' r_2}} + E_{\mathbf{r_1' r_2'}}). \tag{5.21}$$

It then follows also that $(\mathcal{H}_{\overline{f}}, \gamma_{\overline{f}}) := (J_F \mathcal{H}_f, -(J_F \gamma_f J_F)|_{J_F \mathcal{H}_f})$ is the even $\mathcal{A}^{\mathrm{even}}$-bimodule with signed multiplicty matrix

$$\mu_{\overline{f}} = -\mu_f^T = N(-E_{\mathbf{r_2 r_1}} + E_{\mathbf{r_2 r_1'}} + E_{\mathbf{r_2' r_1}} - E_{\mathbf{r_2' r_1'}}).$$

Now, for $\mathcal{C}$ a unital $*$-subalgebra of $\mathcal{A}^{\mathrm{even}}$, let us call a Dirac operator $D \in \mathcal{D}_0(\mathcal{C}, \mathcal{H}_F, \gamma_F, J_F)$ *off-diagonal* if it does not commute with ϵ_F, or equivalently [3, §4] if $[D, \mathcal{Z}(\mathcal{A})] \neq \{0\}$. If $\mathcal{D}_1(\mathcal{C}, \mathcal{H}_F, \gamma_F, J_F, \epsilon_F) \subseteq \mathcal{D}_0(\mathcal{C}, \mathcal{H}_F, \gamma_F, J_F)$ is the subspace consisting of Dirac operators anti-commuting with ϵ_F, then, in fact,

$$\mathcal{D}_0(\mathcal{C}, \mathcal{H}_F, \gamma_F, J_F) = \mathcal{D}_0(\mathcal{C}, \mathcal{H}_F, \gamma_F, J_F, \epsilon_F) \oplus \mathcal{D}_1(\mathcal{C}, \mathcal{H}_F, \gamma_F, J_F, \epsilon_F), \tag{5.22}$$

as can be seen from writing

$$D = \frac{1}{2}\{D, \epsilon_F\}\epsilon_F + \frac{1}{2}[D, \epsilon_F]\epsilon_F$$

for $D \in \mathcal{D}_0(\mathcal{C}, \mathcal{H}_F, \gamma_F, J_F)$. Thus, non-zero off-diagonal Dirac operators exist for $(\mathcal{H}_F, \gamma_F, J_F, \epsilon_F)$ as an S^0-real $\mathcal{C}$-bimodule if and only if

$$\mathcal{D}_1(\mathcal{C}, \mathcal{H}_F, \gamma_F, J_F, \epsilon_F) \neq \{0\}.$$

Our goal is to generalise Theorem 4.1 in [3, §4] and characterise subalgebras of $\mathcal{A}^{\text{even}}$ of maximal dimension admitting off-diagonal Dirac operators.

The following result is the first step in this direction:

PROPOSITION 5.17 ([3, Lemma 4.2]). *A unital $*$-subalgebra $\mathcal{C} \subseteq \mathcal{A}^{\text{even}}$ admits off-diagonal Dirac operators if and only if there exists some partial unitary $T \in \mathcal{L}(\mathbb{C}^{r_1} \oplus \mathbb{C}^{r_1'} \oplus \mathbb{C}^{r_2} \oplus \mathbb{C}^{r_2'})$ with support contained in one of $\mathbb{C}^{r_1}$ or $\mathbb{C}^{r_1'}$ and range contained in one of $\mathbb{C}^{r_2}$ or $\mathbb{C}^{r_2'}$, such that*

$$\mathcal{C} \subseteq \mathcal{A}(T) := \{a \in \mathcal{A}^{\text{even}} \mid [a, T] = [a^*, T] = 0\}.$$

PROOF. First note that the map $\mathcal{D}_1(\mathcal{C}, \mathcal{H}_F, \gamma_F, J_F, \epsilon_F) \to \mathcal{L}^1_{\mathcal{C}}(\mathcal{H}_f, \mathcal{H}_{\overline{f}})$ given by $D \mapsto P_{-i}DP_i$ is an isomorphism, so that $\mathcal{C}$ admits off-diagonal Dirac operators if and only if $\mathcal{L}^1_{\mathcal{C}}(\mathcal{H}_f, \mathcal{H}_{\overline{f}}) \neq \{0\}$. Since a map $S \in \mathcal{L}(\mathcal{H}_f, \mathcal{H}_{\overline{f}})$ satisfies the generalised order one condition for $\mathcal{C}$ if and only if $\rho_{\overline{f}}(c)S - S\rho_f(C)$ is left $\mathcal{C}$-linear for all $c \in \mathcal{C}$, $\mathcal{C}$ admits off-diagonal Dirac operators only if

$$\{S \in \mathcal{L}^{\mathrm{L}}_{\mathcal{C}}(\mathcal{H}_f, \mathcal{H}_{\overline{f}}) \mid -\gamma_{\overline{f}}S = S\gamma_f\} \neq \{0\},$$

or equivalently,

$$\mathcal{C} \subseteq \mathcal{A}_S := \{a \in \mathcal{C}^{\text{even}} \mid \lambda_{\overline{f}}(a)S = S\lambda(a),\ \lambda_{\overline{f}}(a^*)S = S\lambda(a^*)\}$$

for some non-zero $S \in \mathcal{L}(\mathcal{H}_f, \mathcal{H}_{\overline{f}})$ such that $-\gamma_{\overline{f}}S = S\gamma_f$.

Now, let $S \in \mathcal{L}(\mathcal{H}_f, \mathcal{H}_{\overline{f}})$ be non-zero and such that $-\gamma_{\overline{f}}S = S\gamma_f$. Then, the support of S must have non-zero intersection with one of $(\mathcal{H}_F)_{\mathbf{r}_1\mathbf{r}_2}$ or $(\mathcal{H}_F)_{\mathbf{r}_1'\mathbf{r}_2'}$, and the range of S must have non-zero intersection with one of $(\mathcal{H}_F)_{\mathbf{r}_2\mathbf{r}_1}$ or $(\mathcal{H}_F)_{\mathbf{r}_2'\mathbf{r}_1'}$. Thus, $S^{\gamma\delta}_{\alpha\beta} \neq 0$ for some $(\alpha, \beta) \in \{(\mathbf{r}_1, \mathbf{r}_2), (\mathbf{r}_1', \mathbf{r}_2')\}$ and $(\gamma, \delta) \in \{(\mathbf{r}_2, \mathbf{r}_1), (\mathbf{r}_2', \mathbf{r}_1')\}$, so that $\mathcal{A}_S \subseteq \mathcal{A}_{S^{\gamma\delta}_{\alpha\beta}}$. Let us now write

$$S^{\gamma\delta}_{\alpha\beta} = \sum_i A_i \otimes B_i$$

for non-zero $A_i \in M_{n_\gamma \times n_\alpha}(\mathbb{C})$ and for linearly independent $B_i \in M_{Nn_\delta \times Nn_\beta}(\mathbb{C})$. Then for all $a \in \mathcal{A}^{\text{even}}$,

$$\lambda_{\overline{f}}(a)S - S\lambda_f(a) = \sum_i (\lambda_\gamma(a)A_i - A_i\lambda_\alpha(a)) \otimes B_i,$$

so by linear independence of the B_i, $a \in \mathcal{A}_{S^{\gamma\delta}_{\alpha\beta}}$ if and only if for each i,

$$\lambda_\gamma(a)A_i = A_i\lambda_\alpha(a), \quad \lambda_\gamma(a^*)A_i = A_i\lambda_\alpha(a),$$

and hence

$$\mathcal{A}(S) \subseteq \mathcal{A}_{S^{\gamma\delta}_{\alpha\beta}} \subseteq \mathcal{A}(T_0) := \{a \in \mathcal{A}^{\text{even}} \mid [a, T_0] = 0, [a^*, T_1] = 0\}$$

for $T_0 = A_1$, say, viewing T_0 and the elements of $\mathcal{A}^{\text{even}}$ as operators on $\mathbb{C}^{r_1} \oplus \mathbb{C}^{r_1'} \oplus \mathbb{C}^{r_2} \oplus \mathbb{C}^{r_2'}$. However, if $T_0 = PT$ is the polar decomposition of T_0 into a positive operator P on $\mathbb{C}^{n_\gamma}$ and a partial isometry $T : \mathbb{C}^{n_\alpha} \to \mathbb{C}^{n_\gamma}$, it follows that $a \in \mathcal{A}^{\text{even}}$ commutes with T_0 only if it commutes with T, and hence $\mathcal{A}_0 \subseteq \mathcal{A}(T_0) \subseteq \mathcal{A}(T)$, proving the one direction of the claim.

Now suppose that $\mathcal{C} = \mathcal{A}(T)$ for a suitable partial isometry T, which we view as a partial isometry $\mathbb{C}^{n_{\alpha_0}} \to \mathbb{C}^{n_{\gamma_0}}$ for some $\alpha_0 \in \{\mathbf{r}_1, \mathbf{r}_1'\}$, $\gamma_0 \in \{\mathbf{r}_2, \mathbf{r}_2'\}$. Then for any non-zero $\Upsilon \in M_N(\mathbb{C})$, we can define an element $S(\Upsilon) \in \mathcal{L}_{\mathcal{C}}^{\text{LR}}(\mathcal{H}_f, \mathcal{H}_{\overline{f}})$ by setting

$$S(\Upsilon)_{\alpha\beta}^{\gamma\delta} = \begin{cases} T \otimes \Upsilon \otimes T^* & \text{if } \alpha = \delta = \alpha_0,\ \beta = \gamma = \gamma_0, \\ 0 & \text{otherwise,} \end{cases}$$

which, as noted above, corresponds to a unique non-zero element of the space $\mathcal{D}_1(\mathcal{C}, \mathcal{H}_F, \gamma_F, J_F, \epsilon_F)$, so that $\mathcal{C}$ does indeed admit off-diagonal Dirac operators. □

In light of the above characterisation, it suffices to consider subalgebras $\mathcal{A}(T)$ for partial isometries $T : \mathbb{C}^{r_1} \to \mathbb{C}^{r_2}$, so that

$$\mathcal{A}(T) = \{(a_1, a_2, b_1, b_2) \in \mathcal{A}^{\text{even}} \mid b_1 T = T a_1,\ b_1^* T = T a_1^*\} \tag{5.23}$$

$$\cong \mathcal{A}_0(T) \oplus M_{k_1 r_1'/n}(\mathbb{K}_1) \oplus M_{k_2 r_2'/n}(\mathbb{K}_k), \tag{5.24}$$

where

$$\mathcal{A}_0(T) := \{(a, b) \in M_{k_1 r_1/n}(\mathbb{K}_1) \oplus M_{k_2 r_2/n}(\mathbb{K}_2) \mid bT = Ta,\ b^* T = T a^*\}, \tag{5.25}$$

so that our problem is reduced to that of maximising the dimension of $\mathcal{A}_0(T)$.

It is reasonable to assume that T is, in some sense, compatible with the algebraic structures of $M_{k_1 r_1/n}(\mathbb{K}_1)$ and $M_{k_2 r_2/n}(\mathbb{K}_2)$, so as to minimise the restrictiveness of the defining condition on $\mathcal{A}_0(T)$, and hence maximise the dimension of $\mathcal{A}_0(T)$. It turns out that this notion of compatibility takes the form of the following conditions on T:

(1) The subspace $\operatorname{supp}(T)$ of $\mathbb{C}^{r_1}$ is either a $\mathbb{K}_1$-linear subspace of $\mathbb{C}^{r_1} = \mathbb{K}_1^{k_1 r_1/n}$ or, if $\mathbb{K}_1 = \mathbb{H}$, $\operatorname{supp}(T) = E \oplus \mathbb{C}$ for E an $\mathbb{H}$-linear subspace of $\mathbb{C}^{r_1} = \mathbb{H}^{r_1/2}$;
(2) The subspace $\operatorname{im}(T)$ of $\mathbb{C}^{r_2}$ is either a $\mathbb{K}_2$-linear subspace of $\mathbb{C}^{r_2} = \mathbb{K}_1^{k_2 r_2/n}$ or, if $\mathbb{K}_2 = \mathbb{H}$, $\operatorname{im}(T) = E \oplus \mathbb{C}$ for E an $\mathbb{H}$-linear subspace of $\mathbb{C}^{r_2} = \mathbb{H}^{r_2/2}$.

Now, let $r = \operatorname{rank}(T)$, let $d(r) = \dim_{\mathbb{R}}(\mathcal{A}_0(T))$, and let

$$d_i = \begin{cases} 1 & \text{if } \mathbb{K}_i = \mathbb{R}, \\ 2 & \text{if } \mathbb{K}_i = \mathbb{C}, \\ \frac{1}{2} & \text{if } \mathbb{K}_i = \mathbb{H}. \end{cases}$$

Under these assumptions, then, one can show that

(1) If $\mathbb{K}_1 = \mathbb{K}_2$ or $\mathbb{K}_2 = \mathbb{C}$, and, if $\mathbb{K}_1 = \mathbb{H}$, r is even, then

$$\mathcal{A}_0(T) \cong M_{k_1 r/n}(\mathbb{K}_1) \oplus M_{k_1(r_1 - r)/n}(\mathbb{K}_1) \oplus M_{k_2(r_2 - r)/n}(\mathbb{K}_2), \tag{5.26}$$

and hence

$$d(r) = d_1 r^2 + d_1 (r - r_1)^2 + d_2 (r - r_2)^2;$$

(2) If $(\mathbb{K}_1, \mathbb{K}_2) = (\mathbb{H}, \mathbb{R})$ and r is odd, then

$$\mathcal{A}_0(T) \cong \left(M_{(r-1)/2}(\mathbb{H}) \cap M_{r-1}(\mathbb{R})\right) \oplus \mathbb{R} \oplus M_{(r_2 - r - 1)/2}(\mathbb{H}) \oplus M_{r_1 - r}(\mathbb{R}), \tag{5.27}$$

and hence

$$d(r) = (r-1)^2 + 1 + \frac{1}{2}(r - r_2 + 1)^2 + (r - r_1)^2;$$

(3) If $(\mathbb{K}_1, \mathbb{K}_2) = (\mathbb{H}, \mathbb{C})$ and r is odd, then

$$\mathcal{A}_0(T) \cong M_{(r-1)/2}(\mathbb{H}) \oplus \mathbb{C} \oplus M_{(r_2-r-1)/2}(\mathbb{H}) \oplus M_{r_1-r}(\mathbb{C}), \tag{5.28}$$

and hence

$$d(r) = \frac{1}{2}(r-1)^2 + 2 + \frac{1}{2}(r - r_2 + 1)^2 + 2(r - r_1)^2;$$

(4) If $\mathbb{K}_1 = \mathbb{K}_2 = \mathbb{H}$ and r is odd, then

$$\mathcal{A}_0(T) \cong M_{(r-1)/2}(\mathbb{H}) \oplus \mathbb{C} \oplus M_{(r_1-r-1)/2}(\mathbb{H}) \oplus M_{(r_2-r-1)/2}(\mathbb{H}), \tag{5.29}$$

and hence

$$d(r) = \frac{1}{2}(r-1)^2 + 2 + \frac{1}{2}(r - r_1 + 1)^2 + \frac{1}{2}(r - r_2 + 1)^2. \tag{5.30}$$

The other cases are obtained easily, by symmetry, from the ones listed above.

Now, let R_{max} be the set of all $r \in \{1, \ldots, \min(r_1, r_2)\}$ maximising the value of $d(r)$. By checking case by case, one can arrive at the following generalisation of Theorem 4.1 in [3]:

PROPOSITION 5.18. *Let $T : \mathbb{C}^{r_1} \to \mathbb{C}^{r_2}$ be a partial isometry. Then $\mathcal{A}(T)$ attains maximal dimension only if* $\operatorname{rank}(T) \in R_{max}$*, where $R_{max} = \{1\}$ except in the following cases:*

(1) $(\mathbb{K}_1, \mathbb{K}_2) = (\mathbb{C}, \mathbb{C})$ *and* $(r_1, r_2) = (2, 2)$*, in which case* $R_{max} = \{2\}$*;*
(2) $(\mathbb{K}_1, \mathbb{K}_2) = (\mathbb{C}, \mathbb{C})$ *and* $(r_1, r_2) = (3, 3)$*, in which case* $R_{max} = \{1, 2\}$*;*
(3) $(\mathbb{K}_1, \mathbb{K}_2) = (\mathbb{C}, \mathbb{R})$ *and* $(r_1, r_2) = (2, 2)$*, in which case* $R_{max} = \{1, 2\}$*;*
(4) $(\mathbb{K}_1, \mathbb{K}_2) = (\mathbb{C}, \mathbb{H})$ *and* $(r_1, r_2) = (2, 2)$*, in which case* $R_{max} = \{1, 2\}$*;*
(5) $(\mathbb{K}_1, \mathbb{K}_2) = (\mathbb{R}, \mathbb{C})$ *and* $(r_1, r_2) = (2, 2)$*, in which case* $R_{max} = \{1, 2\}$*;*
(6) $(\mathbb{K}_1, \mathbb{K}_2) = (\mathbb{R}, \mathbb{R})$ *and* $(r_1, r_2) = (2, 2)$*, in which case* $R_{max} = \{2\}$*;*
(7) $(\mathbb{K}_1, \mathbb{K}_2) = (\mathbb{R}, \mathbb{R})$ *and* $(r_1, r_2) = (3, 3)$*, in which case* $R_{max} = \{1, 2\}$*;*
(8) $(\mathbb{K}_1, \mathbb{K}_2) = (\mathbb{R}, \mathbb{H})$ *and* $r_1 = 2$*, in which case* $R_{max} = \{1, 2\}$*;*
(9) $(\mathbb{K}_1, \mathbb{K}_2) = (\mathbb{H}, \mathbb{C})$ *and* $(r_1, r_2) = (2, 2)$*, in which case* $R_{max} = \{1, 2\}$*;*
(10) $(\mathbb{K}_1, \mathbb{K}_2) = (\mathbb{H}, \mathbb{R})$ *and* $r_2 = 2$*, in which case* $R_{max} = \{1, 2\}$*;*
(11) $(\mathbb{K}_1, \mathbb{K}_2) = (\mathbb{H}, \mathbb{H})$ *and* $(r_1, r_2) = (4, 4)$*, in which case* $R_{max} = \{4\}$*;*
(12) $(\mathbb{K}_1, \mathbb{K}_2) = (\mathbb{H}, \mathbb{H})$ *and* $(r_1, r_2) \neq (4, 4)$*, in which case* $R_{max} = \{2\}$*.*

Moreover, if T satisfies the aforementioned compatibility conditions, then $\mathcal{A}(T)$ does indeed attain maximal dimension whenever $\operatorname{rank}(T) \in R_{max}$.

One must carry out the same calculations for the other possibilities for the domain and range of T, but this can be done simply by replacing (r_1, r_2) in the above equations and claims with (r_1, r_2'), (r_1', r_2) and (r_1', r_2'). Thus, one can determine the maximal dimension of a subalgebra of $\mathcal{A}^{\text{even}}$ admitting off-diagonal operators by comparing the maximal values of $\dim_{\mathbb{R}}(\mathcal{A}(T))$ for $T : \mathbb{C}^{r_1} \to \mathbb{C}^{r_2}$, $T : \mathbb{C}^{r_1} \to \mathbb{C}^{r_2'}$, $T : \mathbb{C}^{r_1'} \to \mathbb{C}^{r_2}$, and $T : \mathbb{C}^{r_1'} \to \mathbb{C}^{r_2'}$.

Finally, by means of the discussion above and the fact that $\mathrm{Sp}(n)$ acts transitively on 1-dimensional subspaces of $\mathbb{C}^n$, one can readily check that the real C^*-algebra $\mathcal{A}_F$ and the S^0-real $\mathcal{A}_F$-bimodule $(\mathcal{H}_F, \gamma_F, J_F, \epsilon_F)$ of KO-dimension $6 \bmod 8$ of the NCG Standard Model are uniquely determined, up to inner automorphisms of $\mathcal{A}^{\text{even}}$ and unitary equivalence, by the following choice of inputs:

- $n = 4$;
- $(\mathbb{K}_1, \mathbb{K}_2) = (\mathbb{H}, \mathbb{C})$;
- $g_1 \in M_2(\mathbb{H})$, $g_2 \in M_4(\mathbb{C})$;

- $(r_1, r_2) = (2, 4)$;
- $N = 3$.

The value of N, by construction, corresponds to the number of generations of fermions, whilst the values of n, r_1 and r_2 give rise to the number of species of fermion of each chirality per generation. The significance of the other inputs remains to be seen.

6. Conclusion

As we have seen, the structure theory first developed by Paschke and Sitarz [20] and by Krajewski [18] for finite real spectral triples of KO-dimension 0 mod 8 and satisfying orientability and Poincaré duality can be extended quite fully to the case of arbitrary KO-dimension and without the assumptions of orientability and Poincaré duality. In particular, once a suitable ordering is fixed on the spectrum of a finite-dimensional real C^*-algebra $\mathcal{A}$, the study of finite real spectral triples with algebra $\mathcal{A}$ reduces completely to the study of the appropriate multiplicity matrices and of certain moduli spaces constructed using those matrices. This reduction is what has allowed for the success of Krajewski's diagrammatic approach [18, §4] in the cases dealt with by Iochum, Jureit, Schücker, and Stephan [12–17, 22]. We have also seen how to apply this theory both to the "finite geometries" of the current version of the NCG Standard Model [4, 7, 8] and to Chamseddine and Connes's framework [2, 3] for deriving the same finite geometries.

Dropping the orientability requirement comes at a fairly steep cost, as even bimodules of various sorts generally have fairly intricate moduli spaces of Dirac operators. It would therefore be useful to characterise the precise nature of the failure of orientability (and of Poincaré duality) for the finite spectral triple of the current noncommutative-geometric Standard Model. It would also be useful to generalise and study the physically-desirable conditions identified in the extant literature on finite spectral triples, such as dynamical non-degeneracy [22] and anomaly cancellation [18]. Indeed, it would be natural to generalise Krajewski diagrams [18] and the combinatorial analysis they facilitate [17] to bilateral spectral triples of all types. The paper by Paschke and Sitarz [20] also contains further material for generalisation, namely discussion of the noncommutative differential calculus of a finite spectral triple and of quantum group symmetries. In particular, one might hope to characterise finite spectral triples equivariant under the action or coaction of a suitable Hopf algebra [21, 23].

Finally, as was mentioned earlier, the finite geometry of the current NCG Standard Model fails to be S^0-real. However, this failure is specifically the failure of the Dirac operator D to commute with the S^0-real structure ϵ. The "off-diagonal" part of D does, however, take a very special form; we hope to provide in future work a more geometrical interpretation of this term, which provides for Majorana fermions and for the so-called see-saw mechanism [4].

References

[1] John W. Barrett, *A Lorentzian version of the non-commutative geometry of the standard model of particle physics*, J. Math. Phys. **48** (2007), no. 012303.

[2] Ali H. Chamseddine and Alain Connes, *Conceptual explanation for the algebra in the non-commutative approach to the Standard Model*, Phys. Rev. Lett. **99** (2007), no. 191601.

[3] ———, *Why the standard model*, J. Geom. Phys. **58** (2008), 38–47.

[4] Ali H. Chamseddine, Alain Connes, and Matilde Marcolli, *Gravity and the Standard Model with neutrino mixing*, Adv. Theor. Math. Phys. **11** (2007), 991–1089.

[5] Alain Connes, *Geometry from the spectral point of view*, Lett. Math. Phys. **34** (1995), no. 3, 203–238.

[6] ______, *Noncommutative geometry and reality*, J. Math. Phys. **6** (1995), 6194–6231.

[7] ______, *Noncommutative geometry and the Standard Model with neutrino mixing*, JHEP **11** (2006), no. 81.

[8] Alain Connes and Matilde Marcolli, *Noncommutative Geometry, Quantum Fields and Motives*, Colloquium Publications, vol. 55, American Mathematical Society, Providence, RI, 2007.

[9] George A. Elliott, *Towards a theory of classification*, Adv. in Math. **223** (2010), no. 1, 30–48.

[10] ______, *private conversation*, 2008.

[11] Douglas R. Farenick, *Algebras of Linear Transformations*, Springer, New York, 2000.

[12] Bruno Iochum, Thomas Schücker, and Christoph Stephan, *On a classification of irreducible almost commutative geometries*, J. Math. Phys. **45** (2004), 5003–5041.

[13] Jan-H. Jureit and Christoph A. Stephan, *On a classification of irreducible almost commutative geometries, a second helping*, J. Math. Phys. **46** (2005), no. 043512.

[14] Jan-Hendrik Jureit, Thomas Schücker, and Christoph Stephan, *On a classification of irreducible almost commutative geometries III*, J. Math. Phys. **46** (2005), no. 072303.

[15] Jan-Hendrik Jureit and Christoph A. Stephan, *On a classification of irreducible almost commutative geometries IV*, J. Math. Phys. **49** (2008), 033502.

[16] ______, *On a classification of irreducible almost commutative geometries, V* (2009).

[17] Jan-H. Jureit and Christoph A. Stephan, *Finding the standard model of particle physics, a combinatorial problem*, Comp. Phys. Comm. **178** (2008), 230–247.

[18] Thomas Krajewski, *Classification of finite spectral triples*, J. Geom. Phys. **28** (1998), 1–30.

[19] Bing-Ren Li, *Introduction to Operator Algebras*, World Scientific, Singapore, 1992.

[20] Mario Paschke and Andrzej Sitarz, *Discrete spectral triples and their symmetries*, J. Math. Phys. **39** (1998), 6191–6205.

[21] ______, *The geometry of noncommutative symmetries*, Acta Physica Polonica B **31** (2000), 1897–1911.

[22] Thomas Schücker, *Krajewski diagrams and spin lifts* (2005), available at `arXiv:hep-th/0501181v2`.

[23] Andrzej Sitarz, *Equivariant spectral triples*, Noncommutative Geometry and Quantum Groups (Piotr M. Hajac and Wiesław Pusz, eds.), Banach Center Publ., vol. 61, Polish Acad. Sci., Warsaw, 2003, pp. 231–268.

[24] Gerald W. Schwarz, *Smooth functions invariant under the action of a compact Lie group*, Topology **14** (1975), 63–68.

[25] Christoph A. Stephan, *Almost-commutative geometry, massive neutrinos and the orientability axiom in KO-dimension* 6 (2006), available at `arXiv:hep-th/0610097v1`.

Max Planck Institute for Mathematics, Vivatsgasse 7, 53111 Bonn, Germany

Current address: California Institute of Technology, Department of Mathematics, MC 253-37, Pasadena, CA 91125, U.S.A.

E-mail address: `branimir@caltech.edu`

Tensor representations of the general linear super group

Rita Fioresi

ABSTRACT. We show a correspondence between tensor representations of the super general linear group $GL(m|n)$ and tensor representations of the general linear superalgebra $\mathfrak{gl}(m|n)$ using a functorial approach.

1. Introduction

Supersymmetry is an important mathematical tool in physics that enables to treat on equal grounds the two types of elementary particles: bosons and fermions, whose states are described respectively by commuting and anticommuting functions. It is fundamental to seek a unified treatment for these particles since they do transform into each other. Hence considering only symmetries that keep one type separated from the other is not acceptable. For this reason the symmetries of elementary particles must be described not by groups, but by *supergroups*, which are a natural generalization of groups in the $\mathbf{Z}_2$ graded or *super* setting.

The theory of representations of supergroups has a particular importance since it is attached to the problem of the classification of elementary particles.

As in the classical theory, in order to understand the representations of a supergroup, one must first study the representations of its Lie superalgebra. The representation theory of the general linear superalgebra $\mathfrak{gl}(m|n)$ has been the object of study of many people.

In [**3**] Berele and Regev provide a full account of a class of irreducible representations of $\mathfrak{gl}(m|n)$ that turns out to be linked to certain Young tableaux called *semistandard or superstandard tableaux.* The same result appeares also in [**7**] by Dondi and Jarvis in a slightly different setting. Dondi and Jarvis in fact introduce the notion of *super permutation* and use this definition to motivate the semistandard Young tableaux used for the description of the irreducible representations of the general linear superalgebra.

The results by Berele and Regev were later generalized and deepened by Brini, Regonati and Teolis in [**4**]. In their important work, they develop a unified theory that treats simoultaneosly the super and the classical case, through the powerful method of *virtual variables.*

Another account of this subject is found in [**17**]. Sergeev establishes a correspondence between a class of irreducible tensor representations of $\mathfrak{gl}(m|n)$ and the irreducible representations of a certain finite group, different from the permutation group used both in [**3**] and [**7**].

It is important to remark at this point that the theory of representations of superalgebras and of supergroups has dramatic differences with respect to the classical theory. As we will see, not all representations of the super general linear group and its Lie superalgebra are found as tensor representations. Moreover not all representations are completely reducible over **C**.

In this paper we want to understand how tensor representations of the Lie superalgebra $\mathfrak{gl}(m|n)$ can be naturally associated to the representation of the corresponding group $GL(m|n)$. Using [**3, 7**] we are able then to obtain a full classification of the irreducible tensor representations of the general linear supergroup coming from the natural diagonal action and a correspondence between such representations and representations of the symmetric group. These facts are generally known, however we feel that using the functorial language we can explicitly write the exponential map and construct explicitly the correspondence, not only over **C**, but over an arbitrary field. This helps to deepen the understanding since it provides a bridge between different languages and moreover can be used for applications in algebraic supergeometry.

We will do this using the *functor of points approach*, suggested originally by Kostant and Leites [**13, 14**] and later devoloped by Bernstein (see the notes by Deligne and Morgan in [**6**]). This approach allows to recover the geometric intuition of the problem.

This paper is organized as follows.

In section 2 we review some of the basic definitions of supergeometry. Since we will adopt the functorial language we relate our definitions to the other definitions appearing in the literature.

In section 3 we recall briefly the results obtained indipendently by Berele, Regev and Dondi, Jarvis. These results establish a correspondence between tensor representations of the permutation group and tensor representations of the superalgebra $\mathfrak{gl}(m|n)$. Moreover we show that the tensor representations of the Lie superalgebra $\mathfrak{gl}(m|n)$ do not exhaust all polynomial representations of $\mathfrak{gl}(m|n)$.

Finally in section 4 we discuss tensor representations of the general linear supergroup associated to the representations of $\mathfrak{gl}(m|n)$ described in §3.

Acknowledgements. We wish to thank Prof. V. S. Varadarajan, Prof. A. Brini, Prof. F. Regonati and Prof. I. Dimitrov for helpful comments.

2. Basic definitions

Let k be an algebraically closed field of characteristic 0.

All algebras have to be intended over k.

A *superalgebra* A is a $\mathbf{Z}_2$-graded algebra, $A = A_0 \oplus A_1$, $p(x)$ denotes the parity of an homogeneous element x. A is said to be *commutative* if

$$xy = (-1)^{p(x)p(y)}yx$$

Let (salg) denote the category of commutative superalgebras. We call A^0, *the reduced algebra*, the quotient A/I_{odd}, where I_{odd} is the (two-sided) ideal generated by the odd nilpotents.

The concept of an affine supervariety or more generally an affine superscheme can be defined very effectively through its functor of points.

DEFINITION 2.1. An *affine superscheme* is a representable functor:

$$\begin{array}{cccc} \mathbf{X}: & (\text{salg}) & \longrightarrow & (\text{sets}) \\ & A & \mapsto & X(A) = \text{Hom}(k[X], A). \end{array}$$

If $k[X]^0$ has no nilpotents we say that X is an *affine supervariety.*

From this definition one can see that the category (salg) plays a role in algebraic supergeometry similar to the category of commutative algebras for the ordinary (i.e. non super) algebraic geometry. In particular we have an equivalence of categories between the categories of affine superschemes and commutative superalgebras. (For more details see [**8**], [**5**] ch. 5).

EXAMPLES 2.2. 1. *Affine superspace.* Let $V = V_0 \oplus V_1$ be a finite dimensional super vector space. Define the following functor:

$$\mathbf{V} : (\text{salg}) \longrightarrow (\text{sets}), \qquad \mathbf{V}(A) = (A \otimes V)_0 = A_0 \otimes V_0 \oplus A_1 \otimes V_1$$

This functor is representable and it is represented by:

$$k[V] = Sym(V_0) \otimes \wedge(V_1)$$

where $Sym(V_0)$ is the polynomial algebra over the vector space V_0 and $\wedge(V_1)$ the exterior algebra over the vector space V_1. Let's see this more in detail.

If we choose a graded basis for V, $e_1 \dots e_m, \epsilon_1 \dots \epsilon_n$, with e_i even and ϵ_j odd, then

$$k[V] = k[x_1 \dots x_m, \xi_1 \dots \xi_n],$$

where the latin letters denote commuting indeterminates, while the greek ones anticommuting indeterminates i.e. $\xi_i \xi_j = -\xi_j \xi_i$. In this case V is commonly denoted with $k^{m|n}$ and $m|n$ is called the *superdimension* of V. $\mathbf{V}$ is the *functor of points* of the super vector space V.

In particular, if $V = k^{m|n}$:

$$\begin{aligned} \mathbf{V}(A) = & \ \{(a_1 \dots a_m, \alpha_1 \dots \alpha_n) \ \mid \ a_i \in A_0, \quad \alpha_j \in A_1\} = \\ = & \ \text{Hom}(k[V], A) = \{\phi : k[V] \longrightarrow k \ \mid \ \phi(x_i) = a_i, \quad \phi(\xi_j) = \alpha_j\} \end{aligned}$$

Hence $\mathbf{V}(A) = A_0 \otimes k^m \oplus A_1 \otimes k^n$.

2. *Tensor superspace.* Let V be a finite dimensional super vector space. We define the vector space of r-tensors as:

$$T^r(V) =_{\text{def}} \underbrace{V \otimes V \cdots \otimes V}_{r \text{ times}}$$

$T^r(V)$ is a super vector space, the parity of a monomial element is defined as $p(v_1 \otimes \cdots \otimes v_r) = p(v_1) + \cdots + p(v_r)$. The functor of points of $T^r(V)$ viewed as a supervariety is:

$$\mathbf{T^r(V)}(A) = \mathbf{V}(A) \otimes_A \cdots \otimes_A \mathbf{V}(A)$$

We define the superspace of tensors $T(V)$ as:

$$T(V) = \bigoplus_{r \geq 0} T^r(V)$$

and denote with $\mathbf{T(V)}$ its functor of points.

3. *Supermatrices.* Given a finite dimensional super vector space V of dimension $m|n$, the endomorphisms $\mathrm{End}(V)$ over V is itself a supervector space of dimension $m^2 + n^2|2mn$: $\mathrm{End}(V) = \mathrm{End}(V)_0 \oplus \mathrm{End}(V)_1$, where $\mathrm{End}(V)_0$ are the endomorphisms preserving parity, while $\mathrm{End}(V)_1$ are those reversing parity.

Hence we can define the following functor:

$$\mathbf{End}(V) : (\mathrm{salg}) \longrightarrow (\mathrm{sets}), \qquad \mathbf{End}(V)(A) = (A \otimes \mathrm{End}(V))_0$$

This functor is representable (see (1)). Choosing a graded basis for V, $V = k^{m|n}$, the functor is represented by $k[x_{ij}, y_{kl}, \xi_{kj}, \eta_{il}]$ where $1 \leq i, j \leq m$, $m + 1 \leq k, l \leq m + n$.

In this case:

$$\mathbf{End}(V)(A) = \left\{ \begin{pmatrix} a_{m \times m} & \beta_{m \times n} \\ \gamma_{n \times m} & d_{n \times n} \end{pmatrix} \right\}$$

where a, d and β, γ are block matrices with respectively even and odd entries in A.

DEFINITION 2.3. An *affine supergroup* G is a group valued affine superscheme, i.e. it is a representable functor:

$$\begin{array}{cccc} \mathbf{G}: & (\mathrm{salg}) & \longrightarrow & (\mathrm{groups}) \\ & A & \mapsto & \mathbf{GL}(V)(A) \end{array}$$

It is simple to verify that the superalgebra representing the supergroup $\mathbf{G}$ has an Hopf superalgebra structure. More is true: Given a supervariety $\mathbf{G}$, $\mathbf{G}$ is a supergroup if and only if the algebra representing it $k[G]$ is an Hopf superalgebra.

Let V be a finite dimensional super vector space. We are interested in the *general linear supergroup* $\mathbf{GL}(V)$.

DEFINITION 2.4. We define *general linear supergroup* the group valued functor

$$\begin{array}{cccc} \mathbf{GL}(V): & (\mathrm{salg}) & \longrightarrow & (\mathrm{sets}) \\ & A & \mapsto & \mathbf{GL}(V)(A) \end{array}$$

where $\mathbf{GL}(V)(A)$ is the set of automorphisms of the A-supermodule $A \otimes V$, $A \in (\mathrm{salg})$. More explicitly if $V = k^{m|n}$, the functor $\mathbf{GL}(V)$ commonly denoted $\mathbf{GL}(m|n)$ is defined as the set of automorphisms of $A^{m|n} =_{\mathrm{def}} A \otimes k^{m|n}$ and is given by:

$$\mathbf{GL}(m|n)(A) = \left\{ \begin{pmatrix} a_{m \times m} & \beta_{m \times n} \\ \gamma_{n \times m} & d_{n \times n} \end{pmatrix} \mid a, d \quad \text{invertible} \right\}$$

where a, d and β, γ are block matrices with respectively even and odd entries in A.

This functor is representable and it is represented by the Hopf algebra (see [**9**]):

$$k[x_{ij}, y_{\alpha\beta}, \xi_{i\beta}, \eta_{\alpha j}, z, w]/\big((w \det(x) - 1, z \det(y) - 1\big),$$

$$i, j = 1, \dots m \qquad \alpha, \beta = 1, \dots n.$$

We now would like to introduce the notion of Lie superalgebra using the functorial language. We then see it is equivalent to the more standard definitions (see [**12**] for example).

Definition 2.5. Let $\mathfrak{g}$ be a finite dimensional supervector space. The functor (see Example 2.2 (1)):

$$\mathbf{g} : (\text{salg}) \longrightarrow (\text{sets}), \qquad \mathbf{g}(A) = (A \otimes \mathbf{g})_0$$

is said to be a *Lie superalgebra* if it is Lie algebra valued, i.e. for each A there exists a linear map:

$$[\ ,\ ]_A : \mathbf{g}(A) \times \mathbf{g}(A) \longrightarrow \mathfrak{g}(A)$$

satisfying the antisymmetric property and the Jacobi identity.

Notice that in the same way as the supergroup functor is group valued, the Lie superalgebra functor is Lie algebra valued, i. e. it has values in a *classical category*. The super nature of these functors arises from the different starting category, namely (salg), which allows superalgebras as representing objects.

The usual notion of Lie superalgebra, as defined for example by Kac in [**12**] is equivalent to this functorial definition. Let's recall this definition and see the equivalence with the Definition 2.5 more in detail.

Definition 2.6. Let $\mathfrak{g}$ be a super vector space. We say that a bilinear map

$$[,] : \mathfrak{g} \times \mathfrak{g} \longrightarrow \mathfrak{g}$$

is a *superbracket* if $\forall x, y, z \in \mathfrak{g}$:

$$[x, y] = (-1)^{p(x)p(y)}[y, x]$$

$$[x, [y, z]] + (-1)^{p(x)p(y)+p(x)p(z)}[y, [z, x]] + (-1)^{p(x)p(z)+p(y)p(z)}[z, [x, y]] = 0$$

$(\mathfrak{g}, [,])$, is what in the literature is commonly defined as *Lie superalgebra.*

Observation 2.7. The two concepts of Lie superalgebra $\mathbf{g}$ in the functorial setting and superbracket on a supervector space $(\mathfrak{g}, [,])$ are equivalent.

In fact if we have a Lie superalgebra $\mathbf{g}$ there is always a superspace $\mathfrak{g}$ associated to it together with a superbracket. The superbracket on $\mathfrak{g}$ is given following the *even rules.* (For a complete treatment of even rules see pg 57 [**6**]). Given $v, w \in \mathfrak{g}$, we have that since the Lie bracket on $\mathbf{g}(A)$ is A_0-linear:

$$[a \otimes v, b \otimes w] = ab \otimes z \in (A \otimes \mathfrak{g})_0 = \mathbf{g}(A)$$

Hence we can define the bracket $\{v, w\}$ as the element of $\mathfrak{g}$ such that:

$$z = (-1)^{p(a)p(w)}\{v, w\}$$

i. e. satisfying the relation:

$$[a \otimes v, b \otimes w] = (-1)^{p(b)p(v)}ab \otimes \{v, w\}.$$

We need to check it is a superbracket. Let's see for example the antisymmetry property. Observe first that if $a \otimes v, b \otimes w \in (\mathfrak{g} \otimes A)_0$ must be $p(v) = p(a)$, $p(w) = p(b)$ since $(A \otimes \mathfrak{g})_0 = A_0 \otimes \mathfrak{g}_0 \oplus A_1 \otimes \mathfrak{g}_1$. So we can write:

$$[a \otimes v, b \otimes w] = (-1)^{p(b)p(v)} ab \otimes \{v, w\} = (-1)^{p(v)p(w)} ab \otimes \{v, w\}.$$

On the other hand:

$$\begin{aligned}[b \otimes w, a \otimes v] &= (-1)^{p(a)p(w)} ba \otimes \{w, v\} = \\ &= (-1)^{p(a)p(w)+p(a)p(b)} ab \otimes \{w, v\} = \\ &= (-1)^{2p(w)p(v)} ab \otimes \{w, v\} = ab \otimes \{w, v\}.\end{aligned}$$

Comparing the two expression we get the antisymmetry of the superbracket. For the super Jacobi identity the calculation is the same.

Vice-versa if $(\mathfrak{g}, \{,\})$ is a super vector space with a superbracket, we immediately can define its functor of points $\mathbf{g}$. $\mathbf{g}$ is a Lie superalgebra because we have a bracket on $\mathbf{g}(A)$ defined as

$$[a \otimes v, b \otimes w] = (-1)^{p(b)p(v)} ab \otimes \{v, w\}.$$

The previous calculation worked backwards proves that $[,]$ is a (classical) Lie bracket.

With an abuse of language we will call Lie superalgebra both the supervector space $\mathfrak{g}$ with a superbracket $[,]$ and the functor $\mathbf{g}$ as defined in 2.5.

OBSERVATION 2.8. In [**8**] is given the notion of a Lie super algebra associated to an affine supergroup. In particular it is proven that the Lie superalgebra associated to $\mathbf{GL}(m|n)$ is $\mathbf{End}(k^{m|n})$. We will denote $\mathbf{End}(k^{m|n})$ with $\mathfrak{gl}(m|n)$ as supervector space and with $\mathbf{gl}(m|n)$ as its functor of points. The purpose of this paper does not allow for a full description of such correspondence, all the details and the proofs can be found in [**8**].

3. Summary and observations on results by Berele and Regev

In this section we want to review some of the results in [**3, 7**]. We wish to describe the correspondence between tensor representations of the superalgebra $\mathfrak{gl}(m|n)$ and representations of the permutation group. This correspondence is obtained using the double centralizer theorem. (Note: in [**3**] $\mathfrak{gl}(m|n)$ is denoted by $\mathfrak{pl}$).

Let $V = k^{m|n}$ and let $T(V) = \bigoplus_{r \geq 0} T^r(V)$ be the tensor superspace (see Example 2.2 (2)).

We want to define on $T^r(V)$ two actions: one by S_r the permutation group and the other by the Lie superalgebra $\mathfrak{gl}(m|n)$.

Let $\sigma = (i, j) \in S_r$ and let $\{v_i\}_{1 \leq i \leq m+n}$ be a basis of V ($v_1 \dots v_m$ even elements and $v_{m+1} \dots v_{m+n}$ odd ones). Let's define:

$$(v_1 \otimes \cdots \otimes v_r) \cdot \sigma =_{\text{def}} \epsilon v_{\sigma(1)} \otimes \cdots \otimes v_{\sigma(r)}$$

where $\epsilon = -1$ when v_i and v_j are both odd and $\epsilon = 1$ otherwise. This defines a representation τ_r of S_r in $T^r(V)$. The proof of this fact can be found in [**3**] pp.122-123.

Consider now the action θ_r of the Lie superalgebra $\mathfrak{gl}(m|n)$ on $T^r(V)$ given by derivations:

$$\theta_n(X)(v_1 \otimes \cdots \otimes v_r) =_{\text{def}} \sum_i (-1)^{s(X,i)} v_1 \otimes \cdots \otimes X(v_i) \otimes \cdots \otimes v_r$$

$$g \in \mathfrak{gl}(m|n)(A), \quad v_i \in V(A), \quad A \in (\text{salg})$$

with $s(X,i) = p(X)o(i)$ where $o(i)$ denotes the number of odd elements among $v_1 \dots v_i$.

One can see that this is a Lie superalgebra action i.e. it preserves the superbracket and that it extends to an action θ of $\mathfrak{gl}(m|n)$ on $T(V)$ (this is proved in [**3**] 4.7).

In [**3**] Theorem 4.14 and Remark 4.15 prove the important double centralizer theorem:

THEOREM 3.1. *The algebras $\tau_r(S_r)$ and $\theta_r(\mathfrak{gl}(m|n))$ are each the centralizer of the other in* $\text{End}(T^r(V))$.

This result establishes a one to one correspondence between irreducible tensor representations of S_r occurring in τ_r and those of $\mathfrak{gl}(m|n)$ occurring in θ_r.

These representations are parametrized by partitions λ of the integer r. In [**3**] §3 and §4 is worked out completely the structure of irreducible tensor representations of $\mathfrak{gl}(m|n)$ arising in this way.

We are now interested in the dimensions of such representations.

DEFINITION 3.2. Let $t_1 < \cdots < t_m < u_1 < \cdots < u_n$ be integers and λ a partition of r corresponding to a diagram D_λ. A filling T_λ of D_λ is a *semistandard or superstandard tableau* if

1. The part of T_λ filled with the t's is a tableaux.
2. The t's are non decreasing in rows and strictly increasing in columns.
3. The u's are non decreasing in columns and strictly increasing in rows.

As an example that will turn out to be important later let's look at $m = n = 1$, $t_1 = 1$, $u_1 = 2$ and $r = 2$. We can have only two partitions of r: $\lambda = (2)$, $\lambda = (1,1)$. Each partition admits two fillings:

$$\lambda = (2) \qquad 1\ \ 1 \qquad 1\ \ 2$$

$$\lambda = (1,1) \qquad \begin{matrix} 1 \\ 2 \end{matrix} \qquad \begin{matrix} 2 \\ 2 \end{matrix}$$

By Theorem 3.17, 3.18 and 4.17 in [**3**] we have the following:

THEOREM 3.3. *The irreducible representations of $\mathfrak{gl}(m|n)$ occurring in θ_r are parametrized by partitions of λ of the integer r satisfying the hook condition (i. e. if $\lambda = (\lambda_1, \lambda_2 \dots)$, $\lambda_j \leq n$ for $j > m$, see 2.3 in* [**3**] *for more details). The irreducible representations associated to the shape λ has dimension equal to the number of semistandard tableaux of shape λ.*

OBSERVATION 3.4. This theorem tells us immediately that we have no one dimensional representations of $\mathfrak{gl}(m|n)$ occurring in θ_r, if $n > 0$. In fact one can generalize the Example 3.2 to show that since the odd variables allow repetitions on rows, we always have more than one filling for each shape. However there exists a polynomial (or rational) representation of $\mathfrak{gl}(m|n)$ of dimension one, namely the supertrace ([**1**] pg. 100):

$$\begin{array}{ccc} \mathfrak{gl}(m|n) & \longrightarrow & k \cong \mathrm{End}(k) \\ A = \begin{pmatrix} X & Y \\ Z & W \end{pmatrix} & \mapsto & str(A) =_{\mathrm{def}} tr(X) - tr(W) \end{array}$$

This shows that the tensor representations described in [**3**] do not exhaust all polynomial representations of $\mathfrak{gl}(m|n)$, for $n > 0$.

4. Tensor representations of the general linear supergroup

Let's start by introducing the notion of supergroup and of Lie superalgebra representation from a functorial point of view.

DEFINITION 4.1. Given an affine algebraic supergroup $\mathbf{G}$ we say that $\mathbf{G}$ acts on a super vector space W, if we have a natural transformation:

$$r : \mathbf{G} \longrightarrow \mathbf{End}(W)$$

In other words, if we have for any $A \in$ (salg) a functorial morphism

$$r_A : \mathbf{G}(A) \longrightarrow \mathbf{End}(W)(A).$$

Similarly given a Lie superalgebra $\mathbf{g}$ we say that $\mathbf{g}$ acts on a super vector space W, if we have a natural transformation:

$$t : \mathbf{g} \longrightarrow \mathbf{End}(W)$$

preserving the Lie bracket, that is for any $A \in$ (salg), we have a Lie algebra morphism $t_A : \mathbf{g}(A) \longrightarrow \mathrm{End}(W)(A)$. It is easy to verify that this is equivalent to ask that we have a morphism of Lie superalgebras:

$$T : \mathfrak{g} \longrightarrow \mathrm{End}(W)$$

i.e. a super vector space morphism preserving the superbracket. This agrees with the definition of Lie superalgebra representation in [**3**], which we also recalled in §3.

If $W \cong k^{m|n}$ we can identify $r_A(g)$, $g \in \mathbf{G}(A)$ and $t_A(x)$, $x \in \mathbf{g}(A)$ with matrices in $\mathbf{End}(W)(A)$ (see Example 2.2 (3)).

Let V be a finite dimensional super vector space. Define:

$$\rho_r : \mathbf{GL}(V) \longrightarrow \mathbf{End}(T^r(V))$$

$$\rho_{r,A}(g)(v_1 \otimes \cdots \otimes v_n) =_{\mathrm{def}} g(v_1) \otimes \cdots \otimes g(v_n),$$

$$g \in \mathbf{GL}(V)(A), \quad v_i \in \mathbf{V}(A), \quad A \in \text{(salg)}.$$

This is an action of $\mathbf{GL}(V)$ on $T^r(V)$, that can be easily extended to the whole $T(V)$.

We are also interested in the action θ_r of $\mathfrak{gl}(V)$, the Lie superalgebra of $\mathbf{GL}(V)$ on $T^r(V)$ introduced in Section 3.

Let's assume from now on $V = k^{m|n}$. Let e_{ij} denote an elementary matrix in $\mathfrak{gl}(m|n)$. $\{\mathrm{e}_{ij}\}$ is a graded canonical basis for the supervector space $\mathfrak{gl}(m|n)$, with $p(\mathrm{e}_{ij}) = p(i) + p(j)$, where an index i is even if $1 \leq i \leq m$, odd otherwise.

DEFINITION 4.2. Consider the following functor $\mathbf{E}_{ij}$: (salg)⟶(sets), $1 \leq i \neq j \leq m+n$:

$$\mathbf{E}_{ij}(A) = \{I + x\mathrm{e}_{ij} | x \in A_k, k = p(\mathrm{e}_{ij})\}$$

This is an affine supergroup functor represented by $k^{1|0}$ if $p(i) + p(j)$ is even, by $k^{0|1}$ if it is odd. We call $\mathbf{E}_{ij}$ a *one parameter subgroup functor.*

Define also the (additive) supergroup functor $\mathbf{e}_{ij}$: (salg)⟶(sets):

$$\mathbf{e}_{ij}(A) = \{a \otimes \mathrm{e}_{ij} \quad | \quad a \in A_k, \quad k = p(i) + p(j)\}.$$

THEOREM 4.3. *1. For all $A \in$ (salg), the group $\mathbf{GL}(m|n)(A)$ is generated by:*

$$\mathbf{E}_{ij}(A), \quad i = m+1 \dots m+n, \quad j = 1 \dots m$$

$$\mathbf{E}_{kl}(A), \quad k = 1 \dots m, \quad l = m+1 \dots m+n$$

$$\mathbf{GL}(m)(A_0) \times \mathbf{GL}(n)(A_0) = \left\{ \begin{pmatrix} X & 0 \\ 0 & W \end{pmatrix} | X, W \textit{ with entries in } A_0 \right\}$$

where $\mathbf{GL}(m)$ denotes the group functor associated with the classical general linear group.

2. The Lie superalgebra $\mathbf{gl}(m|n)$ is generated by the supergroup functors $\mathbf{e}_{ij}$.

Proof. (2) is immediate. For (1) it is enough to prove the given generators generate the following (see [**19**] pg. 117):

$$\begin{pmatrix} X & 0 \\ 0 & W \end{pmatrix} \quad \begin{pmatrix} I & Y \\ 0 & I \end{pmatrix} \quad \begin{pmatrix} I & 0 \\ Z & I \end{pmatrix}$$

where X, W have entries in A_0 and Y, Z have entries in A_1. This is immediate. □

Consider now the action of the (non super) group $\mathbf{GL}(m) \times \mathbf{GL}(n)$ on the ordinary vector space $V = V_0 \oplus V_1$ and also the action of its Lie algebra $\mathfrak{gl}(m) \times \mathfrak{gl}(n)$ on the same space. We can build the diagonal action ρ^0 of $\mathbf{GL}(m) \times \mathbf{GL}(n)$ on the space of tensors $T(V)$ (again V is viewed disregarding the grading) and also the usual action θ^0 by derivation of $\mathfrak{gl}(m) \times \mathfrak{gl}(n)$ on the same space.

LEMMA 4.4.

$$< \rho^0(GL(m) \times GL(n)) >=< \theta^0(\mathfrak{gl}(m) \times \mathfrak{gl}(n)) >$$

where $< S >$ denotes the subalgebra generated by the set S inside End(V) *the endomorphism of the ordinary vector space $V = k^{m+n}$.*

Proof. This is a consequence of a classical result, see for example [**11**] 8.2. □

THEOREM 4.5.

$$< \rho_{r,A}(\mathbf{GL}(m|n)(A)) >_A = < \theta_{r,A}(\mathfrak{gl}(m|n)(A)) >_A \qquad A \in (\text{salg}).$$

where $< S >_A$ *denotes the subalgebra generated by the set* S *inside* $\mathbf{End}(m|n)(A)$.

Proof. By Theorem 4.3 and Lemma 4.4 it is enough to show that:

$$\rho_{r,A}(\mathbf{E}_{ij}(A)) \in < \theta_{r,A}(\mathbf{gl}(m|n)(A)) >_A,$$

$$\theta_{r,A}(\mathbf{e}_{ij}(A)) \in < \rho_{r,A}(\mathbf{GL}(m|n)(A)) >_A$$

with $i = m+1 \dots m+n$, $j = 1 \dots m$ or $k = 1 \dots m$, $l = m+1 \dots m+n$.

Let D_{ij} be the derivation corresponding to the elementary matrix e_{ij}. So we have that $\mathbf{e}_{ij}(A) = \alpha_1 \otimes D_{ij}$, $\alpha \in A_1$. We claim that

$$\theta_{r,A}(\mathbf{e}_{ij}(A)) = 1(A) - \rho_{r,A}(\mathbf{E}_{ij})(A)$$

This is a simple calculation. □

COROLLARY 4.6. *There is a one to one correspondence between the irreducible representations of* S_r *and the irreducible representations of* $\mathbf{GL}(m|n)$ *occurring in* ρ_r.

OBSERVATION 4.7. By Corollary 4.6 and Theorem 3.3 we have that also the irreducible representations occuring in ρ_r of $\mathbf{GL}(m|n)$ are parametrized by partitions of the integer r. However by Observation 3.4 we have that there is no one dimensional irreducible representation hence also for $GL(m|n)$ we miss an important representation, namely the Berezinian:

$$\begin{array}{ccc} \mathbf{GL}(m|n)(A) & \longrightarrow & A \cong \mathbf{End}(k)(A) \\ \begin{pmatrix} X & Y \\ Z & W \end{pmatrix} & \mapsto & det(W)^{-1} det(X - YW^{-1}Z) \end{array}$$

This shows that the tensor representations of $\mathbf{GL}(m|n)$ do not exhaust all polynomial representations of $\mathbf{GL}(m|n)$, for $n > 0$.

The Berezinian representation has been described by Deligne and Morgan in [**6**] pg 60, in a natural way as an action of $\mathbf{GL}(m|n)$ on Ext group

$$\mathrm{Ext}^m_{\mathrm{Sym}^*(V^*)}(A, \mathrm{Sym}^*(V^*)).$$

Ext plays the same role as the antisymmetric tensors in this super setting.

References

[1] F. A. Berezin, *Introduction to superanalysis.* Edited by A. A. Kirillov. D. Reidel Publishing Company, Dordrecht (Holland) (1987).

[2] A. Baha Balantekin, I. Bars *Dimension and character formula for Lie supergroups*, J. Math. Phy., **22**, 1149-1162, (1981).

[3] A. Berele, A. Regev *Hook Young Diagrams with applications to combinatorics and to representations of Lie superalgebras*, Adv. Math., **64**, 118-175, (1987).

[4] A. Brini, F. Regonati, A. Teolis *The method of virtual variables and representations of Lie superalgebras.* Clifford algebras (Cookeville, TN, 2002), 245–263, Prog. Math. Phys., 34, Birkhauser Boston, MA, (2004). *Combinatorics and representation theory of Lie superalgebras over letterplace superalgebras.* Li, Hongbo (ed.) et al., Computer algebra and geometric algebra with applications. 6th international workshop, IWMM 2004, Shanghai, China, Berlin: Springer. Lecture Notes in Computer Science 3519, 239-257 (2005).

[5] L. Caston, R. Fioresi, *Mathematical Foundations of Supersymmetry,* xxx.lanl.gov, 0710.5742v1, 2007.

[6] P. Deligne and J. Morgan, *Notes on supersymmetry (following J. Bernstein),* in "Quantum fields and strings. A course for mathematicians", Vol 1, AMS, (1999).

[7] P. H. Dondi, P. D. Jarvis *Diagram and superfields tecniques in the classical superalgebras,* J. Phys. A, Math. Gen **14**, 547-563, (1981).

[8] R. Fioresi, M. A. Lledo *On Algebraic Supergroups, Coadjoint Orbits and their Deformations,* Comm. Math. Phy. **245**, no. 1, 177-200, (2004).

[9] R. Fioresi, *On algebraic supergroups and quantum deformations,* math.QA/0111113, J. Algebra Appl. **2**, no. 4, 403–423, (2003).

[10] R. Fioresi, *Supergroups, quantum supergroups and their homogeneous spaces.* Euroconference on Brane New World and Noncommutative Geometry (Torino, 2000). Modern Phys. Lett. A **16** 269–274 (2001).

[11] G. James, A. Kerber *The representation theory of the symmetric group,* Encyclopedia of Mathematics and its applications Vol. 16, Addison Wesley, (1981).

[12] V. Kac *Lie superalgebras* Adv. in Math. **26**, 8-26, (1977).

[13] B. Kostant, *Graded manifolds, Graded Lie theory and prequantization.* Lecture Notes in Math. **570** (1977).

[14] D. A. Leites, *Introduction to the theory of supermanifolds.* Russian Math. Survey. **35**:1 1-64 (1980).

[15] Yu. Manin, *Gauge field theory and complex geometry.* Springer Verlag, (1988).

[16] Yu. Manin, *Topics in non commutative geometry.* Princeton University Press, (1991).

[17] A. N. Sergeev *The tensor algebra of the identity representation as a module over the Lie superalgebras* $\mathfrak{gl}(m|n)$ *and* $Q(n)$, Math. USSR Sbornik, **51**, no. 2, (1985).

[18] M. Scheunert, R. B. Zhang *The general linear supergroup and its Hopf superalgebra of regular functions,* J. Algebra **254**, no. 1, 44-83, (2002).

[19] V. S. Varadarajan *Supersymmetry for mathematicians: an Introduction,* AMS, (2004).

DIPARTIMENTO DI MATEMATICA, UNIVERSITÀ DI BOLOGNA, PIAZZA DI PORTA S. DONATO 5, 40127 BOLOGNA, ITALY

E-mail address: fioresi@dm.unibo.it

Quantum duality principle for quantum Grassmannians

Rita Fioresi and Fabio Gavarini

Abstract. The quantum duality principle (QDP) for homogeneous spaces gives four recipes to obtain, from a quantum homogeneous space, a dual one, in the sense of Poisson duality. One of these recipes fails (for lack of the initial ingredient) when the homogeneous space we start from is not a quasi-affine variety. In this work we solve this problem for the quantum Grassmannian, a key example of quantum projective homogeneous space, providing a suitable analogue of the QDP recipe.

1. Introduction

In the theory of quantum groups, the geometrical objects that one takes into consideration are affine algebraic Poisson groups and their infinitesimal counterparts, namely Lie bialgebras. By "quantization" of either of these, one means a suitable one-parameter deformation of one of the Hopf algebras associated with them. They are respectively the algebra of regular function $\mathcal{O}(G)$, for a Poisson group G, and the universal enveloping algebra $U(\mathfrak{g})$, for a Lie bialgebra $\mathfrak{g}$. Deformations of $\mathcal{O}(G)$ are called *quantum function algebras* (QFA), and are often denoted with $\mathcal{O}_q(G)$, while deformations of $U(\mathfrak{g})$ are called *quantum universal enveloping algebras* (QUEA), denoted with $U_q(\mathfrak{g})$.

The quantum duality principle (QDP), after its formulation in [**9, 10, 11**], provides a recipe to get a QFA out of a QUEA, and vice-versa. This involves a change of the underlying geometric object, according to Poisson duality, in the following sense. Starting from a QUEA over a Lie bialgebra $\mathfrak{g} = \mathit{Lie}(G)$, one gets a QFA for a dual Poisson group G^*. Starting instead from a QFA over a Poisson group G, one gets a QUEA over the dual Lie bialgebra $\mathfrak{g}^*$.

In [**3**], this principle is extended to the wider context of homogeneous Poisson G–spaces. One describes these spaces, in global or in infinitesimal terms, using suitable subsets of $\mathcal{O}(G)$ or of $U(\mathfrak{g})$. Indeed, each homogeneous G–space M can be realized as G/K for some closed subgroup K of G (this amounts to fixing a point in M: it is shown in [**3**], §1.2, how to select such a point). Thus we can deal with either the space or the subgroup. Now, K can be coded in infinitesimal terms by $U(\mathfrak{k})$, where $\mathfrak{k} := \mathit{Lie}(K)$, and in global terms by $\mathcal{I}(K) := \{ \varphi \in \mathcal{O}(G) \,|\, \varphi(K) = 0 \}$,

2000 *Mathematics Subject Classification.* Primary 20G42, 14M15; Secondary 17B37, 17B62.
Key words and phrases. Quantum Grassmann Varieties.
Partially supported by the University of Bologna, funds for selected research topics.

the defining ideal of K. Instead, G/K can be encoded infinitesimally by $U(\mathfrak{g})\,\mathfrak{k}$ and globally by $\mathcal{O}(G/K) \equiv \mathcal{O}(G)^K$, the algebra of K–invariants in $\mathcal{O}(G)$. Note that $U(\mathfrak{g})/U(\mathfrak{g})\,\mathfrak{k}$ identifies with the set of left-invariant differential operators on G/K, or the set of K–invariant, left-invariant differential operators on G.

These constructions *all* make sense in formal geometry, i.e. when dealing simply with formal groups and formal homogeneous spaces, as in [**3**]. Instead, if one looks for *global* geometry, then one construction might fail, namely the description of G/K via its function algebra $\mathcal{O}(G/K) = \mathcal{O}(G)^K$. In fact, this makes sense — i.e., $\mathcal{O}(G)^K$ is enough to describe G/K — if and only if the variety G/K is *quasi-affine*. In particular, this is not the case if G/K is projective, like, for instance, when G/K is a Grassmann variety.

By "quantization" of the homogeneous space G/K one means any quantum deformation (in suitable sense) of any one of the four algebraic objects mentioned before which describe either G/K or K. Moreover one requires that given an infinitesimal or a global quantization for the group G, denoted by $U_q(\mathfrak{g})$ or $\mathcal{O}_q(G)$ respectively, the quantization of the homogeneous space admits a $U_q(\mathfrak{g})$–action or a $\mathcal{O}_q(G)$–coaction respectively, which yields a quantum deformation of the algebraic counterpart of the G–action on G/K.

The QDP for homogeneous G–spaces (cf. [**3**]) starts from an infinitesimal (global) quantization of a G–space, say G/K, and provides a global (infinitesimal) quantization for the Poisson dual G^*–space. The latter is $G^*/K^\perp$ (with $Lie\left(K^\perp\right) = \mathfrak{k}^\perp$, the orthogonal subspace — with respect to the natural pairing between $\mathfrak{g}$ and its dual space $\mathfrak{g}^*$ — to $\mathfrak{k}$ inside $\mathfrak{g}^*$). In particular, the principle gives a concrete recipe

$$\mathcal{O}_q(G/K) \multimap\!\!\rightsquigarrow \mathcal{O}_q(G/K)^\vee =: U_q(\mathfrak{k}^\perp)$$

in which the right-hand side is a quantization of $U(\mathfrak{k}^\perp)$.

However, this recipe makes no sense when $\mathcal{O}_q(G/K)$ is not available. In the non-formal setting this is the case whenever G/K is not quasi-affine, e.g. when it is projective.

In this paper we show how to solve this problem in the special case of the Grassmann varieties, taking G as the general linear group and $K = P$ a maximal parabolic subgroup. We adapt the basic ideas of the original QDP recipe to these new ingredients, and we obtain a new recipe

$$\mathcal{O}_q(G/P) \multimap\!\!\rightsquigarrow \widehat{\mathcal{O}_q(G/P)}^\vee$$

which perfectly makes sense, and yields the same kind of result as predicted by the QDP for the quasi-affine case. In particular, $\widehat{\mathcal{O}_q(G/P)}^\vee$ is a quantization of $U(\mathfrak{p}^\perp)$, obtained through a $(q-1)$–adic completion process.

Our construction goes as follows.

First, we consider the embedding of the Grassmannian G/P (where $G := GL_n$ or $G := SL_n$, and P is a parabolic subgroup of G) inside a projective space, given by Plücker coordinates. This will give us the first new ingredient:

$$\mathcal{O}(G/P) := \text{ring of homogeneous coordinates on } G/P \quad .$$

Many quantizations $\mathcal{O}_q(G/P)$ of $\mathcal{O}(G/P)$ already exist in the literature (see, e.g., [**6, 12, 13**]). All these quantizations, which are equivalent, come together with a quantization of the natural G–action on G/P.

In the original recipe (see [**3**]) $\mathcal{O}_q(G/K) \multimap\!\!\dashrightarrow \mathcal{O}_q(G/K)^\vee$ of the QDP (when G/K is quasi affine) we need to look at a neighborhood of the special point eK (where $e \in G$ is the identity), and at a quantization of it. Therefore, we shall replace the projective variety G/P with such an affine neighborhood, namely the big cell of G/P. This amounts to realize the algebra of regular functions on the big cell as a "homogeneous localization" of $\mathcal{O}(G/P)$, say $\mathcal{O}^{\,loc}(G/P)$, by inverting a suitable element. We then do the same at the quantum level, via the inversion of a suitable almost central element in $\mathcal{O}_q(G/P)$ — which lifts the previous one in $\mathcal{O}(G/P)$. The result is a quantization $\mathcal{O}_q^{\,loc}(G/P)$ of the coordinate ring of the big cell.

Hence we are able to *define* $\mathcal{O}_q(G/P)^\vee := \mathcal{O}_q^{\,loc}(G/P)^\vee$, where the right-hand side is given by the original QDP recipe applied to the big cell as an affine variety (we can forget any group action at this step). By the very construction, this $\mathcal{O}_q(G/P)^\vee$ should be a quantization of $U(\mathfrak{p}^\perp)$ (as an algebra). Indeed, we prove that this is the case, so we might think at $\mathcal{O}_q(G/P)^\vee$ as a quantization (of infinitesimal type) of the variety $G^*/P^\perp$. On the other hand, the construction does not ensure that $\mathcal{O}_q(G/P)^\vee$ also admits a quantization of the G^*–action on $G^*/P^\perp$ (just like the big cell is not a G–space). As a last step, we look at $\mathcal{O}_q\widehat{(G/P)}^\vee$, the $(q-1)$–adic completion of $\mathcal{O}_q(G/P)^\vee$. Of course, it is again a quantization of $U(\mathfrak{p}^\perp)$ (as an algebra). But in addition, it admits a coaction of the $(q-1)$–adic completion of $\mathcal{O}_q(G)^\vee$ — which is a quantization of $U(\mathfrak{g}^*)$. This coaction yields a quantization of the infinitesimal G^*–action on $G^*/P^\perp$. Therefore, in a nutshell, $\mathcal{O}_q\widehat{(G/P)}^\vee$ is a quantization of $G^*/P^\perp$ *as a homogeneous G^*–space*, in the sense explained above.

Notice that our arguments could be applied to any *projective* homogeneous G–space X, *up to having the initial data to start with.* Namely, one needs an embedding of X inside a projective space, a quantization (compatible with the G–action) of the ring of homogeneous coordinates of X (w.r.t. such an embedding), and a quantization of a suitable open dense affine subset of X. This program is carried out in detail in a separate work (see [**2**]).

Finally, this paper is organized as follows.

In section 2 we fix the notation, and we describe the Manin deformations of the general linear group (as a Poisson group), and of its Lie bialgebra, together with its dual. In section 3 we briefly recall results concerning the constructions of the quantum Grassmannian $\mathcal{O}_q(G/P)$ and its quantum big cell $\mathcal{O}_q^{\,loc}(G/P)$. These are known results, treated in detail in [**6, 7**]. Finally, in section 4 we extend the original QDP to build $\mathcal{O}_q(G/P)^\vee$, and we show that its $(q-1)$–adic completion is a quantization of the homogeneous G^*–space $G^*/P^\perp$ dual to the Grassmannian G/P.

Acknowledgements

The first author wishes to thank the Dipartimento di Matematica "Tor Vergata", and in particular Prof. V. Baldoni and Prof. E. Strickland, for the warm hospitality during the period in which this paper was written.

Both authors also thank D. Parashar and M. Marcolli for their kind invitation to the workshop "Quantum Groups and Noncommutative Geometry" held at MPIM in Bonn during August 6–8, 2007.

2. The Poisson Lie group $GL_n(\Bbbk)$ and its quantum deformation

Let $\Bbbk$ be any field of characteristic zero.

In this section we want to recall the construction of a quantum deformation of the Poisson Lie group $GL_n := GL_n(\Bbbk)$. We will also describe explicitly the bialgebra structure of its Lie algebra $\mathfrak{gl}_n := \mathfrak{gl}_n(\Bbbk)$ in a way that fits our purposes, that is to obtain a quantum duality principle for the Grassmann varieties for GL_n (see §4).

Let $\Bbbk_q = \Bbbk\left[q, q^{-1}\right]$ (where q is an indeterminate), the ring of Laurent polynomials over q, and let $\Bbbk(q)$ be the field of rational functions in q.

Definition 2.1. The *quantum matrix algebra* is defined as

$$\mathcal{O}_q(M_{m\times n}) \;=\; \Bbbk_q\big\langle \{x_{ij}\}_{1\le i\le m}^{1\le j\le n} \big\rangle \Big/ I_M$$

where the x_{ij}'s are non commutative indeterminates, and I_M is the two-sided ideal generated by the *Manin relations*

$$\begin{aligned} x_{ij}\,x_{ik} &= q\,x_{ik}\,x_{ij}\,, \qquad x_{ji}\,x_{ki} = q\,x_{ki}\,x_{ji} \qquad \forall\; j<k \\ x_{ij}\,x_{kl} &= x_{kl}\,x_{ij} \qquad\qquad \forall\; i<k\,,\, j>l \;\text{ or }\; i>k\,,\, j<l \\ x_{ij}\,x_{kl} - x_{kl}\,x_{ij} &= \left(q-q^{-1}\right) x_{kj}\,x_{il} \qquad \forall\; i<k\,,\, j<l \end{aligned}$$

Warning: sometimes these relations appear with q exchanged with q^{-1}.

For simplicity we will denote $\mathcal{O}_q(M_{n\times n})$ with $\mathcal{O}_q(M_n)$.

There is a coalgebra structure on $\mathcal{O}_q(M_n)$, given by

$$\Delta(x_{ij}) \;=\; \sum_{k=1}^{n} x_{ik}\otimes x_{kj} \quad , \qquad \epsilon(x_{ij}) \;=\; \delta_{ij} \qquad\qquad (1\le i\,,j\le n)$$

The *quantum general linear group* and the *quantum special linear group* are defined in the following way:

$$\mathcal{O}_q(GL_n) := \mathcal{O}_q(M_n)[T]\Big/\big(TD_q-1\,,1-TD_q\big)$$
$$\mathcal{O}_q(SL_n) := \mathcal{O}_q(M_n)\Big/\big(D_q-1\big)$$

where $D_q := \sum_{\sigma\in S_n}(-q)^{\ell(\sigma)}\,x_{1\,\sigma(1)}\cdots x_{n\,\sigma(n)}$ is a central element, called the *quantum determinant.*

Note: We use the same letter to denote the generators x_{ij} of $\mathcal{O}_q(M_{m\times n})$, of $\mathcal{O}_q(GL_n)$ and of $\mathcal{O}_q(SL_n)$: the context will make clear where they sit.

The algebra $\mathcal{O}_q(GL_n)$ is a quantization of the algebra $\mathcal{O}(GL_n)$ of regular functions on the affine algebraic group GL_n, in the following sense: $\mathcal{O}_q(GL_n)/(q-1)\,\mathcal{O}_q(GL_n)$ is isomorphic to $\mathcal{O}(GL_n)$ as a Hopf algebra (over the field $\Bbbk$). Similarly, $\mathcal{O}_q(SL_n)$ is a quantization of the algebra $\mathcal{O}(SL_n)$ of regular functions on SL_n. Both $\mathcal{O}_q(GL_n)$ and $\mathcal{O}_q(SL_n)$ are Hopf algebras, that is, they also have the antipode. For more details on these constructions see for example [**1**], pg. 215.

By general theory, $\mathcal{O}(GL_n)$ inherits from $\mathcal{O}_q(GL_n)$ a Poisson bracket, which makes it into a Poisson Hopf algebra, so that GL_n becomes a Poisson group. We want to describe now its Poisson bracket. Recall that

$$\mathcal{O}(GL_n) \;=\; \Bbbk\big[\{\,\bar{x}_{ij}\,\}_{i,j=1,\dots,n}\big][t]\Big/\big(t\,d-1\big)$$

where $d := det\big(\bar{x}_{i,j}\big)_{i,j=1,\dots,n}$ is the usual determinant. Setting $\bar{x} = \pi(x)$ for $\pi : \mathcal{O}_q(GL_n) \longrightarrow \mathcal{O}(GL_n)$, the Poisson structure is given (as usual) by

$$\big\{\bar{a}\,,\bar{b}\big\} \;:=\; (q-1)^{-1}\,(a\,b-b\,a)\Big|_{q=1} \qquad\qquad \forall\;\bar{a}\,,\bar{b}\in\mathcal{O}(GL_n)\;.$$

In terms of generators, we have

$$\big\{\bar{x}_{ij}\,,\bar{x}_{ik}\big\} \;=\; \bar{x}_{ij}\,\bar{x}_{ik} \quad \forall\; j<k\,, \qquad \big\{\bar{x}_{ij}\,,\bar{x}_{\ell k}\big\} \;=\; 0 \qquad \forall\; i<\ell\,,k<j$$

$$\big\{\bar{x}_{ij}\,,\bar{x}_{\ell j}\big\} \;=\; \bar{x}_{ij}\,\bar{x}_{\ell j} \quad \forall\; i<\ell\,, \qquad \big\{\bar{x}_{ij}\,,\bar{x}_{\ell k}\big\} \;=\; 2\,\bar{x}_{ij}\,\bar{x}_{\ell k} \quad \forall\; i<\ell\,,j<k$$

$$\big\{d^{-1},\bar{x}_{ij}\big\} \;=0\,, \qquad \big\{d\,,\bar{x}_{ij}\big\} \;=0 \qquad \forall\; i,j=1,\dots,n\,.$$

As GL_n is a Poisson Lie group, its Lie algebra $\mathfrak{gl}_n$ has a Lie bialgebra structure (see [**1**], pg. 24). To describe it, let us denote with E_{ij} the elementary matrices, which form a basis of $\mathfrak{gl}_n$. Define ($\forall\, i=1,\dots,n-1\,,\; j=1,\dots,n$)

$$e_i := E_{i,i+1}\;, \quad g_j := E_{j,j}\;, \quad f_i := E_{i+1,i}\;, \quad h_i := g_i - g_{i+1}$$

Then $\big\{\,e_i\,,\,f_i\,,\,g_j \;\big|\; i=1,\dots,n-1,\, j=1,\dots,n\,\big\}$ is a set of Lie algebra generators of $\mathfrak{gl}_n$, and a Lie cobracket is defined on $\mathfrak{gl}_n$ by

$$\delta(e_i) \;=\; h_i\wedge e_i \;\;, \quad \delta(g_j) \;=\; 0 \;\;, \quad \delta(f_i) \;=\; h_i\wedge f_i \qquad\qquad \forall\; i,j.$$

This cobracket makes $\mathfrak{gl}_n$ itself into a *Lie bialgebra*: this is the so-called *standard* Lie bialgebra structure on $\mathfrak{gl}_n$. It follows immediately that $U(\mathfrak{gl}_n)$ is a co-Poisson Hopf algebra, whose co-Poisson bracket is the (unique) extension of the Lie cobracket of $\mathfrak{gl}_n$ while the Hopf structure is the standard one.

Similar constructions hold for the group SL_n. One simply drops the generator d^{-1}, imposes the relation $d=1$, in the description of $\mathcal{O}(SL_n)$, and replaces the g_s's with the h_i's ($i=1,\dots,n$) when describing $\mathfrak{sl}_n$.

Since $\mathfrak{gl}_n$ is a Lie bialgebra, its dual space $\mathfrak{gl}_n^*$ admits a Lie bialgebra structure, dual to the one of $\mathfrak{gl}_n$. Let $\big\{\,\mathrm{E}_{ij} := E_{ij}^* \;\big|\; i,j=1,\dots,n\,\big\}$ be the basis of $\mathfrak{gl}_n^*$ dual to the basis of elementary matrices for $\mathfrak{gl}_n$. As a Lie algebra, $\mathfrak{gl}_n^*$ can be realized as the subset of $\mathfrak{gl}_n \oplus \mathfrak{gl}_n$ of all pairs

$$\left(\begin{pmatrix} -m_{11} & 0 & \cdots & 0 \\ m_{21} & -m_{22} & \cdots & 0 \\ \vdots & \vdots & \vdots & \vdots \\ m_{n-1,1} & m_{n-1,2} & \cdots & 0 \\ m_{n,1} & m_{n,2} & \cdots & -m_{n,n} \end{pmatrix} , \begin{pmatrix} m_{11} & m_{12} & \cdots & m_{1,n-1} & m_{1,n} \\ 0 & m_{22} & \cdots & m_{2,n-1} & m_{2,n} \\ \vdots & \vdots & \vdots & \vdots & \vdots \\ 0 & 0 & \cdots & m_{n-1,n-1} & m_{n-1,n} \\ 0 & 0 & \cdots & 0 & m_{n,n} \end{pmatrix} \right)$$

with its natural structure of Lie subalgebra of $\mathfrak{gl}_n \oplus \mathfrak{gl}_n$. In fact, the elements E_{ij} correspond to elements in $\mathfrak{gl}_n \oplus \mathfrak{gl}_n$ in the following way:

$$\mathrm{E}_{ij} \cong (E_{ij}, 0) \ \forall i>j , \quad \mathrm{E}_{ij} \cong (-E_{ij}, +E_{ij}) \ \forall i=j , \quad \mathrm{E}_{ij} \cong (0, E_{ij}) \ \forall i<j$$

Then the Lie bracket of $\mathfrak{gl}_n^*$ is given by

$$[\mathrm{E}_{i,j}, \mathrm{E}_{h,k}] = \delta_{j,h} \mathrm{E}_{i,k} - \delta_{k,i} \mathrm{E}_{h,j} , \quad \forall\ i \leq j, h \leq k \ \text{ and } \forall\ i>j, h>k$$

$$[\mathrm{E}_{i,j}, \mathrm{E}_{h,k}] = \delta_{k,i} \mathrm{E}_{h,j} - \delta_{j,h} \mathrm{E}_{i,k} , \quad \forall\ i=j, h>k \ \text{ and } \forall\ i>j, h=k$$

$$[\mathrm{E}_{i,j}, \mathrm{E}_{h,k}] = 0 , \quad \forall\ i<j, h>k \ \text{ and } \forall\ i>j, h<k$$

Note that the elements $(1 \leq i \leq n-1,\ 1 \leq j \leq n)$

$$\mathrm{e}_i = e_i^* = \mathrm{E}_{i,i+1} , \qquad \mathrm{f}_i = f_i^* = \mathrm{E}_{i+1,i} , \qquad \mathrm{g}_j = g_j^* = \mathrm{E}_{jj}$$

are Lie algebra generators of $\mathfrak{gl}_n^*$. In terms of them, the Lie bracket reads

$$[\mathrm{e}_i, \mathrm{f}_j] = 0 , \qquad [\mathrm{g}_i, \mathrm{e}_j] = \delta_{ij} \mathrm{e}_i , \qquad [\mathrm{g}_i, \mathrm{f}_j] = \delta_{ij} \mathrm{f}_j \qquad \forall\ i, j .$$

On the other hand, the Lie cobracket structure of $\mathfrak{gl}_n^*$ is given by

$$\delta(\mathrm{E}_{i,j}) = \sum_{k=1}^{n} \mathrm{E}_{i,k} \wedge \mathrm{E}_{k,j} \qquad \forall\ i, j = 1, \dots, n$$

where $x \wedge y := x \otimes y - y \otimes x$.

Finally, all these formulæ also provide a presentation of $U(\mathfrak{gl}_n^*)$ as a co-Poisson Hopf algebra.

A similar description holds for $\mathfrak{sl}_n^* = \mathfrak{gl}_n^* \big/ Z(\mathfrak{gl}_n^*)$, where $Z(\mathfrak{gl}_n^*)$ is the centre of $\mathfrak{gl}_n^*$, generated by $\mathfrak{l}_n := \mathrm{g}_1 + \cdots + \mathrm{g}_n$. The construction is immediate by looking at the embedding $\mathfrak{sl}_n \hookrightarrow \mathfrak{gl}_n$.

3. The quantum Grassmannian and its big cell

In this section we want to briefly recall the construction of a quantum deformation of the Grassmannian of r–spaces inside an n–dimensional vector space and its big cell, as they appear in [**6, 7**]. The quantum Grassmannian ring will be obtained as a quantum homogeneous space, namely its deformation will come together with a deformation of the natural coaction of the function algebra of the general linear group on it. The deformation will also depend on a specific embedding (the Plücker one) of the Grassmann variety into a projective space. This deformation is very natural, in fact it embeds into the deformation of its big cell ring. Let's see explicitly these constructions.

Let $G := GL_n$, and let P and P_1 be the standard parabolic subgroups

$$P := \left\{ \begin{pmatrix} A & B \\ 0 & C \end{pmatrix} \right\} \subset GL_n \quad , \qquad P_1 := P \bigcap SL_n$$

where A is a square matrix of size r, with $0 < r < n$.

DEFINITION 3.1. The *quantum Grassmannian coordinate ring* $\mathcal{O}_q(G/P)$ with respect to the Plücker embedding is the subalgebra of $\mathcal{O}_q(GL_n)$ generated by the quantum minors (called *quantum Plücker coordinates*)

$$D^I = D^{i_1 \dots i_r} := \sum_{\sigma \in \mathcal{S}_r} (-q)^{\ell(\sigma)} x_{i_1 \sigma(1)} x_{i_2 \sigma(2)} \cdots x_{i_r \sigma(r)}$$

for every ordered r–tuple of indices $I = \{i_1 < \cdots < i_r\}$.

Remark: Equivalently, $\mathcal{O}_q(G/P)$ may be defined in the same way but with $\mathcal{O}_q(SL_n)$ instead of $\mathcal{O}_q(GL_n)$.

The algebra $\mathcal{O}_q(G/P)$ is a quantization of the Grassmannian G/P in the usual sense: the $\Bbbk$–algebra $\mathcal{O}_q(G/P) \big/ (q-1)\, \mathcal{O}_q(G/P)$ is isomorphic to $\mathcal{O}(G/P)$, the algebra of homogeneous coordinates of G/P with respect to the Plücker embedding. In addition, $\mathcal{O}_q(G/P)$ has an important property w.r.t. $\mathcal{O}_q(G)$, given by the following result:

PROPOSITION 3.2.

$$\mathcal{O}_q(G/P) \bigcap (q-1)\, \mathcal{O}_q(G) = (q-1)\, \mathcal{O}_q(G/P)$$

Proof. By Theorem 3.5 in [**13**], we have that certain products of minors $\{p_i\}_{i \in I}$ form a basis of $\mathcal{O}_q(G/P)$ over $\Bbbk_q$. Thus, a generic element in $\mathcal{O}_q(G/P) \bigcap (q-1)\, \mathcal{O}_q(G)$ can be written as

$$\textstyle\sum_{i \in I} \alpha_i \, p_i = (q-1)\, \phi \tag{3.1}$$

for some $\phi \in \mathcal{O}_q(G)$. Moreover, the specialization map

$$\pi_G : \mathcal{O}_q(G) \relbar\joinrel\twoheadrightarrow \mathcal{O}_q(G) \big/ (q-1)\, \mathcal{O}_q(G) = \mathcal{O}(G)$$

maps $\{p_i\}_{i \in I}$ onto a basis $\{\pi_G(p_i)\}_{i \in I}$ of $\mathcal{O}(G/P)$, the latter being a subalgebra of $\mathcal{O}(G)$. Therefore, applying π_G to (3.1) we get $\sum_{i \in I} \overline{\alpha_i}\, \pi_G(p_i) = 0$, where $\overline{\alpha_i} := \alpha_i \mod (q-1)\, \Bbbk_q$, for all $i \in I$. This forces $\alpha_i \in (q-1)\, \Bbbk_q$ for all i, by the linear independence of the $\pi_G(p_i)$'s, whence the claim. □

An immediate consequence of Proposition 3.2 is that the canonical map

$$\mathcal{O}_q(G/P) \big/ (q-1)\, \mathcal{O}_q(G/P) \longrightarrow \mathcal{O}_q(G) \big/ (q-1)\, \mathcal{O}_q(G)$$

is *injective*. Therefore, the specialization map

$$\pi_{G/P} : \mathcal{O}_q(G/P) \relbar\joinrel\twoheadrightarrow \mathcal{O}_q(G/P) \big/ (q-1)\, \mathcal{O}_q(G/P)$$

coincides with the restriction to $\mathcal{O}_q(G/P)$ of the specialization map

$$\pi_G : \mathcal{O}_q(G) \relbar\joinrel\twoheadrightarrow \mathcal{O}_q(G) \big/ (q-1)\, \mathcal{O}_q(G) \quad .$$

Moreover — from a geometrical point of view — the key consequence of this property is that P *is a coisotropic subgroup* of the Poisson group G. This implies the existence of a well defined Poisson structure on the algebra $\mathcal{O}(G/P)$, inherited from the one in $\mathcal{O}(G)$.

OBSERVATION 3.3. The quantum deformation $\mathcal{O}_q(G/P)$ comes naturally equipped with a coaction of $\mathcal{O}_q(GL_n)$ — or, similarly, of $\mathcal{O}_q(SL_n)$ — on it, obtained by restricting the comultiplication Δ. This reads

$$\begin{array}{rccc} \Delta\big|_{\mathcal{O}_q(G/P)} : & \mathcal{O}_q(G/P) & \longrightarrow & \mathcal{O}_q(G) \otimes \mathcal{O}_q(G/P) \\ & D^I & \mapsto & \sum_K D^I_K \otimes D^K \end{array}$$

where, for any $I = (i_1 \dots i_r)$, $K = (k_1 \dots k_r)$, with $1 \leq i_1 < \cdots < i_r \leq n$, $1 \leq k_1 < \cdots < k_r \leq n$, we denote by D^I_K the *quantum minor*

$$D^I_K \;\equiv\; D^{i_1 \dots i_r}_{k_1 \dots k_r} \;:=\; \sum_{\sigma \in \mathcal{S}_r} (-q)^{\ell(\sigma)}\, x_{i_1\, k_{\sigma(1)}}\, x_{i_2\, k_{\sigma(2)}} \cdots x_{i_r\, k_{\sigma(r)}} \quad .$$

This provides a quantization of the natural coaction of $\mathcal{O}(G)$ onto $\mathcal{O}(G/P)$.

The ring $\mathcal{O}_q(G/P)$ is fully described in [**6**] in terms of generators and relations. We refer the reader to this work for further details.

We now turn to the construction of the quantum big cell ring.

DEFINITION 3.4. Let $I_0 = (1 \dots r)$, $D_0 := D^{I_0}$. Define

$$\mathcal{O}_q(G)\big[D_0^{-1}\big] \;:=\; \mathcal{O}_q(G)[T] \Big/ \big(T\, D_0 - 1 \,, D_0\, T - 1\big)$$

Moreover, we define the *big cell ring* $\mathcal{O}_q^{\,loc}(G/P)$ to be the $\Bbbk_q$–subalgebra of $\mathcal{O}_q(G)\big[D_0^{-1}\big]$ generated by the elements

$$t_{ij} \;:=\; (-q)^{r-j}\, D^{1 \dots \widehat{j} \dots r\, i}\, D_0^{-1} \qquad\qquad \forall\; i,j \;:\; 1 \leq j \leq r < i \leq n$$

(see [**7**] for more details).

As in the commutative setting, we have the following result:

PROPOSITION 3.5. $\mathcal{O}_q^{\,loc}(G/P) \cong \mathcal{O}_q(G/P)\big[D_0^{-1}\big]_{proj}$, *where the right-hand side denotes the degree-zero component of the quotient ring* $\mathcal{O}_q(G/P)[T] \Big/ \big(T D_0 - 1 \,, D_0\, T - 1\big)$.

Proof. In the classical setting, the analogous result is proved by this argument: one uses the so-called "straightening relations" to get rid of the extra minors (see, for example, [**4**], §2). Here the argument works essentially the same, using the *quantum straightening* (or *Plücker*) *relations* (see [**6**], §4, [**13**], formula (3.2)(c) and Note I, Note II). □

REMARK 3.6. As before, we have that

$$\mathcal{O}_q^{\,loc}(G/P) \;\bigcap\; (q-1)\, \mathcal{O}_q^{\,loc}(G) \;=\; (q-1)\, \mathcal{O}_q^{\,loc}(G/P)$$

This can be easily deduced from Proposition 3.2, taking into account Proposition 3.5. As a consequence, the map

$$\mathcal{O}_q^{\,loc}(G/P) \Big/ (q-1)\, \mathcal{O}_q^{\,loc}(G/P) \;\longrightarrow\; \mathcal{O}_q^{\,loc}(G) \Big/ (q-1)\, \mathcal{O}_q^{\,loc}(G)$$

is *injective*, so that the specialization map

$$\pi^{loc}_{G/P} \colon \mathcal{O}_q^{loc}(G/P) \longrightarrow\!\!\!\!\to \mathcal{O}_q^{loc}(G/P) \big/ (q-1)\, \mathcal{O}_q^{loc}(G/P)$$

coincides with the restriction of the specialization map

$$\pi^{loc}_{G} : \mathcal{O}_q^{loc}(G) \longrightarrow\!\!\!\!\to \mathcal{O}_q^{loc}(G) \big/ (q-1)\, \mathcal{O}_q^{loc}(G) \quad .$$

The following proposition gives a description of the algebra $\mathcal{O}_q^{loc}(G/P)$:

PROPOSITION 3.7. *The big cell ring is isomorphic to a matrix algebra, via the map*

$$\begin{array}{ccc} \mathcal{O}_q^{loc}(G/P) & \longrightarrow & \mathcal{O}_q\big(M_{(n-r)\times r}\big) \\ t_{ij} & \mapsto & x_{ij} \end{array} \qquad \forall\ 1 \leq j \leq r < i \leq n$$

In particular, the generators t_{ij}'s satisfy the Manin relations.

Proof. See [**7**], Proposition 1.9. □

4. The Quantum Duality Principle for quantum Grassmannians

The quantum duality principle (QDP), originally due to Drinfeld [**5**] and later formalized in [**9**] and extended in [**10, 11**] by Gavarini, is a functorial recipe to obtain a quantum group starting from a given one. The main ingredients are the "Drinfeld functors", which are equivalences between the category of QFA's and the category of QUEA's. Ciccoli and Gavarini extended this principle to the setting of homogeneous spaces. More precisely, in [**3**] they developed the QDP for homogeneous spaces in the *local setting*, i.e. for quantum groups of formal type (where topological Hopf algebras are taken into account). If one tries to find a global version of the QDP for non quasi-affine homogeneous spaces, then problems arise from the very beginning, as explained in §1. The case of *projective* homogeneous spaces has been solved in [**2**], where the original version of the Drinfeld-like functor for which the (global) QDP recipe should fail is suitably modified.

In this section, we apply the general recipe for projective homogeneous spaces to the Grassmannian G/P. The result is a quantization of the homogeneous space *dual* (in the sense of Poisson duality, see [**3**]) to G/P, just as the QDP recipe predicts in the setting of [**3**].

We begin recalling the Drinfeld functor ${}^\vee : QFA \longrightarrow QUEA$.

DEFINITION 4.1. Let G be an affine algebraic group over $\Bbbk$, and $\mathcal{O}_q(G)$ a quantization of its function algebra. Let J be the augmentation ideal of $\mathcal{O}_q(G)$, i.e. the kernel of the counit $\epsilon : \mathcal{O}_q(G) \longrightarrow \Bbbk$. We define

$$\mathcal{O}_q(G)^\vee := \big\langle (q-1)^{-1} J \big\rangle = \sum_{n=0}^{\infty} (q-1)^{-n} J^n \quad \big(\subset \mathcal{O}_q(G) \otimes_{\Bbbk_q} \Bbbk(q) \big) \, .$$

It turns out that $\mathcal{O}_q(G)^\vee$ is a quantization of $U(\mathfrak{g}^*)$, where $\mathfrak{g}^*$ is the dual Lie bialgebra to the Lie bialgebra $\mathfrak{g} = \mathit{Lie}(G)$. So $\mathcal{O}_q(G)^\vee$ is a QUEA, and an infinitesimal quantization for any Poisson group G^* dual to G, i.e. such that $\mathit{Lie}\big(G^*\big) \cong \mathfrak{g}^*$ as Lie bialgebras. Moreover, the association $\mathcal{O}_q(G) \mapsto \mathcal{O}_q(G)^\vee$ yields a functor from QFA's to QUEA's (see [**10, 11**] for more details).

REMARK 4.2. Let $G = GL_n$. Then $\mathcal{O}_q(G)^\vee$ is generated, as a unital subalgebra of $\mathcal{O}_q(G) \otimes_{\Bbbk_q} \Bbbk(q)$, by the elements

$$\mathcal{D}_- := (q-1)^{-1}\left(D_q^{-1} - 1\right), \qquad \chi_{ij} := (q-1)^{-1}\left(x_{ij} - \delta_{ij}\right) \quad \forall\, i,j = 1,\dots,n$$

where the x_{ij}'s are the generators of $\mathcal{O}_q(G)$. As $x_{ij} = \delta_{ij} + (q-1)\,\chi_{ij} \in \mathcal{O}_q(G)^\vee$, we have an obvious embedding of $\mathcal{O}_q(G)$ into $\mathcal{O}_q(G)^\vee$.

In the same spirit — mimicking the construction in [**3**] — we now want to define $\mathcal{O}_q(G/P)^\vee$ when G/P is the Grassmannian.

Let $G = GL_n$, and let P be the maximal parabolic subgroup of §3.

DEFINITION 4.3. Let ϵ' be the natural extension to $\mathcal{O}_q^{\,loc}(G/P)$ of the restriction to $\mathcal{O}_q(G/P)$ of the counit of $\mathcal{O}_q(G)$, and let $J_{G/P}^{\,loc} := Ker(\epsilon')$. We define (as a subset of $\mathcal{O}_q^{\,loc}(G/P) \otimes_{\Bbbk_q} \Bbbk(q)$)

$$\mathcal{O}_q(G/P)^\vee := \left\langle (q-1)^{-1} J_{G/P}^{\,loc} \right\rangle = \sum_{n=0}^{\infty} (q-1)^{-n} \left(J_{G/P}^{\,loc}\right)^n .$$

It is worth pointing out that $\mathcal{O}_q(G/P)^\vee$ is *not* a "quantum homogeneous space" for $\mathcal{O}_q(G)^\vee$ in any natural way, i.e. it does not admit a coaction of $\mathcal{O}_q(G)^\vee$. This is a consequence of the fact that there is no natural coaction of $\mathcal{O}_q(G)$ on $\mathcal{O}_q^{\,loc}(G/P)$. Now we examine this more closely.

Since $\mathcal{O}_q(G/P)^\vee$ is not contained in $\mathcal{O}_q(G)^\vee$, we cannot have a $\mathcal{O}_q(G)^\vee$ coaction induced by the coproduct. This would be the case if $\mathcal{O}_q(G/P)^\vee$ were a (one-sided) *coideal* of $\mathcal{O}_q(G)^\vee$; but this is not true because $\mathcal{O}_q^{\,loc}(G/P)$ is not a (right) coideal of $\mathcal{O}_q(G)$. This reflects the geometrical fact that the big cell of G/P is not a G–space itself. Nevertheless, we shall find a way around this problem simply by *enlarging* $\mathcal{O}_q(G/P)^\vee$ and $\mathcal{O}_q(G)^\vee$, i.e. by taking their $(q-1)$–adic completion (which will not affect their behavior at $q = 1$).

To begin, we provide a concrete description of $\mathcal{O}_q(G/P)^\vee$:

PROPOSITION 4.4.

$$\mathcal{O}_q(G/P)^\vee = \Bbbk_q\left\langle \{\mu_{ij}\}_{i=r+1,\dots,n}^{j=1,\dots,r} \right\rangle \Big/ I_M$$

where $\mu_{ij} := (q-1)^{-1}\, t_{ij}$ *(for all i and j),* I_M *is the ideal of the Manin relations among the* μ_{ij}*'s, and* $t_{ij} = (-q)^{r-j}\, D^{1\dots\widehat{j}\dots r\, i}\, D_0^{-1}$ *(for all i and j).*

Proof. Trivial from definitions and Proposition 3.7. □

We now explain the relation between $\mathcal{O}_q(G/P)^\vee$ and $\mathcal{O}_q(G)^\vee$. The starting point is the following special property:

PROPOSITION 4.5.

$$\mathcal{O}_q(G/P)^\vee \bigcap\, (q-1)\,\mathcal{O}_q(G)^\vee\left[D_0^{-1}\right] = (q-1)\,\mathcal{O}_q(G/P)^\vee$$

Proof. It is the same as for Proposition 3.2. □

REMARK 4.6. As a direct consequence of Proposition 4.5, the canonical map

$$\mathcal{O}_q(G/P)^\vee \Big/ (q-1)\,\mathcal{O}_q(G/P)^\vee \longrightarrow \mathcal{O}_q(G)^\vee\big[D_0^{-1}\big] \Big/ (q-1)\,\mathcal{O}_q(G)^\vee\big[D_0^{-1}\big]$$

is in fact *injective*: therefore, the specialization map

$$\pi_{G/P}^\vee : \mathcal{O}_q(G/P)^\vee \longrightarrow\!\!\!\!\rightarrow \mathcal{O}_q(G/P)^\vee \Big/ (q-1)\,\mathcal{O}_q(G/P)^\vee$$

coincides with the restriction to $\mathcal{O}_q(G/P)^\vee$ of the specialization map

$$\pi_G^\vee : \mathcal{O}_q(G)^\vee\big[D_0^{-1}\big] \longrightarrow\!\!\!\!\rightarrow \mathcal{O}_q(G)^\vee\big[D_0^{-1}\big] \Big/ (q-1)\,\mathcal{O}_q(G)^\vee\big[D_0^{-1}\big] \quad .$$

From now on, let $\widehat{A}$ denote the $(q-1)$–adic completion of any $\Bbbk_q$–algebra A. Note that $\widehat{A}$ and A have the same specialization at $q=1$, i.e. $A/(q-1)\,A$ and $\widehat{A}/(q-1)\,\widehat{A}$ are canonically isomorphic. When $A = \mathcal{O}_q(G)$, note also that $\widehat{\mathcal{O}_q(G)}$ is naturally a complete topological Hopf $\Bbbk_q$–algebra.

The next result shows why it is relevant to introduce such completions.

LEMMA 4.7. *$\mathcal{O}_q(G)^\vee\big[D_0^{-1}\big]$ naturally embeds into $\widehat{\mathcal{O}_q(G)}^\vee$.*

Proof. By remark 4.2 we have that $\mathcal{O}_q(G)^\vee$ is generated by the elements (for all $i,j=1,\dots,n$)

$$\mathcal{D}_- := (q-1)^{-1}\big(D_q^{-1}-1\big)\,, \qquad \chi_{ij} := (q-1)^{-1}\big(x_{ij}-\delta_{ij}\big)$$

inside $\mathcal{O}_q(G)\otimes_{\Bbbk_q}\Bbbk(q)$. On the other hand, observe that

$$x_{ij} = (q-1)\,\chi_{i,j} \in (q-1)\,\mathcal{O}_q(G)^\vee \qquad \forall\ i \neq j$$

and

$$x_{\ell\ell} = 1+(q-1)\,\chi_{\ell\ell} \in \big(1+(q-1)\,\mathcal{O}_q(G)^\vee\big) \qquad \forall\ \ell\,.$$

Then, if we expand explicitly the q–determinant $D_0 := D^{I_0}$, we immediately see that $D_0 \in \big(1+(q-1)\,\mathcal{O}_q(G)^\vee\big)$ as well. Thus D_0 is invertible in $\widehat{\mathcal{O}_q(G)}^\vee$, and so the natural immersion $\mathcal{O}_q(G)^\vee \hookrightarrow \widehat{\mathcal{O}_q(G)}^\vee$ canonically extends to an immersion $\mathcal{O}_q(G)^\vee\big[D_0^{-1}\big] \hookrightarrow \widehat{\mathcal{O}_q(G)}^\vee$. □

COROLLARY 4.8.

(a) The specializations at $q=1$ of $\mathcal{O}_q(G)^\vee$, $\mathcal{O}_q(G)^\vee\big[D_0^{-1}\big]$ and $\widehat{\mathcal{O}_q(G)}^\vee$ are canonically isomorphic. More precisely, the chain

$$\mathcal{O}_q(G)^\vee \hookrightarrow \mathcal{O}_q(G)^\vee\big[D_0^{-1}\big] \hookrightarrow \widehat{\mathcal{O}_q(G)}^\vee$$

of canonical embeddings induces at $q=1$ a chain of isomorphisms.

(b) $\mathcal{O}_q(G/P)^\vee$ embeds into $\widehat{\mathcal{O}_q(G)}^\vee$ via the chain of embeddings

$$\mathcal{O}_q(G/P)^\vee \hookrightarrow \mathcal{O}_q(G)^\vee\big[D_0^{-1}\big] \hookrightarrow \widehat{\mathcal{O}_q(G)}^\vee$$

(c) $$\mathcal{O}_q(G/P)^\vee \bigcap\, (q-1)\,\widehat{\mathcal{O}_q(G)}^\vee = (q-1)\,\mathcal{O}_q(G/P)^\vee \quad .$$

Proof. Part *(a)* and *(b)* are trivial, and *(c)* follows from them. □

Notice that part *(c)* of Corollary 4.8 also implies that

$$\mathcal{O}_q(G/P)^\vee\Big|_{q=1} := \mathcal{O}_q(G/P)^\vee \Big/ (q-1)\,\mathcal{O}_q(G/P)^\vee$$

is a subalgebra of

$$\widehat{\mathcal{O}_q(G)^\vee}\Big|_{q=1} = \mathcal{O}_q(G)^\vee\Big|_{q=1} := \mathcal{O}_q(G)^\vee \Big/ (q-1)\,\mathcal{O}_q(G)^\vee \cong U(\mathfrak{g}^*)$$

just because the specialization map

$$\pi^\vee_{G/P} : \mathcal{O}_q(G/P)^\vee \longrightarrow\!\!\!\!\!\rightarrow \mathcal{O}_q(G/P)^\vee \Big/ (q-1)\,\mathcal{O}_q(G/P)^\vee$$

coincides with the restriction to $\mathcal{O}_q(G/P)^\vee$ of the specialization map

$$\widehat{\pi^\vee_G} : \widehat{\mathcal{O}_q(G)^\vee} \longrightarrow\!\!\!\!\!\rightarrow \widehat{\mathcal{O}_q(G)^\vee} \Big/ (q-1)\,\widehat{\mathcal{O}_q(G)^\vee} \quad .$$

Now we want to see what is $\mathcal{O}_q(G/P)^\vee\Big|_{q=1}$ inside $U(\mathfrak{gl}_n{}^*)$. In other words, we want to understand what is the space that $\mathcal{O}_q(G/P)^\vee$ is quantizing.

Proposition 4.9.

$$\mathcal{O}_q(G/P)^\vee\Big|_{q=1} = U(\mathfrak{p}^\perp)$$

as a subalgebra of $\mathcal{O}_q(G)^\vee\Big|_{q=1} = U(\mathfrak{gl}_n{}^*)$, *where* $\mathfrak{p}^\perp$ *is the orthogonal subspace to* $\mathfrak{p} := \mathit{Lie}(P)$ *inside* $\mathfrak{gl}_n{}^*$.

Proof. Thanks to the previous discussion, it is enough to show that

$$\pi^\vee_G\Big(\mathcal{O}_q(G/P)^\vee\Big) = U(\mathfrak{p}^\perp) \subseteq U(\mathfrak{gl}_n{}^*) = \mathcal{O}_q(G)^\vee\Big|_{q=1} \quad .$$

To do this, we describe the isomorphism $\mathcal{O}_q(G)^\vee\Big|_{q=1} \cong U(\mathfrak{gl}_n{}^*)$ (cf. [**8**]). First, recall that $\mathcal{O}_q(G)^\vee$ is generated by the elements (see Remark 4.2)

$$\mathcal{D}_- := (q-1)^{-1}\left(D_q^{-1}-1\right), \qquad \chi_{ij} := (q-1)^{-1}\left(x_{ij}-\delta_{ij}\right)$$

(for all $i,j=1,\dots,n$) inside $\mathcal{O}_q(G) \otimes_{\Bbbk_q} \Bbbk(q)$. In terms of these generators, the isomorphism reads

$$\mathcal{O}_q(G)^\vee\Big|_{q=1} \longrightarrow U(\mathfrak{gl}_n{}^*)$$
$$\overline{\mathcal{D}_-} \mapsto -(\mathrm{E}_{1,1}+\cdots+\mathrm{E}_{n,n}), \qquad \overline{\chi_{i,j}} \mapsto \mathrm{E}_{i,j} \qquad \forall\ i,j\,.$$

where we used notation $\overline{X} := X \mod (q-1)\,\mathcal{O}_q(G)^\vee$. Indeed, from $\overline{\chi_{i,j}} \mapsto \mathrm{E}_{i,j}$ and $(q-1)^{-1}\left(D_q-1\right) \in \mathcal{O}_q(G)^\vee$, one gets $\overline{D_q} \mapsto 1$ and $\overline{(q-1)^{-1}\left(D_q-1\right)} \mapsto \mathrm{E}_{1,1}+\cdots+\mathrm{E}_{n,n}$. Moreover, the relation $D_q\,D_q^{-1}=1$ in $\mathcal{O}_q(G)$ implies $D_q\,\mathcal{D}_- = -(q-1)^{-1}\left(D_q-1\right)$ in $\mathcal{O}_q(G)^\vee$, so $\overline{\mathcal{D}_-} \mapsto -(\mathrm{E}_{1,1}+\cdots+\mathrm{E}_{n,n})$ as claimed (cf. [**8**], §3, or [**10**], §7). In other words, the specialization $\pi^\vee_G : \mathcal{O}_q(G)^\vee \longrightarrow\!\!\!\!\!\rightarrow U(\mathfrak{gl}_n{}^*)$ is given by

$$\pi^\vee_G(\mathcal{D}_-) = -(\mathrm{E}_{1,1}+\cdots+\mathrm{E}_{n,n}), \qquad \pi^\vee_G(\chi_{i,j}) = \mathrm{E}_{i,j} \qquad \forall\ i,j\,.$$

If we look at $\widehat{\mathcal{O}_q(G)^\vee}$, things are even simpler. Since

$$D_q \in \left(1+(q-1)\,\mathcal{O}_q(G)^\vee\right) \subset \left(1+(q-1)\,\widehat{\mathcal{O}_q(G)^\vee}\right),$$

then $D_q^{-1} \in \left(1+(q-1)\,\widehat{\mathcal{O}_q(G)^\vee}\right)$, and the generator $\mathcal{D}_-$ can be dropped. The specialization map $\widehat{\pi_{G/P}^\vee}$ of course is still described by formulæ as above.

Now let's compute $\pi_{G/P}^\vee\left(\mathcal{O}_q(G/P)^\vee\right) = \widehat{\pi_G^\vee}\left(\mathcal{O}_q(G/P)^\vee\right)$. Recall that $\mathcal{O}_q(G/P)^\vee$ is generated by the μ_{ij}'s, with

$$\mu_{ij} \,:=\, (q-1)^{-1}\,t_{ij} \,=\, (q-1)^{-1}\,(-q)^{r-j}\,D^{1\dots\widehat{j}\dots r\,i}\,D_0^{-1}$$

for $i=r+1,\dots,n$, and $j=1,\dots,r$; thus we must compute $\widehat{\pi_G^\vee}(\mu_{ij})$.

By definition, for every $i \neq j$ the element $x_{ij}=(q-1)\,\chi_{ij}$ is mapped to 0 by $\widehat{\pi_G^\vee}$. Instead, for each ℓ the element $x_{\ell\ell}=1+(q-1)\,\chi_{\ell\ell}$ is mapped to 1 (by $\widehat{\pi_G^\vee}$ again). But then, expanding the q–determinants one easily finds — much like in the proof of Lemma 4.7 — that

$$\widehat{\pi_G^\vee}\left((q-1)^{-1}D^{1\dots\widehat{j}\dots r\,i}\right) \,=\, \left((q-1)^{-1}\sum_{\sigma\in\mathcal{S}_r}(-q)^{\ell(\sigma)}\,x_{1\,\sigma(1)}\cdots x_{r\,\sigma(r)}\right) =$$
$$= \widehat{\pi_G^\vee}\left((q-1)^{-1}\sum_{\sigma\in\mathcal{S}_r}(-q)^{\ell(\sigma)}\prod\nolimits_{k=1}^{r}\left(\delta_{k\,\sigma(k)}+(q-1)\,\chi_{k\,\sigma(k)}\right)\right)$$

The only term in $(q-1)$ in the expansion of $D^{1\dots\widehat{j}\dots r\,i}$ comes from the product

$$(1+(q-1)\chi_{1\,1})\cdots(1+(q-1)\chi_{r\,r})(q-1)\chi_{i\,j} \equiv (q-1)\chi_{i\,j} \mod (q-1)^2\mathcal{O}(G/P)$$

Therefore, from the previous analysis we get

$$\widehat{\pi_G^\vee}\left((q-1)^{-1}\,D^{1\dots\widehat{j}\dots r\,i}\right) \,=\, \widehat{\pi_G^\vee}\,(\chi_{i,j}) \,=\, \mathrm{E}_{i,j}$$
$$\widehat{\pi_G^\vee}\,(D_0) \,=\, \widehat{\pi_G^\vee}(1) \,=\, 1\,, \qquad \widehat{\pi_G^\vee}\,(D_0^{-1}) \,=\, \widehat{\pi_G^\vee}(1) \,=\, 1$$

so in the end $\widehat{\pi_G^\vee}(\mu_{ij}) = (-1)^{r-j}\,\mathrm{E}_{i,j}$, for all $1 \leq j \leq r < i \leq n$.

The outcome is that $\pi_{G/P}^\vee\left(\mathcal{O}_q(G/P)^\vee\right) \,=\, U(\mathfrak{h})$, where

$$\mathfrak{h} \,:=\, \mathit{Span}\left(\left\{\,\mathrm{E}_{i,j}\,\middle|\,r+1 \leq i \leq n,\; 1 \leq j \leq r\,\right\}\right)\,.$$

On the other hand, from the very definitions and our description of $\mathfrak{gl}_n{}^*$ one easily finds that $\mathfrak{h}=\mathfrak{p}^\perp$, for $\mathfrak{p}:=\mathit{Lie}(P)$. The claim follows. □

Proposition 4.9 claims that $\mathcal{O}_q(G/P)^\vee$ is a quantization of $U(\mathfrak{p}^\perp)$, i.e. it is a unital $\Bbbk_q$–algebra whose semiclassical limit is $U(\mathfrak{p}^\perp)$. Now, the fact that $U(\mathfrak{p}^\perp)$ describes (infinitesimally) a homogeneous space for G^* is encoded in algebraic terms by the fact that it is a (left) coideal of $U(\mathfrak{g}^*)$; in other words, $U(\mathfrak{p}^\perp)$ is a (left) $U(\mathfrak{g}^*)$–comodule w.r.t. the restriction of the coproduct of $U(\mathfrak{g}^*)$. Thus, for $\mathcal{O}_q(G/P)^\vee$ to be a quantization of $U(\mathfrak{p}^\perp)$ *as a homogeneous space* we need also a quantization of this fact: namely, we would like $\mathcal{O}_q(G/P)^\vee$ to be a left coideal of $\mathcal{O}_q(G)^\vee$, our quantization of $U(\mathfrak{g}^*)$. But this makes no sense at all, as $\mathcal{O}_q(G/P)^\vee$ is not even a subset of $\mathcal{O}_q(G)^\vee$!

This problem leads us to enlarge a bit our quantizations $\mathcal{O}_q(G/P)^\vee$ and $\mathcal{O}_q(G)^\vee$: we take their $(q-1)$–adic completions, namely $\widehat{\mathcal{O}_q(G/P)^\vee}$ and $\widehat{\mathcal{O}_q(G)^\vee}$. While not affecting their behavior at $q=1$ (i.e., their semiclassical limits are the same), this operation solves the problem. Indeed, $\widehat{\mathcal{O}_q(G)^\vee}$ is big enough to contain $\mathcal{O}_q(G/P)^\vee$, by Corollary 4.8*(b)*. Then, as $\widehat{\mathcal{O}_q(G)^\vee}$ is a topological Hopf algebra, inside it we must look at the closure of $\mathcal{O}_q(G/P)^\vee$. Thanks to Corollary 4.8*(c)* (which means, roughly, that an Artin-Rees lemma holds), the latter is nothing but $\widehat{\mathcal{O}_q(G/P)^\vee}$. Finally, next result tells us that $\widehat{\mathcal{O}_q(G/P)^\vee}$ is a left coideal of $\widehat{\mathcal{O}_q(G)^\vee}$, as expected.

PROPOSITION 4.10. *$\widehat{\mathcal{O}_q(G/P)^\vee}$ is a left coideal of $\widehat{\mathcal{O}_q(G)^\vee}$.*

Proof. Recall that the coproduct $\widehat{\Delta}$ of $\widehat{\mathcal{O}_q(G)^\vee}$ takes values in the *topological* tensor product $\widehat{\mathcal{O}_q(G)^\vee}\,\widehat{\otimes}\,\widehat{\mathcal{O}_q(G)^\vee}$, which by definition is the $(q-1)$–adic completion of the *algebraic* tensor product $\widehat{\mathcal{O}_q(G)^\vee}\otimes\widehat{\mathcal{O}_q(G)^\vee}$. Our purpose then is to show that this coproduct $\widehat{\Delta}$ maps $\widehat{\mathcal{O}_q(G/P)^\vee}$ in the topological tensor product $\widehat{\mathcal{O}_q(G)^\vee}\,\widehat{\otimes}\,\widehat{\mathcal{O}_q(G/P)^\vee}$.

By construction, the coproduct of $\mathcal{O}_q(G)^\vee$, hence of $\widehat{\mathcal{O}_q(G)^\vee}$ too, is induced by that of $\mathcal{O}_q(G)$, say $\Delta : \mathcal{O}_q(G) \longrightarrow \mathcal{O}_q(G)\otimes\mathcal{O}_q(G)$. Now, the latter can be uniquely (canonically) extended to a coassociative algebra morphism

$$\widetilde{\Delta} : \mathcal{O}_q(G)\big[D_{I_0}^{-1}\big] \longrightarrow \mathcal{O}_q(G)\big[D_{I_0}^{-1}\big]\,\widetilde{\otimes}\,\mathcal{O}_q(G)\big[D_{I_0}^{-1}\big]$$

where $\widetilde{\otimes}$ is the $J_\otimes$–adic completion of the algebraic tensor product, with

$$J_\otimes := J\otimes\mathcal{O}_q(G) + \mathcal{O}_q(G)\otimes J\ , \qquad J := Ker\big(\epsilon_{\mathcal{O}_q(G)}\big)\ .$$

In fact, since $\Delta(D_0) = D_0\otimes D_0 + \sum_{K\neq I_0} D_K^{I_0}\otimes D^K$, one easily computes

$$\widetilde{\Delta}(D_0) = \Big(1+\sum_{K\neq I_0} D_K^{I_0}\,D_0^{-1}\otimes D^K D_0^{-1}\Big)(D_0\otimes D_0)$$

$$\widetilde{\Delta}(D_0^{-1}) = (D_0\otimes D_0)^{-1}\Big(1+\sum_{K\neq I_0} D_K^{I_0}\,D_0^{-1}\otimes D^K D_0^{-1}\Big)^{-1}$$

$$= \Big(D_0^{-1}\otimes D_0^{-1}\Big)\sum_{n\geq 0}(-1)^n\Big(\sum_{K\neq I_0} D_K^{I_0}\,D_0^{-1}\otimes D^K D_0^{-1}\Big)^n$$

Let's now look at the restriction $\widetilde{\Delta}_r$ of $\widetilde{\Delta}$ to $\mathcal{O}_q^{loc}(G/P)$. We have

$$\widetilde{\Delta}_r(t_{ij}) = \widetilde{\Delta}_r\big(D^{1\dots\widehat{j}\dots r\,i}\,D_0^{-1}\big) = \widetilde{\Delta}\big(D^{1\dots\widehat{j}\dots r\,i}\big)\cdot\widetilde{\Delta}(D_0)^{-1} =$$

$$= \Big(\sum_L D_L^{1\dots\widehat{j}\dots r\,i}\,D_0^{-1}\otimes D^L D_0^{-1}\Big)\cdot\sum_{n\geq 0}(-1)^n\Big(\sum_{K\neq I_0} D_K^{I_0}\,D_0^{-1}\otimes D^K D_0^{-1}\Big)^n$$

Now, by Proposition 3.5 we know that each product $D^L D_{I_0}^{-1}$ is a combination of the t_{ij}'s. Hence the formula above shows that $\widetilde{\Delta}_r$ maps $\mathcal{O}_q^{loc}(G/P)$ into $\mathcal{O}_q(G)\big[D_0^{-1}\big]\,\widetilde{\otimes}\,\mathcal{O}_q^{loc}(G/P)$.

By scalar extension, $\widetilde{\Delta}$ uniquely extends to a map defined on the $\Bbbk(q)$–vector space $\Bbbk(q) \otimes_{\Bbbk_q} \mathcal{O}_q(G)\big[D_0^{-1}\big]$, which we still call $\widetilde{\Delta}$. Its restriction to the similar scalar extension of $\mathcal{O}_q^{\,loc}(G/P)$ clearly coincides with the scalar extension of $\widetilde{\Delta}_r$, hence we call it $\widetilde{\Delta}_r$ again. Finally, the restriction of $\widetilde{\Delta}$ to $\mathcal{O}_q(G)^\vee\big[D_0^{-1}\big]$ and of $\widetilde{\Delta}_r$ to $\mathcal{O}_q\big(G/P\big)^\vee$ both coincide — by construction — with the proper restrictions of the coproduct of $\widehat{\mathcal{O}_q(G)^\vee}$ (cf. Corollary 4.8).

In the end, we are left to compute $\widetilde{\Delta}_r(\mu_{ij})$. The computation above gives

$$\widehat{\Delta}(\mu_{ij}) \;=\; \widetilde{\Delta}_r(\mu_{ij}) \;=\; (q-1)^{-1}\,\widetilde{\Delta}_r(t_{ij}) \;=$$
$$= (q-1)^{-1}\sum_L D_L^{1\dots\widehat{j}\dots r\,i} D_0^{-1} \otimes D^L D_0^{-1} \cdot \sum_{n\geq 0}(-1)^n\bigg(\sum_{K\neq I_0} D_K^{I_0} D_0^{-1} \otimes D^K D_0^{-1}\bigg)^n$$

Now, each left-hand side factor above belongs to $\widehat{\mathcal{O}_q(G)^\vee} \,\widehat{\otimes}\, \widehat{\mathcal{O}_q\big(G/P\big)^\vee}$, because either $D^L \in J_{G/P}^{\,loc}$ (if $L \neq I_0$, with notation of §4.3), or $D_L^{1\dots\widehat{j}\dots r\,i} \in J$ (if $L = I_0$, with $J := Ker\big(\epsilon_{\mathcal{O}_q(G)}\big)$). On right-hand side instead we have

$$D^K \in J_{G/P}^{\,loc} \subseteq (q-1)\,\mathcal{O}_q\big(G/P\big)^\vee \;, \qquad D_K^{I_0} \in J \subseteq (q-1)\,\mathcal{O}_q(G)^\vee$$

whence — as $D_0^{-1} \in \widehat{\mathcal{O}_q(G)^\vee}$ and $D_0^{-1} \in \widehat{\mathcal{O}_q\big(G/P\big)^\vee}$ — we get

$$\sum_{K\neq I_0} D_K^{I_0} D_0^{-1} \otimes D^K D_0^{-1} \;\in\; (q-1)^2\, \widehat{\mathcal{O}_q(G)^\vee} \,\widehat{\otimes}\, \widehat{\mathcal{O}_q\big(G/P\big)^\vee}$$

so that $\displaystyle\sum_{n\geq 0}(-1)^n\bigg(\sum_{K\neq I_0} D_K^{I_0} D_0^{-1} \otimes D^K D_0^{-1}\bigg)^n \in \widehat{\mathcal{O}_q(G)^\vee} \,\widehat{\otimes}\, \widehat{\mathcal{O}_q\big(G/P\big)^\vee}$

The final outcome is $\widehat{\Delta}(\mu_{ij}) \in \widehat{\mathcal{O}_q(G)^\vee} \,\widehat{\otimes}\, \widehat{\mathcal{O}_q\big(G/P\big)^\vee}$ for all i and all j. As the μ_{ij}'s topologically generate $\widehat{\mathcal{O}_q\big(G/P\big)^\vee}$, this proves the claim. □

In the end, we get the main result of this paper.

THEOREM 4.11. *$\widehat{\mathcal{O}_q\big(G/P\big)^\vee}$ is a quantum homogeneous G^*–space, which is indeed an infinitesimal quantization of the homogeneous G^*–space $\mathfrak{p}^\perp$.*

Proof. Just collect the previous results. By Proposition 4.9 and by the fact that $\widehat{\mathcal{O}_q\big(G/P\big)^\vee}\Big|_{q=1} = \mathcal{O}_q\big(G/P\big)^\vee\Big|_{q=1}$ we have that the specialization of $\widehat{\mathcal{O}_q\big(G/P\big)^\vee}$ is $U\big(\mathfrak{p}^\perp\big)$. Moreover we saw that $\widehat{\mathcal{O}_q\big(G/P\big)^\vee}$ is a subalgebra, and left coideal, of $\widehat{\mathcal{O}_q(G)^\vee}$. Finally, we have

$$\widehat{\mathcal{O}_q\big(G/P\big)^\vee} \bigcap\, (q-1)\,\widehat{\mathcal{O}_q(G)^\vee} \;=\; (q-1)\,\widehat{\mathcal{O}_q\big(G/P\big)^\vee}$$

as an easy consequence of Corollary 4.8 *(c)*. Therefore, $\widehat{\mathcal{O}_q\big(G/P\big)^\vee}$ is a quantum homogeneous space, in the usual sense. As $\widehat{\mathcal{O}_q(G)^\vee}$ is a quantization of $\mathfrak{g}^*$, we have that $\widehat{\mathcal{O}_q\big(G/P\big)^\vee}$ is in fact a quantum homogeneous space for G^*; of course, this is a quantization of *infinitesimal* type. □

REMARK 4.12. All these computations can be repeated, step by step, taking $G = SL_n$ and $P = P_1$.

References

[1] V. Chari, A. Pressley, *Quantum Groups*, Cambridge Univ. Press, Cambridge (1994).
[2] N. Ciccoli, R. Fioresi, F. Gavarini, *Quantum Duality Principle for Projective Homogeneous Spaces*, J. Noncommut. Geom. **2** (2008), 449–496.
[3] N. Ciccoli, F. Gavarini, *A quantum duality principle for coisotropic subgroups and Poisson quotients*, Adv. Math. **199** (2006), 104–135.
[4] C. De Concini, D. Eisenbud, C. Procesi, *Young Diagrams and Determinantal Varieties*, Invent. Math. **56** (1980), 129–165.
[5] V. G. Drinfeld, *Quantum groups*, Proc. Intern. Congress of Math. (Berkeley, 1986) (1987), 798–820.
[6] R. Fioresi, *Quantum deformation of the Grassmannian manifold*, J. Algebra **214** (1999), 418–447.
[7] R. Fioresi, *A deformation of the big cell inside the Grassmannian manifold* $G(r,n)$, Rev. Math. Phys. **11** (1999), 25–40.
[8] F. Gavarini, *Quantum function algebras as quantum enveloping algebras*, Comm. Algebra **26** (1998), 1795–1818.
[9] F. Gavarini, *The quantum duality principle*, Ann. Inst. Fourier (Grenoble) **52** (2002), 809–834.
[10] F. Gavarini, *The global quantum duality principle: theory, examples, and applications*, 120 pages, see `http://arxiv.org/abs/math.QA/0303019` (2003).
[11] F. Gavarini, *The global quantum duality principle*, J. Reine Angew. Math. **612** (2007), 17–33.
[12] V. Lakshmibai, N. Reshetikhin, *Quantum flag and Schubert schemes*, Contemp. Math. **134**, Amer. Math. Soc., Providence, RI (1992), 145–181.
[13] E. Taft, J. Towber, *Quantum deformation of flag schemes and Grassmann schemes, I. A q-deformation of the shape-algebra for* GL(n), J. Algebra **142** (1991), 1–36.

DIPARTIMENTO DI MATEMATICA, UNIVERSITÀ DI BOLOGNA, PIAZZA DI PORTA S. DONATO 5, I-40127 BOLOGNA, ITALY
E-mail address: `fioresi@dm.unibo.it`

DIPARTIMENTO DI MATEMATICA, UNIVERSITÀ DI ROMA "TOR VERGATA", VIA DELLA RICERCA SCIENTIFICA 1, I-00133 ROMA, ITALY
E-mail address: `gavarini@mat.uniroma2.it`

Some remarks on the action of quantum isometry groups

Debashish Goswami

ABSTRACT. We give some new sufficient conditions on a spectral triple to ensure that the quantum group of orientation and volume preserving isometries defined in [6] has a C^*-action on the underlying C^* algebra.

1. Introduction

Taking motivation from the work of Wang, Banica, Bichon and others (see [20], [21], [2], [3], [4], [22] and references therein), we have given a definition of quantum isometry group based on a 'Laplacian' in [11], and then followed it up by a formulation of 'quantum group of orientation preserving isometries' in [6] (see also [7], [9], [8], [5] for many explicit computations). The main result of [6] is that given a spectral triple (of compact type) $(\mathcal{A}^\infty, \mathcal{H}, D)$ and a positive unbounded operator R commuting with D, there is a universal object in the category of compact quantum groups which have a unitary representation (say U) on the Hilbert space $\mathcal{H}$ w.r.t. which D is equivariant and the normal (co)-action α_U of the quantum group obtained by conjugation by the unitary representation leaves the weak closure of $\mathcal{A}^\infty$ invariant and preserves a canonical functional τ_R called 'R-twisted volume form' described in [6]. The Woronowicz subalgebra of this universal quantum group generated by the 'matrix elements' of $\alpha_U(a)$, $a \in \mathcal{A}^\infty$ is called 'the quantum group of orientation and R-twisted volume preserving isometries' and denoted by $QISO_R^+(D)$.

However, it is not clear from the definition and construction of this quantum group whether α_U is a C^* -action of $QISO_R^+(D)$ on the C^* algebra generated by $\mathcal{A}^\infty$ (in the sense of Woronowicz and Podles). The problem is that to prove the existence of a universal object in [6] we had to make use of the Hilbert space and the strong operator topology coming from it. This is the reason why we demanded only the stability of the von Neumann algebra generated by $\mathcal{A}^\infty$ in the definition of an isometric and orientation preserving quantum group action on the spectral triple $(\mathcal{A}^\infty, \mathcal{H}, D)$. In a sense, we worked in a suitable category of 'measurable' actions and could prove the existence of a universal object there. In the classical situation, i.e. for isometric orientation preserving group actions on Riemannian spin manifolds, the apparently weaker condition of measurability turns out to be equivalent to a topological, in fact smooth action, thanks to the Sobolev's theorem

The author gratefully acknowledges support obtained from the Indian National Academy of Sciences through the grants for a project on 'Noncommutative Geometry and Quantum Groups'.

(see the proof of the converse part of Theorem 2.2 in [**6**]).The analogous question in the noncommutative situation is to see whether α_U (which is a-priori only a normal action) is a C^* action, and we shall show that under some assumptions which very much resemble the conditions for the classical Sobolev's theorem, we can indeed answer this question in the affirmative. In fact, we have already given a number of sufficient conditions for ensuring C^*-action in [**6**], and the conditions given in the present article add to this list, strengthening our belief that the action of quantum group of orientation preserving isometries is in general a C^* action.

2. Preliminaries

2.1. Generalities on quantum groups and their action. We review some basic facts about quantum groups (see, e.g. [**25**], [**24**], [**1**], [**19**] and references therein). A compact quantum group (to be abbreviated as CQG from now on) is given by a pair $(\mathcal{S}, \Delta)$, where $\mathcal{S}$ is a unital separable C^* algebra equipped with a unital $*$-homomorphism $\Delta : \mathcal{S} \to \mathcal{S} \otimes \mathcal{S}$ (where $\otimes$ denotes the injective tensor product) satisfying
(ai) $(\Delta \otimes id) \circ \Delta = (id \otimes \Delta) \circ \Delta$ (co-associativity), and
(aii) the linear span of $\Delta(\mathcal{S})(\mathcal{S} \otimes 1)$ and $\Delta(\mathcal{S})(1 \otimes \mathcal{S})$ are norm-dense in $\mathcal{S} \otimes \mathcal{S}$.
It is well-known (see [**25**], [**24**]) that there is a canonical dense $*$-subalgebra $\mathcal{S}_0$ of $\mathcal{S}$, consisting of the matrix coefficients of the finite dimensional unitary (co)-representations (to be defined shortly) of $\mathcal{S}$, and maps $\epsilon : \mathcal{S}_0 \to \mathbb{C}$ (co-unit) and $\kappa : \mathcal{S}_0 \to \mathcal{S}_0$ (antipode) defined on $\mathcal{S}_0$ which make $\mathcal{S}_0$ a Hopf $*$-algebra.

We say that the compact quantum group $(\mathcal{S}, \Delta)$ (co)-acts on a unital C^* algebra $\mathcal{B}$, if there is a unital $*$-homomorphism (called an action) $\alpha : \mathcal{B} \to \mathcal{B} \otimes \mathcal{S}$ satisfying the following :
(bi) $(\alpha \otimes id) \circ \alpha = (id \otimes \Delta) \circ \alpha$, and
(bii) the linear span of $\alpha(\mathcal{B})(1 \otimes \mathcal{S})$ is norm-dense in $\mathcal{B} \otimes \mathcal{S}$.
It is known (see [**16**]) that the above is equivalent to the existence of a dense unital $*$-subalgebra $\mathcal{B}_0$ of $\mathcal{B}$ on which the action α is an algebraic action of the Hopf algebra $\mathcal{S}_0$, i.e. α maps $\mathcal{B}_0$ into $\mathcal{B}_0 \otimes_{\mathrm{alg}} \mathcal{S}_0$ and also $(\mathrm{id} \otimes \epsilon) \circ \alpha = \mathrm{id}$ on $\mathcal{B}_0$.

Such an action will be called a C^* or topological action, to distinguish it from a normal action on a von Neumann algebra, which we briefly mention later.

DEFINITION 2.1. *A unitary (co) representation of a compact quantum group (S, Δ) on a Hilbert space $\mathcal{H}$ is a map U from $\mathcal{H}$ to the Hilbert $\mathcal{S}$ module $\mathcal{H} \otimes \mathcal{S}$ such that the element $\widetilde{U} \in \mathcal{M}(\mathcal{K}(\mathcal{H}) \otimes \mathcal{S})$ given by $\widetilde{U}(\xi \otimes b) = U(\xi)(1 \otimes b)$ $(\xi \in \mathcal{H}, b \in \mathcal{S})$ is a unitary satisfying*

$$(\mathrm{id} \otimes \Delta)\widetilde{U} = \widetilde{U}_{(12)}\widetilde{U}_{(13)},$$

where for an operator $X \in \mathcal{B}(\mathcal{H}_1 \otimes \mathcal{H}_2)$ we have denoted by X_{12} and X_{13} the operators $X \otimes I_{\mathcal{H}_2} \in \mathcal{B}(\mathcal{H}_1 \otimes \mathcal{H}_2 \otimes \mathcal{H}_2)$, and $\Sigma_{23} X_{12} \Sigma_{23}$ respectively (Σ_{23} being the unitary on $\mathcal{H}_1 \otimes \mathcal{H}_2 \otimes \mathcal{H}_2$ which flips the two copies of $\mathcal{H}_2$).

Given a unitary representation U we shall denote by α_U the $*$-homomorphism $\alpha_U(X) = \widetilde{U}(X \otimes 1)\widetilde{U}^*$ for $X \in \mathcal{B}(\mathcal{H})$. For a not necessarily bounded, densely defined (in the weak operator topology) linear functional τ on $\mathcal{B}(\mathcal{H})$, we say that α_U preserves τ if α_U maps a suitable (weakly) dense $*$-subalgebra (say $\mathcal{D}$) in the domain of τ into $\mathcal{D} \otimes_{\mathrm{alg}} \mathcal{S}$ and $(\tau \otimes \mathrm{id})(\alpha_U(a)) = \tau(a)1_{\mathcal{S}}$ for all $a \in \mathcal{D}$. When

τ is bounded and normal, this is equivalent to $(\tau \otimes \mathrm{id})(\alpha_U(a)) = \tau(a)1_{\mathcal{S}}$ for all $a \in \mathcal{B}(\mathcal{H})$.

We say that a (possibly unbounded) operator T on $\mathcal{H}$ commutes with U if $T \otimes I$ (with the natural domain) commutes with $\widetilde{U}$. Sometimes such an operator will be called U-equivariant.

We also need to consider von Neumann algebraic quantum group of compact type. This is given by a von Neumann algebra $\mathcal{M}$ equipped with a normal coassociative coproduct and also a faithful normal state (Haar state) invariant under the coproduct. We refer to [**18**], [**15**] for more details, in fact for a locally compact von Neumann algebraic quantum group, of which those of compact type form a very special and relatively simple class. We shall actually be concerned with the canonical von Neumann algebraic compact quantum group coming from the GNS representation of a CQG (w.r.t. the Haar state). Let us mention that the Haar state (say h) on a CQG $\mathcal{S}$ is not necessarily faithful, but it is faithful on the dense $*$-algebra $\mathcal{S}_0$ mentioned before. Let $\rho_h : \mathcal{S} \to \mathcal{B}(L^2(h))$ be the GNS representation. It is easily seen that the coproduct on $\mathcal{S}$ is implemented by the canonical left regular unitary representation, say U, on this space, and then α_U can be used as the definition of a normal coassociative coproduct on the von Neumann algebra $\mathcal{M} = \rho_h(\mathcal{S})'' \subseteq \mathcal{B}(L^2(h))$. Since the Haar state is a vector state in $L^2(h)$ and hence normal invariant state, it is clear that $\mathcal{M}$ is a compact type von Neumann algebraic quantum group.

We remark here that the definition of unitary representation as well as (co)-action has a natural analogue for such von Neumann algebraic quantum groups, and it can be shown that any such representation decomposes into direct sum of irreducible ones, and that any irreducible representation of a compact Hopf von Neumann algebra is finite dimensional. The proofs of these facts are almost the same as the proof in the C^* case, with the only difference being that the norm topology must be replaced by appropriate weak or strong operator topology. Moreover, we remark that given a C^*-action $\alpha : \mathcal{A} \to \mathcal{A} \otimes \mathcal{S}$ of the CQG $\mathcal{S}$ on a separable unital C^* algebra embedded in $\mathcal{B}(\mathcal{H})$ for some Hilbert space $\mathcal{H}$, such that the action α is implemented by some unitary representation U of the CQG $\mathcal{S}$ on $\mathcal{H}$, i.e. $\alpha = \alpha_U$, we can canonically construct a normal action of the von Neumann algebraic compact quantum group $\mathcal{M}$ as follows. First, replace the action α by $\alpha_h = (\mathrm{id} \otimes \rho_h) \circ \alpha$, which is an action of $\rho_h(\mathcal{S})$, and note that $\alpha_h(a) = U_h(a \otimes 1)U_h^*$ for $a \in \mathcal{A}$, where $U_h := (\mathrm{id} \otimes \rho_h)(U)$, a unitary representation of $\rho_h(\mathcal{S})$ on $\mathcal{H}$. Now we use the right hand side of the above to extend the definition of α_h to the whole of $\mathcal{B}(\mathcal{H})$, in particular on $\mathcal{N} := \mathcal{A}'' \subseteq \mathcal{B}(\mathcal{H})$, and observe that this indeed gives a normal action of $\mathcal{M} = \rho_h(\mathcal{A})''$ on $\mathcal{N}$.

2.2. Quantum group of orientation and volume preserving isometries. Next we give an overview of the definition of quantum isometry groups as in [**6**].

DEFINITION 2.2. *A quantum family of orientation preserving isometries of the spectral triple $(\mathcal{A}^\infty, \mathcal{H}, D)$ (of compact type) is given by a pair $(\mathcal{S}, U)$ where $\mathcal{S}$ is a separable unital C^*-algebra and U is a linear map from $\mathcal{H}$ to $\mathcal{H} \otimes \mathcal{S}$ such that $\widetilde{U}$ given by $\widetilde{U}(\xi \otimes b) = U(\xi)(1 \otimes b)$ ($\xi \in \mathcal{H}$, $b \in \mathcal{S}$) extends to a unitary element of $\mathcal{M}(\mathcal{K}(\mathcal{H}) \otimes \mathcal{S})$ satisfying the following*
(i) for every state ϕ on $\mathcal{S}$ we have $U_\phi D = DU_\phi$, where $U_\phi := (\mathrm{id} \otimes \phi)(\widetilde{U})$;

(ii) $(\mathrm{id} \otimes \phi) \circ \alpha_U(a) \in (\mathcal{A}^\infty)'' \ \forall a \in \mathcal{A}^\infty$ *for every state* ϕ *on* $\mathcal{S}$, *where* $\alpha_U(x) := \widetilde{U}(x \otimes 1)\widetilde{U}^*$ *for* $x \in \mathcal{B}(\mathcal{H})$.

In case the C^**-algebra* $\mathcal{S}$ *has a coproduct* Δ *such that* $(\mathcal{S}, \Delta)$ *is a compact quantum group and* U *is a unitary representation of* $(\mathcal{S}, \Delta)$ *on* $\mathcal{H}$, *we say that* $(\mathcal{S}, \Delta)$ *acts by orientation-preserving isometries on the spectral triple.*

Consider the category $\mathbf{Q}$ with the object-class consisting of all quantum families of orientation preserving isometries $(\mathcal{S}, U)$ of the given spectral triple, and the set of morphisms $\mathrm{Mor}((\mathcal{S}, U), (\mathcal{S}', U'))$ being the set of unital $*$-homomorphisms $\Phi : \mathcal{S} \to \mathcal{S}'$ satisfying $(\mathrm{id} \otimes \Phi)(U) = U'$. We also consider another category $\mathbf{Q}'$ whose objects are triplets $(\mathcal{S}, \Delta, U)$, where $(\mathcal{S}, \Delta)$ is a compact quantum group acting by orientation preserving isometries on the given spectral triple, with U being the corresponding unitary representation. The morphisms are the homomorphisms of compact quantum groups which are also morphisms of the underlying quantum families of orientation preserving isometries. The forgetful functor $F : \mathbf{Q}' \to \mathbf{Q}$ is clearly faithful, and we can view $F(\mathbf{Q}')$ as a subcategory of $\mathbf{Q}$.

Unfortunately, in general $\mathbf{Q}'$ or $\mathbf{Q}$ will not have a universal object, as discussed in [**6**]. We have to restrict to a subcategory described below to get a universal object in general, though in some cases. Fix a positive, possibly unbounded, operator R on $\mathcal{H}$ which commutes with D and consider the weakly dense $*$-subalgebra $\mathcal{E}_D$ of $\mathcal{B}(\mathcal{H})$ generated by the rank-one operators of the form $|\xi >< \eta|$ where ξ, η are eigenvectors of D. Define $\tau_R(x) = Tr(Rx),\ x \in \mathcal{E}_D$.

DEFINITION 2.3. *A quantum family of orientation preserving isometries* $(\mathcal{S}, U)$ *of* $(\mathcal{A}^\infty, \mathcal{H}, D)$ *is said to preserve the* R*-twisted volume, (simply said to be volume-preserving if* R *is understood) if one has* $(\tau_R \otimes \mathrm{id})(\alpha_U(x)) = \tau_R(x) 1_{\mathcal{S}}$ *for all* $x \in \mathcal{E}_D$, *where* $\mathcal{E}_D$ *and* τ_R *are as above.*

If, furthermore, the C^**-algebra* $\mathcal{S}$ *has a coproduct* Δ *such that* $(\mathcal{S}, \Delta)$ *is a CQG and* U *is a unitary representation of* $(\mathcal{S}, \Delta)$ *on* $\mathcal{H}$, *we say that* $(\mathcal{S}, \Delta)$ *acts by (*R*-twisted) volume and orientation-preserving isometries on the spectral triple.*

We shall consider the categories $\mathbf{Q}_R$ *and* $\mathbf{Q}'_R$ *which are the full subcategories of* $\mathbf{Q}$ *and* $\mathbf{Q}'$ *respectively, obtained by restricting the object-classes to the volume-preserving quantum families.*

The following result is proved in [**6**].

THEOREM 2.4. *The category* $\mathbf{Q}_R$ *of quantum families of volume and orientation preserving isometries has a universal (initial) object, say* $(\widetilde{\mathcal{G}}, U_0)$. *Moreover,* $\widetilde{\mathcal{G}}$ *has a coproduct* Δ_0 *such that* $(\widetilde{\mathcal{G}}, \Delta_0)$ *is a compact quantum group and* $(\widetilde{\mathcal{G}}, \Delta_0, U_0)$ *is a universal object in the category* $\mathbf{Q}'_0$. *The representation* U_0 *is faithful.*

The Woronowicz subalgebra of $\tilde{\mathcal{G}}$ generated by elements of the form $\{< (\xi \otimes 1), \alpha_0(a)(\eta \otimes 1) >_{\tilde{\mathcal{G}}},\ a \in \mathcal{A}^\infty\}$, where $\alpha_0 \equiv \alpha_{U_0}$ and $< \cdot, \cdot >_{\tilde{\mathcal{G}}}$ denotes the $\tilde{\mathcal{G}}$-valued inner product of the Hilbert module $\mathcal{H} \otimes \tilde{\mathcal{G}}$, is called the quantum group of orientation and volume preserving isometries, and denoted by $QISO_R^+(D)$.

3. C^*-action of $QISO_R^+(D)$

It is not clear from the definition and construction of $QISO_R^+(D)$ whether the C^* algebra $\mathcal{A}$ generated by $\mathcal{A}^\infty$ is stable under $\alpha_0 := \alpha_{U_0}$ in the sense that $(\mathrm{id} \otimes \phi) \circ \alpha_0$ maps $\mathcal{A}$ into $\mathcal{A}$ for every ϕ. Moreover, even if $\mathcal{A}$ is stable, the question remains whether α_0 is a C^*-action of the CQG $QISO_R^+(D)$. Although we could not yet decide whether the general answer to the above two questions are affirmative, we have given a number of sufficient conditions for it in [**6**]. In fact, those conditions already cover all classical compact Riemannian manifolds, and many noncommutative ones as well. In what follows, we shall provide yet another set of sufficient conditions, which will be valid for many interesting spectral triples constructed from Lie group actions on C^* algebras.

Suppose that there are compact Lie groups $\tilde{G}$, G with a (finite) covering map $\gamma : \tilde{G} \to G$ (which is group homomorphism), such that the following hold:
(a) There is a strongly continuous (w.r.t. the norm-topology) action of G on $\mathcal{A}$, i.e. $g \mapsto \beta_g(a)$ is norm-continuous for $a \in cla$.
(b) there exists a strongly continuous unitary representation $V_{\tilde{g}}$ of $\tilde{G}$ on $\mathcal{H}$ which commutes with D and R, and we also have $V_{\tilde{g}} a V_{\tilde{g}}^{-1} = \beta_g(a)$, where $a \in \mathcal{A}, \tilde{g} \in \tilde{G}$, and $g = \gamma(\tilde{g})$.
(c) In the decomposition of the G-action β on $\mathcal{A}$ into irreducible subspaces, each irreducible representation of G occurs with at most finite (including zero) multiplicity.
It follows from (b) that we can extend β_g to the von Neumann algebra $\mathcal{A}''$ as a strongly continuous (w.r.t. SOT) action of G. Indeed, for $g \in G$, $a \in \mathcal{A}''$, we can choose any $\tilde{g}$ such that $\gamma(\tilde{g}) = g$, and also choose a bounded net a_n from $\mathcal{A}$ converging in the S.O.T. to a. Clearly, $\beta_g(a_n) = V_{\tilde{g}} a_n V_{\tilde{g}}^{-1}$ converges in S.O.T., and the limit defines $\beta_g(a)$.

Since G is a Lie group, we choose a basis of its Lie algebra, say $\{\chi_1, ..., \chi_N\}$. Each χ_i induces closable derivation on $\mathcal{A}$ (w.r.t. the norm topology) as well as on $\mathcal{A}''$ (w.r.t. SOT) which will be denoted by δ_i and $\tilde{\delta}_i$ respectively. We do have natural Frechet spaces $\mathcal{E}_1 := \bigcap_{n \geq 1, 1 \leq i_j \leq N} \mathrm{Dom}(\delta_{i_1} ... \delta_{i_n})$ and $\mathcal{E}_2 := \bigcap_{n \geq 1, 1 \leq i_j \leq N} \mathrm{Dom}(\tilde{\delta}_{i_1} ... \tilde{\delta}_{i_n})$, and we assume that
(d) $\mathcal{E}_1$ and $\mathcal{E}_2$ coincide with $\mathcal{A}^\infty$.

REMARK 3.1. *We have a feeling that in (c), it may be enough to require only the finite dimensionality of the identity representation. Indeed, if the action is ergodic, i.e. the identity representation has multiplicity one, it is known (see* [**14**]*) that (c) does hold, and in fact, by Lemma 8.1.20 of* [**13**] *(see also* [**12**]*), one can also deduce (d). This strongly indicates that if there is a generalization of the results obtained in* [**14**] *for actions which have finite multiplicity of the identity representation, then one may relax the condition (d), and also weaken (c). However, we are not aware of any such generalization of* [**14**].

Let $\hat{G}$ denote the (countable) set of equivalence classes of irreducible representations of G, and let $\mathcal{V}_\pi$ denote the (finite dimensional by assumption) subspace of $\mathcal{A}''$ which is the range of the 'spectral projection' P_π corresponding to π, namely $x \mapsto \int_G c_\pi(g) \beta_g(x) dg$, c_π being the character of π. It is easy to see from the assumptions made that elements of $\mathcal{V}_\pi$ actually belong to $\mathcal{A}^\infty$, and clearly, this subspace coincides with the range of the restriction of P_π on $\mathcal{A}$. Thus, in particular, the linear span of $\mathcal{V}_\pi$, $\pi \in \hat{G}$, is norm-dense in $\mathcal{A}$ as well.

Now we have the following:

THEOREM 3.2. *Under the above assumptions, $\mathcal{A}$ has the C^*-action of $\mathcal{G} = QISO_R^+(D)$ given by the restriction of α_0 on $\mathcal{A}$.*

Proof: As discussed in the previous section, we consider the reduced CQG $\rho_h(\mathcal{G})$ where ρ_h is the GNS representation of the Haar state, and denote the von Neumann algebraic quantum group (of compact type) $\rho_h(\mathcal{G})''$ by $\mathcal{M}$. Clearly, α_0 extends to a normal action of $\mathcal{M}$ on $\mathcal{A}'' \subseteq \mathcal{B}(\mathcal{H})$, given by $\alpha_0(x) = U_0(x \otimes 1)U_0^* \subseteq \mathcal{A}'' \otimes \mathcal{B}(L^2(h))$ for $x \in \mathcal{A}''$, which decomposes into finite dimensional irreducible subspaces of $\mathcal{A}''$, say $\{\mathcal{A}_i, i \in I\}$ (I some index set).

We claim that G can be identified with a quantum subgroup of $\mathcal{G}$, in the sense that $C(G)$ is identified with a quotient of $\mathcal{G}$ by some Woronowicz ideal. To see this, note that by assumption, $V_{\tilde{g}}$ ($\forall \tilde{g} \in \tilde{G}$) commutes with D and $\tau_R(V_{\tilde{g}} x V_{\tilde{g}}^{-1}) = \mathrm{Tr}(R V_{\tilde{g}} x V_{\tilde{g}}^{-1}) = \mathrm{Tr}(V_{\tilde{g}} R x V_{\tilde{g}}^{-1}) = \tau_R(x)$, for all $x \in \mathcal{E}_D$, since $V_{\tilde{g}}$ is unitary commuting with R. Moreover, $V_{\tilde{g}} \mathcal{A} V_{\tilde{g}}^{-1} \subseteq \mathcal{A}''$. This implies, by the universality of $\tilde{\mathcal{G}}$, that there is a Woronowicz ideal $\tilde{\mathcal{I}}$ of $\tilde{\mathcal{G}}$ such that $C(\tilde{G}) \cong \tilde{\mathcal{G}}/\tilde{\mathcal{I}}$. Since $C(G)$ is a Woronowicz subalgebra of $C(\tilde{G})$ in the obvious way, via the injective $*$-homomorphism which sends $f \in C(G)$ to $f \circ \gamma \in C(\tilde{G})$, our claim follows.

it is clear that each $\mathcal{A}_i$ is G-invariant, i.e. $\beta_g(\mathcal{A}_i) \subseteq \mathcal{A}_i$ for all g.

Thus, we can further decompose $\mathcal{A}_i$ into G-irreducible subspaces, say $\mathcal{A}_i^\pi$, $\pi \in \hat{G}$. Clearly, $\mathcal{A}_i^\pi \subseteq \mathcal{V}_\pi \subseteq \mathcal{A}^\infty$. Thus, the restriction of α_0 to the linear span of all the $\mathcal{A}_i^\pi$'s, say $\mathcal{V}$ (which is already a $*$-algebra by the general theory of CQG), is a Hopf algebraic action of $\mathcal{M}$. However, by definition of $\mathcal{G} = QISO_R^+(D)$ it is clear that $\alpha_0|_{\mathcal{A}_i^\pi}$ is actually a Hopf algebraic action of the CQG $\rho_h(\mathcal{G})$. Moreover, since $\mathcal{A}_i^\pi$ is finite dimensional, the 'matrix coefficients' of $\alpha_0|_{\mathcal{A}_i^\pi}$ must come from the Hopf algebra $\mathcal{G}_0$ mentioned in the previous section, on which ρ_h is faithful, and we thus see, by identifying $\rho_h(\mathcal{G}_0)$ with $\mathcal{G}_0$, that $\alpha_0(\mathcal{A}_i^\pi) \subseteq \mathcal{A}_i^\pi \otimes_{\mathrm{alg}} \mathcal{G}_0$.

Now it suffices to prove the norm-density of the subspace $\mathcal{V}$ in $\mathcal{A}$. To this end, we note that, by the weak density of $\mathcal{V}$ in $\mathcal{A}''$, the range $\mathcal{V}_\pi$ of $P_\pi|_{\mathcal{A}''}$ is the weak closure of $P_\pi(\mathcal{V})$. But P_π being finite dimensional, the range must coincide with $P_\pi(\mathcal{V})$, which is nothing but the linear span of those (finitely many) $\mathcal{A}_i^\pi$ which are nonzero , i.e. for which the irreducible of type π occurs in the decomposition of $\mathcal{A}_i$. It follows that $\mathcal{V}$ contains $\mathcal{V}_\pi$ for each π, hence (by the norm-density of $\mathrm{Span}\{\mathcal{V}_\pi, \pi \in \hat{G}\}$ in $\mathcal{A}$) $\mathcal{V}$ is norm-dense in $\mathcal{A}$. □

Examples:

We shall mention two classes of spectral triples which satisfy our assumptions.

(I) The assumptions of the above theorem are valid (with $R = I$) in case the given spectral triple obtained from an ergodic action of a compact Lie group G on the underlying C^* algebra $\mathcal{A}$. Let G have a Lie algebra basis $\{\chi_1, \ldots, \chi_N\}$ and an ergodic G-action on $\mathcal{A}$. It is well-known (see [**14**]) that $\mathcal{A}$ has a canonical faithful G-invariant trace, say τ, and if we embed $\mathcal{A}$ in the corresponding GNS space $L^2(\tau)$, the operator $i\delta_j$ (where δ_j is as before) extends to a self adjoint operator on $L^2(\tau)$. Taking $\mathcal{H} = L^2(\tau) \otimes \mathbb{C}^n$, where n is the smallest positive integer such that the Clifford algebra of dimension N admits a faithful representation as $n \times n$ matrices, we consider $D = \sum_j i\delta_j \otimes \gamma_j$, γ_j being the Clifford matrices. As we have already remarked, the smooth algebra $\mathcal{A}^\infty$ corresponding to the G-action on $\mathcal{A}$ satisfies our assumption (d), and it has the natural representation $a \mapsto a \otimes 1_{\mathbb{C}^n}$ on $\mathcal{H}$. If the operator D has a self-adjoint extension with compact resolvent, it is clear that $(\mathcal{A}^\infty, \mathcal{H}, D)$ gives a spectral triple satisfying all the assumptions (a)-(d) with $\tilde{G} = G$, $R = I$ and the unitary representation V being $\beta_g \otimes I_{\mathbb{C}^n}$, where β_g is the given ergodic action of G on $\mathcal{A}$, extended naturally as a unitary representation on

$L^2(\tau)$. It may be mentioned that the standard spectral triple on the noncommutative tori arises in this way.
(II) Another interesting class of examples satisfying assumptions (a)-(d) come from those classical spectral triples for which the action of the group of orientation preserving Riemannian isometries on $C(M)$ (where M denotes the underlying manifold) is such that any irreducible representation occurs with at most finite multiplicities in its decomposition. It is easy to see that the classical spheres and tori are indeed such manifolds.

Acknowledgement:
The author would like to thank the anonymous referee for constructive suggestions and comments which led to improvement of the paper. In particular, Remark 3.1 is motivated by referee's comments.

References

[1] Baaj, S. and Skandalis, G.: Unitaires multiplicatifs et dualite pour les produits croises de C^*-algebres, Ann. Sci. Ecole Norm. Sup. (4) **26** (1993), no. 4, 425–488.
[2] Banica, T.: Quantum automorphism groups of small metric spaces, Pacific J. Math. **219**(2005), no. 1, 27–51.
[3] Banica, T.: Quantum automorphism groups of homogeneous graphs, J. Funct. Anal. **224**(2005), no. 2, 243–280.
[4] Bichon, J.: Quantum automorphism groups of finite graphs, Proc. Amer. Math. Soc. **131**(2003), no. 3, 665–673.
[5] Bhowmick, J.: Quantum Isometry Group of the n tori, preprint (2008), arXiv 0803.4434.
[6] Bhowmick, J. and Goswami, D.: Quantum Group of Orientation preserving Riemannian Isometries, preprint (2008), arXiv 0806.3687.
[7] Bhowmick, J., Goswami, D. and Skalski, A.: Quantum Isometry Groups of 0- Dimensional Manifolds , preprint(2008), arXiv 0807.4288., to appear in Trans. A. M. S.
[8] Bhowmick, J. and Goswami, D.: Quantum isometry groups : examples and computations, Comm. Math. Phys. **285** (2009), no.2, 421–444..
[9] Bhowmick, J. and Goswami, D.: Quantum isometry groups of the Podles sphere, preprint (2008)
[10] Connes, A.: "Noncommutative Geometry", Academic Press, London-New York (1994).
[11] Goswami, D.: Quantum Group of isometries in Classical and Non Commutative Geometry, Comm. Math. Phys. **285** (2009), no. 1, 141–160.
[12] Goswami, D.: On Equivariant Embedding of Hilbert C^* modules, to appear in Proc. Ind. Acad. Sci. (Math. Sci.), available at arXiv:0704.0110.
[13] Goswami, D. and Sinha, K. B.: "Quantum Stochastic Calculus and Noncommutative Geometry", Cambridge Tracts in Mathematics **169**, Cambridge University Press, U.K., 2007.
[14] Hoegh-Krohn, R, Landstad, M. B., Stormer, E.: Compact ergodic groups of automorphisms, Ann. of Math. (2) **114** (1981), no. 1, 75–86.
[15] Kustermans, J. and Vaes, S.: Locally compact quantum groups in the von Neumann algebraic setting, Math. Scand. **92** (2003), no. 1, 68–92.
[16] Podles, P.: Symmetries of quantum spaces, Subgroups and quotient spaces of quantum SU(2) and SO(3) groups, Comm. Math. Phys. **170**(1995), no. 1, 1–20.
[17] Soltan, P. M.: Quantum families of maps and quantum semigroups on finite quantum spaces, preprint, arXiv:math/0610922.
[18] Vaes, S.: The unitary implementation of a locally compact quantum group action, J. Funct. Anal. **180** (2001), no. 2, 426–480.
[19] Maes, A. and Van Daele, A.: Notes on compact quantum groups, Nieuw Arch. Wisk. (4)**16**(1998), no. 1-2, 73–112.
[20] Wang, S.: Free products of compact quantum groups, Comm. Math. Phys. **167** (1995), no. 3, 671–692.
[21] Wang, S.: Quantum symmetry groups of finite spaces, Comm. Math. Phys. **195**(1998), 195–211.

[22] Wang, S.: Structure and isomorphism classification of compact quantum groups $A_u(Q)$ and $B_u(Q)$, J. Operator Theory **48** (2002), 573–583.

[23] Wang, S. : Ergodic actions of universal quantum groups on operator algebras. Comm. Math. Phys. 203 (1999), no. 2, 481–498.

[24] Woronowicz, S. L.: Compact matrix pseudogroups, Comm. Math. Phys. **111**(1987), no. 4, 613–665.

[25] Woronowicz, S. L.: "Compact quantum groups", pp. 845–884 in Symétries quantiques (Quantum symmetries) (Les Houches, 1995), edited by A. Connes et al., Elsevier, Amsterdam, 1998.

[26] Woronowicz, S. L.: Pseudogroups, pseudospaces and Pontryagin duality, Proceedings of the International Conference on Mathematical Physics, Lausane (1979), Lecture Notes in Physics **116**, pp. 407-412.

Stat-Math Unit, Kolkata Centre, Indian Statistical Institute, 203, B. T. Road, Kolkata 700 108, India

Generic Hopf Galois extensions

Christian Kassel

ABSTRACT. In previous joint work with Eli Aljadeff we attached a generic Hopf Galois extension $\mathcal{A}_H^\alpha$ to each twisted algebra ${}^\alpha H$ obtained from a Hopf algebra H by twisting its product with the help of a cocycle α. The algebra $\mathcal{A}_H^\alpha$ is a flat deformation of ${}^\alpha H$ over a "big" central subalgebra $\mathcal{B}_H^\alpha$ and can be viewed as the noncommutative analogue of a versal torsor in the sense of Serre. After surveying the results on $\mathcal{A}_H^\alpha$ obtained with Aljadeff, we establish three new results: we present a systematic method to construct elements of the commutative algebra $\mathcal{B}_H^\alpha$, we show that a certain important integrality condition is satisfied by all finite-dimensional Hopf algebras generated by grouplike and skew-primitive elements, and we compute $\mathcal{B}_H^\alpha$ in the case where H is the Hopf algebra of a cyclic group.

1. Introduction

In this paper we deal with associative algebras ${}^\alpha H$ obtained from a Hopf algebra H by twisting its product by a cocycle α. This class of algebras, which for simplicity we call *twisted algebras*, coincides with the class of so-called *cleft Hopf Galois extensions* of the ground field; classical Galois extensions and strongly group-graded algebras belong to this class. As has been stressed many times (see, e.g., [**22**]), Hopf Galois extensions can be viewed as noncommutative analogues of principal fiber bundles (also known as G-torsors), for which the rôle of the structural group is played by a Hopf algebra. Hopf Galois extensions abound in the world of quantum groups and of noncommutative geometry. The problem of constructing systematically Hopf Galois extensions of a given algebra for a given Hopf algebra and of classifying them up to isomorphism has been addressed in a number of papers over the last fifteen years; let us mention [**4, 5, 10, 12, 13, 14, 15, 16, 17, 19, 20, 21**]. This list is far from being exhaustive, but gives a pretty good idea of the activity on this subject.

A new approach to this problem was recently considered in [**2**]; this approach mixes commutative algebra with techniques from noncommutative algebra such as *polynomial identities*. In particular, in that paper Eli Aljadeff and the author

---

2000 *Mathematics Subject Classification.* Primary (16W30, 16S35, 16R50) Secondary (13B05, 13B22, 16E99, 58B32, 58B34, 81R50).

Key words and phrases. Hopf algebra, Galois extension, twisted product, generic, cocycle, integrality.

Partially funded by ANR Project BLAN07-3_183390.

attached two "universal algebras" $\mathcal{U}_H^\alpha$, $\mathcal{A}_H^\alpha$ to each twisted algebra ${}^\alpha H$. The algebra $\mathcal{U}_H^\alpha$, which was built out of polynomial identities satisfied by ${}^\alpha H$, was the starting point of *loc. cit.* In the present paper we concentrate on the second algebra $\mathcal{A}_H^\alpha$ and survey the results obtained in [**2**] from the point of view of this algebra. In addition, we present here two new results, namely Theorem 7.2 and Proposition 8.1, as well as a computation in Subsection 4.3.

The algebra $\mathcal{A}_H^\alpha$ is a "generic" version of ${}^\alpha H$ and can be seen as a kind of universal Hopf Galois extension. To construct $\mathcal{A}_H^\alpha$ we introduce the *generic cocycle* cohomologous to the original cocycle α and we consider the commutative algebra $\mathcal{B}_H^\alpha$ generated by the values of the generic cocycle and of its convolution inverse. Then $\mathcal{A}_H^\alpha$ is a cleft H-Galois extension of $\mathcal{B}_H^\alpha$. We call $\mathcal{A}_H^\alpha$ the *generic Galois extension* and $\mathcal{B}_H^\alpha$ the *generic base space.* They satisfy the following remarkable properties.

Any "form" of ${}^\alpha H$ is obtained from $\mathcal{A}_H^\alpha$ by a specialization of $\mathcal{B}_H^\alpha$. Conversely, under an additional integrality condition, any central specialization of $\mathcal{A}_H^\alpha$ is a form of ${}^\alpha H$. Thus, the set of algebra morphisms $\mathrm{Alg}(\mathcal{B}_H^\alpha, K)$ parametrizes the isomorphism classes of K-forms of ${}^\alpha H$ and $\mathcal{A}_H^\alpha$ can be viewed as the noncommutative analogue of a *versal deformation space* or a *versal torsor* in the sense of Serre (see [**11**, Chap. I]). We believe that such versal deformation spaces are of interest and deserve to be computed for many Hopf Galois extensions. Even when the Hopf algebra H is a group algebra, in which case our theory simplifies drastically, not many examples have been computed (see [**1, 3**] for results in this case).

Our approach also leads to the emergence of new interesting questions on Hopf algebras such as Question 7.1 below. We give a positive answer to this question for a class of Hopf algebras that includes the finite-dimensional ones that are generated by grouplike and skew-primitive elements.

Finally we present a new systematic way to construct elements of the generic base space $\mathcal{B}_H^\alpha$. These elements are the images of certain universal noncommutative polynomials under a certain tautological map. In the language of polynomial identities, these noncommutative polynomials are central identities.

The paper is organized as follows. In Section 2 we recall the concept of a Hopf Galois extension and discuss the classification problem for such extensions. In Section 3 we define Hopf algebra cocycles and the twisted algebras ${}^\alpha H$. We construct the generic cocycle and the generic base space $\mathcal{B}_H^\alpha$ in Section 4; we also compute $\mathcal{B}_H^\alpha$ when H is the Hopf algebra of a cyclic group. In Section 5 we illustrate the theory with a nontrivial, still not too complicated example, namely with the four-dimensional Sweedler algebra. In Section 6 we define the generic Hopf Galois extension $\mathcal{A}_H^\alpha$ and state its most important properties. Some results of Section 6 hold under a certain integrality condition; in Section 7 we prove that this condition is satisfied by a certain class of Hopf algebras. In Section 8 we present the above-mentioned general method to construct elements of $\mathcal{B}_H^\alpha$. The contents of Subsection 4.3 and of Sections 7 and 8 are new.

We consistently work over a fixed field k, over which all our constructions will be defined. As usual, unadorned tensor symbols refer to the tensor product of k-vector spaces. All algebras are assumed to be associative and unital, and all algebra morphisms preserve the units. We denote the unit of an algebra A by 1_A, or simply by 1 if the context is clear. The set of algebra morphisms from an algebra A to an algebra B will be denoted by $\mathrm{Alg}(A, B)$.

2. Principal fiber bundles and Hopf Galois extensions

2.1. Hopf Galois extensions. A principal fiber bundle involves a group G acting, say on the right, on a space X such that the map

$$X \times G \to X \times_Y X\,;\ (x,g) \mapsto (x,xg)$$

is an isomorphism (in the category of spaces under consideration). Here Y represents some version of the quotient space X/G and $X \times_Y X$ the fiber product.

In a purely algebraic setting, the group G is replaced by a Hopf algebra H with coproduct $\Delta : H \to H \otimes H$, coünit $\varepsilon : H \to k$, and antipode $S : H \to H$. In the sequel we shall make use of the Heyneman-Sweedler sigma notation (see [**23**, Sect. 1.2]): we write

$$\Delta(x) = \sum_{(x)} x_{(1)} \otimes x_{(2)}$$

for the coproduct of $x \in H$ and

$$\Delta^{(2)}(x) = \sum_{(x)} x_{(1)} \otimes x_{(2)} \otimes x_{(3)}$$

for the iterated coproduct $\Delta^{(2)} = (\Delta \otimes \mathrm{id}_H) \circ \Delta = (\mathrm{id}_H \otimes \Delta) \circ \Delta$, and so on.

The G-space X is replaced by an algebra A carrying the structure of an H-comodule algebra. Recall that an algebra A is an *H-comodule algebra* if it has a right H-comodule structure whose coaction $\delta : A \to A \otimes H$ is an algebra morphism.

The space of *coïnvariants* of an H-comodule algebra A is the subspace A^H of A defined by

$$A^H = \{a \in A \,|\, \delta(a) = a \otimes 1\}\,.$$

The subspace A^H is a subalgebra and a subcomodule of A. We then say that $A^H \subset A$ is an *H-extension* or that A is an H-extension of A^H. An H-extension is called *central* if A^H lies in the center of A.

An H-extension $B = A^H \subset A$ is said to be *H-Galois* if A is faithfully flat as a left B-module and the linear map $\beta : A \otimes_B A \to A \otimes H$ defined for a, $b \in A$ by

$$a \otimes b \mapsto (a \otimes 1_H)\,\delta(b)$$

is bijective. For a survey of Hopf Galois extensions, see [**18**, Chap. 8].

EXAMPLE 2.1. The group algebra $H = k[G]$ of a group G is a Hopf algebra with coproduct, coünit, and antipode respectively given for all $g \in G$ by

$$\Delta(g) = g \otimes g\,, \quad \varepsilon(g) = 1\,, \quad S(g) = g^{-1}\,.$$

This is a pointed Hopf algebra. It is well known (see [**7**, Lemma 4.8]) that an H-comodule algebra A is the same as a G-graded algebra

$$A = \bigoplus_{g \in G} A_g\,.$$

The coaction $\delta : A \to A \otimes H$ is given by $\delta(a) = a \otimes g$ for $a \in A_g$ and $g \in G$. We have $A^H = A_e$, where e is the neutral element of G. Such a algebra is an H-Galois extension of A_e if and only if the product induces isomorphisms

$$A_g \otimes_{A_e} A_h \cong A_{gh} \qquad (g, h \in G)\,.$$

2.2. (Uni)versal extensions. An isomorphism $f : A \to A'$ of H-Galois extensions is an isomorphism of the underlying H-comodule algebras, i.e., an algebra morphism satisfying

$$\delta \circ f = (f \otimes \mathrm{id}_H) \circ \delta\,.$$

Such an isomorphism necessarily sends A^H onto A'^H.

For any Hopf algebra H and any commutative algebra B, let $\mathrm{CGal}_H(B)$ denote the set of isomorphism classes of central H-Galois extensions of B. It was shown in [**13**, Th. 1.4] (see also [**14**, Prop. 1.2]) that any morphism $f : B \to B'$ of commutative algebras induces a functorial map

$$f_* : \mathrm{CGal}_H(B) \to \mathrm{CGal}_H(B')$$

given by $f_*(A) = B' \otimes_B A$ for all H-Galois extensions A of B. The set $\mathrm{CGal}_H(B)$ is also a contravariant functor in H (see [**14**, Prop. 1.3]), but we will not make use of this fact here.

Recall that principal fiber bundles are classified as follows: there is a principal G-bundle $EG \to BG$, called the universal G-bundle such that any principal G-bundle $X \to Y$ is obtained from pulling back the universal one along a continuous map $f : Y \to BG$, which is unique up to homotopy.

By analogy, a *(uni)versal H-Galois extension* would be a central H-Galois extension $\mathcal{B}_H \subset \mathcal{A}_H$ such that for any commutative algebra B and any central H-Galois extension A of B there is a (unique) morphism of algebras $f : \mathcal{B}_H \to B$ such that $f_*(\mathcal{A}_H) \cong A$. In other words, the map

$$\mathrm{Alg}(\mathcal{B}_H, B) \to \mathrm{CGal}_H(B)\,;\; f \mapsto f_*(\mathcal{A}_H)$$

would be surjective (bijective). We have no idea if such (uni)versal H-Galois extensions exist for general Hopf algebras.

In the sequel, we shall only consider the case where $B = k$ and the H-Galois extensions of k are *cleft*. Such extensions coincide with the twisted algebras ${}^{\alpha}H$ introduced in the next section. To such an H-Galois extension we shall associate a central H-Galois extension $\mathcal{B}_H^{\alpha} \subset \mathcal{A}_H^{\alpha}$, such that the functor $\mathrm{Alg}(\mathcal{B}_H^{\alpha}, -)$ parametrizes the "forms" of ${}^{\alpha}H$. In this way we obtain an H-Galois extension that is versal for a family of H-Galois extensions close to ${}^{\alpha}H$ in some appropriate étale-like Grothendieck topology.

3. Twisted algebras

The definition of the twisted algebras ${}^{\alpha}H$ uses the concept of a cocycle, which we now recall.

3.1. Cocycles. Let H be a Hopf algebra and B a commutative algebra. We use the following terminology. A bilinear map $\alpha : H \times H \to B$ is a *cocycle of H with values in B* if

$$\sum_{(x),(y)} \alpha(x_{(1)}, y_{(1)})\, \alpha(x_{(2)}y_{(2)}, z) = \sum_{(y),(z)} \alpha(y_{(1)}, z_{(1)})\, \alpha(x, y_{(2)}z_{(2)})$$

for all $x, y, z \in H$. In the literature, what we call a cocycle is often referred to as a "left 2-cocycle."

A bilinear map $\alpha : H \times H \to B$ is said to be *normalized* if

(3.1) $$\alpha(x, 1_H) = \alpha(1_H, x) = \varepsilon(x)\, 1_B$$

for all $x \in H$.

Two cocycles $\alpha, \beta : H \times H \to B$ are said to be *cohomologous* if there is an invertible linear map $\lambda : H \to B$ such that

$$\beta(x,y) = \sum_{(x),(y)} \lambda(x_{(1)})\, \lambda(y_{(1)})\, \alpha(x_{(2)}, y_{(2)})\, \lambda^{-1}(x_{(3)}y_{(3)}) \tag{3.2}$$

for all $x, y \in H$. Here "invertible" means invertible with respect to the convolution product and $\lambda^{-1} : H \to B$ denotes the inverse of λ. We write $\alpha \sim \beta$ if α, β are cohomologous cocycles. The relation $\sim$ is an equivalence relation on the set of cocycles of H with values in B.

3.2. Twisted product. Let H be a Hopf algebra, B a commutative algebra, and $\alpha : H \times H \to B$ be a normalized cocycle with values in B. From now on, all cocycles are assumed to be invertible with respect to the convolution product.

Let u_H be a copy of the underlying vector space of H. Denote the identity map from H to u_H by $x \mapsto u_x$ $(x \in H)$.

We define the *twisted algebra* $B \otimes {}^{\alpha}H$ as the vector space $B \otimes u_H$ equipped with the associative product given by

$$(b \otimes u_x)(c \otimes u_y) = \sum_{(x),(y)} bc\, \alpha(x_{(1)}, y_{(1)}) \otimes u_{x_{(2)}y_{(2)}} \tag{3.3}$$

for all b, $c \in B$ and x, $y \in H$. Since α is a normalized cocycle, $1_B \otimes u_1$ is the unit of $B \otimes {}^{\alpha}H$.

The algebra $A = B \otimes {}^{\alpha}H$ is an H-comodule algebra with coaction

$$\delta = \mathrm{id}_B \otimes \Delta : A = B \otimes H \to B \otimes H \otimes H = A \otimes H\,.$$

The subalgebra of coïnvariants of $B \otimes {}^{\alpha}H$ coincides with $B \otimes u_1$. Using (3.1) and (3.3), it is easy to check that this subalgebra lies in the center of $B \otimes {}^{\alpha}H$.

It is well known that each twisted algebra $B \otimes {}^{\alpha}H$ is a central H-Galois extension of B. Actually, the class of twisted algebra coincides with the class of so-called central cleft H-Galois extensions; see [**6**], [**9**], [**18**, Prop. 7.2.3].

An important special case of this construction occurs when $B = k$ is the ground field and $\alpha : H \times H \to k$ is a cocycle of H with values in k. In this case, we simply call α a cocycle of H. Then the twisted algebra $k \otimes {}^{\alpha}H$, which we henceforth denote by ${}^{\alpha}H$, coincides with u_H equipped with the associative product

$$u_x\, u_y = \sum_{(x),(y)} \alpha(x_{(1)}, y_{(1)})\, u_{x_{(2)}y_{(2)}}$$

for all x, $y \in H$. The twisted algebras of the form ${}^{\alpha}H$ coincide with the so-called *cleft H-Galois objects*, which are the cleft H-Galois extensions of the ground field k. We point out that for certain Hopf algebras H all H-Galois objects are cleft, e.g., if H is finite-dimensional or is a pointed Hopf algebra.

When $H = k[G]$ is the Hopf algebra of a group as in Example 2.1, then a G-graded algebra $A = \bigoplus_{g \in G} A_g$ is an H-Galois object if and only if $A_g A_h = A_{gh}$ for all $g, h \in G$ and $\dim A_g = 1$ for all $g \in G$. Such an H-extension is cleft and thus isomorphic to ${}^{\alpha}H$ for some normalized invertible cocycle α.

3.3. Isomorphisms of twisted algebras. By [**6, 8**] there is an isomorphism of H-comodule algebras between the twisted algebras ${}^{\alpha}H$ and ${}^{\beta}H$ if and only if the cocycles α and β are cohomologous in the sense of (3.2). It follows that the set of isomorphism classes of cleft H-Galois objects is in bijection with the set of cohomology classes of invertible cocycles of H.

When the Hopf algebra H is cocommutative, then the convolution product of two cocycles is a cocycle and the set of cohomology classes of invertible cocycles of H is a group. This applies to the case $H = k[G]$; in this case the group of cohomology classes of invertible cocycles of $k[G]$ is isomorphic to the cohomology group $H^2(G, k^\times)$ of the group G with values in the group $k^\times$ of invertible elements of k.

In general, the convolution product of two cocycles is *not* a cocycle and thus the set of cohomology classes of invertible cocycles is not a group. One of the *raisons d'être* of the constructions presented here and in [**2**] lies in the lack of a suitable cohomology group governing the situation. We come up instead with the generic Galois extension defined below.

4. The generic cocycle

Let H be a Hopf algebra and $\alpha : H \times H \to k$ an invertible normalized cocycle.

4.1. The cocycle σ. Our first aim is to construct a "generic" cocycle of H that is cohomologous to α.

We start from the equation (3.2)

$$\beta(x,y) = \sum_{(x),(y)} \lambda(x_{(1)})\,\lambda(y_{(1)})\,\alpha(x_{(2)}, y_{(2)})\,\lambda^{-1}(x_{(3)}y_{(3)})$$

expressing that a cocycle β is cohomologous to α, and the equation

$$\sum_{(x)} \lambda(x_{(1)})\,\lambda^{-1}(x_{(2)}) = \sum_{(x)} \lambda^{-1}(x_{(1)})\,\lambda(x_{(2)}) = \varepsilon(x)\,1 \tag{4.1}$$

expressing that the linear form λ is invertible with inverse λ^{-1}. To obtain the generic cocycle, we proceed to mimic (3.2), replacing the scalars $\lambda(x), \lambda^{-1}(x)$ respectively by symbols t_x, t_x^{-1} satisfying (4.1).

Let us give a meaning to the symbols t_x, t_x^{-1}. To this end we pick another copy t_H of the underlying vector space of H and denote the identity map from H to t_H by $x \mapsto t_x$ ($x \in H$).

Let $S(t_H)$ be the symmetric algebra over the vector space t_H. If $\{x_i\}_{i\in I}$ is a basis of H, then $S(t_H)$ is isomorphic to the polynomial algebra over the indeterminates $\{t_{x_i}\}_{i\in I}$.

By [**2**, Lemma A.1] there is a unique linear map $x \mapsto t_x^{-1}$ from H to the field of fractions $\operatorname{Frac} S(t_H)$ of $S(t_H)$ such that for all $x \in H$,

$$\sum_{(x)} t_{x_{(1)}}\, t^{-1}_{x_{(2)}} = \sum_{(x)} t^{-1}_{x_{(1)}}\, t_{x_{(2)}} = \varepsilon(x)\,1\,. \tag{4.2}$$

Equation (4.2) is the symbolic counterpart of (4.1).

Mimicking (3.2), we define a bilinear map

$$\sigma : H \times H \to \operatorname{Frac} S(t_H)$$

with values in the field of fractions $\operatorname{Frac} S(t_H)$ by the formula

$$\sigma(x,y) = \sum_{(x),(y)} t_{x_{(1)}}\, t_{y_{(1)}}\, \alpha(x_{(2)}, y_{(2)})\, t^{-1}_{x_{(3)}y_{(3)}} \tag{4.3}$$

for all $x, y \in H$. The bilinear map σ is a cocycle of H with values in $\operatorname{Frac} S(t_H)$; by definition, it is cohomologous to α. We call σ the *generic cocycle attached to* α.

The cocycle α being invertible, so is σ, with inverse σ^{-1} given for all $x, y \in H$ by

$$\sigma^{-1}(x,y) = \sum_{(x),(y)} t_{x_{(1)}y_{(1)}} \, \alpha^{-1}(x_{(2)}, y_{(2)}) \, t^{-1}_{x_{(3)}} \, t^{-1}_{y_{(3)}} \,, \tag{4.4}$$

where α^{-1} is the inverse of α.

In the case where $H = k[G]$ is the Hopf algebra of a group, the generic cocycle and its inverse have the following simple expressions:

$$\sigma(g,h) = \alpha(g,h) \, \frac{t_g \, t_h}{t_{gh}} \quad \text{and} \quad \sigma^{-1}(g,h) = \frac{1}{\alpha(g,h)} \, \frac{t_{gh}}{t_g \, t_h} \tag{4.5}$$

for all $g, h \in G$.

4.2. The generic base space. Let $\mathcal{B}_H^\alpha$ be the subalgebra of $\operatorname{Frac} S(t_H)$ generated by the values of the generic cocycle σ and of its inverse σ^{-1}. For reasons that will become clear in Section 6, we call $\mathcal{B}_H^\alpha$ the *generic base space.*

Since $\mathcal{B}_H^\alpha$ is a subalgebra of the field $\operatorname{Frac} S(t_H)$, it is a domain and the transcendence degree of the field of fractions of $\mathcal{B}_H^\alpha$ cannot exceed the dimension of H.

In the case where H is finite-dimensional, $\mathcal{B}_H^\alpha$ is a finitely generated algebra. One can obtain a presentation of $\mathcal{B}_H^\alpha$ by generators and relations using standard monomial order techniques of commutative algebra.

4.3. A computation. Let $H = k[\mathbb{Z}]$ be the Hopf algebra of the group $\mathbb{Z}$ of integers. We write $\mathbb{Z}$ multiplicatively and identify its elements with the powers x^m of a variable x ($m \in \mathbb{Z}$).

We take α to be the trivial cocycle, i.e., $\alpha(g,h) = 1$ for all $g, h \in \mathbb{Z}$ (this is no restriction since $H^2(\mathbb{Z}, k^\times) = 0$). In this case the symmetric algebra $S(t_H)$ coincides with the polynomial algebra $k[t_m \,|\, m \in \mathbb{Z}]$. Set $y_m = t_m / t_1^m$ for each $m \in \mathbb{Z}$. We have $y_1 = 1$ and $y_0 = t_0$.

By (4.5), the generic cocycle is given by

$$\sigma(x^m, x^n) = \frac{t_m t_n}{t_{m+n}}$$

for all $m, n \in \mathbb{Z}$. This can be reformulated as

$$\sigma(x^m, x^n) = \frac{y_m y_n}{y_{m+n}} \,.$$

The inverse of σ is given by

$$\sigma^{-1}(x^m, x^n) = \frac{1}{\sigma(x^m, x^n)} = \frac{y_{m+n}}{y_m y_n} \,.$$

A simple computation yields the following expressions of y_m in the values of σ and σ^{-1}:

$$y_m = \begin{cases} \sigma^{-1}(x^{m-1}, x) \, \sigma^{-1}(x^{m-2}, x) \cdots \sigma^{-1}(x, x) & \text{if } m \geq 2 \,, \\ 1 & \text{if } m = 1 \,, \\ \sigma(x^0, x^0) & \text{if } m = 0 \,, \\ \sigma(x^m, x^{-m}) \sigma(x^{-m-1}, x) \sigma(x^{-m-2}, x) \cdots \sigma(x, x) \sigma(x^0, x^0) & \text{if } m \leq -1 \,. \end{cases}$$

It follows that the elements $y_m^{\pm 1}$ belong to $\mathcal{B}_H^\alpha$ for all $m \in \mathbb{Z} - \{1\}$ and generate this algebra. It is easy to check that the family $(y_m)_{m\neq 1}$ is algebraically independent, so that $\mathcal{B}_H^\alpha$ is the Laurent polynomial algebra

$$\mathcal{B}_H^\alpha = k[\, y_m^{\pm 1} \,|\, m \in \mathbb{Z} - \{1\}\,] \,.$$

We deduce the algebra isomorphism

$$k[t_m^{\pm 1} \,|\, m \in \mathbb{Z}] \cong \mathcal{B}_H^\alpha [t_1^{\pm 1}] \,. \tag{4.6}$$

If in the previous computations we replace $\mathbb{Z}$ by the cyclic group $\mathbb{Z}/N$, where N is some integer $N \geq 2$, then the algebra $\mathcal{B}_H^\alpha$ is again a Laurent polynomial algebra:

$$\mathcal{B}_H^\alpha = k[y_0^{\pm 1}, y_2^{\pm 1}, \dots, y_N^{\pm 1}]$$

where $y_0, y_2, \dots, y_{N-1}$ are defined as above and $y_N = t_0/t_1^N$. In this case, the algebra $k[t_0^{\pm 1}, t_1^{\pm 1}, \dots, t_{N-1}^{\pm 1}]$ is an integral extension of $\mathcal{B}_H^\alpha$:

$$k[t_0^{\pm 1}, t_1^{\pm 1}, \dots, t_{N-1}^{\pm 1}] \cong \mathcal{B}_H^\alpha [t_1]/(t_1^N - y_0/y_N) \,.$$

5. The Sweedler algebra

We now illustrate the constructions of Section 4 on Sweedler's four-dimensional Hopf algebra. We assume in this section that the characteristic of the ground field k is different from 2.

The *Sweedler algebra* H_4 is the algebra generated by two elements x, y subject to the relations

$$x^2 = 1 \,, \quad xy + yx = 0 \,, \quad y^2 = 0 \,.$$

It is four-dimensional. As a basis of H_4 we take the set $\{1, x, y, z\}$, where $z = xy$.

The algebra H_4 carries the structure of a Hopf algebra with coproduct, coünit, and antipode given by

$$\begin{array}{rclrcl}
\Delta(1) & = & 1 \otimes 1 \,, & \Delta(x) & = & x \otimes x \,, \\
\Delta(y) & = & 1 \otimes y + y \otimes x \,, & \Delta(z) & = & x \otimes z + z \otimes 1 \,, \\
\varepsilon(1) & = & \varepsilon(x) = 1 \,, & \varepsilon(y) & = & \varepsilon(z) = 0 \,, \\
S(1) & = & 1 \,, & S(x) & = & x \,, \\
S(y) & = & z \,, & S(z) & = & -y \,.
\end{array}$$

By definition, the symbols t_x and t_x^{-1} satisfy the equations

$$\begin{array}{rclrcl}
t_1 \, t_1^{-1} & = & 1 \,, & t_x \, t_x^{-1} & = & 1 \,, \\
t_1 \, t_y^{-1} + t_y \, t_x^{-1} & = & 0 \,, & t_x \, t_z^{-1} + t_z \, t_1^{-1} & = & 0 \,.
\end{array}$$

Hence,

$$t_1^{-1} = \frac{1}{t_1} \,, \quad t_x^{-1} = \frac{1}{t_x} \,, \quad t_y^{-1} = -\frac{t_y}{t_1 t_x} \,, \quad t_z^{-1} = -\frac{t_z}{t_1 t_x} \,.$$

Masuoka [**15**] showed that any cleft H_4-Galois object has, up to isomorphism, the following presentation:

$${}^\alpha H_4 = k\langle\, u_x, u_y \,|\, u_x^2 = a \,, \;\; u_x u_y + u_y u_x = b \,, \;\; u_y^2 = c \,\rangle$$

for some scalars a, b, c with $a \neq 0$. To indicate the dependence on the parameters a, b c, we denote ${}^\alpha H_4$ by $A_{a,b,c}$.

It is easy to check that the center of $A_{a,b,c}$ is trivial for all values of a, b, c. Moreover, the algebra $A_{a,b,c}$ is simple if and only $b^2 - 4ac \neq 0$. If $b^2 - 4ac = 0$, then $A_{a,b,c}$ is isomorphic as an algebra to H_4; the latter is not semisimple since the two-sided ideal generated by y is nilpotent.

The generic cocycle σ attached to α has the following values:

$$\begin{aligned}
\sigma(1,1) &= \sigma(1,x) = \sigma(x,1) = t_1\,, \\
\sigma(1,y) &= \sigma(y,1) = \sigma(1,z) = \sigma(z,1) = 0\,, \\
\sigma(x,x) &= at_x^2 t_1^{-1}, \\
\sigma(y,y) &= \sigma(z,y) = -\sigma(y,z) = (at_y^2 + bt_1t_y + ct_1^2)\, t_1^{-1}, \\
\sigma(x,y) &= -\sigma(x,z) = (at_xt_y - t_1t_z)\, t_1^{-1}, \\
\sigma(y,x) &= \sigma(z,x) = (bt_1t_x + at_xt_y + t_1t_z)\, t_1^{-1}, \\
\sigma(z,z) &= -(t_z^2 + bt_xt_z + act_x^2)\, t_1^{-1},
\end{aligned}$$

The values of the inverse σ^{-1} are equal to the values of σ possibly divided by positive powers of t_1 and of $\sigma(x,x) = at_x^2 t_1^{-1}$.

By definition, $\mathcal{B}_{H_4}^{\alpha}$ is the subalgebra of $\operatorname{Frac} S(t_{H_4})$ generated by the values of σ and σ^{-1}. If we set

$$\begin{aligned}
&E = t_1\,, \quad R = a\,t_x^2\,, \quad S = a\,t_y^2 + b\,t_1t_y + c\,t_1^2\,, \\
&T = t_x\,(2a\,t_y + b\,t_1)\,, \quad U = a\,t_x^2\,(2\,t_z + b\,t_x)\,,
\end{aligned} \tag{5.1}$$

then we can reformulate the above (nonzero) values of σ as follows:

$$\begin{aligned}
&\sigma(1,1) = \sigma(1,x) = \sigma(x,1) = E\,, \\
&\sigma(x,x) = \frac{R}{E}\,, \\
&\sigma(y,y) = \sigma(z,y) = -\sigma(y,z) = \frac{S}{E}\,, \\
&\sigma(x,y) = -\sigma(x,z) = \frac{RT - EU}{2ER}\,, \\
&\sigma(y,x) = \sigma(z,x) = \frac{RT + EU}{2ER}\,, \\
&\sigma(z,z) = \frac{a\,U^2 - (b^2 - 4ac)\,R^3}{4a\,ER^2}\,.
\end{aligned}$$

From the previous equalities we conclude that $E^{\pm 1}$, $R^{\pm 1}$, S, T, U belong to $\mathcal{B}_{H_4}^{\alpha}$ and that they generate it as an algebra.

In [**2**, Sect. 10] we obtained the following presentation of $\mathcal{B}_{H_4}^{\alpha}$ by generators and relations.

THEOREM 5.1. *We have*

$$\mathcal{B}_{H_4}^{\alpha} \cong k[E^{\pm 1}, R^{\pm 1}, S, T, U]/(P_{a,b,c})\,,$$

where

$$P_{a,b,c} = T^2 - 4RS - \frac{b^2 - 4ac}{a}\, E^2 R\,.$$

It follows from the previous theorem that the algebra morphisms from $\mathcal{B}_{H_4}^{\alpha}$ to a field K containing k are in one-to-one correspondence with the quintuples $(e,r,s,t,u) \in K^5$ verifying $e \neq 0$, $r \neq 0$, and the equation

$$t^2 - 4rs = \frac{b^2 - 4ac}{a}\, e^2 r\,. \tag{5.2}$$

In other words, the set of K-points of $\mathcal{B}_{H_4}^{\alpha}$ is the hypersurface of equation (5.2) in $K^{\times} \times K^{\times} \times K \times K \times K$.

6. The generic Galois extension

As in Section 4, we consider a Hopf algebra H and an invertible normalized cocycle $\alpha : H \times H \to k$. Let ${}^{\alpha}H$ be the corresponding twisted algebra.

6.1. The algebra $\mathcal{A}_H^{\alpha}$. By the definition of the commutative algebra $\mathcal{B}_H^{\alpha}$ given in Section 4.2, the generic cocycle σ takes its values in $\mathcal{B}_H^{\alpha}$. Therefore we may apply the construction of Section 3.2 and consider the twisted algebra

$$\mathcal{A}_H^{\alpha} = \mathcal{B}_H^{\alpha} \otimes {}^{\sigma}H \, .$$

The product of $\mathcal{A}_H^{\alpha}$ is given for all b, $c \in \mathcal{B}_H^{\alpha}$ and x, $y \in H$ by

$$(b \otimes u_x)(c \otimes u_y) = \sum_{(x),(y)} bc\,\sigma(x_{(1)}, y_{(1)}) \otimes u_{x_{(2)}y_{(2)}} \, .$$

We call $\mathcal{A}_H^{\alpha}$ the *generic Galois extension* attached to the cocycle α.

The subalgebra of coïnvariants of $\mathcal{A}_H^{\alpha}$ is equal to $\mathcal{B}_H^{\alpha} \otimes u_1$; this subalgebra is central in $\mathcal{A}_H^{\alpha}$. Therefore, $\mathcal{A}_H^{\alpha}$ is a central cleft H-Galois extension of $\mathcal{B}_H^{\alpha}$.

By [**2**, Prop. 5.3], there is an algebra morphism $\chi_0 : \mathcal{B}_H^{\alpha} \to k$ such that

$$\chi_0\big(\sigma(x,y)\big) = \alpha(x,y) \quad \text{and} \quad \chi_0\big(\sigma^{-1}(x,y)\big) = \alpha^{-1}(x,y)$$

for all $x, y \in H$. Consider the maximal ideal $\mathfrak{m}_0 = \operatorname{Ker}(\chi_0 : \mathcal{B}_H^{\alpha} \to k)$ of $\mathcal{B}_H^{\alpha}$. According to [**2**, Prop. 6.2], there is an isomorphism of H-comodule algebras

$$\mathcal{A}_H^{\alpha}/\mathfrak{m}_0\, \mathcal{A}_H^{\alpha} \cong {}^{\alpha}H \, .$$

Thus, $\mathcal{A}_H^{\alpha}$ is a *flat deformation* of ${}^{\alpha}H$ over the commutative algebra $\mathcal{B}_H^{\alpha}$.

Certain properties of ${}^{\alpha}H$ lift to the generic Galois extension $\mathcal{A}_H^{\alpha}$ such as the one recorded in the following result of [**2**], where $\operatorname{Frac} \mathcal{B}_H^{\alpha}$ stands for the field of fractions of $\mathcal{B}_H^{\alpha}$.

THEOREM 6.1. *Assume that the ground field k is of characteristic zero and the Hopf algebra H is finite-dimensional. If the algebra ${}^{\alpha}H$ is simple (resp. semisimple), then so is*

$$\operatorname{Frac} \mathcal{B}_H^{\alpha} \otimes_{\mathcal{B}_H^{\alpha}} \mathcal{A}_H^{\alpha} = \operatorname{Frac} \mathcal{B}_H^{\alpha} \otimes {}^{\sigma}H \, .$$

6.2. Forms. We have just observed that ${}^{\alpha}H \cong \mathcal{A}_H^{\alpha}/\mathfrak{m}_0\, \mathcal{A}_H^{\alpha}$ for some maximal ideal $\mathfrak{m}_0$ of $\mathcal{B}_H^{\alpha}$. We may now wonder what can be said of the other *central specializations* of $\mathcal{A}_H^{\alpha}$, that is of the quotients $\mathcal{A}_H^{\alpha}/\mathfrak{m}\, \mathcal{A}_H^{\alpha}$, where $\mathfrak{m}$ is an arbitrary maximal ideal of $\mathcal{B}_H^{\alpha}$. To answer this question, we need the following terminology.

Let $\beta : H \times H \to K$ be a normalized invertible cocycle with values in a field K containing the ground field k. We say that the twisted H-comodule algebra $K \otimes {}^{\beta}H$ is a *K-form* of ${}^{\alpha}H$ if there is a field L containing K and an L-linear isomorphism of H-comodule algebras

$$L \otimes_K (K \otimes {}^{\beta}H) \cong L \otimes_k {}^{\alpha}H \, .$$

We now state two theorems relating forms of ${}^{\alpha}H$ to central specializations of the generic Galois extension $\mathcal{A}_H^{\alpha}$. For proofs, see [**2**, Sect. 7].

THEOREM 6.2. *For any K-form $K \otimes {}^{\beta}H$ of ${}^{\alpha}H$, where $\beta : H \times H \to K$ is a normalized invertible cocycle with values in an extension K of k, there exist an algebra morphism $\chi : \mathcal{B}_H^{\alpha} \to K$ and a K-linear isomorphism of H-comodule algebras*

$$K_{\chi} \otimes_{\mathcal{B}_H^{\alpha}} \mathcal{A}_H^{\alpha} \cong K \otimes {}^{\beta}H \, .$$

Here K_χ stands for K equipped with the $\mathcal{B}_H^\alpha$-module structure induced by the algebra morphism $\chi : \mathcal{B}_H^\alpha \to K$. We have

$$K_\chi \otimes_{\mathcal{B}_H^\alpha} \mathcal{A}_H^\alpha \cong \mathcal{A}_H^\alpha / \mathfrak{m}_\chi \, \mathcal{A}_H^\alpha \, ,$$

where $\mathfrak{m}_\chi = \operatorname{Ker}(\chi : \mathcal{B}_H^\alpha \to K)$.

There is a converse to Theorem 6.2; it requires an additional condition.

THEOREM 6.3. *If* $\operatorname{Frac} S(t_H)$ *is integral over the subalgebra* $\mathcal{B}_H^\alpha$, *then for any field* K *containing* k *and any algebra morphism* $\chi : \mathcal{B}_H^\alpha \to K$, *the* H*-comodule* K*-algebra* $K_\chi \otimes_{\mathcal{B}_H^\alpha} \mathcal{A}_H^\alpha = \mathcal{A}_H^\alpha / \mathfrak{m}_\chi \, \mathcal{A}_H^\alpha$ *is a* K*-form of* ${}^\alpha H$.

It follows that if $\operatorname{Frac} S(t_H)$ is integral over $\mathcal{B}_H^\alpha$, then the map

$$\begin{array}{ccc} \operatorname{Alg}(\mathcal{B}_H^\alpha, K) & \longrightarrow & K\text{-}\operatorname{Forms}({}^\alpha H) \\ \chi & \longmapsto & K_\chi \otimes_{\mathcal{B}_H^\alpha} \mathcal{A}_H^\alpha = \mathcal{A}_H^\alpha / \mathfrak{m}_\chi \, \mathcal{A}_H^\alpha \end{array}$$

is a *surjection* from the set of algebra morphisms $\mathcal{B}_H^\alpha \to K$ to the set of isomorphism classes of K-forms of ${}^\alpha H$. Thus the set $\operatorname{Alg}(\mathcal{B}_H^\alpha, K)$ parametrizes the K-forms of ${}^\alpha H$. Using terminology of singularity theory, we say that the Galois extension $\mathcal{B}_H^\alpha \subset \mathcal{A}_H^\alpha$ is a *versal deformation space* for the forms of ${}^\alpha H$ (we would call this space universal if the above surjection was bijective).

By Theorem 6.1, the central localization $\operatorname{Frac} \mathcal{B}_H^\alpha \otimes_{\mathcal{B}_H^\alpha} \mathcal{A}_H^\alpha$ is a simple algebra if the algebra ${}^\alpha H$ is simple. Under the integrality condition above, we have the following related result (see [**2**, Th. 7.4]).

THEOREM 6.4. *If* $\operatorname{Frac} S(t_H)$ *is integral over* $\mathcal{B}_H^\alpha$ *and if the algebra* ${}^\alpha H$ *is simple, then* $\mathcal{A}_H^\alpha$ *is an Azumaya algebra with center* $\mathcal{B}_H^\alpha$.

This means that $\mathcal{A}_H^\alpha / \mathfrak{m} \, \mathcal{A}_H^\alpha$ is a simple algebra for any maximal ideal $\mathfrak{m}$ of $\mathcal{B}_H^\alpha$. For instance, any full matrix algebra with entries in a commutative algebra is Azumaya.

EXAMPLE 6.5. For the Sweedler algebra H_4, we proved in [**2**, Sect. 10] that $\mathcal{A}_{H_4}^\alpha$ is given as an algebra by

$$\mathcal{A}_{H_4}^\alpha \cong \mathcal{B}_{H_4}^\alpha \langle X, Y \rangle / (X^2 - R \, , \, Y^2 - S \, , \, XY + YX - T) \, , \tag{6.1}$$

where $\mathcal{B}_{H_4}^\alpha$ is as in Theorem 5.1 and the elements R, S, T of $\mathcal{B}_{H_4}^\alpha$ are defined by (5.1). As an $\mathcal{B}_{H_4}^\alpha$-module, $\mathcal{A}_{H_4}^\alpha$ is free with basis $\{1, X, Y, XY\}$.

7. The integrality condition

In view of Theorem 6.3 it is natural to ask the following question.

QUESTION 7.1. *Under which condition on the pair* (H, α) *is* $\operatorname{Frac} S(t_H)$ *integral over the subalgebra* $\mathcal{B}_H^\alpha$*?*

Question 7.1 has a negative answer in the case where $H = k[\mathbb{Z}]$ and α is the trivial cocycle. Indeed, it follows from (4.6) that $\operatorname{Frac} S(t_H)$ is then a pure transcendental extension (of degree one) of the field of fractions of $\mathcal{B}_H^\alpha$.

We give a positive answer in the following important case.

THEOREM 7.2. *Let* H *be a Hopf algebra generated as an algebra by a set* Σ *of grouplike and skew-primitive elements such that the grouplike elements of* Σ *are of finite order and generate the group of grouplike elements of* H *and such that each skew-primitive element of* Σ *generates a finite-dimensional subalgebra of* H. *Then* $\operatorname{Frac} S(t_H)$ *is integral over the subalgebra* $\mathcal{B}_H^\alpha$ *for every cocycle* α *of* H.

Theorem 7.2 implies a positive answer to Question 7.1 for any finite-dimensional Hopf algebra generated by grouplike and skew-primitive elements. It is conjectured that all finite-dimensional *pointed* Hopf algebras are generated by grouplike and skew-primitive elements; if this conjecture holds, then Question 7.1 has a positive answer for any finite-dimensional Hopf algebra that is pointed.

Recall that $g \in H$ is *grouplike* if $\Delta(g) = g \otimes g$; it then follows that $\varepsilon(g) = 1$. The inverse of a grouplike element and the product of two grouplike elements are grouplike. An element $x \in H$ is *skew-primitive* if

$$\Delta(x) = g \otimes x + x \otimes h \tag{7.1}$$

for some grouplike elements $g, h \in H$; this implies $\varepsilon(x) = 0$. The product of a skew-primitive element by a grouplike element is skew-primitive.

In order to prove Theorem 7.2, we need the following lemma.

LEMMA 7.3. *If $x^{[1]}, \dots, x^{[n]}$ are elements of H, then*

$$\begin{aligned} t_{x^{[1]}\cdots x^{[n]}} = \sum_{x^{[1]},\dots,x^{[n]}} & \sigma^{-1}(x^{[1]}_{(1)} \cdots x^{[n-1]}_{(1)}, x^{[n]}_{(1)}) \times \\ & \times \sigma^{-1}(x^{[1]}_{(2)} \cdots x^{[n-2]}_{(2)}, x^{[n-1]}_{(2)}) \cdots \sigma^{-1}(x^{[1]}_{(n-1)}, x^{[2]}_{(n-1)}) \times \\ & \times t_{x^{[1]}_{(n)}} t_{x^{[2]}_{(n)}} \cdots t_{x^{[n-1]}_{(3)}} t_{x^{[n]}_{(2)}} \times \\ & \times \alpha(x^{[1]}_{(n+1)}, x^{[2]}_{(n+1)}) \cdots \alpha(x^{[1]}_{(2n-2)} x^{[2]}_{(2n-2)} \cdots x^{[n-2]}_{(6)}, x^{[n-1]}_{(4)}) \times \\ & \times \alpha(x^{[1]}_{(2n-1)} x^{[2]}_{(2n-1)} \cdots x^{[n-1]}_{(5)}, x^{[n]}_{(3)}) . \end{aligned}$$

PROOF. We prove the formula by induction on n. When $n = 2$ it reduces to

$$t_{xy} = \sum_{(x),(y)} \sigma^{-1}(x_{(1)}, y_{(1)})\, t_{x_{(2)}}\, t_{y_{(2)}}\, \alpha(x_{(3)}, y_{(3)}) \tag{7.2}$$

for $x, y \in H$. Let us first prove (7.2). By (4.4) the right-hand side of (7.2) is equal to

$$\begin{aligned} & \sum_{(x),(y)} t_{x_{(1)}y_{(1)}}\, \alpha^{-1}(x_{(2)}, y_{(2)})\, t^{-1}_{x_{(3)}} \underbrace{t^{-1}_{y_{(3)}}\, t_{y_{(4)}}}\, t_{x_{(4)}}\, \alpha(x_{(5)}, y_{(5)}) \\ = & \sum_{(x),(y)} t_{x_{(1)}y_{(1)}}\, \alpha^{-1}(x_{(2)}, y_{(2)})\, \underbrace{t^{-1}_{x_{(3)}}\, t_{x_{(4)}}}\, \alpha(x_{(5)}, y_{(3)}) \\ = & \sum_{(x),(y)} t_{x_{(1)}y_{(1)}}\, \underbrace{\alpha^{-1}(x_{(2)}, y_{(2)})\, \alpha(x_{(3)}, y_{(3)})} \\ = & \sum_{(x),(y)} t_{x_{(1)}y_{(1)}}\, \varepsilon(x_{(2)})\, \varepsilon(y_{(2)}) \\ = & \sum_{(x),(y)} t_{x_{(1)}y_{(1)}}\, \varepsilon(x_{(2)}y_{(2)}) = t_{xy} . \end{aligned}$$

Let us assume that Lemma 7.3 holds for all n-tuples of H and consider a sequence $(x^{[1]}, x^{[2]}, \dots, x^{[n+1]})$ of $n+1$ elements of H. By the induction hypothesis and by (7.2), $t_{x^{[1]}x^{[2]}\cdots x^{[n+1]}} = t_{(x^{[1]}x^{[2]})\cdots x^{[n+1]}}$ is equal to

$$\sum_{x^{[1]},x^{[2]},\dots,x^{[n+1]}} \sigma^{-1}\big((x^{[1]}_{(1)}x^{[2]}_{(1)})\cdots x^{[n]}_{(1)}, x^{[n+1]}_{(1)}\big) \times$$
$$\times\, \sigma^{-1}\big((x^{[1]}_{(2)}x^{[2]}_{(2)})\cdots x^{[n-1]}_{(2)}, x^{[n]}_{(2)}\big)\cdots \sigma^{-1}\big(x^{[1]}_{(n-1)}x^{[2]}_{(n-1)}, x^{[3]}_{(n-1)}\big) \times$$
$$\times\, t_{x^{[1]}_{(n)}x^{[2]}_{(n)}}\, t_{x^{[3]}_{(n)}} \cdots t_{x^{[n]}_{(3)}}\, t_{x^{[n+1]}_{(2)}} \times$$
$$\times\, \alpha\big(x^{[1]}_{(n+1)}x^{[2]}_{(n+1)}, x^{[3]}_{(n+1)}\big)\cdots$$
$$\cdots \alpha\big((x^{[1]}_{(2n-2)}\, x^{[2]}_{(2n-2)})\, x^{[3]}_{(2n-2)} \cdots x^{[n-1]}_{(6)}, x^{[n]}_{(4)}\big) \times$$
$$\times\, \alpha\big((x^{[1]}_{(2n-1)}\, x^{[2]}_{(2n-1)})\, x^{[3]}_{(2n-1)} \cdots x^{[n]}_{(5)}, x^{[n+1]}_{(3)}\big)$$
$$= \sum_{x^{[1]},x^{[2]},\dots,x^{[n+1]}} \sigma^{-1}\big(x^{[1]}_{(1)}x^{[2]}_{(1)}\cdots x^{[n]}_{(1)}, x^{[n+1]}_{(1)}\big)\,\sigma^{-1}\big(x^{[1]}_{(2)}x^{[2]}_{(2)}\cdots x^{[n-1]}_{(2)}, x^{[n]}_{(2)}\big)\cdots$$
$$\cdots \sigma^{-1}\big(x^{[1]}_{(n-1)}x^{[2]}_{(n-1)}, x^{[3]}_{(n-1)}\big)\,\sigma^{-1}\big(x^{[1]}_{(n)}, x^{[2]}_{(n)}\big) \times$$
$$\times\, t_{x^{[1]}_{(n+1)}}\, t_{x^{[2]}_{(n+1)}}\, t_{x^{[3]}_{(n)}} \cdots t_{x^{[n]}_{(3)}}\, t_{x^{[n+1]}_{(2)}} \times$$
$$\times\, \alpha\big(x^{[1]}_{(n+2)}, x^{[2]}_{(n+2)}\big)\,\alpha\big(x^{[1]}_{(n+3)}x^{[2]}_{(n+3)}, x^{[3]}_{(n+1)}\big)\cdots$$
$$\cdots \alpha\big(x^{[1]}_{(2n)}\, x^{[2]}_{(2n)}\, x^{[3]}_{(2n-2)} \cdots x^{[n-1]}_{(6)}, x^{[n]}_{(4)}\big) \times$$
$$\times\, \alpha\big(x^{[1]}_{(2n+1)}\, x^{[2]}_{(2n+1)}\, x^{[3]}_{(2n-1)} \cdots x^{[n]}_{(5)}, x^{[n+1]}_{(3)}\big),$$

which is the desired formula for $n+1$ elements. □

PROOF OF THEOREM 7.2. Let A be the integral closure of $\mathcal{B}^{\alpha}_{H}$ in $\operatorname{Frac} S(t_H)$. To prove the theorem it suffices to establish that each generator t_z of $S(t_H)$ belongs to A.

We start with the unit of H. By [**2**, Lemma 5.1], $t_1 = \sigma(1,1)$. Thus t_1 belongs to $\mathcal{B}^{\alpha}_{H}$, hence to A.

Let g be a grouplike element of the generating set Σ. By hypothesis, there is an integer $n \geq 2$ such that $g^n = 1$. We apply Lemma 7.3 to $x^{[1]} = \cdots = x^{[n]} = g$. Since any iterated coproduct $\Delta^{(p)}$ applied to g yields

$$\Delta^{(p)}(g) = g \otimes g \otimes \cdots \otimes g\,, \tag{7.3}$$

where the right-hand side is the tensor product of p copies of g, we obtain

$$t_{g^n} = \sigma^{-1}(g^{n-1}, g)\,\sigma^{-1}(g^{n-2}, g)\cdots\sigma^{-1}(g,g)\, t_g^n \times \\ \times\, \alpha(g,g)\cdots\alpha(g^{n-2},g)\,\alpha(g^{n-1},g)\,. \tag{7.4}$$

Since the values of an invertible cocycle on grouplike elements are invertible elements, since $t_{g^n} = t_1$, and since $\sigma^{-1}(g,h) = 1/\sigma(g,h)$ for all grouplike elements g, h, Formula (7.4) implies

$$t_g^n = t_1\, \frac{\sigma(g^{n-1},g)\,\sigma(g^{n-2},g)\cdots\sigma(g,g)}{\alpha(g,g)\cdots\alpha(g^{n-2},g)\,\alpha(g^{n-1},g)}\,.$$

The right-hand side belongs to $\mathcal{B}^{\alpha}_{H}$. It follows that t_g is in A for each grouplike element of Σ.

Since the grouplike elements of Σ are of finite order and generate the group of grouplike elements of H, any grouplike element g of H is a product $g = g^{[1]}\cdots g^{[n]}$

of grouplike elements of Σ for which we have just established that $t_{g^{[1]}}, \ldots, t_{g^{[n]}}$ belong to A. It then follows from Lemma 7.3 and (7.3) that

$$t_{g^{[1]}\cdots g^{[n]}} = \kappa(g^{[1]}, \ldots, g^{[n]})\, t_{g^{[1]}}\, t_{g^{[2]}} \cdots t_{g^{[n-1]}}\, t_{g^{[n]}}\,,$$

where $\kappa(g^{[1]}, \ldots, g^{[n]})$ is the invertible element of $\mathcal{B}_H^\alpha$ given by

$$\kappa(g^{[1]}, \ldots, g^{[n]}) = \frac{\alpha(g^{[1]}, g^{[2]}) \cdots \alpha(g^{[1]}\, g^{[2]} \cdots g^{[n-2]}, g^{[n-1]})\, \alpha(g^{[1]}\, g^{[2]} \cdots g^{[n-1]}, g^{[n]})}{\sigma(g^{[1]}\, g^{[2]} \cdots g^{[n-1]}, g^{[n]})\, \sigma(g^{[1]}\, g^{[2]} \cdots g^{[n-2]}, g^{[n-1]}) \cdots \sigma(g^{[1]}, g^{[2]})}\,.$$

Therefore, $t_g \in A$ for every grouplike element of H.

We next show that t_x belongs to A for every skew-primitive element x of Σ. It is easy to check that if x satisfies (7.1), then for all $p \geq 2$,

$$\Delta^{(p)}(x) = g^{\otimes p} \otimes x + \sum_{i=1}^{p-1} g^{\otimes(p-i)} \otimes x \otimes h^{\otimes i} + x \otimes h^{\otimes p}\,. \tag{7.5}$$

Thus the iterated coproduct of any skew-primitive element x is a sum of tensor product of elements, all of which are grouplike, except for exactly one, which is x. It then follows from Lemma 7.3 and (7.5) that for each $n \geq 1$ the element t_{x^n} is a linear combination with coefficients in $\mathcal{B}_H^\alpha$ of monomials of the form $t_{g_1} t_{g_2} \cdots t_{g_{n-p}} t_x^p$, where $0 \leq p \leq n$ and $g_1, \ldots, g_{n-p}$ are grouplike elements. It is easily checked that in this linear combination there is a unique monomial of the form t_x^n whose coefficient is the invertible element of $\mathcal{B}_H^\alpha$

$$\sigma^{-1}(g^{n-1}, g)\, \sigma^{-1}(g^{n-2}, g) \cdots \sigma^{-1}(g, g)\, \alpha(h, h) \cdots \alpha(h^{n-2}, h)\, \alpha(h^{n-1}, h)\,.$$

Since t_g belongs to A for any grouplike element $g \in H$, it follows that, for all $n \geq 1$, the element t_{x^n} is a polynomial of degree n in t_x with coefficients in A. By hypothesis, there are scalars $\lambda_1, \ldots, \lambda_{n-1}, \lambda_n \in k$ for some positive integer n such that

$$x^n + \lambda_1 x^{n-1} + \cdots + \lambda_{n-1} x + \lambda_n = 0\,.$$

Therefore, t_x satisfies a degree n polynomial equation with coefficients in the integral closure A and with highest-degree coefficient equal to 1. This proves that $t_x \in A$.

To complete the proof, it suffices to check that t_z belongs to A for any product z of grouplike or skew-primitive elements $x^{[1]}, \ldots, x^{[n]}$ such that $t_{x^{[1]}}, \ldots, t_{x^{[n]}}$ belong to A. It follows from Lemma 7.3, (7.3), and (7.5) that t_z is a linear combination with coefficients in $\mathcal{B}_H^\alpha$ of products of the variables $t_{x^{[1]}}, \ldots, t_{x^{[n]}}$ and of variables of the form t_g, where g is grouplike. Since these monomials belong to A, so does t_z. □

8. How to construct elements of $\mathcal{B}_H^\alpha$

In the example considered in Section 5 we reformulated the values of the generic cocycle in terms of certain rational fractions E, R, S, T, U. The aim of this last section is to explain how we found these fractions by presenting a general systematic way of producing elements of $\mathcal{B}_H^\alpha$ for an arbitrary Hopf algebra. To this end we introduce a new set of symbols.

8.1. The symbols X_x. Let H be a Hopf algebra and X_H a copy of the underlying vector space of H; we denote the identity map from H to X_H by $x \mapsto X_x$ for all $x \in H$.

Consider the tensor algebra $T(X_H)$ of the vector space X_H over the ground field k:

$$T(X_H) = \bigoplus_{r \geq 0} X_H^{\otimes r} .$$

If $\{x_i\}_{i \in I}$ is a basis of H, then $T(X_H)$ is the free noncommutative algebra over the set of indeterminates $\{X_{x_i}\}_{i \in I}$.

The algebra $T(X_H)$ is an H-comodule algebra equipped with the coaction $\delta : T(X_H) \to T(X_H) \otimes H$ given for all $x \in H$ by

$$\delta(X_x) = \sum_{(x)} X_{x_{(1)}} \otimes x_{(2)} . \tag{8.1}$$

8.2. Coïnvariant elements of $T(X_H)$. Let us now present a general method to construct coïnvariant elements of $T(X_H)$. We need the following terminology.

Given an integer $n \geq 1$, an *ordered partition* of $\{1, \ldots, n\}$ is a partition $\underline{I} = (I_1, \ldots, I_r)$ of $\{1, \ldots, n\}$ into disjoint nonempty subsets $I_1, \ldots, I_r$ such that $i < j$ for all $i \in I_k$ and $j \in I_{k+1}$ $(1 \leq k \leq r-1)$.

If $x[1], \ldots, x[n]$ are n elements of H and if $I = \{i_1 < \cdots < i_p\}$ is a subset of $\{1, \ldots, n\}$, we set $x[I] = x[i_1] \cdots x[i_p] \in H$. If $\underline{I} = (I_1, \ldots, I_r)$ is an ordered partition of $\{1, \ldots, n\}$, then clearly $x[I_1] \cdots x[I_r] = x[1] \cdots x[n]$.

Now let $x[1], \ldots, x[n]$ be n elements of H and $\underline{I} = (I_1, \ldots, I_r)$, $\underline{J} = (J_1, \ldots, J_s)$ be ordered partitions of $\{1, \ldots, n\}$. We consider the following element of $T(X_H)$:

$$P_{x[1],\ldots,x[n];\underline{I},\underline{J}} = \sum_{(x[1]),\ldots,(x[n])} X_{x[I_1]_{(1)}} \cdots X_{x[I_r]_{(1)}} X_{S(x[J_s]_{(2)})} \cdots X_{S(x[J_1]_{(2)})} . \tag{8.2}$$

The element $P_{x[1],\ldots,x[n];\underline{I},\underline{J}}$ is an homogeneous element of $T(X_H)$ of degree $r+s$. Observe that $P_{x[1],\ldots,x[n];\underline{I},\underline{J}}$ is linear in each variable $x[1], \ldots, x[n]$.

We have the following generalization of [**2**, Lemma 2.1].

PROPOSITION 8.1. *Each element $P_{x[1],\ldots,x[n];\underline{I},\underline{J}}$ of $T(X_H)$ is coïnvariant.*

PROOF. By (8.1), $\delta(P_{x[1],\ldots,x[n];\underline{I},\underline{J}})$ is equal to

$$\begin{aligned}
&\sum_{(x[1]),\ldots,(x[n])} X_{x[I_1]_{(1)}} \cdots X_{x[I_r]_{(1)}} X_{S(x[J_s]_{(4)})} \cdots X_{S(x[J_1]_{(4)})} \\
&\qquad\qquad \otimes x[I_1]_{(2)} \cdots x[I_r]_{(2)} \, S(x[J_s]_{(3)}) \cdots S(x[J_1]_{(3)}) \\
&= \sum_{(x[1]),\ldots,(x[n])} X_{x[I_1]_{(1)}} \cdots X_{x[I_r]_{(1)}} X_{S(x[J_s]_{(4)})} \cdots X_{S(x[J_1]_{(4)})} \\
&\qquad\qquad \otimes x[1]_{(2)} \cdots x[n]_{(2)} \, S(x[n]_{(3)}) \cdots S(x[1]_{(3)}) \\
&= \sum_{(x[1]),\ldots,(x[n])} X_{x[I_1]_{(1)}} \cdots X_{x[I_r]_{(1)}} X_{S(x[J_s]_{(3)})} \cdots X_{S(x[J_1]_{(3)})} \\
&\qquad\qquad \otimes \varepsilon(x[1]_{(2)}) \cdots \varepsilon(x[n]_{(2)}) \\
&= \sum_{(x[1]),\ldots,(x[n])} X_{x[I_1]_{(1)}} \cdots X_{x[I_r]_{(1)}} X_{S(x[J_s]_{(2)})} \cdots X_{S(x[J_1]_{(2)})} \otimes 1 .
\end{aligned}$$

Therefore, $\delta(P_{x[1],\dots,x[n];\underline{I},\underline{J}}) = P_{x[1],\dots,x[n];\underline{I},\underline{J}} \otimes 1$ and the conclusion follows. □

As special cases of the previous proposition, the following elements of $T(X_H)$ are coïnvariant for all $x, y \in H$:

$$P_x = P_{x;(\{1\}),(\{1\})} = \sum_{(x)} X_{x_{(1)}} X_{S(x_{(2)})} \tag{8.3}$$

and

$$P_{x,y} = P_{x,y;(\{1\},\{2\}),(\{1,2\})} = \sum_{(x),(y)} X_{x_{(1)}} X_{y_{(1)}} X_{S(x_{(2)}y_{(2)})} \,. \tag{8.4}$$

8.3. The generic evaluation map. As in Section 4, let H be a Hopf algebra, $\alpha : H \times H \to k$ a normalized invertible cocycle, and ${}^{\alpha}H$ the corresponding twisted algebra.

Consider the algebra morphism $\mu_\alpha : T(X_H) \to S(t_H) \otimes {}^{\alpha}H$ defined for all $x \in H$ by

$$\mu_\alpha(X_x) = \sum_{(x)} t_{x_{(1)}} \otimes u_{x_{(2)}} \,.$$

The morphism μ_α possesses the following properties (see [**2**, Sect. 4]).

PROPOSITION 8.2. *(a) The morphism $\mu_\alpha : T(X_H) \to S(t_H) \otimes {}^{\alpha}H$ is an H-comodule algebra morphism.*

(b) If the ground field k is infinite, then for every H-comodule algebra morphism $\mu : T(X_H) \to {}^{\alpha}H$, there is a unique algebra morphism $\chi : S(t_H) \to k$ such that

$$\mu = (\chi \otimes \mathrm{id}) \circ \mu_\alpha \,.$$

In other words, any H-comodule algebra morphism $\mu : T(X_H) \to {}^{\alpha}H$ is obtained by specialization from μ_α. For this reason we call μ_α the *generic evaluation map* for ${}^{\alpha}H$.

Now we have the following result (see [**2**, Sect. 8]).

PROPOSITION 8.3. *If $P \in T(X_H)$ is coïnvariant, then $\mu_\alpha(P)$ belongs to $\mathcal{B}_H^{\alpha}$.*

It follows that the image $\mu_\alpha(P_{x[1],\dots,x[n];\underline{I},\underline{J}})$ of all coïnvariant elements defined by (8.2) belong to $\mathcal{B}_H^{\alpha}$. This provides a systematic way to produce elements of $\mathcal{B}_H^{\alpha}$.

EXAMPLE 8.4. When $H = H_4$ is the Sweedler algebra, it is easy to check that the elements R, S, T, U of (5.1) are obtained in this way: we have

$$R = \mu_\alpha(P_x)\,, \quad T = \mu_\alpha(P_{y-z})\,, \quad U = \mu_\alpha(P_{x,z})\,, \quad ES = \mu_\alpha(P_{y,y})\,,$$

where $\{1, x, y, z\}$ is the basis of H_4 defined in Section 5 and P_x, P_{y-z}, $P_{x,z}$, and $P_{y,y}$ are special cases of the noncommutative polynomials defined by (8.3) and (8.4).

REMARK 8.5. In [**2**] we developped a theory of *polynomial identities* for H-comodule algebras. This theory applies in particular to the twisted algebras ${}^{\alpha}H$. We established that the H-identities of ${}^{\alpha}H$, as defined in *loc. cit.*, are exactly the elements of $T(X_H)$ that lie in the kernel of the generic evaluation map μ_α. Thus the H-comodule algebra

$$\mathcal{U}_H^{\alpha} = T(X_H)/\operatorname{Ker}\mu_\alpha$$

plays the rôle of a *universal comodule algebra*. We also constructed an H-comodule algebra morphism $\mathcal{U}_H^{\alpha} \to \mathcal{A}_H^{\alpha}$; under certain conditions this map turns the generic Galois extension $\mathcal{A}_H^{\alpha}$ into a *central localization* of the universal comodule algebra $\mathcal{U}_H^{\alpha}$ (see [**2**, Sect. 9] for details).

References

1. E. Aljadeff, D. Haile, M. Natapov, *Graded identities of matrix algebras and the universal graded algebra*, arXiv:0710.5568, Trans. Amer. Math. Soc. (2009), in press.
2. E. Aljadeff, C. Kassel, *Polynomial identities and noncommutative versal torsors*, Adv. Math. 218 (2008), 1453–1495, doi:10.1016/j.aim.2008.03.014.
3. E. Aljadeff, M. Natapov, *On the universal G-graded central simple algebra*, Actas del XVI Coloquio Latinoamericano de Álgebra (eds. W. Ferrer Santos, G. Gonzalez-Sprinberg, A. Rittatore, A. Solotar), Biblioteca de la Revista Matemática Iberoamericana, Madrid 2007.
4. T. Aubriot, *On the classification of Galois objects over the quantum group of a nondegenerate bilinear form*, Manuscripta Math. 122 (2007), 119–135.
5. J. Bichon, *Galois and bigalois objects over monomial non-semisimple Hopf algebras*, J. Algebra Appl. 5 (2006), 653–680.
6. R. J. Blattner, M. Cohen, S. Montgomery, *Crossed products and inner actions of Hopf algebras*, Trans. Amer. Math. Soc. 298 (1986), 671–711.
7. R. J. Blattner, S. Montgomery, *A duality theorem for Hopf module algebras*, J. Algebra 95 (1985), 153–172.
8. Y. Doi, *Equivalent crossed products for a Hopf algebra*, Comm. Algebra 17 (1989), 3053–3085.
9. Y. Doi, M. Takeuchi, *Cleft comodule algebras for a bialgebra*, Comm. Algebra 14 (1986), 801–817.
10. Y. Doi, M. Takeuchi, *Quaternion algebras and Hopf crossed products*, Comm. Algebra 23 (1995), 3291–3325.
11. S. Garibaldi, A. Merkurjev, J.-P. Serre, *Cohomological invariants in Galois cohomology*, Univ. Lecture Ser. 28, Amer. Math. Soc., Providence, RI, 2003.
12. R. Günther, *Crossed products for pointed Hopf algebras*, Comm. Algebra 27 (1999), 4389–4410.
13. C. Kassel, *Quantum principal bundles up to homotopy equivalence*, The Legacy of Niels Henrik Abel, The Abel Bicentennial, Oslo, 2002, O. A. Laudal, R. Piene (eds.), Springer-Verlag 2004, 737–748 (see also arXiv:math.QA/0507221).
14. C. Kassel, H.-J. Schneider, *Homotopy theory of Hopf Galois extensions*, Ann. Inst. Fourier (Grenoble) 55 (2005), 2521–2550.
15. A. Masuoka, *Cleft extensions for a Hopf algebra generated by a nearly primitive element*, Comm. Algebra 22 (1994), 4537–4559.
16. A. Masuoka, *Cocycle deformations and Galois objects for some cosemisimple Hopf algebras of finite dimension*, New trends in Hopf algebra theory (La Falda, 1999), 195–214, Contemp. Math., 267, Amer. Math. Soc., Providence, RI, 2000.
17. A. Masuoka, *Abelian and non-abelian second cohomologies of quantized enveloping algebras*, J. Algebra 320 (2008), 1–47.
18. S. Montgomery, *Hopf algebras and their actions on rings*, CBMS Conf. Series in Math., vol. 82, Amer. Math. Soc., Providence, RI, 1993.
19. F. Panaite, F. Van Oystaeyen, *Clifford-type algebras as cleft extensions for some pointed Hopf algebras*, Comm. Algebra 28 (2000), 585–600.
20. P. Schauenburg, *Hopf bi-Galois extensions*, Comm. Algebra 24 (1996), 3797–3825.
21. P. Schauenburg, *Galois objects over generalized Drinfeld doubles, with an application to* $u_q(\mathfrak{sl}_2)$, J. Algebra 217 (1999), 584–598.
22. H.-J. Schneider, *Principal homogeneous spaces for arbitrary Hopf algebras*, Israel J. Math. 72 (1990), 167–195.
23. M. Sweedler, *Hopf algebras*, W. A. Benjamin, Inc., New York, 1969.

Université de Strasbourg & CNRS, Institut de Recherche Mathématique Avancée, 7 rue René Descartes, 67084 Strasbourg, France

E-mail address: `kassel@math.u-strasbg.fr`

URL: `www-irma.u-strasbg.fr/~kassel/`

Quantizing the Moduli Space of Parabolic Higgs Bundles

Avijit Mukherjee

ABSTRACT. We consider the moduli space $\mathcal{M}_H^s$ of stable parabolic Higgs bundles (of rank 2 for simplicity) over a compact Riemann surface of genus $g > 1$. This is a smooth variety over $\mathbb{C}$, equipped with a holomorphic symplectic form Ω_H. Any symplectic form is known to admit a quantization, but in general the quantization is not unique. We fix a projective structure $\mathcal{P}$ on X. Using $\mathcal{P}$ we show that there is a *canonical* quantization of Ω_H on a certain Zariski open dense subset $\mathcal{U} \subset \mathcal{M}_H^s$, once a projective structure $\mathcal{P}$ on X has been specified.

1. Introduction

In this talk, we shall outline a quantization scheme for a certain smooth variety (over $\mathbb{C}$) which admits a (holomorphic) symplectic form. By *quantization* we shall refer to the *Moyal-Weyl quantization* [**1**]. The smooth variety in question is taken to be the moduli space $\mathcal{M}_H^s$ of stable parabolic Higgs bundles (of rank 2 for simplicity, and) of fixed degree over a compact Riemann surface of genus $g > 1$. It is well-known that $\mathcal{M}_H^s$ is a smooth, irreducible quasi-projective complex variety.

A natural holomorphic symplectic form Ω_H on $\mathcal{M}_H^s$ is known to exist and its construction has been described in details in [**8**],[**7**]. The symplectic form defines a Poisson structure on $\mathcal{O}_{\mathcal{M}_H^s}$, the sheaf of complex valued algebraic functions on $\mathcal{M}_H^s$, which is then amenable to a quantization procedure. However, given a symplectic structure, the space of all possible quantizations of that symplectic structure is *infinite-dimensional*, and hence non-unique. In this talk, we address the question: is there a *canonical* quantization of Ω_H? To obtain an affirmative answer, it is seen that one has to fix some additional datum, a projective structure $\mathcal{P}$ on X. Once this is fixed/chosen, there *does* exist a *preferred*, and hence in that sense canonical, quantization of the symplectic form Ω_H on a certain Zariski open dense subset of $\mathcal{M}_H^s$.

The present talk is a report of that pursuit. We shall here outline the basic ideas and the strategy of the proof, skipping most of the details, which have appeared in [**5**]. Though in this work, we have restricted ourselves to the case of rank 2 parabolic vector bundles, the generalization of our results and conclusions to higher rank cases is straight-forward. (The restriction is motivated by reasons of notational and computational simplicity.)

The author would like to thank the MPIM–Bonn, and the organizers of the Workshop for providing the opportunity of presenting these results.

In the next section, we begin by recalling the relevant notions and definitions in this context.

2. Preliminaries

Let us first recall the basic definitions and also fix our notations. Unless otherwise stated, X will always denote a compact, connected Riemann surface of genus $g > 1$, or in other words, X is a smooth, projective curve (over $\mathbb{C}$) of genus $g > 1$.

2.1. Parabolic bundles. Let X be a compact, connected Riemann surface of genus g, with $g \geq 2$. We fix a finite subset of points

$$S := \{s_1, s_2, \cdots, s_n\} \subset X\,.$$

(Parabolic structures can also be defined for the cases of $g = 0, 1$. For these cases, if $g = 0$, we would take $n \geq 4$; and for $g = 1$, we take $n \geq 1$. However in our present problem, we only consider $g \geq 2$.) Also we shall, for computational convenience and simplicity, consider the case of parabolic vector bundles of rank 2 and arbitrary (but fixed) degree, on X.

A *parabolic vector bundle of rank 2* over X, with parabolic structure over S, consists of the following datum[**14**]:

(1) a holomorphic vector bundle E of rank 2 over X;
(2) for each point $s \in S$, a line $F_s \subset E_s$ of the fiber E_s;
(3) for each point $s \in S$, a real number λ_s (*parabolic weights*) with $0 < \lambda_s < 1$.

For a parabolic vector bundle $E_* := (E, \{F_s\}, \{\lambda_s\})$ as above, the *parabolic degree* is defined to be

$$\text{par-deg}(E_*) := \deg(E) + \sum_{s \in S} \lambda_s$$

where $\deg(E)$ is the degree of E [**14**]. The parabolic vector bundle E_* is called *stable* if for every holomorphic line subbundle L of E, the inequality

$$\deg(L) + \sum_{s \in S'} \lambda_s < \frac{\text{par-deg}(E_*)}{2} \tag{2.1}$$

where $S' := \{s \in S \,|\, L_s = F_s\}$ [**14**]. If a weaker condition, namely

$$\deg(L) + \sum_{s \in S'} \lambda_s \leq \frac{\text{par-deg}(E_*)}{2} \tag{2.2}$$

is valid, then E_* is called *semistable* [**14**].

2.2. Parabolic Higgs bundles. Let K_X denote the holomorphic cotangent bundle of X. A Higgs structure of a parabolic vector bundle $E_* := (E, \{F_s\}, \{\lambda_s\})$ is a holomorphic section

$$\theta \in H^0(X, \text{End}(E) \otimes K_X \otimes \mathcal{O}_X(S)) \tag{2.3}$$

with the property that for each $s \in S$, the image of the homomorphism

$$\theta(s) : E_s \longrightarrow (E \otimes K_X \otimes \mathcal{O}_X(S))_s$$

is contained in the subspace $F_s \otimes (K_X \otimes \mathcal{O}_X(S))_s \subset (E \otimes K_X \otimes \mathcal{O}_X(S))_s$ and $\theta(s)(F_s) = 0$ [**11**], [**8**], [**9**]. In other words, $\theta(s)$ is nilpotent with respect to the flag $0 \subset F_s \subset E_s$.

A parabolic Higgs bundle (E_*, θ) as above is called *stable* if for every line subbundle L of E with

$$\theta(L) \subseteq L \otimes K_X \otimes \mathcal{O}_X(S) \subset E \otimes K_X \otimes \mathcal{O}_X(S)$$

the inequality (2.1) is satisfied. If the weaker inequality (2.2) is valid, then (E_*, θ) is called *semistable.* In particular, if E_* is stable (respectively, semistable), then (E_*, θ) is stable (respectively, semistable). However, (E_*, θ) can be stable (respectively, semistable) without E_* being stable (respectively, semistable).
It is known that the moduli space of stable parabolic Higgs bundles is an irreducible, smooth, quasiprojective variety of dimension $8g_X - 6 + 2n$. This moduli space will be denoted by $\mathcal{M}_H^s$. There is a natural inclusion of the total space of the holomorphic cotangent bundle of the moduli space of stable parabolic bundles on X into $\mathcal{M}_H^s$. Given a stable parabolic bundle E_*, the holomorphic cotangent space at the point E_* of the moduli space of stable parabolic bundles is the space of all sections θ as in the definition of Higgs structure in (2.3) [**7**],[**8**]. Therefore, we have a tautological map from the holomorphic cotangent bundle of the moduli space of stable parabolic bundles to $\mathcal{M}_H^s$. The image of this map is a Zariski open subset. The moduli space of semistable parabolic Higgs bundles will be denoted by $\mathcal{M}_H^{ss}$. It is an irreducible, normal, quasiprojective variety, and $\mathcal{M}_H^s$ is a Zariski open smooth subvariety of it.

2.3. Symplectic structure on the moduli space. The natural symplectic form on the total space of the cotangent bundle of the moduli space of stable parabolic bundles extends to a symplectic form on the moduli space $\mathcal{M}_H^s$ of stable parabolic Higgs bundles. The symplectic form on $\mathcal{M}_H^s$ is described in [**8**],[**7**]. A crucial ingredient in the construction of this symplectic form is the observation that:

THEOREM 2.1. *The tangent space $T_{(E_*,\theta)}\mathcal{M}_H^s$ of the variety $\mathcal{M}_H^s$ at the point represented by the parabolic Higgs bundle (E_*, θ) is identified with the hypercohomology $\mathbb{H}^1(C_\bullet)$ [**8**],*

where, the complex $C_\bullet$ is the 2-term deformation complex of a Higgs bundle defined by:

$$C_\bullet \ : \ C_0 \ = \ \mathrm{End}(E) \stackrel{[-,\theta]}{\longrightarrow} C_1 \ = \ K_X \otimes \mathrm{End}(E),$$

with $\mathrm{End}(E)$ at the 0-th position, and if $\theta = dz \otimes A$ in a local trivialization z, and s is a local section of $\mathrm{End}(E)$, then $[s, \theta] = dz \otimes (sA - As)$.

In fact it turns out that this holomorphic symplectic form is itself exact and there exists a one-form Ψ on the variety $\mathcal{M}_H^s$, such that: $\Omega_H \ = \ d\Psi$. The two–form $\mathrm{d}\Psi$ gives our desired symplectic form on $\mathcal{M}_H^s$. The restriction of $\mathrm{d}\Psi$ to the Zariski open subset of $\mathcal{M}_H^s$ defined by the total space of the cotangent bundle of the moduli space of parabolic bundles coincides with the canonical symplectic structure on cotangent bundles. (See [**8**], [**9**].). We shall henceforth denote this symplectic structure on $\mathcal{M}_H^s$ by Ω_H. This symplectic form defines a *Poisson* structure on $\mathcal{O}_{\mathcal{M}_H^s}$, the sheaf of holomorphic (= $\mathbb{C}$-valued algebraic) functions on $\mathcal{M}_H^s$.

Definition: A *quantization* of Ω_H is a 1-parameter family of associative algebra structures on $\mathcal{O}_{\mathcal{M}_H^s}$, deforming the abelian algebraic structure (defined by pointwise multiplication) with the infinitesimal deformations of the pointwise multiplication structure being governed by the Poisson structure. Any symplectic structure admits a quantization, but there is as such no unique (canonical) quantization! That is, the space of all quantizations of a symplectic structure is infinite-dimensional. The main theme of this talk is to present the observation that a *preferred* or a canonical quantization scheme does however exist, once a projective structure on X has been chosen/fixed.

THEOREM 2.2. *Let X be a compact Riemann surface (of genus ≥ 2) equipped with a (chosen) projective structure $\mathcal{P}$. There there exists a certain Zariski open dense subset $U \subset \mathcal{M}_H^s$, such that the projective structure $\mathcal{P}$ gives a canonical quantization of the symplectic structure Ω_H over U* [**5**].

(This subset U will be defined in the following section.)

2.4. A few useful facts about Projective Structures on X. A projective atlas on X is defined by giving a covering of X by holomorphic coordinate charts $\{U_\alpha, \phi_\alpha\}_{\alpha \in A}$, for some set A, and where for every α,

$$U_\alpha \subset X \qquad \text{and} \quad \phi_\alpha : U_\alpha \longrightarrow V \subset \mathbb{C}$$

is a bi-holomorphism such that $\phi_\beta \cdot \phi_\alpha^{-1}$ is the restriction of a Möbius transformation on the image of $\phi_\alpha(U_\alpha \bigcap U_\beta)$ in $\mathbb{C}$. A projective structure on X is an equivalence class of projective atlases. Recall that the *Uniformization Theorem* states that if, $\widetilde{X} \longrightarrow X$ is the universal cover of a Riemann surface X, then $\widetilde{X}$ is bi-holomorphic to $\mathbb{C}$, or $\mathbb{P}^1$ or $\mathbb{H}$. Moreover it is known that $\mathrm{Aut}(\mathbb{C}/\mathbb{P}^1/\mathbb{H}) \subset$ Möbius group. Hence any Riemann surface admits a natural projective structure. Further, it is known that the space of all projective structures on X is an affine space for $H^0\left(X, {K_X}^{\otimes 2}\right)$, the space of quadratic differentials where $\dim_{\mathbb{C}} H^0\left(X, {K_X}^{\otimes 2}\right) = 3g_X - 3$

3. Quantization

We start with a brief recapitulation of the relevant ideas and notions. Let M denote a complex manifold, TM and T^*M its respective holomorphic tangent and cotangent bundles. Let ω be a holomorphic symplectic form on M, and let $\tau : T^*M \longrightarrow TM$ be the holomorphic isomorphism defined by ω.

$$\textit{i.e.,} \qquad \tau^{-1}(v) \cdot u = \omega(u, v), \qquad \text{for} \quad u, v \in T_x M, \ x \in M$$

Let f and g be any two holomorphic functions defined on an open subset $U \subset M$. The map sending the pair,

$$(f, g) \longmapsto \{f, g\} := \omega(\tau(df), \tau(dg))$$

defines a holomorphic Poisson structure on $\mathcal{H}(M)$, the space of all locally defined holomorphic functions on M. The Poisson bracket thus provides the commutative algebra $\mathcal{H}(M)$, with the structure of a Lie algebra satisfying the Leibnitz rule:

$$\{fg, k\} = f\{g, k\} + g\{f, k\}$$

Let us define a formal parameter h, and

$$\mathcal{A}(M) := \mathcal{H}(M)[[h]] \quad = \text{ space of all formal Taylor series in } h.$$

Given two elements $f, g \in \mathcal{A}(M)$ (*i.e.*, $f := \sum_{j=0}^{\infty} h^j f_j$, where $f_j \in \mathcal{H}(M)$, and $g := \sum_{j=0}^{\infty} h^j g_j$, where $g_j \in \mathcal{H}(M)$), a *Quantization* of the Poisson structure is an assignment of an associative algebra operation $\star$ on $\mathcal{A}(M)$, for every pair, $(f,g) \in \mathcal{A}(M)$, defined by,

$$(f,g) \longmapsto (f \star g) := \sum_{j=0}^{\infty} h^j \phi_j, \qquad \phi_i \in \mathcal{H}(M).$$

satisfying the following rules: [**1**],[**13**]
(a) Each $\phi_i \in \mathcal{H}(M)$ is some polynomial (independent of f and g) in derivatives (of arbitrary order) of $\{f_i\}$ and $\{g_i\}$.
(b) $\phi_0 = f_0 g_0$
(c) $1 \star f = f \star 1 = f$, for every $f \in \mathcal{H}(M)$
(d) $f \star g - g \star f = \sqrt{-1}\, h\, \{f_0, g_0\} + h^2 \beta$, where $\beta \in \mathcal{A}(M)$ depends on the choice of f and g.

3.1. Moyal-Weyl Quantization. Let V be a complex vector space, Θ a constant symplectic form on V and let $\mathcal{H}(V)$ denote the space of all locally defined holomorphic functions on V, equipped with the Poisson structure.

Let

$$\triangle : V \longrightarrow V \times V$$

denote the diagonal homomorphism defined by $v \longmapsto (v,v)$. Then it is known that there exists a differential operator

$$D: \mathcal{H}(V \times V) \longrightarrow \mathcal{H}(V \times V)$$

with constant coefficients, such that for any pair $(f,g) \in \mathcal{H}(V)$,

$$\{f,g\} = \triangle^* D(f \otimes g)$$

where $f \otimes g$ is the function on $V \times V$ defined by $(u,v) \longmapsto f(u)g(v)$. Moreover D is unique. The *Moyal-Weyl algebra* is then defined by

$$f \star g = \triangle^* \exp\big(\sqrt{-1}\, hD/2\big)(f \otimes g) \in \mathcal{A}(V)$$

The Moyal-Weyl algebra can be extended to a multiplication operation on $\mathcal{A}(V)$ using bilinearity with respect to h,

$$f \star g = \sum_{i,j=0}^{\infty} h^{i+j} (f_i \star g_j) \in \mathcal{A}(V)$$

THEOREM 3.1. *The $\star$ operation converts $\mathcal{A}(V)$ into an associative algebra that quantizes the symplectic structure Θ* [**1**],[**13**].

In this present work, we shall always mean the above *Moyal-Weyl quantization* scheme whenever we refer to quantization. Note that if $\mathrm{Sp}(V)$ denotes the group of all linear automorphisms of V preserving the symplectic form Θ, then the differential operator D commutes with the diagonal action of $\mathrm{Sp}(V)$ on $V \times V$, from which it readily follows that:

$$(f \circ G) \star (g \circ G) = (f \star g) \circ G, \quad \text{for any } G \in \mathrm{Sp}(V), \text{ and } f,g \in \mathcal{A}(V)$$

For X, a projective algebraic curve, let $\mathcal{Z} := K_X \setminus 0_X$ be the complement of the zero section of the total space of the holomorphic cotangent bundle. The total space of K_X has a natural algebraic symplectic structure. (The total space of K_X

has a tautological one-form on it and its exterior derivative defines this symplectic structure). Let

$$\Theta_0 \in H^0(\mathcal{Z}, \Omega^2_{\mathcal{Z}})$$

be the algebraic symplectic form on $\mathcal{Z}$, obtained by restricting the natural symplectic form on the total space of K_X. Then it is known, that

THEOREM 3.2. *Given a projective structure on X, there exists a natural quantization of the symplectic form Θ_0 on the symplectic manifold $\mathcal{Z}$ [3].*

4. Quantization of Ω_H

In this section we will consider the quantization problem for the moduli space of stable parabolic Higgs bundles. Our Main Theorem (which stems from an extension of the ideas captured by main theorem proved in [3]) is:

THEOREM 4.1. *Let X be a compact Riemann surface equipped with a projective structure $\mathcal{P}$. The projective structure $\mathcal{P}$ then gives a (natural/canonical) quantization of a certain Zariski open dense subset (to be defined later) $\mathcal{U} \subset \mathcal{M}^s_H$ equipped with the symplectic form Ω_H.*

The construction of this Zariski open dense subset $\mathcal{U} \subset \mathcal{M}^s_H$ will be explained below.
Fix one of the points $s_1 \in S$, over which the parabolic structure is defined. Then we have the following lemma.

LEMMA 4.2. *There is a non-empty Zariski open dense subset $U_1 \subset \mathcal{M}^s_H$, such that for all parabolic Higgs bundles (E_*, θ) in U_1,*
(a) dim $H^0(X, E \otimes \mathcal{O}_X((g_X - 1/2 - d/2)s_1)) = 1$, *if the degree d of E is odd.*
(b) dim $\{\beta \in H^0(X, E \otimes \mathcal{O}_X((g_X - d')s_1)) \mid \beta(s_1) \subset F_{s_1}\} = 1$, *where $d' = d/2$ (when the degree d of E is even),*
and $F_{s_1} \subset E_{s_1}$ is the line defining the quasi-parabolic structure over the fixed point s_1 [7].

Fix a positive integer δ, such that $2(\delta + g_X) - 1 > \max\{0, 6(g_X - 1) + \#S\}$, and then set,

$$\begin{aligned} \delta_0 &:= 2(\delta + g_X) - 1, && \text{if } d \text{ is } odd, \\ \delta_0 &:= 2(\delta + g_X), && \text{when } d \text{ is } even. \end{aligned}$$

Let $\mathcal{M}^{\delta_0}_H$ denote the moduli space of all rank two, stable parabolic Higgs bundles (E_*, θ), where E_* is a parabolic vector bundle of rank 2 and degree δ_0 with parabolic weights λ_{s_1} at $s_1 \in S$ (as fixed earlier) and θ, a Higgs structure on E_*. We can, in addition, introduce and construct a *Moduli Space* $\mathcal{M}^{\delta_0}_T$ of *Triples* ([6],[7]) of the form (E_*, θ, σ), where $(E_*, \theta) \in \mathcal{M}^{\delta_0}_H$ and $\sigma \in H^0(X, E) \setminus \{0\}$ is a non-zero section of E. The construction of this moduli space of triples has been discussed elsewhere ([6],[7]). Here it suffices to make the observation that the projection (the forgetful map) $p : \mathcal{M}^{\delta_0}_T \longrightarrow \mathcal{M}^{\delta_0}_H$ that sends any $(E_*, \theta, \sigma) \longrightarrow (E_*, \theta)$ is a smooth projective bundle, whose fibre over the point in $\mathcal{M}^{\delta_0}_H$ corresponding to (E_*, θ) is $\mathbb{P}H^0(X, E)$, the projective space of lines in $H^0(X, E)$. The numerical condition on δ_0 that we have assumed then ensures that the dim $H^0(X, E)$ is independent of E.([6],[7])

Let Z denote the total space of the line bundle $K_X \otimes \mathcal{O}_X(S)$. In an earlier work ([**7**]), we have constructed a surjective morphism

$$\mu : \mathcal{M}_T^{\delta_0} \longrightarrow \mathrm{Hilb}^l(Z)$$

where $l = 4g_X + 2\delta + n - 3$ when d is odd and $l = 4g_X + 2\delta + n - 2$ when d is even, with $n = \#S$, and $\mathrm{Hilb}^j(Z)$ as usual denotes the Hilbert scheme of points of length j of Z ([**6**],[**7**]). Now consider any $(E_*, \theta) \in U_1$, where U_1 is the open subset in Lemma 4.2. When d is odd, let $\beta \in H^0(X, E \otimes \mathcal{O}_X((g_X - 1/2 - d/2)s_1))$ be a non-zero section. (For the case, d is even, we analogously choose a $\beta \in H^0(X, E \otimes \mathcal{O}_X((g_X - d/2)s_1))$, with $\beta(s_1) \subset F_{s_1}$). The previous lemma ensures that a non-zero β always exists and any two choices of β differ by the multiplication by an element of $\mathbb{C}^*$. Given a choice of β, we can construct the map:

$$\begin{aligned} \psi : U_1 &\longrightarrow \mathcal{M}_T^{\delta_0} \\ (E_*, \theta) &\longrightarrow (E'_*, \theta, \beta \otimes s_\delta) \end{aligned}$$

where s_δ is the section of the line bundle $\mathcal{O}_X(\delta s_1)$ defined by the constant function 1, and E'_* is the appropiately twisted parabolic bundle,

$$\begin{aligned} E'_* &:= E_* \otimes \mathcal{O}_X((g_X - 1/2 - d/2 + \delta)s_1), \quad \text{when } d \text{ is odd,} \\ E'_* &:= E_* \otimes \mathcal{O}_X((g_X - d/2 + \delta)s_1), \quad \text{when } d \text{ is even} \end{aligned}$$

We note that this twisting ensures that the resulting isomorphism:

$$i_0 : \mathcal{M}_H^s \longrightarrow \mathcal{M}_H^{\delta_0}$$

that sends any pair: $(E_*, \theta) \longrightarrow (E'_*, \theta)$ preserves the symplectic structures of the two moduli spaces.

Let

$$\psi_1 : U_1 \longrightarrow \mathrm{Hilb}^l(Z)$$

be the composition of the two maps defined above. It is known that $\mathrm{Hilb}^l(Z)$ admits a meromorphic symplectic form ([**2**], [**7**]), and we shall denote that by Θ. Now, using this map ψ_1, we shall construct a map from a non-empty Zariski open subset $U_2 \subset U_1$ to $\mathrm{Hilb}^{4g_X+n-3}(Z')$. The details of this construction are explained in [**5**]. Here we just summarize the results. Let $Z' := Z \setminus \{0_{s_1}\} \subset Z$ be the complement of the point. Then

THEOREM 4.3. *There is a subset $U_2 \subset U_1$ and a unique map:*

$$\psi_2 : U_2 \longrightarrow \mathit{Hilb}^{4g_X+n-3}(Z')$$

such that U_2 is non-empty and Zariski open dense subset of U_1 and $\psi_1(y) \cap Z' = \psi_2(y)$. Moreover, it is easy to see that the map ψ_2 is dominant and the meromorphic symplectic form on $\mathit{Hilb}^{4g_X+n-3}(Z')$ pulls back to the symplectic form on U_2.

Finally let $V \subset \mathrm{Hilb}^{4g_X+n-3}(Z')$ be the Zariski open dense subset, corresponding to the *distinct* $4g_X + n - 3$ points of Z' and let $\pi_X : Z' \longrightarrow X$ be the natural projection. Define

$$U := \{y \in V | y \cap (\pi_X^{-1}(S) \cup 0_X) = \emptyset\}$$

where 0_X is the image of the zero section of the line bundle $K_X \otimes \mathcal{O}_X(S)$. Then U is a Zariski open dense subset of $\mathrm{Hilb}^{4g_X+n-3}(Z')$ and the complement of U is a divisor of $\mathrm{Hilb}^{4g_X+n-3}(Z')$. The meromorphic symplectic form Θ on $\mathrm{Hilb}^{4g_X+n-3}(Z)$ will define a symplectic form on the open subset U, as the pole of Θ would be supported on the divisor of $\mathrm{Hilb}^{4g_X+n-3}(Z)$ consisting of all zero-dimensional subschemes with support intersecting $\pi_X^{-1}(S)$. We define a Zariski open subset of U_2 by

$$\mathcal{U} := \psi_2^{-1}(U),$$

then we find that the above Zariski open dense subset $\mathcal{U} \subset \mathcal{M}_H^s$, equipped with the symplectic form Ω_H admits a *natural* quantization. In fact to quantize Ω_H over $\mathcal{U}$, it suffices to quantize Θ over U.

4.1. Quantization of Θ over U : In this section we will sketch the basic features of our quantization scheme for the quantization of Θ over U. We recall that we had defined the variety $\mathcal{Z} := K_X \setminus 0_X$, and p the natural projection $\mathcal{Z} \longrightarrow X$. Set

$$\mathcal{Z}^0 := \mathcal{Z} \setminus p^{-1}(S) \subset \mathcal{Z} \quad \text{and let} \quad \widehat{\mathcal{Z}} \subset \left(\mathcal{Z}^0\right)^{4g_X+n-3}$$

be the Zariski open dense subset of the Cartesian product parametrizing all distinct $4g_X + n - 3$ points of $\mathcal{Z}^0$. Clearly the permutation group Σ of

$$\{1, 2, 3, \cdots, 4g_X + n - 3\}$$

acts freely on $\widehat{\mathcal{Z}}$, which immediately yields the identification:

$$\widehat{\mathcal{Z}}/\Sigma = U.$$

One can then see that the symplectic structure θ_0 on $\mathcal{Z}$, defines/extends to a symplectic structure Θ_m on $\mathcal{Z}^m$, by pointwise multiplication [**2**]. Since the action of Σ preserves the symplectic form on $\mathcal{Z}^{4g_X+n-3}$, the symplectic form on $\widehat{\mathcal{Z}}$ descends to U, in other words U inherits the symplectic form Θ.
Now we have already discussed the fact that fixing a projective structure $\mathcal{P}$ on X, gives us a preferred or canonical quantization of the symplectic structure Θ_0 on $\mathcal{Z}$. Using the pointwise construction and the free action of Σ, this quantization in turn, gives a quantization of the symplectic structure Θ_m on $\mathcal{Z}^m$ for all $m \geq 1$. As the action of the permutation group Σ on $\left(\mathcal{Z}^0\right)^{4g_X+n-3}$ preserves this quantization, the quantization of the symplectic form Θ_{4g_X+n-3} on $\widehat{\mathcal{Z}}$, descends to a quantization of the symplectic variety $\widehat{\mathcal{Z}}/\Sigma$ $(= U)$, hence giving us a quantization of the symplectic structure Θ over U.
Now let $\Pi : \mathcal{U} \longrightarrow U$ be the restriction of the map ψ_2. Then from [[**7**],Theorem 3.2], it follows that $\Pi^*\Theta = \Omega_H$. Therefore to quantize Ω_H over $\mathcal{U}$, it suffices to quantize Θ over U, and this we have just achieved. Therefore we have proved the following theorem:

THEOREM 4.4. *Let X be a compact Riemann surface, equipped with a projective structure $\mathcal{P}$, and let $\mathcal{M}_H^s$ be the moduli space of stable, parabolic Higgs bundles on X. Let Ω_H denote the symplectic form on $\mathcal{M}_H^s$. Then the choice of the projective structure $\mathcal{P}$ gives a canonical quantization of the Zariski open dense subset $\mathcal{U} \subset \mathcal{M}_H^s$, equipped with the symplectic form Ω_H. Moreover the map from the space of all projective structures on X to the space of all quantizations of $(\mathcal{U}, \Omega_H)$ is injective.*

References

[1] F. Bayern, M. Flato, C. Frondsal, A. Lichnerowicz, D. Sternheimer,Deformation Theory and Quantizations I, II. *Ann. Phys.* **111** (1978) 61-151;
B. V. Fedosov, A simple geometrical construction of deformation quantization, *J. Differential Geom.*, **40** (1994) 213-238
A Weinstein, Deformation Quantization, *Séminaire Bourbaki* **46** (789), (1994), 213-238.

[2] A Beauville, Variétés Kähleriennes dont la première class de Chern est nulle, *J. Differential Geom* **18** (1983) 755-782.

[3] D. Ben-Zvi and I. Biswas, A Quantization on Riemann surfaces with projective structure, *Lett. Math. Phys* **54**, (2000), 73-82.

[4] Bradlow, S., Garcìa-Prada, O.: Stable triples, equivariant bundles and dimension reduction. Math. Ann. **304**, 225–252 (1996)

[5] Biswas, I., Mukherjee, A.: Quantization of a Moduli Space of Parabolic Higgs Bundles, *Int. J. Mathematics*, . **15**, (2004) 907-917.

[6] Biswas, I., Mukherjee, A.: Symplectic structures of moduli space of Higgs bundles over a curve and Hilbert scheme of points on the canonical bundle. Commun. Math. Phys. **221**, 293–304 (2001)

[7] Biswas, I., Mukherjee, A.: Symplectic structures on moduli space of parabolic Higgs bundles and Hilbert scheme, *Commun. Math. Phys.* **240**, (2003) 149-159.

[8] Biswas, I., Ramanan, S.: An infinitesimal study of the moduli of Hitchin pairs. Jour. Lond. Math. Soc. **49**, 219–231 (1994)

[9] Faltings, G.: Stable G-bundles and projective connections. Jour. Alg. geom. **2**, 507–568 (1993)

[10] Garcìa-Prada, O.: Dimensional reduction of stable bundles, vortices and stable pairs. Internat. Jour. Math. **5**, 1–52 (1994)

[11] Hitchin, N.: The self–duality equations on a Riemann surface. Proc. Lond. Math. Soc. **55**, 59–126 (1987)

[12] Hitchin, N.: Stable bundles and integrable systems. Duke Math. Jour. **54**, 91–114 (1987)

[13] M De Wilde and P B A Lecomte, Existence of star-products and formal deformations of the Poisson Lie algebra of arbitrary symplectic manifold. *Lett. Math. Phys*, **7** (1983) 487-496

[14] Mehta, V., Seshadri, C.S.: Moduli of vector bundles on curves with parabolic structure. Math. Ann. **248**, 205–239 (1980)

RamaKrishna Mission Vivekananda University, Belur Math, WB 711202 INDIA, and MPIM, Vivatsgasse 7, 53111 Bonn, GERMANY.

E-mail address: avijit@mpim-bonn.mpg.de

Locally compact quantum groups. Radford's S^4 formula.

A. Van Daele

ABSTRACT. Let A be a finite-dimensional Hopf algebra. The left and the right integrals on A are related by means of a distinguished group-like element δ of A. Similarly, there is this element $\widehat{\delta}$ in the dual Hopf algebra $\widehat{A}$. Radford showed that

$$S^4(a) = \delta^{-1}(\widehat{\delta} \triangleright a \triangleleft \widehat{\delta}^{-1})\delta$$

for all a in A where S is the antipode of A and where $\triangleright$ and $\triangleleft$ are used to denote the standard left and right actions of $\widehat{A}$ on A. The formula still holds for multiplier Hopf algebras with integrals (algebraic quantum groups).
In the theory of locally compact quantum groups, an analytical form of Radford's formula can be proven (in terms of bounded operators on a Hilbert space).
In this talk, we do not have the intention to discuss Radford's formula as such, but rather to use it, together with related formulas, for illustrating various aspects of the road that takes us from the theory of Hopf algebras (including compact quantum groups) to multiplier Hopf algebras (including discrete quantum groups) and further to the more general theory of locally compact quantum groups.

1. Introduction

As we have mentioned in the abstract, this note is about different steps along the road from the (purely algebraic) theory of Hopf algebras to the (analytical) theory of locally compact quantum groups. The formula of Radford, under its different forms at each level, is only used to illustrate certain aspects in this development.

In *Section* 1, we start with the simplest case. We take a finite-dimensional Hopf algebra A and we recall Radford's formula for the fourth power of the antipode in this case (see [R]), introducing the terminology that will be used further. We use S for the antipode and δ and $\widehat{\delta}$ for the distinguished group-like elements in A and the dual $\widehat{A}$. We call these the modular elements for reasons we explain later. We are also interested in the *-algebra case and in particular when the underlying algebra is an operator algebra. This means that A can be represented as a *-algebra of operators on a (finite-dimensional) Hilbert space. Then however, the integrals are positive, the modular elements are 1 and $S^2 = \iota$ (the identity map) so that Radford's formula becomes a triviality. We speak about a finite quantum group but in the literature, it is usually called a finite-dimensional Kac algebra (see [E-S]).

In *Section* 2, we first consider the case of a Hopf algebra A, not necessarily finite-dimensional, but with integrals (a co-Frobenious Hopf algebra). Radford's formula in this case was obtained in [B-B-T] where the modular element $\widehat{\delta}$ is seen as a homomorphism from A to $\mathbb{C}$. In this note however, we consider the dual $\widehat{A}$ of this Hopf algebra and describe it as a multiplier Hopf algebra. The element $\widehat{\delta}$ is then an element in the multiplier algebra $M(\widehat{A})$ of $\widehat{A}$.

In the operator algebra framework, we get here (essentially) a compact quantum group (as introduced by Woronowicz in [W2] and [W3]). In this setting, we necessarily have $\delta = 1$, but now it can happen that $\widehat{\delta} \neq 1$ (e.g. for the compact quantum group $SU_q(2)$, see [W1]). We also consider discrete quantum groups. They were first introduced in [P-W] as duals of compact quantum groups. Later they have been studied, as independent objects and independently in [E-R] and [VD2]. In this case of course, $\widehat{\delta} = 1$ while possibly $\delta \neq 1$. Radford's formula gives $S^4(a) = \delta^{-1}a\delta$ for all a in the algebra. In fact, one can define the square root $\delta^{\frac{1}{2}}$ of δ and show that even $S^2(a) = \delta^{-\frac{1}{2}}a\delta^{\frac{1}{2}}$. It is a fundamental formula for discrete quantum groups.

Section 3 is about algebraic quantum groups. We already needed the notion of a multiplier Hopf algebra (see [VD1]) in Section 2 for properly dealing with discrete quantum groups. However, it is only in this section that we introduce the concept. We also look at the case with integrals and then we speak about algebraic quantum groups (cf. [VD3]). For an algebraic quantum group (A, Δ), it is possible to define a dual $(\widehat{A}, \widehat{\Delta})$. It is again an algebraic quantum group. This duality extends the duality of finite-dimensional Hopf algebras (as used in Section 1), as well as the duality between compact and discrete quantum groups (as in Section 2). Also in this more general case, we have the existence of the modular elements δ and $\widehat{\delta}$, now in the multiplier algebras, and Radford's formula is still valid. It seems appropriate to give a proof (or rather sketch it) in this situation because it will follow easily from well-known results in the theory (see [D-VD-W]). As this case is more general than the previous ones (e.g. the finite-dimensional and the co-Frobenius Hopf algebras), this proof is also valid for these earlier cases.

Also here, we consider the *-algebra case and in particular when the integrals are positive. Then, the underlying algebras are operator algebras (now *-algebras of *bounded* operators on a possibly infinite-dimensional Hilbert space). We also have an analytical form of Radford's formula here and it is very close to the form we will obtain in the still more general case of locally compact quantum groups (in Section 4). Observe that now it can happen that both δ and $\widehat{\delta}$ are non-trivial.

It should not come as a surprise that, for *-algebraic quantum groups, we can formulate a form of Radford's result that is similar to the one we will obtain for general locally compact quantum groups. After all, the theory of *-algebraic quantum groups has been a source of inspiration for the development of locally compact quantum groups (as found in [K-V1], [K-V2] and [K-V3]). See e.g. the paper by Kustermans and myself [K-VD] and also the more recent paper entitled *Multiplier Hopf *-algebras with positive integrals: A laboratory for locally compact quantum groups* [VD6].

Finally, in *Section* 4 we briefly discuss the most general and tecnically far more difficult case of a locally compact quantum group. We recall the definition (within the setting of von Neumann algebras) and we explain how the basic ingredients of the analytical form of Radford's result are constructed. About the proof, we have to be very short because this would take us too far. Nevertheless, we say something about it and especially, what kind of similarities there are with the case of algebraic quantum groups. Observe some differences in conventions in this section.

This note contains no new results. It is more like a short survey of various levels, from Hopf algebras to locally compact quantum groups, making a link between the purely algebraic approach to quantum groups and the operator algebra approach. It is well-known that working with operator algebras in this context puts sometimes very severe restrictions on possible results, special cases and examples. Think e.g. of the fact that it forces the square of the antipode to be the identity map in the finite-dimensional case (see Section 1). On the other hand, it also has some nice advantages like the analytic structure of a *-algebraic quantum group (see Section 3). In any case, we are strongly convinced that a fair amount of knowlegde of 'the other side' can be of great help, not only for a basic understanding, but also because it sometimes provides different and handy tools to obtain new results or to treat old results in a better way. We think Radford's formula is a good illustration of this fact. Therefore, with this note, we hope to contribute to increase the interest of algebraists in the analytical aspects and vice versa.

Let us finish this introduction with some notation and conventions, as well as with providing some basic references. More of this will be given throughout the note.

We work with associative algebras over the comlex numbers since we often will also consider an involution on the algebra, making it into an operator algebra. The algebras need not have an identity, but we always assume that the product, as a bilinear map, is non-degenerate. This allows to consider the algebra as a two-sided ideal sitting in the multiplier algebra. If the algebra has a unit, we denote it by 1. This will also be used for the unit in the multiplier algebra. We will systematically use ι for the identity map.

We use A' for the space of all linear functionals on a vector space A and call it the dual space of A. Often, we will consider a suitable subspace of this full dual space. Most of the time, our tensor products are purely algebraic, except in the last section on locally compact quantum groups where we work with von Neumann algebras and von Neumann algebraic tensor products. Unfortunately, some other conventions in Section 4 are also different from those in the earlier sections. This is mainly due to differences between the algebraic approach and the operator algebra approach.

The basic references for Hopf algebras are of course [A] and [S]. For compact quantum groups we have [W2] and [W3], see also [M-VD]. For discrete quantum groups we refer to [P-W], [E-R] and [VD2]. The basic theory of multiplier Hopf algebras is found in [VD1] and when they have integrals, the reference is [VD3]. See also [VD-Z] for a survey paper on the subject. Finally, the general theory of locally compact quantum groups is developed in [K-V1], [K-V2] and [K-V3]. See also [M-N] and [M-N-W] for a different approach and [VD8] for a more recent and simpler treatment of the theory.

Acknowledgements I would like to thank the organizers of the Workshop on Quantum Groups and Noncommutative Geometry (MPIP Bonn, August 2007) for giving me the opportunity to talk about this subject. I am also grateful to P.M. Hajac who drew my attention to the paper by Kaufman and Radford [K-R].

2. Finite quantum groups

Let A be a finite-dimensional Hopf algebra (over the complex numbers) with coproduct Δ, counit ε and antipode S. Let $\widehat{A}$ denote the dual Hopf algebra of A. We will us the *pairing notation.* So, if $a \in A$ and $b \in \widehat{A}$ we write $\langle a, b\rangle$ for the value of b in the element a.

Let φ be a left integral on A. There exists a distinguished group-like element δ in A defined by the formula $\varphi(S(a)) = \varphi(a\delta)$ for all $a \in A$. We will call δ the *modular element* (for reasons we will explain later, in Section 3). Similarly, when $\widehat{\varphi}$ is a left integral on $\widehat{A}$, there is the modular element $\widehat{\delta}$ in $\widehat{A}$ satisfying $\widehat{\varphi}(S(b)) = \widehat{\varphi}(b\widehat{\delta})$ for all $b \in \widehat{A}$.

Now we can state Radford's formula (see [R]):

THEOREM 2.1. *For all $a \in A$, we have*

$$S^4(a) = \delta^{-1}(\widehat{\delta} \triangleright a \triangleleft \widehat{\delta}^{-1})\delta.$$

We use the standard left and right actions of the dual $\widehat{A}$ on A defined by

$$b \triangleright a = \sum_{(a)} a_{(1)}\langle a_{(2)}, b\rangle \qquad \text{and} \qquad a \triangleleft b = \sum_{(a)} a_{(2)}\langle a_{(1)}, b\rangle$$

for $a \in A$ and $b \in \widehat{A}$ (where we use the Sweedler notation).

Later, we will give a proof of this formula in the more general setting of algebraic quantum groups (see Section 3).

Let us also consider the case of a Hopf *-algebra. We assume that A is a *-algebra and that Δ is a *-homomorphism. Then ε is a *-homomorphism but S need not be a *-map. In stead, it is invertible and satisfies $S(a)^* = S^{-1}(a^*)$ for all a. So, it is a *-map if and only if $S^2 = \iota$, the identiy map.

If moreover A is an operator algebra, then there exists a *positive* left integral φ (and conversely). Then necessarily $\varphi(1) > 0$ so that left and right integrals coincide. This implies that $\delta = 1$. One can show that again $\widehat{A}$ will be an operator algebra and so also $\widehat{\delta} = 1$. Radford's formula implies that in this case $S^4 = \iota$. In fact, it follows that already $S^2 = \iota$ and that the integrals are traces. We will give a short argument later in the more general case of a discrete quantum group (see the next section and also Section 3).

In this note, we will call a finite-dimensional Hopf *-algebra with positive integrals a *finite quantum group.* In the literature however, one often calls it a finite-dimensional Kac algebra (see [E-S]).

3. Compact and discrete quantum groups

Now, let A be any Hopf algebra. We do no longer assume that it is finite-dimensional, but we require that it has integrals. Assume also that it has an invertible antipode. Again there exists a unique group-like element δ in A such that $\varphi(S(a)) = \varphi(a\delta)$ for all $a \in A$ when φ is a left integral on A.

The dual space A' is an algebra but no longer a Hopf algebra (in general). However, there still is the distinguished element $\widehat{\delta} \in A'$. It is a homomorphism, it is invertible and Radford's formula is still valid. For all $a \in A$, we have

$$S^4(a) = \delta^{-1}(\widehat{\delta} \triangleright a \triangleleft \widehat{\delta}^{-1})\delta.$$

The actions are defined as before by

$$f \triangleright a = \sum_{(a)} f(a_{(2)})a_{(1)} \qquad \text{and} \qquad a \triangleleft f = \sum_{(a)} f(a_{(1)})a_{(2)}$$

for all $a \in A$ and $f \in A'$.

The proof we plan to give later (for algebraic quantum groups) will also include this case.

If moreover A is a *-algebra and Δ a *-homomorphism, still ε will be a *-homomorphism and $S(a)^* = S^{-1}(a^*)$ for all $a \in A$. And if A is an operator algebra, the left integral is positive, it is also a right integral and so $\delta = 1$.

We agree to use the term *compact quantum group* for this case. Indeed, it is essentially a compact quantum group as defined by Woronowicz in [W3].

Remark that $\widehat{\delta}$ need not be 1 in this case, the integrals need not be traces and $S^2 \neq \iota$ is still possible. The standard example where this happens is the quantum $SU_q(2)$ (see [W1]).

Let us now consider the case of a discrete quantum group. Discrete quantum groups can be obtained as duals of compact quantum groups. Although it is more natural to treat them within the framework of multiplier Hopf algebras (see later), we will briefly consider the case already now (and see why we need to pass to multiplier Hopf algebras).

The following result is part of the motivation for what we will do later.

PROPOSITION 3.1. *Let A be a Hopf algebra with a left integral φ. Define the dual $\widehat{A}$ as the subspace of A' containing all elements of the form $\varphi(\,\cdot\, a)$ with $a \in A$. It is a subalgebra of A'. If we define the coproduct $\widehat{\Delta} : A' \to (A \otimes A)'$ by dualizing the product on A, we find that*

$$\widehat{\Delta}(\widehat{A})(1 \otimes \widehat{A}) \subseteq \widehat{A} \otimes \widehat{A} \qquad \textit{and} \qquad (\widehat{A} \otimes 1)\widehat{\Delta}(\widehat{A}) \subseteq \widehat{A} \otimes \widehat{A}$$

in the algebra $(A \otimes A)'$.

So, we get that $\widehat{\Delta}$ maps $\widehat{A}$ into the multiplier algebra $M(\widehat{A} \otimes \widehat{A})$ (as we will define it later). Moreover, the pair $(\widehat{A}, \widehat{\Delta})$ is a multiplier Hopf algebra (and not a Hopf algebra in general).

If we define $\widehat{\psi}(b) = \varepsilon(a)$ when $b = \varphi(\,\cdot\, a)$, we get a right integral on $\widehat{A}$. This means here that

$$(\widehat{\psi} \otimes \iota)(\widehat{\Delta}(b)(1 \otimes b')) = \widehat{\psi}(b)b'$$

for all $b, b' \in \widehat{A}$. The antipode S leaves $\widehat{A}$ invariant and converts $\widehat{\psi}$ to a left integral $\widehat{\varphi}$ on $\widehat{A}$. The element $\widehat{\delta}$, considered earlier, is in $M(\widehat{A})$ and still satifsies $\widehat{\varphi}(S(b)) = \widehat{\varphi}(b\widehat{\delta})$ for all $b \in \widehat{A}$.
If A is a compact quantum group, it turns out that $\widehat{A}$ is a direct sum of matrix algebras. This takes us to the following definition of a discrete quantum group.

DEFINITION 3.2. *A discrete quantum group is a pair (A, Δ) where A is a direct sum of matrix algebras (with the standard involution), Δ is a coproduct on A and such that there is a counit ε and an antipode S.*

It is not a Hopf algebra (except when it is a finite direct sum), but it is a multiplier Hopf algebra (see further). Indeed, we have $\Delta(A) \subseteq M(A \otimes A)$, the multiplier algebra of $A \otimes A$, but in general $\Delta(A)$ does not belong to $A \otimes A$ itself.

For discrete quantum groups, we can prove (among other things) the following result.

THEOREM 3.3. *There exists a positive left integral φ and a positive group-like element δ in the multiplier algebra $M(A)$ of A defined by $\varphi(S(a)) = \varphi(a\delta)$ for all $a \in A$. This element moreover satisfies*

$$S^2(a) = \delta^{-\frac{1}{2}} a \delta^{\frac{1}{2}}$$

for all a. We also have $\varphi(ab) = \varphi(bS^2(a))$ for all $a, b \in A$ and therefore, the map $a \mapsto \varphi(a\delta^{\frac{1}{2}})$ is a trace on A.

The first formula is a slightly stronger version of Radford's formula for these discrete quantum groups. It can be dualized to get a similar expression for the square S^2 of the antipode of a compact quantum group.

One way to develop discrete quantum groups is by viewing them as duals of compact quantum groups (as done in [P-W]). This however is not the best choice. It is relatively easy to develop the theory of discrete quantum groups (and prove the above results) directly from the definition above. Using the standard trace on each component, one can obtain quickly a formula for both integrals as well as for the modular element. See e.g. [VD2].

It can happen that $\delta \neq 1$ (so that left and right integrals are different). It can also happen that $S^2 \neq \iota$ so that the integrals are not traces. This can of course only happen if $\delta \neq 1$.

The standard example is the dual of the compact quantum group $SU_q(2)$ whose underlying algebra is the direct sum $\oplus_{n=0}^{\infty} M_n$. All objects can easily be given in terms of the deformation parameter q, except for the comultiplication (which is quite complicated), see e.g. [VD4].

On the other hand, if $\delta = 1$ we must have that $S^2 = \iota$ and that the integrals are traces. This generalizes the corresponding result for finite quantum groups as we have seen in Section 1. Observe also that if we have a quantum group that is both discrete and compact, it must be a finite quantum group.

4. Algebraic quantum groups

Discrete and compact quantum groups are special cases of algebraic quantum groups. Also the duality of algebraic quantum groups generalizes the one between discrete and compact quantum groups. We will briefly review this theory. For details, we refer to the literature, see [VD1], [VD3] and [VD-Z].

The basic ingredient is that of a multiplier Hopf algebra:

DEFINITION 4.1. *Let A be an algebra over $\mathbb{C}$, with or without identity, but with a non-degenerate product. A coproduct (or comultiplication) on A is a non-degenerate homomorphism $\Delta : A \to M(A \otimes A)$ (the multiplier algebra of $A \otimes A$), satisfying coassociativity $(\Delta \otimes \iota)\Delta = (\iota \otimes \Delta)\Delta$. The pair (A, Δ) is called a (regular) multiplier Hopf algebra if there exists a counit and an (invertible) antipode. If A is a *-algebra and Δ a *-homomorphism, regularity is automatic and we call it a multiplier Hopf *-algebra.*

There is a lot to say about this definition and we refer to the literature for details. However, it is important to notice that any Hopf (*-)algebra is a multiplier Hopf (*-)algebra and conversely, if the underlying algebra of a multiplier Hopf algebra has an identity, it is actually a Hopf algebra. Also remark that the counit and the antipode are unique.

Next, we consider algebraic quantum groups:

DEFINITION 4.2. *Let (A, Δ) be a regular multiplier Hopf algebra. A left integral is a non-zero linear functional $\varphi : A \to \mathbb{C}$ satisfying left invariance $(\iota \otimes \varphi)\Delta(a) = \varphi(a)1$ in $M(A)$ for all $a \in A$. Similarly, a right integral is defined.*

If a left integral φ exists, also a right integral ψ exists (namely $\psi = \varphi \circ S$). In that case, we call (A, Δ) an *algebraic quantum group.* If moreover (A, Δ) is a multiplier Hopf *-algebra with a positive left integral φ (i.e. such that $\varphi(a^*a) \geq 0$ for all a), then also a positive right integral exists (which is not a trivial result!). In that case, we call (A, Δ) a *-algebraic quantum group.

Remark that the term 'algebraic' does not refer to the possible quantization of algebraic groups, but we use it rather because *-algebraic quantum groups are locally compact quantum groups (considered in the next section) that can be treated with purely algebraic techniques.

Integrals on regular multiplier Hopf algebras are unique (up to a scalar) if they exist. They are faithful in the sense that (for the left integral φ) we have $a = 0$ if either $\varphi(ab) = 0$ for all b or $\varphi(ba) = 0$ for all b. From the uniqueness it follows that there is a constant ν (the *scaling constant*), given by $\varphi(S^2(a)) = \nu\varphi(a)$ for all $a \in A$. It can happen that $\nu \neq 1$ but when A is a *-algebraic quantum group (with positive integrals), we must have $\nu = 1$ (see [DC-VD]).

In general, integrals need not be traces, but there exist automorphisms σ and σ' (called the *modular automorphisms*) satisfying

$$\varphi(ab) = \varphi(b\sigma(a)) \qquad\qquad \psi(ab) = \psi(b\sigma'(a))$$

for all $a, b \in A$ when φ is a left integral and ψ a right integral. The term 'modular' comes from operator algebra theory and the modular automorphism group of a faithful normal state (or semi-finite weight) on a von Neumann algebra (see the next section).

Important for us in this note that focuses on Radford's formula is the *modular element* δ. It is a group-like element in the multiplier algebra $M(A)$ satisfying $\varphi(S(a)) = \varphi(a\delta)$ for all a just as in the case of Hopf algebras with integrals. It can be defined, using the uniqueness of integrals, by the formula $(\varphi \otimes \iota)\Delta(a) = \varphi(a)\delta$ for all a. In this case, the term 'modular' is used because it is related with the modular function for a non-unimodular locally compact group. In fact, also the modular automorphism group in the theory of von Neumann algebras finds its origin in the theory of non-unimodular locally compact groups.

There are many relations among these objects and again, we refer to the literature.

For any algebraic quantum group, we have a dual:

THEOREM 4.3. *Let (A, Δ) be an algebraic quantum group. Define the subspace $\widehat{A}$ of the dual space A' of functionals of the form $\varphi(\,\cdot\, a)$ where $a \in A$. The adjoints of the coproduct and the product on A define a product and a coproduct $\widehat{\Delta}$ on $\widehat{A}$, making $(\widehat{A}, \widehat{\Delta})$ into an algebraic quantum group, called the dual of (A, Δ). A right integral $\widehat{\psi}$ on $\widehat{A}$ is given by the formula $\widehat{\psi}(\omega) = \varepsilon(a)$ when $\omega = \varphi(\,\cdot\, a)$ and $a \in A$. If (A, Δ) is a *-algebraic quantum group, then so is $(\widehat{A}, \widehat{\Delta})$ and $\widehat{\psi}$ as defined before is positive when φ is positive.*

The last statement in the above theorem is a consequence of *Plancherel's formula*. Here it says that $\widehat{\psi}(\widehat{a}^*\widehat{a}) = \varphi(a^*a)$ if $a \in A$ and $\widehat{a} = \varphi(\,\cdot\, a)$, its *Fourier transform*.

Also remark that the dual of $(\widehat{A}, \widehat{\Delta})$ is again (A, Δ).

We will use the pairing notation (as we have already done in Section 1 for a finite-dimensional Hopf algebra and its dual). We also have the standard actions of $\widehat{A}$ on A and of A on $\widehat{A}$. In the first case, we have

$$\langle b \triangleright a, b' \rangle = \langle a, b'b \rangle$$
$$\langle a \triangleleft b, b' \rangle = \langle a, bb' \rangle$$

for all $a \in A$ and $b, b' \in B$. It is not completely obvious that these elements are well-defined in A, but it can be shown. Moreover, these actions are unital. This means that elements of the form $b \triangleright a$ with a in A and b in $\widehat{A}$ span all of A and similarly for the right action. See [Dr-VD] and [Dr-VD-Z].

Let us now first state some of the formulas relating the various objects of (A, Δ) and indicate how they can be proven. We use the notations introduced before.

PROPOSITION 4.4. *Let (A, Δ) be an algebraic quantum group. We have $\sigma \circ S \circ \sigma' = S$ and $\delta\sigma(a) = \sigma'(a)\delta$ and for all a. Also for all $a \in A$ we have*

$$\Delta(\sigma(a)) = (S^2 \otimes \sigma)\Delta(a).$$

The first formulas follow in a straighforward way from the definitions of σ and σ' with $\psi = \varphi \circ S = \varphi(\,\cdot\,\delta)$. For the second one, we use that for all $a, b \in A$,

$$S((\iota \otimes \varphi)(\Delta(a)(1 \otimes b))) = (\iota \otimes \varphi)((1 \otimes a)\Delta(b)),$$

two times in combination with the definition of σ. This last formula itself follows easily from left invariance of φ and the standard properties of the antipode.

We will also need some other properties. We have that the automorphisms S^2, σ and σ' all commute with each other. And we also have that $\sigma(\delta) = \sigma'(\delta) = \frac{1}{\nu}\delta$ where ν is the scaling constant.

Next, we state and prove some of the formulas relating objects of (A, Δ) with objects of the dual $(\widehat{A}, \widehat{\Delta})$.

PROPOSITION 4.5. *Let (A, Δ) be an algebraic quantum group and let $(\widehat{A}, \widehat{\Delta})$ be its dual. We have $\widehat{\delta}^{-1} = \varepsilon \circ \sigma$ where $\widehat{\delta}$ is the modular element of $\widehat{A}$, seen as a homomorphism of A. Also $\sigma(a) = \widehat{\delta}^{-1} \triangleright S^2(a)$ for all $a \in A$.*

PROOF. To prove the first formula, we start with $c \in A$ and we take the element $b = \varphi(\,\cdot\, c)$ in the dual $\widehat{A}$. Because for all a, a' in A we have $\varphi(a'c\sigma(a)) = \varphi(aa'c)$, we get $\varphi(\,\cdot\, c\sigma(a)) = b \triangleleft a$. If we apply $\widehat{\psi}$ we find $\varepsilon(c\sigma(a)) = \widehat{\psi}(b \triangleleft a)$. Because $(\iota \otimes \widehat{\psi})\widehat{\Delta}(b) = \widehat{\psi}(b)\widehat{\delta}^{-1}$ (a formula that can easily be obtained from the definition of $\widehat{\delta}$ by using the antipode), we get $\widehat{\psi}(b \triangleleft a) = \widehat{\psi}(b)\langle a, \widehat{\delta}^{-1}\rangle$. Combining all results and using that $\widehat{\psi}(b) = \varepsilon(c)$, we find the first formula of the proposition.

To obtain the second formula, consider the equation $\Delta(\sigma(a)) = (S^2 \otimes \sigma)\Delta(a)$, obtained in the previous proposition, apply $\iota \otimes \varepsilon$ and use the first formula of this proposition. □

In the proof above, we have used the left action of A on $\widehat{A}$. We also have looked at $\widehat{\delta}^{-1}$ as a linear functional on A by extending the pairing between A and $\widehat{A}$ to $M(\widehat{A})$ in an obvious way. If the quantum group is counimodular, that is if $\widehat{\delta} = 1$, it follows from these results that $\sigma = S^2$. This is the case for discrete quantum groups as we saw in Theorem 2.3.

Now we are ready to give a simple proof of Radford's formula for algebraic quantum groups.

THEOREM 4.6. *Let (A, Δ) be an algebraic quantum group. When δ and $\widehat{\delta}$ are the modular elements in A and its dual $\widehat{A}$, then*

$$S^4(a) = \delta^{-1}(\widehat{\delta} \triangleright a \triangleleft \widehat{\delta}^{-1})\delta$$

for all $a \in A$.

PROOF. ¿From the second formula in Proposition 3.5 we find $\widehat{\delta} \triangleright a = S^2(\sigma^{-1}(a))$. Similarly, or by applying the antipode on this formula, we obtain $a \triangleleft \widehat{\delta}^{-1} = S^2(\sigma'(a))$. If we combine these two formulas with the relation $\sigma'(a) = \delta\sigma(a)\delta^{-1}$ and use that $S^2(\delta) = \delta$, we get Radford's formula. □

The proof we have given can be found in [D-VD-Z] and in [D-VD], where we have generalized this result further to algebraic quantum hypergroups.

Next, let us look at the case of a *-algebraic quantum group. The requirement of positivity of the integrals is quite strong. We have mentioned already that it forces the scaling constant ν to be 1. On the other hand, we end up with an operator algebra and this allows to work on Hilbert spaces and use spectral theory. In this case, we arrive at what is called the *analytic structure* of a *-algebraic quantum group (see [K] and also [DC-VD]). Roughly speaking, it means that powers of S^2, σ, σ' and δ all have analytical extensions to the whole complex plane. More precisely, we get the following result. We only consider S^2 and δ because we focus in this note on Radford's formula.

PROPOSITION 4.7. *Let (A, Δ) be a *-algebraic quantum group. There exists an analytic function $\tau : z \mapsto \tau_z$ on $\mathbb{C}$ such that τ_z is an automorphism of (A, Δ), that $\tau_{z+y} = \tau_z \circ \tau_y$ for all $z, y \in \mathbb{C}$ and so that $S^2 = \tau_{-i}$. Similarly, there is an analytic function $z \mapsto \delta^z$ so that $\delta^z \in M(A)$, that $\delta^{z+y} = \delta^z\delta^y$ for all $z, y \in \mathbb{C}$ and such that $\delta^z = \delta$ for $z = 1$ (justifying the notation).*

Analyticity here is in a strong sense. In the first case, we want $z \mapsto f(\tau_z(a))$ analytic for all $a \in A$ and all $f \in A'$. In the second case, we want e.g. $z \mapsto f(a\delta^z)$ analytic for all $a \in A$ and $f \in A'$. These analytical extensions are unique.

Then, we can get the *analytical form* of Radford's formula. For real numbers, we obtain the following:

THEOREM 4.8. *Let (A, Δ) be a *-algebraic quantum group. Let τ_z and δ^z for $z \in \mathbb{C}$ be defined as in the previous proposition. Consider also $\widehat{\delta}^z \in M(\widehat{A})$ in a similar way. Then, for all $t \in \mathbb{R}$, we have*

$$\tau_{2t}(a) = \delta^{-it}(\widehat{\delta}^{it} \triangleright a \triangleleft \widehat{\delta}^{-it})\delta^{it}$$

for all $a \in A$.

This is the form of Radford's formula that we will be able to generalize to general locally compact quantum groups (see the next section). The result however is true for all complex numbers. In particular, we can take $z = -\frac{i}{2}$. This yields

$$S^2(a) = \delta^{-\frac{1}{2}}(\widehat{\delta}^{\frac{1}{2}} \triangleright a \triangleleft \widehat{\delta}^{-\frac{1}{2}})\delta^{\frac{1}{2}}$$

for all $a \in A$. Indeed, as a consequence of the result in Proposition 3.7, we can also define the square roots $\delta^{\frac{1}{2}}$ and $\widehat{\delta}^{\frac{1}{2}}$ in $M(A)$ and $M(\widehat{A})$ respectively. These are still group-like elements.

We should make a reference to a paper by Kaufman and Radford here [K-R]. They discover the formula with the square roots for Drinfel'd doubles that are ribbon Hopf algebras.

Finally, consider some special cases. If e.g. (A, Δ) is counimodular, this is by definition when left and right integrals on $\widehat{A}$ are the same, so that $\widehat{\delta} = 1$, we find that $S^2(a) = \delta^{-\frac{1}{2}} a \delta^{\frac{1}{2}}$ for all a. This is the formula that we have seen in Theorem 2.3 for discrete quantum groups. They are counimodular because compact quantum groups are unimodular. If (A, Δ) is both unimodular and counimodular, then we must have $S^2 = \iota$. In this case, it follows from Proposition 3.5 that both σ and σ' are trivial. This means that the integrals are traces. This, in particular, applies to the case of finite quantum groups (as in Section 1).

5. Locally compact quantum groups

We start this section with the definition of a locally compact quantum group in the von Neumann algebra setting.

DEFINITION 5.1. *Let M be a von Neumann algebra. A coproduct on M is a normal unital *-homomorphism $\Delta : M \to M \otimes M$, the von Neumann algebraic tensor product, satisfying coassociativity $(\Delta \otimes \iota)\Delta = (\iota \otimes \Delta)\Delta$. If there exist faithful normal semi-finite weights φ and ψ on M that are left, resp. right invariant, then the pair (M, Δ) is called a locally compact quantum group.*

We collect some important remarks about this concept:

REMARKS 5.2. *i) The adapted form of continuity of Δ in the von Neumann algebra setting is expressed in the requirement that the coproduct is normal.*
ii) By this continuity, the *-homomorphisms $\Delta \otimes \iota$ and $\iota \otimes \Delta$ are well-defined from $M \otimes M$ to $M \otimes M \otimes M$ and so coassociativity makes sense.
iii) A weight on a von Neumann algebra is, roughly speaking, an unbounded positive linear functional. It is called semi-finite if it is bounded on enough elements. And again it is called normal if it satisfies the proper continuity.

For the theory of von Neumann algebras and the notions needed above, we refer to the books of Takesaki [T1] and [T2].

The weight φ is called left invariant if $\varphi((\omega \otimes \iota)\Delta(x)) = \varphi(x)\omega(1)$ whenever x is a positive element in the von Neumann algebra with $\varphi(x) < \infty$ and ω is a positive element in the predual M_* of M. Similarly, right invariance of the weight ψ is defined. These weights are unique (up to a scalar) and are called the left and right *Haar weights.* They are of course the analogues of the left and right integrals in the theory of *-algebraic quantum groups.

It is also possible to define locally compact quantum groups in the framework of C*-algebras, but that is somewhat more complicated. In fact, both approaches are equivalent in the sense that they define the same objects. We refer to the original works by Kustermans and Vaes; see [K-V1], [K-V2] and [K-V3]. Independently, the notion was also developed by Masuda, Nakagami and Woronowicz; see [M-N] and [M-N-W]. A more recent and simpler development of the theory can be found in

[VD8] and a discussion on the equivalence of the C*-approach and the von Neumann approach is e.g. given in [VD7].

The basic examples come from a locally compact group G. On the one hand, there is the abelian von Neumann algebra $L^\infty(G)$, defined with respect to the left Haar measure. The coproduct Δ is given as before by $\Delta(f)(p,q) = f(pq)$ when $f \in L^\infty(G)$ and $p,q \in G$. The invariant weights φ and ψ are obtained by integration with respect to the left and right Haar measures on the group. On the other hand, there is the group von Neuman algebra $VN(G)$ generated by the left regular representation λ of the group on the Hilbert space $L^2(G)$. In this case, the coproduct is given by $\Delta(\lambda_p) = \lambda_p \otimes \lambda_p$. The left and right integrals are the same. Formally, we must have $\varphi(\lambda_p) = 0$, except when $p = e$, the identity of the group, but it is not so easy to define this weight properly.

Any multiplier Hopf *-algebra with positive integrals, i.e. a *-algebraic quantum group, gives rise to a locally compact quantum group (see [K-VD]):

THEOREM 5.3. *Let (A, Δ) be a *-algebraic quantum group with left integral φ. Consider the GNS-representation π_φ of A associated with φ. The coproduct on A yields a coproduct on the von Neumann algebra M generated by $\pi_\varphi(A)$ making it into a locally compact quantum group.*

The Haar weights are of course nothing else but the unique normal extensions of the original left and right integrals.

It is an interesting, but open problem to describe those locally compact quantum groups that can arise from *-algebraic quantum groups as above. In the case of locally compact groups, the problem has been solved in [L-VD]. The requirement is that there exists a compact open subgroup. In particular, when G is a totally disconected locally compact group, the two associated locally compact quantum groups are essentially *-algebraic quantum groups. In connection with this problem, let us also observe the following. For any *-algebraic quantum group, the scaling constant ν is necessarily 1 (see [DC-VD]). However, there are examples of locally compact quantum groups where this is not the case (see [VD5]). We will come back to this statement later.

Let us now indicate how the theory of locally compact quantum groups is developed (as e.g. in [VD8]) and focus on the relevant formulas, needed to formulate Radford's result.

So, we start with a locally compact quantum group (M, Δ) with left and right Haar weights φ and ψ as in Definition 4.1. We recall the GNS construction:

PROPOSITION 5.4. *Denote by $\mathcal{N}_\varphi$ the set of elements $x \in M$ so that $\varphi(x^*x) < \infty$. It is a dense left ideal of M and φ has a unique extension (still denoted by φ) to the *-algebra spanned by elements of the form x^*y with $x, y \in \mathcal{N}_\varphi$. There exists a Hilbert space $\mathcal{H}_\varphi$ and an injective linear map $\Lambda_\varphi : \mathcal{N}_\varphi \to \mathcal{H}_\varphi$ with dense range such that $\langle \Lambda_\varphi(x), \Lambda_\varphi(y) \rangle = \varphi(y^*x)$ for all $x, y \in \mathcal{N}_\varphi$. There also exists a faithful, unital and normal *-representation π_φ of M on $\mathcal{H}_\varphi$ given by $\pi_\varphi(y)\Lambda_\varphi(x) = \Lambda_\varphi(yx)$ whenever $x \in \mathcal{N}_\varphi$ and $y \in M$.*

In what follows, we will drop the index φ and use $\mathcal{H}$ and Λ in stead of $\mathcal{H}_\varphi$ and Λ_φ. We will also omit π_φ and assume that M acts directly on the space $\mathcal{H}$.

Next, we recall some results from the Tomita-Takesaki modular theory (see e.g. [T2]):

PROPOSITION 5.5. *There is a closed, conjugate linear, possibly unbounded but densely defined involutive operator T on $\mathcal{H}$ so that $\Lambda(x) \in \mathcal{D}(T)$, the domain of T, for any $x \in \mathcal{N}_\varphi \cap \mathcal{N}_\varphi^*$ and $T\Lambda(x) = \Lambda(x^*)$. If $T = J\nabla^{\frac{1}{2}}$ denotes the polar decomposition of T, then J is a conjugate linear isometric involutive operator and ∇ a positive non-singular self-adjoint operator. If M' denotes the commutant of M, we have $JMJ = M'$. Also $\nabla^{it}M\nabla^{-it} = M$ for all $t \in \mathbb{R}$.*

It follows from this result that we can define a one-parameter group (σ_t) of automorphisms of M, called the *modular automorphism group*, by $\sigma_t(x) = \nabla^{it}x\nabla^{-it}$ for $x \in M$ and $t \in \mathbb{R}$. A similar construction will give the modular automorphisms (σ'_t) associated with the right Haar weight ψ.

Using a proper notion of an analytic extension, one can show that $\varphi(xy) = \varphi(y\sigma_{-i}(x))$ for the appropriate elements x and y. So (σ_{-i}) plays the role of the modular automorphism σ as we have it for *-algebraic quantum groups. Similarly σ'_{-i} plays the role of the modular automorphism σ'. We apologize for the possible confusion caused by the difference in notations used here (and further in this section).

There is also something called the 'relative modular theory' when two weights are considered. If we apply results from this theory to the invariant weights φ and ψ, we find the following:

PROPOSITION 5.6. *There exists a positive non-singular self-adjoint operator δ on the Hilbert space $\mathcal{H}$ such that for all $t \in \mathbb{R}$ we have $\delta^{it} \in M$ and $\psi = \varphi(\delta^{\frac{1}{2}} \cdot \delta^{\frac{1}{2}})$.*

It should be mentioned that it is not so easy to interprete this last formula in a correct way.

When thinking of a *-algebraic quantum group, where we have $\sigma(\delta) = \delta$ (because the scaling constant is trivial), we see that this formula is another form of the one we have for algebraic quantum groups, namely $\psi = \varphi(\,\cdot\,\delta)$. Here, we call δ the *modular operator.*

These are the first main ingredients of the theory. Remark that these objects are only dependent on the weights φ and ψ on the von Neumann algebras M and seem in no way related with the coproduct structure. This is not completely correct as the result in Proposition 4.6 would not be true for any pair of weights.

Next, let us consider the dual locally compact quantum group $(\widehat{M}, \widehat{\Delta})$ with left and right Haar weights $\widehat{\varphi}$ and $\widehat{\psi}$. The precise construction is quite involved but in essence, it is a careful analytic version of the same construction for *-algebraic quantum groups.

The Hilbert space associated with the dual left Haar weight $\widehat{\varphi}$ is identified with $\mathcal{H}$ and the map $\widehat{\Lambda}$ associated with $\widehat{\varphi}$ is defined in such a way that $\widehat{\Lambda}(\widehat{x}) = \Lambda(x)$ when x is an appropriate element in M and $\widehat{x}$ its Fourier transform $\varphi(\,\cdot\,x)$. Remark that a different convention is used in the sense that the dual coproduct is flipped causing, among other things, that the dual right integral $\widehat{\psi}$ is now the dual left integral $\widehat{\varphi}$. This convention is common in the operator algebra approach.

And just as for the original locally compact quantum group (M, Δ), we also have the conjugate linear isometric operator $\widehat{J}$ on $\mathcal{H}$ for the dual $(\widehat{M}, \widehat{\Delta})$ satisfying $\widehat{J}\widehat{M}\widehat{J} = \widehat{M}'$ and the modular automorphisms $(\widehat{\sigma}_t)$ and $(\widehat{\sigma}'_t)$ of $\widehat{M}$, as well as the modular operator $\hat{\delta}$ for the dual.

The *scaling group* can be characterized as follows:

PROPOSITION 5.7. *There exists a one-parameter group of automorpisms* (τ_t) *of* (M, Δ) *such that*

$$\Delta(\sigma_t(x)) = (\tau_t \otimes \sigma_t)\Delta(x)$$
$$\Delta(\sigma'_t(x)) = (\sigma'_t \otimes \tau_{-t})\Delta(x)$$

for all $x \in M$ *and* $t \in \mathbb{R}$. *All the automorphisms in* (σ_t), (σ'_t) *and* (τ_t) *mutually commute.*

Similarly, we have the scaling group $(\widehat{\tau}_t)$ on the dual, characterized by similar formulas.

If we take a proper analytic extension, we see that τ_{-i} is like the square S^2 of the antipode in a *-algebraic quantum group. The first formula replaces $\Delta(\sigma(a))) = (S^2 \otimes \sigma)\Delta(a)$ and the second one is $\Delta(\sigma'(a)) = (\sigma' \otimes S^{-2})\Delta(a)$ for an element a in a *algebraic quantum group.

Again, the proof is technically rather difficult. It essentially uses the polar decomposition of an operator $\Lambda(x) \mapsto \Lambda(S(x)^*)$ where S is the 'antipode', roughly defined by the formula

$$S((\iota \otimes \varphi)(\Delta(x)(1 \otimes y))) = (\iota \otimes \varphi)((1 \otimes x)\Delta(y))$$

for well-chosen elements x and y in the von Neumann algebra M.

There are several relations among the data we have so far:

PROPOSITION 5.8. *When* $x \in M$ *and* $y \in \widehat{M}$ *we have*

$$\begin{aligned} \sigma_t(x) &= \nabla^{it} x \nabla^{-it} & \qquad \tau_t(x) &= \widehat{\nabla}^{it} x \widehat{\nabla}^{-it} \\ \widehat{\sigma}_t(y) &= \widehat{\nabla}^{it} y \widehat{\nabla}^{-it} & \qquad \widehat{\tau}_t(y) &= \nabla^{it} y \nabla^{-it} \end{aligned}$$

for all $t \in \mathbb{R}$.

The formulas on the left were mentioned already but the others are new (and somewhat remarkable). We do not have any counterparts of these equations in the theory of *-algebraic quantum groups. This is not so with the following results.

PROPOSITION 5.9. *There exists a strictly positive number* ν, *called the scaling constant, satisfying*

$$\begin{aligned} \varphi \circ \tau_t &= \nu^{-t} \varphi & \qquad \varphi \circ \sigma'_t &= \nu^t \varphi \\ \psi \circ \tau_t &= \nu^{-t} \psi & \qquad \psi \circ \sigma_t &= \nu^{-t} \psi \end{aligned}$$

for all $t \in \mathbb{R}$.

When extending these formulas analytically to the complex number $-i$, we find e.g. $\varphi \circ S^2 = \nu^i \varphi$ and we see that ν^i turns out to replace the scaling constant as introduced for *-algebraic quantum groups. As mentioned already, in this case, the scaling constant can be non-trivial, see e.g. [VD5].

Also the above result is a consequence of the uniqueness of the invariant weights. And finally, we have some formulas relating δ with the other data:

PROPOSITION 5.10. *We have* $\tau_t(\delta) = \delta$ *and*

$$\sigma_t(\delta) = \nu^t \delta \qquad\qquad \sigma'_t(\delta) = \nu^{-t}\delta$$

for all t. *We also have* $\widehat{J}\delta\widehat{J} = \delta^{-1}$.

Of course, these formulas have to be interpreted (e.g. by looking at powers δ^{is} of δ). There is also a formula for $J\delta J$ but that is more complicated. Similar equations hold for the dual modular operator $\widehat{\delta}$.

Having defined the main objects and the most important formulas, we can now state the analytical form of Radford's formula for locally compact quantum groups (see Theorem 4.20 in [VD8]):

THEOREM 5.11. *Because the left Haar weight is relatively invariant, we can define a one-parameter group of unitary operators, denoted* P^{it}, *by the formula* $P^{it}\Lambda(x) = \nu^{\frac{1}{2}t}\Lambda(\tau_t(x))$ *for all* $x \in \mathcal{N}$. *Then we have*

$$P^{-2it} = \delta^{it}(J\delta^{it}J)\widehat{\delta}^{it}(\widehat{J}\widehat{\delta}^{it}\widehat{J})$$

for all t.

Compare this formula, call it the 'second' formula in what follows, with the formula in Theorem 3.8, which we will call the 'first' one. And assume for the moment that the scaling constant is 1. Change t to $-t$ in the first formula and 'apply' Λ. On the left hand side, we get $P^{-2it}\Lambda(a)$. When we look at the right hand side, first we have left multiplication with δ^{it} in the first formula which we find as the operator δ^{it} in the second formula. Next we have right multiplication with δ^{-it} in the first formula that results in the operator $J\delta^{it}J$ in the second formula. The change in sign comes from the fact that J is conjugate linear and δ self-adjoint. Also remember that J is the unitary part in the polar decomposition of the map $\Lambda(x) \mapsto \Lambda(x^*)$ and the fact that the involution changes the order allows to express right multiplication with elements as operators, using this map. The third factor in the second formula comes from the left action of $\widehat{\delta}^{-it}$. Now, the difference in sign is coming from the difference in conventions about the dual coproduct. Flipping this coproduct causes $\widehat{\delta}$ to be replaced by $\widehat{\delta}^{-1}$. Finally, the right action of $\widehat{\delta}^{it}$ corresponds with the factor $\widehat{J}\widehat{\delta}^{it}\widehat{J}$. We have the same sign here because it is changed two times for reasons explained earlier.

If the scaling constant is not equal to 1, we get an extra factor on the left because $P^{-2it}\Lambda(x) = \nu^{-t}\Lambda(\tau_t(x))$. This factor will also occur on the right hand side because right multiplication with δ^{-it} is not exactly the same as $J\delta^{it}J$. There is a factor $\nu^{\frac{1}{2}t}$ coming from the commutation rules between the modular operator ∇ and δ (as $\sigma(\delta) = \frac{1}{\nu}\delta$ in the case of algebraic quantum groups). Similarly, this scalar will pop up when comparing the right action of $\widehat{\delta}^{it}$ with the factor $\widehat{J}\widehat{\delta}^{it}\widehat{J}$.

So, we see that the two formulas are completely in accordance with each other and that it is justified to call the formula in Theorem 4.11 above the analytical form of Radford's formula for locally compact quantum groups.

Also here, it is interesting to look at some special cases. If e.g. $\widehat{\delta} = 1$, also in this case we have $\sigma_t = \tau_t$ and $\sigma'_t = \tau_{-t}$ (as for discrete quantum groups). If both modular operators are 1, then necessarily the scaling group and the modular automorphisms are trival, causing the Haar weights to be traces and $S^2 = \iota$.

References

[A] E. Abe: *Hopf algebras.* Cambridge University Press (1977).

[B-B-T] M. Beattie, D. Bulacu & B. Torrecillas: *Radford's S^4 formula for co-Frobenius Hopf algebras.* J. Algebra 307 (2007), 330-342.

[DC-VD] K. De Commer & A. Van Daele: *Multiplier Hopf algebras imbedded in C^*-algebraic quantum groups.* Preprint K.U. Leuven (2006). Arxiv math.OA/0611872. To appear in Rocky Mountain Journal of Mathematics.

[D-VD] L. Delvaux & A. Van Daele: *Algebraic quantum hypergroups.* Preprint University of Hasselt and K.U. Leuven (2006). math.RA/0606466

[D-VD-W] L. Delvaux, A. Van Daele & S. Wang: *A note on Radford's S^4 formula.* Preprint University of Hasselt, K.U. Leuven and Nanjing University (2006). Arxiv math.RA/0608096.

[Dr-VD] B. Drabant & A. Van Daele: *Pairing and Quantum double of multiplier Hopf algebras.* Algebras and Representation Theory 4 (2001), 109-132.

[Dr-VD-Z] B. Drabant, A. Van Daele & Y. Zhang: *Actions of multiplier Hopf algebras.* Commun. Alg. 27 (1999), 4117–4172.

[E-R] E.G. Effros & Z.-J. Ruan : *Discrete quantum groups I. The Haar measure.* Int. J. Math. 5 (1994) 681-723.

[E-S] M. Enock & J.-M. Schwartz: *Kac algebras and duality for locally compact groups.* Springer (1992).

[K] J. Kustermans: *The analytic structure of algebraic quantum groups.* J. Algebra 259 (2003), 415–450.

[K-R] L. Kaufman & D. Radford: *A necessary and sufficient condition for a finite-dimensional Drinfel'd double to be a ribbon Hopf algebra.* J. of Alg. 159 (1993), 98-114.

[K-V1] J. Kustermans & S. Vaes: *A simple definition for locally compact quantum groups.* C.R. Acad. Sci. Paris Sér I 328 (1999), 871–876.

[K-V2] J. Kustermans & S. Vaes: *Locally compact quantum groups.* Ann. Sci. École Norm. Sup. (4) (33) (2000), 837–934.

[K-V3] J. Kustermans & S. Vaes: *Locally compact quantum groups in the von Neumann algebra setting.* Math. Scand. 92 (2003), 68–92.

[K-VD] J. Kustermans & A. Van Daele: *C^*-algebraic quantum groups arising from algebraic quantum groups.* Int. J. Math. 8 (1997), 1067–1139.

[L-VD] M.B. Landstad & A. Van Daele: *Groups with compact open subgroups and multiplier Hopf *-algebras.* Expo. Math. 26 (2008), 197–217.

[M-VD] A. Maes & A. Van Daele: *Notes on compact quantum groups.* Nieuw Archief voor Wiskunde, Vierde serie 16 (1998), 73–112.

[M-N] M. Masuda & Y. Nakagami: *A von Neumann algebra framework for the duality of quantum groups.* Publ. RIMS Kyoto **30** (1994), 799–850.

[M-N-W] T. Masuda, Y. Nakagami & S.L. Woronowicz: *A C^*-algebraic framework for the quantum groups.* Int. J. of Math. 14 (2003), 903–1001.

[P-W] P. Podleś & S.L. Woronowicz: *Quantum deformation of Lorentz group.* Commun. Math. Phys. 130 (1990), 381–431.

[R] D. Radford: *The order of the antipode of any finite-dimensional Hopf algebra is finite.* Amer. J. Math. 98 (1976), 333–355.

[S] M. Sweedler: *Hopf algebras.* Benjamin, New-York (1969).

[T1] M. Takesaki: *Theory of Operator Algebras I.* Springer-Verlag, New York (1979).

[T2] M. Takesaki: *Theory of Operator Algebras II.* Springer-Verlag, New York (2001).

[VD1] A. Van Daele: *Multiplier Hopf algebras.* Trans. Am. Math. Soc. 342 (1994), 917–932.

[VD2] A. Van Daele: *Discrete quantum groups.* J. of Alg. 180 (1996), 431-444.

[VD3] A. Van Daele: *An algebraic framework for group duality.* Adv. Math. 140 (1998), 323–366.

[VD4] A. Van Daele: *Quantum groups with invariant integrals.* Proc. Natl. Acad. Sci. USA 97 (2000), 541-546.

[VD5] A. Van Daele: *The Haar measure on some locally compact quantum groups.* Preprint K.U. Leuven (2001). Arxiv math.OA/0109004.

[VD6] A. Van Daele: *Multiplier Hopf *-algebras with positive integrals: A laboratory for locally compact quantum groups.* Irma Lectures in Mathematical and Theoretical Physics 2: Locally compact Quantum Groups and Groupoids. Proceedings of the meeting in Strasbourg on Hopf algebras, quantum groups and their applications (2002). Ed. V. Turaev & L. Vainerman. Walter de Gruyter, (2003), 229–247.

[VD7] A. Van Daele: *Locally compact quantum groups: The von Neumann algebra versus the C^*-algebra approach.* Bulletin of Kerala Mathematics Association, Special issue (2006), 153–177.

[VD8] A. Van Daele: *Locally compact quantum groups. A von Neumann algebra approach.* Preprint K.U. Leuven (2006). Arxiv math.OA/0602212.

[VD-Z] A. Van Daele & Y. Zhang: *A survey on multiplier Hopf algebras.* In 'Hopf algebras and Quantum Groups', eds. S. Caenepeel & F. Van Oyestayen, Dekker, New York (1998), pp. 259–309.

[W1] S.L. Woronowicz: *Twisted $SU(2)$ group. An example of a non-commutative differential calculus.* Publ. RIMS, Kyoto University 23 (1987), 117–181.

[W2] S.L. Woronowicz: *Compact matrix pseudogroups.* Comm. Math. Phys. 111 (1987), 613-665.

[W3] S.L. Woronowicz: *Compact quantum groups.* Quantum symmetries/Symmétries quantiques. Proceedings of the Les Houches summer school 1995, North-Holland, Amsterdam (1998), 845–884.

DEPARTMENT OF MATHEMATICS, K.U. LEUVEN, CELESTIJNENLAAN 200B, B-3001 HEVERLEE (BELGIUM).

E-mail address: `Alfons.VanDaele@wis.kuleuven.be`

Categorical aspects of Hopf algebras

Robert Wisbauer

ABSTRACT. Hopf algebras allow for useful applications, for example in physics. Yet they also are mathematical objects of considerable theoretical interest and it is this aspect which we want to focus on in this survey. Our intention is to present techniques and results from module and category theory which lead to a deeper understanding of these structures. We begin with recalling parts from module theory which do serve our purpose but which may also find other applications. Eventually the notion of Hopf algebras (in module categories) will be extended to Hopf monads on arbitrary categories.

1. Introduction

The author's interest in coalgebraic structures and Hopf algebras arose from the observation that the categories considered in those situations are similar to those in module theory over associative (and nonassociative) rings. At the beginning in the 1960's, the study of coalgebras was to a far extent motivated by the classical theory of algebras over fields; in particular, the finiteness theorem for comodules brought the investigations close to the theory of finite dimensional algebras. Moreover, comodules for coalgebras C over fields can be essentially handled as modules over the dual algebra C^*.

Bringing in knowledge from module theory, coalgebras over commutative rings could be handled and from this it was a short step to extend the theory to *corings* over *non-commutative* rings (e.g. [**BrWi**]). This allows, for example, to consider for bialgebras B over a commuatative ring R, the tensorproduct $B \otimes_R B$ as coring over B and the Hopf bimodules over B as $B \otimes_R B$-comodules. Clearly this was a conceptual simplification of the related theory and the basic idea could be transferred to other situations. Some of these aspects are outlined in this talk.

Since Lawvere's categorification of general algebra, algebras and coalgebras are used as basic notions in universal algebra, logic, and theoretical computer science, for example (e.g. [**AdPo**], [**Gu**], [**TuPl**]).

The categories of interest there are far from being additive. The transfer of Hopf algebras in module categories to Hopf monads in arbitrary categories provides the chance to understand and study this notion in this wider context.

Generalisations of Hopf theory to *monoidal* categories were also suggested in papers by Moerdijk [**Moer**], Loday [**Lod**] and others. Handling these notions in arbitrary categories may also help to a better understanding of their concepts.

Not surprisingly, there is some overlap with the survey talks [**Wi.H**] and [**Wi.G**]. Here a broader point of view is taken and more recent progress is recorded.

2. Algebras

Let R be an associative and commutative ring with unit. Denote by $\mathbb{M}_R$ the category of (right) R-modules.

2.1. **Algebras.** An R-algebra (A, m, e) is an R-module A with R-linear maps, product and unit,

$$m : A \otimes_R A \to A, \quad e : R \to A,$$

satisfying associativity and unitality conditions expressed by commutativity of the diagrams

$$\begin{array}{ccc} A\otimes_R A\otimes_R A & \xrightarrow{m\otimes I} & A\otimes_R A \\ {\scriptstyle I\otimes m}\downarrow & & \downarrow{\scriptstyle m} \\ A\otimes_R A & \xrightarrow{\quad m \quad} & A, \end{array} \qquad \begin{array}{ccccc} A & \xrightarrow{I\otimes e} & A\otimes_R A & \xleftarrow{e\otimes I} & A \\ & {\scriptstyle =}\searrow & \downarrow{\scriptstyle m} & \swarrow{\scriptstyle =} & \\ & & A. & & \end{array}$$

2.2. **Tensorproduct of algebras.** Given two R-algebras (A, m_A, e_A) and (B, m_B, e_B), the tensor product $A \otimes_R B$ can be made an algebra with product

$$m_{A\otimes B} : \ A\otimes_R B\otimes_R A\otimes_R B \xrightarrow{I\otimes\tau\otimes I} A\otimes_R A\otimes_R B\otimes_R B \xrightarrow{m_A\otimes m_B} A\otimes_R B\ ,$$

and unit $e_A \otimes e_B : R \to A \otimes_R B$, for some R-linear map

$$\tau : B \otimes_R A \to A \otimes_R B.$$

inducing commutative diagrams

$$\begin{array}{ccccc} B\otimes_R B\otimes_R A & & \xrightarrow{\qquad m_B\otimes I \qquad} & & B\otimes_R A \\ {\scriptstyle I\otimes\tau}\downarrow & & & & \downarrow{\scriptstyle \tau} \\ B\otimes_R A\otimes_R B & \xrightarrow{\tau\otimes I} & A\otimes_R B\otimes B & \xrightarrow{I\otimes m_B} & A\otimes_R B, \end{array} \qquad \begin{array}{ccc} A & \xrightarrow{e_B\otimes I} & B\otimes_R A \\ & {\scriptstyle I\otimes e_B}\searrow & \downarrow{\scriptstyle \tau} \\ & & A\otimes_R B, \end{array}$$

and similar diagrams derived from the product m_A and unit e_A of A.

It is easy to see that the canonical twist map

$$\text{tw} : A \otimes_R B \to B \otimes_{R}, A, \quad a \otimes b \to b \otimes a,$$

satisfies the conditions on τ and this is widely used to define a product on $A \otimes_R B$. However, there are many other such maps of interest.

These kind of conditions can be readily transferred to functors on arbitrary categories and in this context they are known as *distributive laws* (e.g. [**Be**], [**Wi.A**]).

3. Category of A-modules

Let A be an associative R-algebra with unit.

3.1. **A-modules.** A left A-module M is an R-module with an R-linear map $\rho_M : A \otimes_R M \to M$ with commutative diagrams

$$\begin{array}{ccc} A \otimes_R A \otimes_R M & \xrightarrow{I \otimes \rho_M} & A \otimes_R M \\ {\scriptstyle m\otimes I}\downarrow & & \downarrow{\scriptstyle \rho_M} \\ A \otimes_R M & \xrightarrow{\rho_M} & A, \end{array} \qquad \begin{array}{ccc} M & \xrightarrow{e \otimes I} & A \otimes_R M \\ & \searrow^{=} & \downarrow{\scriptstyle \rho_M} \\ & & M. \end{array}$$

The category ${}_A\mathbb{M}$ of (unital) left A-modules is a Grothendieck category with A a finitely generated projective generator.

Properties of (the ring, module) A are reflected by properties of the category ${}_A\mathbb{M}$. These interdependencies were studied under the title *homological classification of rings.*

To use such techniques for the investigation of the structure of a right A-module M, one may consider the smallest Grothendieck (full) subcategory of ${}_A\mathbb{M}$ which contains M. For this purpose recall that an A-module N is called

M-generated if there is an epimorphism $M^{(\Lambda)} \to N$, Λ an index set, and
M-subgenerated if N is a submodule of an M-generated module.

3.2. **The category $\sigma[M]$.** For any A-module M, denote by $\sigma[M]$ the full subcategory of ${}_A\mathbb{M}$ whose objects are all M-subgenerated modules. This is the smallest Grothendieck category containing M. Thus it shares many properties with ${}_A\mathbb{M}$, however it need not contain neither a projective nor a finitely generated generator. For example, one may think of the category of abelian torsion groups which is just the subcategory $\sigma[\mathbb{Q}/\mathbb{Z}]$ of ${}_\mathbb{Z}\mathbb{M}$ (without non-zero projective objects).

In general, M need not be a generator in $\sigma[M]$. A module $N \in \sigma[M]$ with $\sigma[N] = \sigma[M]$ is said to be a *subgenerator* in $\sigma[M]$. Of course, M is a subgenerator in $\sigma[M]$ (by definition). The notion of a subgenerator also plays a prominent role in the categories considered for coalgebraic structures (e.g. 4.2, 5.3).

An A-module N is a subgenerator in ${}_A\mathbb{M}$ if and only if A embeds in a finite direct sum of copies of N, i.e. $A \hookrightarrow N^k$, for some $k \in \mathbb{N}$. Such modules are also called *cofaithful.*

The notion of singularity in ${}_A\mathbb{M}$ can be transferred to $\sigma[M]$: A module $N \in \sigma[M]$ is called *singular in $\sigma[M]$* or *M-singular* if $N \simeq L/K$ for $L \in \sigma[A]$ and $K \subset L$ an essential submodule.

3.3. **Trace functor.** The *inclusion functor* $\sigma[P] \to {}_A\mathbb{M}$ has a *right adjoint* $\mathcal{T}^M : {}_A\mathbb{M} \to \sigma[M]$, sending $X \in {}_A\mathbb{M}$ to

$$\mathcal{T}^M(X) := \sum\{f(N) \mid N \in \sigma[M],\ f \in \mathrm{Hom}_A(N, X)\}.$$

3.4. **Functors determined by $P \in {}_A\mathbb{M}$.** Given any A-module P with $S = \mathrm{End}_A(P)$, there is an adjoint pair of functors

$$P \otimes_S - : {}_S\mathbb{M} \to {}_A\mathbb{M}, \quad \mathrm{Hom}_A(P, -) : {}_A\mathbb{M} \to {}_S\mathbb{M},$$

with (co)restriction

$$P \otimes_S - : {}_S\mathbb{M} \to \sigma[P], \quad \mathrm{Hom}_A(P, -) : \sigma[P] \to {}_S\mathbb{M}.$$

and functorial isomorphism

$$\mathrm{Hom}_A(P \otimes_S X, Y) \to \mathrm{Hom}_S(X, \mathrm{Hom}_A(P, Y)),$$

$$\text{unit} \quad \eta_X : X \to \mathrm{Hom}_A(P, P \otimes_S X), \quad x \mapsto [p \mapsto p \otimes x];$$

$$\text{counit} \quad \varepsilon_Y : P \otimes_S \mathrm{Hom}_A(P, Y) \to Y, \quad p \otimes f \mapsto f(p).$$

These functors determine an *equivalence of categories* if and only if η and ε are natural isomorphisms.

In any category $\mathbb{A}$, an object $G \in \mathbb{A}$ is said to be a *generator* provided the functor $\mathrm{Mor}_{\mathbb{A}}(G, -) : \mathbb{A} \to \mathrm{Ens}$ is faithful. It is a property of Grothendieck categories that these functors are even fully faithful ([**Nast**, III, Teoremă 9.1]).

Let $P \in {}_A\mathbb{M}$, $S = \mathrm{End}_A(P)$. Then P is a right S-module and there is a canonical ring morphism

$$\phi : A \to B = \mathrm{End}_S(M), \quad a \mapsto [m \mapsto am].$$

P is called *balanced* provided ϕ is an isomorphism.

3.5. P **as generator in** ${}_A\mathbb{M}$. *The following are equivalent:*

(a) *P is a generator in ${}_A\mathbb{M}$;*

(b) *$\mathrm{Hom}_A(P, -) : {}_A\mathbb{M} \to {}_S\mathbb{M}$ is (fully) faithful;*

(c) *$\varepsilon : P \otimes_S \mathrm{Hom}_A(P, N) \to N$ is surjective (bijective), $N \in {}_A\mathbb{M}$;*

(d) *P is balanced and P_S is finitely generated and projective.*

Note that the equivalence of (a) and (d) goes back to Morita [**Mor**]. It need not hold in more general situations. In [**Wi.G**, 2.6] it is shown:

3.6. P **as generator in** $\sigma[P]$. *The following are equivalent:*

(a) *P is a generator in $\sigma[P]$;*

(b) *$\mathrm{Hom}_A(P, -) : \sigma[P] \to {}_S\mathbb{M}$ is (fully) faithful;*

(c) *$\varepsilon_N : P \otimes_S \mathrm{Hom}_A(P, N) \to N$ is sur-(bi-)jective, $N \in \sigma[P]$;*

(d) *$\phi : A \to B$ is dense, P_S is flat and*
ε_V is an isomorphism for all injectives $V \in \sigma[P]$.

The elementary notions sketched above lead to interesting characterisations of Azumaya R-algebras (R a commutative ring) when applied to A considered as an (A, A)-bimodule, or - equivalently - as a module over $A \otimes_R A^o$.

In this situation we have for any $A \otimes_R A^o$-module M,

$$\mathrm{Hom}_{A \otimes_R A^o}(A, M) = Z(M) = \{m \in M \mid am = ma \text{ for all } a \in A\},$$

and $\mathrm{End}_{A \otimes_R A^o}(A) \simeq Z(A)$, the center of A.

3.7. **Azumaya algebras.** *Let A be a central R-algebra, that is $Z(A) = R$. Then the following are equivalent:*

(a) *A is a (projective) generator in ${}_{A \otimes_R A^o}\mathbb{M}$;*

(b) *$A \otimes_R A^o \simeq \mathrm{End}_R(A)$ and A_R is finitely generated and projective;*

(c) *$\mathrm{Hom}_{A \otimes_R A^o}(A, -) : {}_{A \otimes_R A^o}\mathbb{M} \to \mathbb{M}_R$ is (fully) faithful;*

(d) *$A \otimes_R - : \mathbb{M}_R \to {}_{A \otimes_R A^o}\mathbb{M}$ is an equivalence;*

(e) *$\mu : A \otimes_R A^o \to A$ splits in ${}_{A \otimes_R A^o}\mathbb{M}$ (A is R-separable).*

The preceding result can also be formulated for not necessarily associative algebras by referring to the following.

3.8. **Multiplication algebra.** Let A be a (non-associative) R-algebra with unit. Then any $a \in A$ induces R-linear maps

$$L_a : A \to A,\ x \mapsto ax; \quad R_a : A \to A,\ x \mapsto xa.$$

The *multiplication algebra* of A is the (associative) subalgebra

$$M(A) \subset \mathrm{End}_R(A) \text{ generated by } \{L_a, R_a \mid a \in A\}.$$

Then A is a left module over $M(A)$ generated by 1_A (in general not projective) and $\mathrm{End}_{M(A)}(A)$ is isomorphic to the center of A. By $\sigma[{}_{M(A)}A]$, or $\sigma[A]$ for short, we denote the full subcategory of ${}_{M(A)}\mathbb{M}$ subgenerated by A. (For algebras A without unit these notions are slightly modified, e.g. [**Wi.B**]).

This setting allows to define *Azumaya* also for non-associative algebras (e.g. [**Wi.B**, 24.8]).

3.9. **Azumaya algebras.** *Let A be a central R-algebra with unit. Then the following are equivalent:*

(a) *A is a (projective) generator in ${}_{M(A)}\mathbb{M}$;*

(b) *$M(A) \simeq \mathrm{End}_R(A)$ and A_R is finitely generated and projective;*

(c) *$\mathrm{Hom}_{M(A)}(A, -) : {}_{M(A)}\mathbb{M} \to \mathbb{M}_R$ is (fully) faithful;*

(d) *$A \otimes_R - : \mathbb{M}_R \to {}_{M(A)}\mathbb{M}$ is an equivalence.*

The fact that the generator property of A as $A \otimes_R A^o$-module implies projectivity is a consequence of the commutativity of the corresponding endomorphism ring ($=Z(A)$).

Restricting to the subcategory $\sigma[A]$ we obtain

3.10. **Azumaya rings.** *Let A be a central R-algebra with unit. Then the following are equivalent:*

(a) *A is a (projective) generator in $\sigma[{}_{M(A)}A]$;*

(b) *$M(A)$ is dense in $\mathrm{End}_R(A)$ and A_R is faithfully flat;*

(c) *$\mathrm{Hom}_{M(A)}(A, -) : \sigma[{}_{M(A)}A] \to \mathbb{M}_R$ is (fully) faithful;*

(d) *$A \otimes_R - : \mathbb{M}_R \to \sigma[{}_{M(A)}A]$ is an equivalence.*

For any algebra A, central localisation is possible with respect to the maximal (or prime) ideals of the center $Z(A)$ and also with respect to central idempotents of A.

3.11. **Pierce stalks.** Let A be a (non-associative) algebra and denote by $B(A)$ the set of central idempotents of A which form a Boolean ring. Denote by $\mathcal{X}$ the set of all maximal ideals of $B(A)$. For any $x \in \mathcal{X}$, the set $B(A) \setminus x$ is a multiplicatively closed subset of (the center) of A and we can form the ring of fractions $A_x = AS^{-1}$. These are called the *Pierce stalks* of A (e.g. [**Wi.B**, Section 18]). They may be applied for local-global characterisations of algebraic structures, for example (see [**Wi.B**, 26.8], [**Wi.M**]):

3.12. **Pierce stalks of Azumaya rings.** *Let A be a central (non-associative) R-algebra with unit. Then the following are equivalent:*

(a) *A is an Azumaya algebra;*

(b) *A is finitely presented in $\sigma[A]$ and for every $x \in \mathcal{X}$, A_x is an Azumaya ring;*

(c) *for every $x \in \mathcal{X}$, A_x is an Azumaya ring with center R_x.*

Considering the (A, A)-bimodules for an associative ring A may be regarded as an extension of the module theory over commutative rings to non-commutative rings. Using the multiplication algebra $M(A)$ we can even handle non-associative algebras A. In particular, we can describe a kind of central localisation of semiprime algebras A. This may help to handle notions in non-commutative geometry.

One problem in transferring localisation techniques from semiprime commutative rings to semiprime non-commutative rings is that the latter need not be non-singular as one-sided modules. To guarentee this, additional assumptions on the ring are required (e.g. Goldie's theorem). This is not the case if we consider A in the category $\sigma[A]$.

A module $N \in \sigma[A]$ is called *A-singular* if $N \simeq L/K$ for $L \in \sigma[A]$ and $K \subset L$ an essential $M(A)$-submodule (see 3.2). The following is shown in [**Wi.B**, Section 32].

3.13. **Central closure of semiprime algebras.** *Let A be a semiprime R-algebra and $\widehat{A}$ the injective hull of A in $\sigma[{}_{M(A)}A]$. Then*

(i) *A is non-singular in $\sigma[{}_{M(A)}A]$.*

(ii) $\mathrm{End}_{M(A)}(\widehat{A})$ *is a regular, selfinjective, commutative ring, called the* extended centroid.

(iii) $\widehat{A} = A\,\mathrm{Hom}_{M(A)}(A, \widehat{A}) = A\,\mathrm{End}_{M(A)}(\widehat{A})$ *and allows for a ring structure (for $a, b \in A$, $\alpha, \beta \in \mathrm{End}_{M(A)}(\widehat{A})$),*

$$(a\alpha) \cdot (b\beta) := ab\,\alpha\beta.$$

This is the (Martindale) central closure *of A.*

(iv) *$\widehat{A}$ is a simple ring if and only if A is strongly prime (as an $M(A)$-module).*

Not surprisingly - the above results applied to $A = \mathbb{Z}$ yield the rationals $\mathbb{Q}$ as the (self-)injective hull of the integers $\mathbb{Z}$.

A semiprime ring A is said to be *strongly prime (as $M(A)$-module)* if its central closure is a simple ring, and an ideal $I \subset A$ is called *strongly prime* provided the factor ring A/I is strongly prime.

Using this notion, an associative ring A is defined to be a *Hilbert ring* if any strongly prime ideal of A is the intersection of maximal ideals. This is the case if and only if for all $n \in \mathbb{N}$, every maximal ideal $\mathcal{J} \subset A[X_1, \ldots, X_n]$ contracts to a maximal ideal of A or - equivalently - $A[X_1, \ldots, X_n]/\mathcal{J}$ is finitely generated as an $A/\mathcal{J} \cap A$-module (liberal extension). This yields a natural noncommutative version of Hilbert's Nullstellensatz (see [**KaWi**]).

The techniques considered in 3.13 were extended in Lomp [**Lomp**] to study the action of Hopf algebras on algebras.

4. Coalgebras and comodules

The module theory sketched in the preceding section provides useful techniques for the investigation of coalgebras and comodules. In this section R will denote a commutative ring.

4.1. **Coalgebras.** An *R-coalgebra* is a triple (C, Δ, ε) where C is an R-module with R-linear maps

$$\Delta : C \to C \otimes_R C, \quad \varepsilon : C \to R,$$

satisfying coassociativity and counitality conditions.

The tensor product $C \otimes_R D$ of two R-coalgebras C and D can be made to a coalgebra with a similar procedure as for algebras. For this a suitable linear map $\tau' : C \otimes_R D \to D \otimes_R C$ is needed leading to the corresponding commutative diagrams (compare 2.2).

The dual R-module $C^* = \mathrm{Hom}_R(C, R)$ has an associative ring structure given by the *convolution product*

$$f * g = (g \otimes f) \circ \Delta \ \text{ for } f, g \in C^*,$$

with unit ε.

Replacing $g \otimes f$ by $f \otimes g$ (as done in the literature) yields a multiplication opposite to the one given before. This does not do any harm but has some effect on the formalism considered later on.

4.2. **Comodules.** A *left comodule* over a coalgebra C is a pair (M, ϱ^M) where M is an R-module with an R-linear map (coaction)

$$\varrho^M : M \to C \otimes_R M$$

satisfying compatibility and counitality conditions.

A morphism between C-comodules M, N is an R-linear map $f : M \to N$ with $\varrho^N \circ f = (I \otimes f) \circ \varrho^M$. The set (group) of these morphisms is denoted by $\mathrm{Hom}^C(M, N)$.

The category ${}^C\mathbb{M}$ of left C-comodules is additive, with coproduct and cokernels - but not necessarily with kernels.

The functor $C \otimes_R - : \mathbb{M}_R \to {}^C\mathbb{M}$ is right adjoint to the forgetful functor ${}^C\mathbb{M} \to \mathbb{M}_R$, that is, there is an isomorphism

$$\mathrm{Hom}^C(M, C \otimes_R X) \to \mathrm{Hom}_R(M, X), \quad f \mapsto (\varepsilon \otimes I) \circ f,$$

and from this it follows that

$$\mathrm{End}^C(C) \simeq \mathrm{Hom}_R(C, R) = C^*,$$

which is a ring morphism - or antimorphism depending on the choice for the convolution product (see 4.1).

C is a subgenerator in ${}^C\mathbb{M}$, since any C-comodule leads to a diagram

$$\begin{array}{ccccc} & & R^{(\Lambda)} & & C \otimes_R R^{(\Lambda)} \xrightarrow{\simeq} C^{(\Lambda)} \\ & & \downarrow{\scriptstyle h} & & \downarrow{\scriptstyle I\otimes h} \\ 0 & \longrightarrow & M & \xrightarrow{\varrho^M} & C \otimes_R M, \end{array}$$

where h is an epimorphism for some index set Λ.

Monomorphisms in ${}^C\mathbb{M}$ need not be injective maps and - as a consequence - generators G in ${}^C\mathbb{M}$ need not be flat modules over their endomorphism rings and the functor $\mathrm{Hom}^C(G, -) : {}^C\mathbb{M} \to \mathbf{Ab}$ need not be full.

All monomorphisms in ${}^C\mathbb{M}$ are injective maps if and only if C is flat as an R-module. In this case ${}^C\mathbb{M}$ has kernels.

There is a close relationship between comodules and modules.

4.3. **C-comodules and C^*-modules.** Any C-comodule $\varrho^M : M \to C \otimes_R M$ is a C^*-module by the action

$$\tilde{\varrho}^M : C^* \otimes M \xrightarrow{I \otimes \varrho^M} C^* \otimes C \otimes M \xrightarrow{ev \otimes I} M .$$

For any $M, N \in {}^C\mathbb{M}$, $\mathrm{Hom}^C(M, N) \subset \mathrm{Hom}_{C^*}(M, N)$ and hence there is a faithful functor

$$\Phi : {}^C\mathbb{M} \to {}_{C^*}\mathbb{M}, \quad (M, \varrho^M) \mapsto (M, \tilde{\varrho}^M)$$

To make Φ a full functor, the morphism (natural in $Y \in \mathbb{M}_R$)

$$\alpha_Y : C \otimes_R Y \to \mathrm{Hom}_R(C^*, Y), \quad c \otimes y \mapsto [f \mapsto f(c)y],$$

has to be injective for all $Y \in \mathbb{M}_R$ (*α-condition*, see [**BrWi**, 4.3]):

4.4. **${}^C\mathbb{M}$ a full module subcategory.** *The following are equivalent:*

(a) $\Phi : {}^C\mathbb{M} \to {}_{C^*}\mathbb{M}$ *is a full functor;*

(b) $\Phi : {}^C\mathbb{M} \to \sigma[{}_{C^*}C]$ $(\subset {}_{C^*}\mathbb{M})$ *is an equivalence;*

(c) α_Y *is injective for all* $Y \in \mathbb{M}_R$*;*

(d) C_R *is locally projective.*

This observation shows that under the given conditions the investigation of the category of comodule reduces to the study of C^*-modules, more precisely, the study of the category $\sigma[{}_{C^*}C]$ (see [**BrWi**], [**Wi.F**]).

As a special case we have (see [**BrWi**, 4.7]):

4.5. **${}^C\mathbb{M}$ a full module category.** *The following are equivalent:*

(a) $\Phi : {}^C\mathbb{M} \to {}_{C^*}\mathbb{M}$ *is an equivalence;*

(b) α *is an isomorphism;*

(c) C_R *is finitely generated and projective.*

4.6. **Natural morphism.** Applying $\mathrm{Hom}_R(X, -)$ to the morphism α_Y leads to the morphism, natural in $X, Y \in \mathbb{M}_R$,

$$\tilde{\alpha}_{X,Y} : \mathrm{Hom}_R(X, C \otimes_R Y) \to \mathrm{Hom}_R(X, \mathrm{Hom}_R(C^*, Y)) \xrightarrow{\simeq} \mathrm{Hom}_R(C^* \otimes_R X, Y).$$

If α_Y is a monomorphism, then $\alpha_{X,Y}$ is a monomorphism,
if α_Y is an isomorphism, then $\alpha_{X,Y}$ is an isomorphism, $X, Y \in \mathbb{M}_R$.

The latter means that the monad $C^* \otimes_R -$ and the comonad $C \otimes_R -$ form an adjoint pair of endofunctors on $\mathbb{M}_R$, while the former condition means a weakened form of adjunction.

It is known (from category theory) that, for the monad $C^* \otimes_R -$, the right adjoint $\mathrm{Hom}_R(C^*, -)$ is a comonad and the category ${}_{C^*}\mathbb{M}$ is equivalent to the category $\mathbb{M}^{\mathrm{Hom}_R(C^*,-)}$ of $\mathrm{Hom}_R(C^*, -)$-comodules (e.g. [**BöBrWi**, 3.5]).

Thus $\alpha : C \otimes - \to \mathrm{Hom}_R(C^*, -)$ may be considered as a comonad morphism yielding a functor

$$\tilde{\Phi} : {}^C\mathbb{M} \longrightarrow \mathbb{M}^{\mathrm{Hom}_R(C^*,-)},$$
$$M \to C \otimes_R M \longmapsto M \to C \otimes_R M \xrightarrow{\alpha_M} \mathrm{Hom}_R(C^*, M).$$

As noticed in 4.4 and 4.5, this functor is fully faithful if and only if α is injective; it is an equivalence provided α is a natural isomorphism.

5. Bialgebras and Hopf algebras

Combining algebras and coalgebras leads to the notion of

5.1. **Bialgebras.** An R-bialgebra is an R-module B carrying an algebra structure (B, m, e) and a coalgebra structure (B, Δ, ε) with compatibility conditions which can be expressed in two (equivalent) ways

(a) $m : B \otimes_R B \to B$ and $e : R \to B$ are coalgebra morphisms;

(b) $\Delta : B \to B \otimes_R B$ and $\varepsilon : B \to R$ are algebra morphisms.

To formulate this, an algebra and a coalgebra structure is needed on the tensor-product $B \otimes_R B$ as defined in 2.2 and 4.1 (with the twist tw map taken for τ). The twist map (or a braiding) can be avoided at this stage by referring to an *entwining map*

$$\psi : B \otimes_R B \to B \otimes_R B,$$

which allows to express compatibility between algebra and coalgebra structure by commutativity of the diagram (e.g. [**BöBrWi**, 8.1])

$$\begin{array}{ccccc}
B \otimes_R B & \xrightarrow{m} & B & \xrightarrow{\Delta} & B \otimes_R B \\
{\scriptstyle \Delta \otimes I_B}\downarrow & & & & \uparrow{\scriptstyle m \otimes I_B} \\
B \otimes_R B \otimes_R B & & \xrightarrow{I_B \otimes \psi} & & B \otimes_R B \otimes_R B.
\end{array}$$

In the standard situation this entwining is derived from the twist map as

$$\psi = (m \otimes I) \circ (I \otimes \mathrm{tw}) \circ (\delta \otimes I) : B \otimes_R B \to B \otimes_R B,\ a \otimes b \mapsto a_{\underline{1}} \otimes b a_{\underline{2}}.$$

This is a special case of 6.12 (see also [**BöBrWi**, 8.1]).

5.2. **Hopf modules.** *Hopf modules* for a bialgebra B are R-modules M with a B-module and a B-comodule structure

$$\rho_M : B \otimes_R M \to M, \quad \rho^M : M \to B \otimes_R M,$$

satisfying the compatibility condition

$$\rho^M(bm) = \Delta(b) \cdot \rho^M(m), \quad \text{for } b \in B,\ m \in M.$$

Here we use that - due to the algebra map Δ - the tensor product $N \otimes_R M$ of two B-modules can be considered as a left B-module via the diagonal action

$$b \cdot (m \otimes n) = \Delta(b)(m \otimes n) = \sum b_{\underline{1}} n \otimes b_{\underline{2}} m.$$

This makes the category ${}_B\mathbb{M}$ monoidal.

If the compatibility between m and Δ is expressed by an entwining map $\psi : B \otimes_R B \to B \otimes_R B$ (see 5.1), then the Hopf modules are characterised by commutativity of the diagram

$$\begin{array}{ccccc}
B \otimes_R M & \xrightarrow{\rho_M} & M & \xrightarrow{\rho^M} & B \otimes_R M \\
{\scriptstyle I \otimes \rho^M}\downarrow & & & & \uparrow{\scriptstyle I \otimes \rho_M} \\
B \otimes_R B \otimes_R M & & \xrightarrow{\psi \otimes I} & & B \otimes_R B \otimes_R M.
\end{array}$$

5.3. **Category of Hopf modules.** Morphisms between two B-Hopf modules M and N are R-linear maps $f : M \to N$ which are B-module as well as B-comodule morphisms. With these morphisms, the Hopf modules form an additive category, we denote it by ${}_B^B\mathbb{M}$. Certainly B is an object in ${}_B^B\mathbb{M}$, but in general it is neither a generator nor a subgenerator.

As mentioned above, $B \otimes_R B$ has a (further) left B-module structure induced by Δ, we denote the resulting module by $B \otimes^b B$. It is not difficult to see that $B \otimes^b B$ is an object in ${}_B^B\mathbb{M}$ and is a subgenerator in this category (e.g. [**BrWi**, 14.5]).

Similarly, one may keep the trivial B-module structure on $B \otimes_R B$ but introduce a new comodule structure on it. This is again a Hopf module, denoted by $B \otimes^c B$, and is also a subgenerator in ${}_B^B\mathbb{M}$ (e.g. [**BrWi**, 14.5]).

As for comodules, monomorphisms in ${}_B^B\mathbb{M}$ need not be injective maps unless B is flat as an R-module.

If B is locally projective as an R-module, the comodule structure of the Hopf modules may be considered as B^*-module structure and their module and comodule structures yield a structure as module over the smash product $B\#B^*$. In this case, ${}_B^B\mathbb{M}$ is isomorphic to $\sigma[{}_{B\#B^*}B \otimes^b B]$, the full subcategory of ${}_{B\#B^*}\mathbb{M}$ subgenerated by $B \otimes^b B$ (or $B \otimes^c B$) (e.g. [**BrWi**, 14.15]).

5.4. **Comparison functor.** For any R-bialgebra B, there is a *comparison functor*

$$\phi_B^B : {}_R\mathbb{M} \to {}_B^B\mathbb{M}, \quad X \mapsto (B \otimes_R X, m \otimes I_X, \Delta \otimes I_X),$$

which is full and faithful since, by module and comodule properties, for any $X, Y \in \mathbb{M}_R$,

$$\mathrm{Hom}_B^B(B \otimes_R X, B \otimes_R Y) \simeq \mathrm{Hom}_R^B(X, B \otimes_R Y) \simeq \mathrm{Hom}_R(X, Y),$$

with the trivial B-comodule structure on X. In particular, $\mathrm{End}_B^B(B) \simeq R$.

5.5. **The bimonad** $\mathrm{Hom}_R(B, -)$**.** As mentioned in 4.6, for a monad (comonad) $B \otimes_R -$, the right adjoint functor $\mathrm{Hom}_R(B, -)$, we denote it by $[B, -]$, is a comonad (monad).

An entwining $\psi : B \otimes_R B \to B \otimes_R B$ may be seen as an entwining between the monad $B \otimes_R -$ and the comonad $B \otimes_R -$,

$$\tilde{\psi} : B \otimes_R B \otimes_R - \to B \otimes_R B \otimes_R -$$

and this induces an entwining between the Hom-functors (see [**BöBrWi**, 8.2])

$$\widehat{\psi} : [B, [B, -]] \to [B, [B, -]].$$

This allows to define $[B, -]$-*Hopf modules* (similar to 5.2), the category $\mathbb{M}_{[B,-]}^{[B,-]}$, and a comparison functor (with obvious notation)

$$\phi_{[B,-]}^{[B,-]} : {}_R\mathbb{M} \to \mathbb{M}_{[B,-]}^{[B,-]}, \quad X \mapsto ([B, X], \Delta_X^*, m_X^*).$$

5.6. **Antipode.** For any bialgebra B, a *convolution product* can be defined on the R-module $\mathrm{End}_R(B)$ by putting, for $f, g \in \mathrm{End}_R(B)$, (compare 4.1)

$$f * g = m \circ (f \otimes g) \circ \Delta.$$

This makes $(\mathrm{End}_R(B), *)$ an R-algebra with identity $e \circ \varepsilon$.

An *antipode* is an $S \in \operatorname{End}_R(B)$ which is inverse to the identity map I_B of B with respect to $*$, that is $S * I_B = e \circ \varepsilon = I_B * S$ or - explicitely -

$$m \circ (S \otimes I_B) \circ \Delta = e \circ \varepsilon = m \circ (I_B \otimes S) \circ \Delta.$$

If B has an antipode it is called a *Hopf algebra.*

The existence of an antipode is equivalent to the canonical map

$$\gamma : \; B \otimes_R B \xrightarrow{\delta \otimes I} B \otimes_R B \xrightarrow{I \otimes m} B \otimes_R B$$

being an isomorphism (e.g. [**BrWi**, 15.2]).

The importance of the antipode is clear by the (see [**BöBrWi**, 8.11])

5.7. **Fundamental Theorem.** *For any R-bialgebra B, the following are equivalent:*

(a) *B is a Hopf algebra (i.e. has an antipode);*

(b) $\phi^B_B : {}_R\mathbb{M} \to {}^B_B\mathbb{M}$ *is an equivalence;*

(c) $\phi^{[B,-]}_{[B,-]} : {}_R\mathbb{M} \to \mathbb{M}^{[B,-]}_{[B,-]}$ *is an equivalence;*

(d) $\operatorname{Hom}^B_B(B,-) : {}^B_B\mathbb{M} \to {}_R\mathbb{M}$ *is full and faithful.*

If B_R is flat then (a)-(d) *are equivalent to:*

(e) *B is a generator in ${}^B_B\mathbb{M}$.*

Recall that for B_R locally projective, ${}^B_B\mathbb{M}$ is equivalent to $\sigma[{}_{B\#B^*}B \otimes^b B]$ and thus we have:

5.8. **Corollary.** *Let B be an R-bialgebra with B_R locally projective. Then the following are equivalent:*

(a) *B is a Hopf algebra;*

(b) *B is a subgenerator in ${}^B_B\mathbb{M}$ and $B\#B^*$ is dense in* $\operatorname{End}_R(B)$*;*

(c) *B is a generator in ${}^B_B\mathbb{M}$.*

These characterisations are very similar to those of Azumaya rings (see 3.10). This indicates, for example, that Pierce stalks may also be applied to characterise (properties of) Hopf algebras.

The notion of *bialgebras* addresses one functor with algebra and coalgebra structures. More general, one may consider relationships between distinct algebras and coalgebras:

5.9. **Entwined algebras and coalgebras.** Given an R-algebra (A, m, e) and an R-coalgebra (C, Δ, ε), an *entwining* (between monad $A \otimes_R -$ and comonad $C \otimes_R -$) is an R-linear map

$$\psi : A \otimes_R C \to C \otimes_R A,$$

inducing certain commutative diagrams. This notion was introduced in Brzeziński and Majid [**BrMa**] and is a special case of a mixed *distributive law* (see 6.5). *Entwined modules* are defined as R-modules M which are modules (M, ϱ_M) and comodules (M, ϱ^M), inducing commutativity of the diagram (e.g. [**BrWi**, 32.4])

$$\begin{array}{ccccc}
A \otimes M & \xrightarrow{\varrho_M} & M & \xrightarrow{\varrho^M} & C \otimes M \\
\Big\downarrow{\scriptstyle I_A \otimes \varrho^M} & & & & \Big\uparrow{\scriptstyle I \otimes \varrho_M} \\
A \otimes C \otimes M & & \xrightarrow{\psi \otimes I} & & C \otimes A \otimes M.
\end{array}$$

With morphisms which are A-module as well as C-comodule maps, the entwined modules form a category denoted by ${}^C_A\mathbb{M}$.

$C \otimes_R A$ is naturally a right A-module and ψ can be applied to define a left A-module structure on it,

$$a \cdot (c \otimes b) = \psi(a, c)b, \text{ for } a, b \in A,\ c \in C.$$

Moreover, a coproduct can be defined on $C \otimes_R A$, making $C \otimes_R A$ an A-*coring*, a notion which extends the notion of R-coalgebras to non-commutative base rings A. The category ${}^C_A\mathbb{M}$ of entwining modules can be considered as ${}^{C\otimes A}\mathbb{M}$, the category of left comodules over the coring $C \otimes_R A$ (e.g. [**BrWi**, 32.6]).

To get a comparison functor as in 5.4, we have to require that A is an object in ${}^C_A\mathbb{M}$; this is equivalent to the existence of a *grouplike element* in the A-coring $C \otimes_R A$ (e.g. [**BrWi**, 28.1 and 23.16]).

5.10. **Galois corings.** Let (A, C) be an entwined pair of an algebra A and a coalgebra C. Assume that A is an entwined module by $\varrho^A : A \to C \otimes_A A$. Then there is a comparison functor

$$\phi^C_A : \mathbb{M}_R \to {}^C_A\mathbb{M} : X \mapsto (A \otimes_R X, m \otimes I, \varrho^A \otimes I),$$

which is left adjoint to the (coinvariant) functor $\mathrm{Hom}^C_A(A, -) : {}^C_A\mathbb{M} \to \mathbb{M}_R$.

Moreover, $B = \mathrm{Hom}^C_A(A, A)$ is a subring of A, $\mathrm{Hom}^C_A(A, C \otimes_R A) \simeq A$, and evaluation yields a (canonical) map

$$\gamma : A \otimes_B A \to C \otimes_R A.$$

Now $C \otimes_R A$ is said to be a *Galois A-coring* provided γ is an isomorphism (e.g. [**BrWi**, 28.18]). This describes coalgebra-Galois extensions or non-commutative principal bundles. If - in this case - A_B is a faithfully flat module, then the functor

$$\mathbb{M}_B \to {}^C_A\mathbb{M} : Y \mapsto (A \otimes_B Y, m \otimes I, \varrho^A \otimes I)$$

is an equivalence of categories.

This extends the fundamental theorem for Hopf algebras to entwined structures: If $A = C = H$ is a Hopf algebra, then (H, H) is an entwining, $B = R$, and the resulting γ is an isomorphsism if and only if H has an antipode (see 5.6).

6. General categories

As seen in the preceding sections, the notions of algebras, coalgebras, and Hopf algebras are all buit up on the tensor product. Hence a first step to generalisation is to consider monoidal categories $(\mathbb{V}, \otimes, \mathbb{I})$. For example, entwining structures in such categories are considered in Mesablishvili [**Me**]. Furthermore, opmonoidal monads T on $\mathbb{V}$ were considered by Bruguières and Virelizier (in [**BruVir**, 2.3]) which may be considered as an entwining of the monad T with the comonad $- \otimes T(\mathbb{I})$. The *generalised bialgebras* in Loday [**Lod**], defined as Schur functors (on vector spaces) with a monad structure (operads) and a specified coalgebra structure, may also be seen as a generalisation of entwining structures [**Lod**, 2.2.1].

However, algebras and coalgebras also show up in more general categories as considered in universal algebra, theoretical computer science, logic, etc. (e.g. Gumm [**Gu**], Turi and Plotkin [**TuPl**], Adámek and Porst [**AdPo**]). It is of some interest to understand how the notion of Hopf algebras can be transferred to these settings. In what follows we consider an arbitrary category $\mathbb{A}$.

6.1. **Monads on $\mathbb{A}$.** A *monad on* $\mathbb{A}$ is a triple (F, m, e) with a functor $F : \mathbb{A} \to \mathbb{A}$ and natural transformations

$$m : FF \to F, \quad e : I_{\mathbb{A}} \to F,$$

inducing commutativity of certain diagrams (as for algebras, see 2.1).

F-modules are defined as $X \in \mathrm{Obj}(\mathbb{A})$ with morphisms $\varrho_X : F(X) \to X$ and certain commutative diagrams (as for the usual modules, see 3.1).

The catgegory of F-modules is denoted by $\mathbb{A}_F$. The free functor

$$\phi_F : \mathbb{A} \to \mathbb{A}_F, \; X \mapsto (F(X), m_X)$$

is left adjoint to the forgetful functor $U_F : \mathbb{A}_F \to \mathbb{A}$ by the isomorphism, for $X \in \mathbb{A}$, $Y \in \mathbb{A}_F$,

$$\mathrm{Mor}_{\mathbb{A}_F}(F(X), Y) \to \mathrm{Mor}_{\mathbb{A}}(X, U_F(Y)), \quad f \mapsto f \circ e_X.$$

6.2. **Comonads on $\mathbb{A}$.** A *comonad on* $\mathbb{A}$ is a triple (G, δ, ε) with a functor $G : \mathbb{A} \to \mathbb{A}$ and natural transformations

$$\delta : G \to GG, \quad \varepsilon : G \to I_{\mathbb{A}},$$

satisfying certain commuting diagrams (reversed to the module case).

$\mathbb{G}$*-comodules* are objects $X \in \mathrm{Obj}(\mathbb{A})$ with morphisms $\varrho^X : X \to G(X)$ in $\mathbb{A}$ and certain commutative diagrams.

The category of G-comodules is denoted by $\mathbb{A}^G$. The free functor

$$\phi^G : \mathbb{A} \to \mathbb{A}^G, \; X \mapsto (G(X), \delta_X)$$

is right adjoint to the forgetful functor $U^G : \mathbb{A}^G \to \mathbb{A}$ by the isomorphism, for $X \in \mathbb{A}^G$, $Y \in \mathbb{A}$,

$$\mathrm{Mor}_{\mathbb{A}^G}(X, G(Y)) \to \mathrm{Mor}_{\mathbb{A}}(U^G(X), Y), \quad f \mapsto \varepsilon_Y \circ f.$$

Monads and comonads are closely related with

6.3. **Adjoint functors.** A pair of functors $L : \mathbb{A} \to \mathbb{B}$, $R : \mathbb{B} \to \mathbb{A}$ is said to be *adjoint* if there is an isomorphism, natural in $X \in \mathbb{A}$, $Y \in \mathbb{B}$,

$$\mathrm{Mor}_{\mathbb{B}}(L(X), Y) \xrightarrow{\simeq} \mathrm{Mor}_{\mathbb{A}}(X, R(Y)),$$

also described by natural transformations $\eta : I_{\mathbb{A}} \to RL$, $\varepsilon : LR \to I_{\mathbb{B}}$. This implies

$$\text{a monad } (RL, R\varepsilon_L, \eta) \text{ on } \mathbb{A}\,, \quad \text{a comonad } (LR, L\eta_R, \varepsilon) \text{ on } \mathbb{B}.$$

L is full and faithful if and only if $\varepsilon : GF \to I_{\mathbb{A}}$ is an isomorphism.

L is an equivalence (with inverse R) if and only if ε and η are natural isomorphisms.

6.4. **Lifting properties.** Compatibility between endofunctors $F, G : \mathbb{A} \to \mathbb{A}$ can be described by *lifting properties.* For this, let $F : \mathbb{A} \to \mathbb{A}$ be a monad and $G : \mathbb{A} \to \mathbb{A}$ any functor on $\mathbb{A}$ and consider the diagram

$$\begin{array}{ccc} \mathbb{A}_F & \xrightarrow{\overline{G}} & \mathbb{A}_F \\ {\scriptstyle U_F}\downarrow & & \downarrow{\scriptstyle U_F} \\ \mathbb{A} & \xrightarrow{G} & \mathbb{A}. \end{array}$$

If a $\overline{G}$ exists making the diagram commutative it is called a *lifting of G*. The questions arising are:

(i) does a lifting $\overline{G}$ exist ?

(ii) if G is a monad - is $\overline{G}$ again a monad (monad lifting)?
(iii) if G is a comonad - is $\overline{G}$ also a comonad (comonad lifting)?

For R-algebras A and B, (i) together with (ii) may be compared with the definition of an algebra structure on $A \otimes_R B$ and leads to diagrams similar to those in 2.1.

For an R-algebras A and an R-coalgebra C, (i) together with (iii) corresponds to the entwinings considered in 5.9.

We formulate this in the general case (e.g. [**Wi.A**, 5.3]).

6.5. **Mixed distributive law (entwining).** *Let (F, m, e) be a monad and (G, δ, ε) a comonad. Then a comonad lifting $\overline{G} : \mathbb{A}_F \to \mathbb{A}_F$ exists if and only if there is a natural transformations*

$$\lambda : FG \to GF$$

inducing commutativity of the diagrams

$$\begin{array}{ccccc} FFG & & \xrightarrow{m_G} & & FG \\ {\scriptstyle F\lambda}\downarrow & & & & \downarrow{\scriptstyle \lambda} \\ FGF & \xrightarrow{\lambda_F} & GFF & \xrightarrow{Gm} & GF, \end{array} \qquad \begin{array}{ccccc} FG & \xrightarrow{F\delta} & FGG & \xrightarrow{\lambda_G} & GFG \\ {\scriptstyle \lambda}\downarrow & & & & \downarrow{\scriptstyle G\lambda} \\ GF & & \xrightarrow{\delta_F} & & GGF, \end{array}$$

$$\begin{array}{ccc} G & \xrightarrow{e_G} & FG \\ & {\scriptstyle Ge}\searrow & \downarrow{\scriptstyle \lambda} \\ & & GF, \end{array} \qquad \begin{array}{ccc} FG & \xrightarrow{F\varepsilon} & F \\ {\scriptstyle \lambda}\downarrow & \nearrow{\scriptstyle \varepsilon_F} & \\ GF. & & \end{array}$$

Entwining is also used to express compatibility for an endofunctor which is a monad as well as a comonad. Notice that the diagrams in 6.5 either contain the product m *or* the coproduct δ, the unit e *or* the counit ε. Additional conditions are needed for adequate compatibility.

6.6. **(Mixed) bimonad.** An endofunctor $B : \mathbb{A} \to \mathbb{A}$ is said to be a *(mixed) bimonad* if it is

(i) a monad (B, m, e) with $e : I \to B$ a comonad morphism,
(ii) a comonad (B, δ, ε) with $\varepsilon : B \to I$ a monad morphism,
(iii) with an entwining functorial morphism $\psi : BB \to BB$,
(iv) with a commutative diagram

$$\begin{array}{ccccc} BB & \xrightarrow{m} & B & \xrightarrow{\delta} & BB \\ {\scriptstyle B\delta}\downarrow & & & & \uparrow{\scriptstyle Bm} \\ BBB & & \xrightarrow{\psi_B} & & BBB. \end{array}$$

6.7. **(Mixed) B-bimodules.** For a bimonad B on $\mathbb{A}$, *(mixed) bimodules* are defined as B-modules and B-comodules X satisfying the pentagonal law

$$\begin{array}{ccccc} B(X) & \xrightarrow{\varrho_X} & X & \xrightarrow{\varrho^X} & B(X) \\ {\scriptstyle B(\varrho^X)}\downarrow & & & & \uparrow{\scriptstyle B(\varrho_X)} \\ BB(X) & & \xrightarrow{\psi_X} & & BB(X). \end{array}$$

B-bimodule morphisms are B-module as well as B-comodule morphisms. We denote the category of B-bimodules by $\mathbb{A}^B_B$.

There is a comparison functor (compare 5.4)

$$\phi^B_B : \mathbb{A} \to \mathbb{A}^B_B, \quad A \longmapsto [BB(A) \overset{\mu_A}{\to} B(A) \overset{\delta_A}{\to} BB(A)],$$

which is full and faithful by the isomorphisms, functorial in $X, X' \in \mathbb{A}$,

$$\mathrm{Mor}^B_B(B(X), B(X')) \simeq \mathrm{Mor}_B(B(X), X') \simeq \mathrm{Mor}_{\mathbb{A}}(X, X').$$

In particular, this implies $\mathrm{End}^B_B(B(X)) \simeq \mathrm{End}_{\mathbb{A}}(X)$, for any $X \in \mathbb{A}$.

Following the pattern in 5.6 we define an

6.8. **Antipode.** Let B be a bimonad. An *antipode* of B is a natural transformation $S : B \to B$ leading to commutativity of the diagram

$$\begin{array}{ccccc} B & \xrightarrow{\varepsilon} & I & \xrightarrow{e} & B \\ \delta\big\downarrow & & & & \big\uparrow m \\ BB & & \overset{S_B}{\underset{BS}{\rightrightarrows}} & & BB \end{array}$$

We call B a *Hopf bimonad* provided it has an antipode.

As for Hopf algebras (see 5.6) we observe that the canonical natural transformation

$$\gamma : BB \xrightarrow{\delta_B} BBB \xrightarrow{Bm} BB$$

is an isomorphism if and only if B has an antipode (e.g. [**BrWi**, 15.1]).

The Fundamental Theorem for Hopf algebras states that the existence of an antipode is equivalent to the comparison functor being an equivalence (see 5.7). To get a corresponding result in our general setting we have to impose slight conditions on the base category and on the functor (see [**MeWi**, 5.6]):

6.9. **Fundamental Theorem for bimonads.** *Let B be a bimonad on the category $\mathbb{A}$ and assume that $\mathbb{A}$ admits colimits or limits and B preserves them. Then the following are equivalent:*

(a) *B is a Hopf bimonad (see 6.8);*

(b) *$\gamma = Bm \cdot \delta B : BB \to BB$ is a natural isomorphism;*

(c) *$\gamma' = mB \cdot B\delta B : BB \to BB$ is a natural isomorphism;*

(d) *the comparison functor $\phi^B_B : \mathbb{A} \to {}^B_B\mathbb{A}$ is an equivalence.*

Recall that for an R-module B, the tensor functor $B \otimes_R -$ has a right adjoint and we have observed in 5.5 that a bialgebra structure on B can be transferred to the adjoint $\mathrm{Hom}_R(B, -)$.

As shown in [**MeWi**, 7.5], this applies for general bimonads provided they have a right adjoint:

6.10. **Adjoints of bimonads.** *Let B be an endofunctor of $\mathbb{A}$ with right adjoint $R : \mathbb{A} \to \mathbb{A}$. Then B is a bimonad (with antipode) if and only if R is a bimonad (with antipode).*

As a special case we have that for any R-Hopf algebra H, the functor $\mathrm{Hom}_R(H,-)$ is a Hopf monad on $\mathbb{M}_R$. This is not a tensor functor unless H_R is finitely generated and projective.

As pointed out in 5.1, no twist map (or braiding) is needed on the base category to formulate the compatibility conditions for bialgebras (and bimonads). There may exist a kind of braiding relations for bimonads based on distributive laws.

6.11. **Double entwinings.** Let B be an endofunctor on the category $\mathbb{A}$ with a monad structure $\underline{B}=(B,m,e)$ and a comonad structure $\overline{B}=(B,\delta,\varepsilon)$.

A natural transformation $\tau : BB \to BB$ is said to be a *double entwining* provided

(i) τ is a mixed distributive law from the monad $\underline{B}$ to the comonad $\overline{B}$;

(ii) τ is a mixed distributive law from the comonad $\overline{B}$ to the monad $\underline{B}$.

6.12. **Induced bimonad.** Let $\tau : BB \to BB$ be a double entwining with commutative diagrams

$$\begin{array}{ccc} BB & \xrightarrow{B\varepsilon} & B \\ {\scriptstyle m}\downarrow & & \downarrow{\scriptstyle \varepsilon} \\ B & \xrightarrow{\varepsilon} & 1, \end{array} \qquad \begin{array}{ccc} 1 & \xrightarrow{e} & B \\ {\scriptstyle e}\downarrow & & \downarrow{\scriptstyle \delta} \\ B & \xrightarrow[eB]{} & BB, \end{array} \qquad \begin{array}{ccc} 1 & \xrightarrow{e} & B \\ & \searrow^{=} & \downarrow{\scriptstyle \varepsilon} \\ & & 1, \end{array}$$

$$\begin{array}{ccccc} BB & \xrightarrow{m} & B & \xrightarrow{\delta} & BB \\ {\scriptstyle \delta\delta}\downarrow & & & & \uparrow{\scriptstyle mm} \\ BBBB & & \xrightarrow[B\tau B]{} & & BBBB. \end{array}$$

Then the composite

$$\overline{\tau} : BB \xrightarrow{\delta B} BBB \xrightarrow{B\tau} BBB \xrightarrow{mB} BB$$

is a *mixed distributive law* from the monad $\underline{B}$ to the comonad $\overline{B}$ making $(B,m,e,\delta,\varepsilon,\overline{\tau})$ a bimonad (see 6.6).

It is obvious that for any bimonad B, the product BB is again a monad as well as a comonad.

BB is also a *bimonad* provided τ satisfies the *Yang-Baxter equation*, that is, commutativity of the diagram

$$\begin{array}{ccccc} BBB & \xrightarrow{\tau B} & BBB & \xrightarrow{B\tau} & BBB \\ {\scriptstyle B\tau}\downarrow & & & & \downarrow{\scriptstyle \tau B} \\ BBB & \xrightarrow[\tau B]{} & BBB & \xrightarrow[B\tau]{} & BBB. \end{array}$$

If this holds, then BB is a bimonad with

product $\overline{m}: BBBB \xrightarrow{B\tau B} BBBB \xrightarrow{mm} BB$,

coproduct $\overline{\delta}: BB \xrightarrow{\delta\delta} BBBB \xrightarrow{B\tau B} BBBB$,

entwining $\overline{\overline{\tau}}: BBBB \xrightarrow{B\tau B} BBBB \xrightarrow{\tau\tau} BBBB \xrightarrow{B\tau B} BBBB$.

Finally, if τ is a double entwining satisfying the Yang-Baxter equation and $\tau^2=1$, then an *opposite bimonad* B^{op} can be defined for B with

$$\text{product} \quad m \cdot \tau : BB \xrightarrow{\tau} BB \xrightarrow{m} B,$$
$$\text{coproduct} \quad \tau \cdot \delta : B \xrightarrow{\delta} BB \xrightarrow{\tau} BB.$$

If B has an antipode S, then $S : B^{\rm op} \to B$ is a bimonad morphism provided that

$$\tau \cdot BS = SB \text{ and } \tau \cdot BS = SB.$$

In the classical theory of Hopf algebras, the category $\mathbb{M}_R$ of R-modules over a commutative ring R (or vector spaces) is taken as category $\mathbb{A}$ and tensor functors $B \otimes_R -$ are considered (which have right adjoints $\mathrm{Hom}_R(B, -)$). Here the Fundamental theorem for bimonads 6.9 implies that for Hopf algebras 5.7. The twist map provides a braiding on $\mathbb{M}_R$ and this induces a double entwining on the tensor functor $B \otimes_R -$.

We conclude with a non-additive example of our notions.

6.13. **Endofunctors on Set.** On the category **Set** of sets, any set G induces an endofunctor

$$G \times - : \mathbf{Set} \to \mathbf{Set}, \quad X \mapsto G \times X,$$

which has a right adjoint

$$\mathrm{Map}(G, -) : \mathbf{Set} \to \mathbf{Set}, \quad X \mapsto \mathrm{Map}(G, X).$$

Recall (e.g. from [**Wi.A**, 5.19]) that

(1) $G \times -$ is a monad if and only if G is a monoid;

(2) $G \times -$ is comonad with coproduct $\delta : G \to G \times G$, $g \mapsto (g, g)$;

(3) there is an entwining morphism

$$\psi : G \times G \to G \times G, \quad (g, h) \mapsto (gh, g).$$

Thus for any monoid G, $G \times -$ is a bimonad and

Hopf monads on Set. *For a bimonad $G \times -$, the following are equivalent:*

(a) $G \times -$ *is a Hopf monad;*

(b) $\mathrm{Mor}(G, -)$ *is a Hopf monad;*

(c) G *is a group.*

Here we also have a double entwining given by the twist map

$$\tau : G \times G \times - \mapsto G \times G \times -, \quad (a, b, -) \mapsto (b, a, -).$$

6.14. **Remarks.** After reporting about bialgebras and the compatibilty of their algebra and coalgebra part, we considered the entwining of distinct algebras and coalgebras (see 5.9). Similarly, one may try to extend results for bimonads to the entwining of a monad F and a distinct comonad G on a category $\mathbb{A}$ and to head for a kind of *Fundamental Theorem*, that is, an equivalence between the category $\mathbb{A}_F^G$ and, say, a module category over some coinvariants. For this one has to extend the notion of (co)modules over rings to (co)actions of (co)monads on functors and to introduce the notion of *Galois functors*. Comparing with 5.10, a crucial question is when F allows for a G-coaction. For this a *grouplike natural transformation* $I \to G$ is needed. In cooperation with B. Mesablishvili the work on these problems is still in progress.

References

[AdPo] Adámek, J. and and Porst, H.E., From varieties of algebras to covarieties of coalgebras, Electronic Notes in Theoretical Computer Science 44.1 (2001)

[Be] Beck, J., *Distributive laws*, in Seminar on Triples and Categorical Homology Theory, B. Eckmann (ed.), Springer LNM 80, 119-140 (1969)

[BöBrWi] Böhm, G., Brzeziński, T. and Wisbauer, R., *Monads and comonads in module categories*, arXiv:0804.1460 (2008)

[BruVir] Bruguières, A. and Virelizier, A., *Hopf monads*, Adv. Math. 215(2), 679-733 (2007)

[BrMa] Brzeziński, T. and Majid, Sh., *Comodule bundles*, Commun. Math. Physics 191, 467-492 (1998)

[BrWi] Brzeziński, T. and Wisbauer, R., *Corings and Comodules*, London Math. Soc. Lecture Note Series 309, Cambridge University Press (2003)

[EiMo] Eilenberg, S. and Moore, J.C., *Adjoint functors and triples*, Ill. J. Math. 9, 381-398 (1965)

[Gu] Gumm, H.P., *Universelle Coalgebra*, in: Allgemeine Algebra, Ihringer, Th., Berliner Stud. zur Math., Band 10, 155-207, Heldermann Verlag (2003)

[KaWi] Kaučikas, A. and Wisbauer, R., *Noncommutative Hilbert Rings*, J. Algebra Appl. 3(4), 437-444 (2004)

[Lod] Loday, J.-L., *Generalized bialgebras and triples of operads*, arXiv:math/0611885

[Lomp] Lomp, Ch., *A central closure construction for certain algebra extensions. Applications to Hopf actions*, J. Pure Appl. Algebra 198 (1-3), 297-316 (2005)

[Me] Mesablishvili, B., *Entwining Structures in Monoidal Categories*, J. Algebra 319(6), 2496-2517 (2008)

[MeWi] Mesablishvili, B. and Wisbauer, R., *Bimonads and Hopf monads on categories*, arXiv:math.QA/0710.1163 (2007)

[Moer] Moerdijk, I., Monads on tensor categories, J. Pure Appl. Algebra 168(2-3), 189-208 (2002)

[Mor] Morita,K., *Duality for modules and its applications to the theory of rings with minimum condition*, Sci. Rep. Tokyo Kyoiku Daigaku, Sect. A, 6, 83-142 (1958)

[Nast] Năstăsescu, C., Rings. Modules. Categories. (Inele. Module. Categorii.). (Romanian) Bucuresti: Editura Academiei Republicii Socialiste Romania (1976)

[TuPl] Turi, D. and Plotkin, G., *Towards a mathematical operational Semantics*, Proceedings 12th Ann. IEEE Symp. on Logic in Computer Science, LICS'97, Warsaw, Poland (1997)

[Wi.F] Wisbauer, R., *Foundations of module and ring theory. A handbook for study and research*, Algebra, Logic and Appl., Gordon and Breach, Philadelphia (1991)

[Wi.M] Wisbauer, R., *Modules and structure presheaves for arbitrary rings*, Adv. Math., Beijing 20, No.1, 15-23 (1991).

[Wi.B] Wisbauer, R., *Modules and Algebras: Bimodule Structure and Group Actions on Algebras*, Pitman Mono. PAM 81, Addison Wesley, Longman (1996)

[Wi.A] Wisbauer, R., *Algebras versus coalgebras*, Appl. Categor. Struct. 16(1-2) (2008), 255-295.

[Wi.H] Wisbauer, R., *Hopf monads on categories*, Jain, S.K. (ed.) et al., Noncommutative rings, group rings, diagram algebras and their applications. Intern. conf. University of Madras 2006; American Mathematical Society, Contemporary Mathematics 456, 219-230 (2008).

[Wi.G] Wisbauer, R., *Generators in module and comodule categories*, Proc. of the Faith-Osofsky conference, Zanesville (2007).

Mathematical Institute, Heinrich Heine University, 40225 Düsseldorf, Germany
E-mail address: `wisbauer@math.uni-duesseldorf.de`

Laplacians and gauged Laplacians on a quantum Hopf bundle

Alessandro Zampini

ABSTRACT. This paper presents an analysis of the set of connections and covariant derivatives on a U(1) quantum Hopf bundle on the standard quantum sphere S^2_q, whose total space algebra $SU_q(2)$ is equipped with the 3d left covariant differential calculus by Woronowicz. The introduction of a Hodge duality on both $\Omega(SU_q(2))$ and on $\Omega(S^2_q)$ allows for the study of Laplacians and of gauged Laplacians.

This paper is dedicated to Sergio Albeverio, on the occasion of his 70th birthday.

1. Introduction

This paper is focussed on the analysis of a class of Hall Hamiltonians in the noncommutative set up. It is intended as a survey of the general formulation of quantum principal bundles, and as a description of a specific procedure to introduce, on both the total space and the base space of a quantum Hopf bundle, a set of Laplacian operators and to couple them with gauge connections. It also presents a detailed formulation of the classical Hopf bundle. The emphasis in the presentation of structures from differential geometry will be given to their algebraic aspects extended to the noncommutative setting.

Classical Hall Hamiltonians are gauged Laplace operators acting on the space of sections of the vector bundles associated to the principal bundles $\pi : G \to G/\mathrm{K}$ over homogeneous spaces (with G semisimple and K compact) and can be constructed in terms of the Casimir operators of G and K. With (ρ, V) a representation of K, one has the identification of sections of the associated vector bundle $E = G \times_{\rho(\mathrm{K})} V$ with equivariant maps from G to V, $\Gamma(G/\mathrm{K}, E) \simeq C^\infty(G, V)_{\rho(\mathrm{K})} \subset C^\infty(G) \otimes V$. Given a connection on G one has a covariant derivative ∇ on $\Gamma(G/\mathrm{K}, E)$, so that the gauged Laplacian operator is $\Delta^E = (\nabla\nabla^* + \nabla^*\nabla) = \star\nabla\star\nabla$, where the dual ∇^* is defined from the metric induced on the homogeneous space basis G/K by the Cartan-Killing metric on G, or equivalently the Hodge duality comes from the induced metric on G/K. If the connection is the canonical one, given by the orthogonal splitting of the Lie algebra $\mathfrak{g}$ of G in terms of the Lie algebra $\mathfrak{k}$ of the gauge group and of its orthogonal complement, then the gauged Laplacian operator can be cast in terms of the quadratic Casimirs of $\mathfrak{g}$ and $\mathfrak{k}$:

$$\Delta^E = (\Delta^G \otimes 1 - 1 \otimes C_{\mathfrak{k}})|_{C^\infty(G,V)_{\rho(\mathrm{K})}} = (C_{\mathfrak{g}} \otimes 1 - 1 \otimes C_{\mathfrak{k}})|_{C^\infty(G,V)_{\rho(\mathrm{K})}} \quad (1.1)$$

The above formula [4] simplifies the diagonalisation of the gauged Laplacian, and has important applications in the study of the heat kernel expansion and index theorems on principal bundles.

The natural evolution is to develop models of the Hall effect on noncommutative spaces whose symmetries are described in terms of quantum groups. In [24] the first model of 'excitations moving on a quantum 2-sphere' in the field of a magnetic monopole has been studied. It is described by quantum principal U(1)-bundle over a quantum sphere S^2_q having as a total space the manifold of the quantum group $SU_q(2)$ [6]. The natural associated line bundles are classified by the winding number $n \in \mathbb{Z}$: equipped $SU_q(2)$ with the three dimensional left covariant calculus from Woronowicz [38], the gauge monopole connection is studied and a gauged Laplacian acting on sections of the associated bundle is completely diagonalised. That paper presents a first generalisation of the relation (1.1). Its most interesting aspect is that the corresponding energies are not invariant under the exchange monopole/antimonopole, namely the spectrum of the gauged Laplacian is not invariant under the inversion of the direction of the magnetic field, a manifestation of the phenomenon usually referred to as 'quantisation removes degeneracy'. A parallel study of the relation (1.1) is presented in [11], where Laplacians on a quantum projective plane are gauged via the monopole connection.

The analysis in [24] embodies two specific starting points. The first one is that the quantum Casimir C_q for the universal envelopping algebra $\mathcal{U}_q(\mathfrak{su}(2))$ dual to $SU_q(2)$ – thus playing the quantum role of the classical envelopping algebra dual to the classical Lie group – *is* a quadratic operator in the generators of $\mathcal{U}_q(\mathfrak{su}(2))$ acting on $SU_q(2)$, but *can not* be cast in the form of a whatever rank polynomial in the left invariant generators of the left invariant three dimensional differential calculus by Woronowicz, so to say in the basis of natural left invariant derivations associated to this differential calculus. The second starting point is given by the studies performed in [26]. In that paper a $\star$-Hodge operator on the exterior algebra on the Podleś sphere S^2_q – coming from the differential two dimensional calculus induced on S^2_q by the three dimensional calculus on $SU_q(2)$ – had been introduced, so to make it possible the definition of a Laplacian operator on S^2_q.

This paper develops the analysis started in [24], and describes another generalisation of the relation (1.1) to the setting of the same quantum Hopf bundle. A family of compatible $\star$-Hodge structures on the exterior algebras $\Omega(SU_q(2))$ and $\Omega(S^2_q)$, depending on a set of real parameters, are introduced, giving the corresponding Laplacians $\square_{SU_q(2)} = \star d \star d : \mathcal{A}(SU_q(2)) \mapsto \mathcal{A}(SU_q(2))$, and $\square_{S^2_q} = \star d \star d : \mathcal{A}(S^2_q) \mapsto \mathcal{A}(S^2_q)$. The connections on the principal bundle allows for a gauging of the Laplacian $\square_{S^2_q}$ on each associated line bundle. When $\square_{S^2_q}$ is gauged into $\square_{D_0}$ via the monopole connection, one finds

$$q^{2n}\square_{D_0} = (\square_{SU_q(2)} + \gamma X_z X_z), \tag{1.2}$$

where the integer $n \in \mathbb{Z}$ specifies the value of the monopole charge. This is the relation generalising the first equality in (1.1): the role of the quadratic Casimir of the gauge group algebra is played by $\gamma X_z X_z \triangleright$, with X_z the vertical derivation of the fibration, and $\gamma \in \mathbb{R}_+$ appears in this formulation as a parametrisation for a set of compatible $\star$-Hodge structures giving Laplacians satisfying the same relation (1.2).

This paper begins with an exposition of the classical Hopf bundle $\pi : S^3 \to S^2$. Section 2 presents a global – i.e. charts independent – description of the differential calculi on both the Lie group manifold $SU(2) \simeq S^3$ and on the homogeneous space $S^2 = S^3/\mathrm{U}(1)$, and introduces on the exterior algebras $\Omega(S^3)$ and $\Omega(S^2)$ the Hodge duality structures coming from a Cartan-Killing type metric on the Lie algebra $\mathfrak{su}(2)$, in order to define Laplacian operators. The principal bundle structure is described in terms of a well known principal bundle atlas. The aim of the section is to explicitly compute for such a specific Hopf bundle, following the classical approach from differential geometry, the main structures which will be generalised to the quantum setting. A more general and complete analysis of a noncommutative geometry approach to the differential geometry of principal and quantum bundles is in [**3**].

Section 3 describes the quantum formulation [**6**] of the principal bundle having $\mathcal{A}(\mathrm{SU}_q(2))$ as total space algebra, $\mathcal{A}(\mathrm{S}^2_q)$ as base manifold algebra and $\mathcal{A}(\mathrm{U}(1))$ as gauge group algebra, with the differential calculus on $\mathrm{SU}_q(2)$ given the 3d left-covariant calculus introduced by Woronowicz [**37, 38**].

Section 4 presents a $\star$-Hodge duality on $\Omega(\mathrm{SU}_q(2))$, allowing for the definition of a Laplacian operator. The Hodge duality is introduced following [**22**]; section 5 describes an evolution of this approach, giving a $\star$-Hodge duality structure on $\Omega(\mathrm{S}^2_q)$, and analysing its compatibility with the one on $\Omega(\mathrm{SU}_q(2))$.

Section 6 provides a complete explicit description of the set of connections on this specific realisation of the quantum Hopf bundle, and of the main properties of the covariant derivative operators on each associated line bundle. The emphasis is on the domain of the covariant derivative operators – the set of horizontal coequivariant elements of the bundle – which appears here as the quantum counterpart of the classical forms also called tensorial forms. Section 7 studies the coupling of the Laplacian operator on $\Omega(\mathrm{S}^2_q)$ to the gauge connections.

Section 8 applies to the commutative algebras $\{\mathcal{A}(\mathrm{SU}_q(2)), \mathcal{A}(S^2), \mathcal{A}(\mathrm{U}(1))\}$ the formalism developed in the quantum setting, in order to recover the structure of the classical Hopf bundle from an algebraic perspective. Section 9 closes the paper with an evolution of section 6, describing how a covariant derivative operator can be defined on $\Omega(\mathrm{SU}_q(2))$, the whole exterior algebra on the total space $\mathrm{SU}_q(2)$ of the quantum Hopf bundle, following the formalism developed in [**12, 13**].

2. The classical Hopf bundle

The first formulation of what are nowadays known as Hopf fibrations is contained in [**18, 19**] in terms of projecting spheres to spheres of lower dimensions: it came also as a geometric formulation of the Dirac's model of magnetic monopole [**10**]. The following lines are intended as a concise introduction to the formalism of fiber – and principal – bundles, aimed to set the notations that will be used in this paper: excellent textbooks – like for example [**20, 28**] – deeply and extensively describe this subject.

With $\pi : \mathcal{P} \to \mathcal{M}$ a smooth surjective map from a manifold $\mathcal{P}$ to a manifold $\mathcal{M}$, $(\mathcal{P}, \mathcal{M}, \pi)$ is a fibre bundle with typical fibre $\mathcal{F}$ over $\mathcal{M}$ if there is a fibre bundle atlas with charts (U_i, λ_i), where U_i is an open covering of $\mathcal{M}$ and the diffeomorphisms $\lambda_i : \pi^{-1}(U_i) \to U_i \times \mathcal{F}$ are such that $\pi : \pi^{-1}(U_i) \to U_i$ is the composition of λ_i with the projection onto the first factor in $U_i \times \mathcal{F}$. The manifold $\mathcal{P}$ is called the total space of the bundle, the manifold $\mathcal{M}$ is the base of the bundle. From the definition it follows that $\pi^{-1}(m)$ is diffeomorphic to $\mathcal{F}$ – the fibre of the bundle –

for any $m \in \mathcal{M}$. For any $f \in \mathcal{F}$ one has $\lambda_i \circ \lambda_j^{-1}(m, f) = (m, \lambda_{ij}(m, f))$ where $\lambda_{ij} : (U_i \cap U_j) \times \mathcal{F} \to \mathcal{F}$ is smooth and $\lambda_{ij}(m, \ )$ belongs to the group $\mathrm{Diff}(\mathcal{F})$ of diffeomorphisms of the fibre $\mathcal{F}$ for each $m \in U_i \cap U_j$. The mappings λ_{ij} are called the transition functions of the bundle, and satisfy the cocycle condition $\lambda_{ij}(m, \ ) \circ \lambda_{jk}(m, \ ) = \lambda_{ik}(m, \ )$ for $m \in U_i \cap U_j \cap U_k$, with $\lambda_{ii}(m, \ ) = id_{\mathcal{F}}$ for $m \in U_i$.

A fibre bundle $(\mathcal{P}, \mathcal{M}, \pi)$ is called a vector bundle if its typical fibre $\mathcal{F}$ is a vector space and if the trivialisation diffeomorphisms λ_i give transition functions λ_{ij} which are invertible linear maps, elements in $\mathrm{GL}(\mathcal{F})$ for any $m \in \mathcal{M}$. A principal bundle $(\mathcal{P}, \mathrm{K}, [\mathcal{M}], \pi)$ with structure group K is a fibre bundle $(\mathcal{P}, \mathcal{M}, \pi)$ with typical fibre K and transition functions $\lambda_{ij}(m, \ ) \in \mathrm{Aut}(\mathrm{K})$ which give the left translation of the group K on itself. On the total space of a principal bundle there is also a right action of the Lie group K – that is $\mathrm{r}_{k'}(\mathrm{r}_k(p)) = \mathrm{r}_{kk'}(p)$ for any $p \in \mathcal{P}$ and $k, k' \in \mathrm{K}$ – such that $\pi(\mathrm{r}_k(p)) = \pi(p)$, and such that the action is free and transitive. The base $\mathcal{M}$ of the bundle can be identified with the quotient $\mathcal{P}/\mathrm{K}$ with respect to such a right action.

Given G a Lie group and $\mathrm{K} \subset G$ a closed Lie subgroup of it, the group manifold G is the total space manifold of a principal bundle $(G, \mathrm{K}, G/\mathrm{K}, \pi)$ with base space G/K - the space of left cosets - and typical fiber given by the structure or gauge group K, so that the bundle projection $\pi : G \to G/\mathrm{K}$ is the canonical projection. The right principal action of the gauge group K on G is given as $\mathrm{r}_k(g) = gk$ for any $k \in \mathrm{K}$ and $g \in G$. This action trivially satisfies the requirements of being free and transitive. If $\mathfrak{k}$ is the Lie algebra of the group K, the fundamental vector field $X_\tau \in \mathfrak{X}(G)$ associated to $\tau \in \mathfrak{k}$ is defined as the infinitesimal generator of the right principal action $\mathrm{r}_{\exp s\tau}(g) = g \exp s\tau$ of the one parameter subgroup $\exp s\tau \subset \mathrm{K}$: the mapping $\tau \in \mathfrak{k} \to \{X_\tau\} \in \mathfrak{X}(G)$ is a Lie algebra isomorphism between $\mathfrak{k}$ and the set of fundamental vector fields $\{X_\tau\}$. A differential form $\phi \in \Omega(G)$ is called horizontal if $i_{X_\tau}\phi = 0$ for any fundamental vector field X_τ.

If $\rho : \mathrm{K} \to \mathrm{GL}(W)$ is a finite dimensional representation of K on the vector space W, the associated vector bundle to G is the vector bundle whose total space is $\mathcal{E} = G \times_{\rho(\mathrm{K})} W$, having typical fiber W. It is defined as the quotient of the product $G \times W$ by the equivalence relation $(\mathrm{r}_k(g) = gk; w) \sim (g; \rho(k) \cdot w)$ for any choice of $g \in G$, $k \in \mathrm{K}$ and $w \in W$: $(\mathcal{E}, G/\mathrm{K}, \pi_{\mathcal{E}})$ is a fibre bundle with a projection $\pi_{\mathcal{E}} : \mathcal{E} \to G/\mathrm{K}$ which is consistently defined on the quotient as $\pi_{\mathcal{E}}[g, w]_{\rho(\mathrm{K})} = \pi(g)$ from the principal bundle projection π.

With $\mathrm{r}_k^* : \Omega(G) \to \Omega(G)$ the action of K on the exterior algebra $\Omega(G)$ induced as a pull-back of the right action r_k of K on G, the $\rho(\mathrm{K})$-equivariant r-forms of the principal bundle are W-valued forms on G defined as:

$$\Omega^r(G, W)_{\rho(\mathrm{K})} = \{\phi \in \Omega^r(G, W) = \Omega^r(G) \otimes W : \mathrm{r}_k^*(\phi) = \rho^{-1}(k)\phi\}. \tag{2.1}$$

A section of the associated bundle $\mathcal{E}$ is an element in $\Gamma(G/\mathrm{K}, \mathcal{E})$, namely a map $\sigma : G/\mathrm{K} \to \mathcal{E}$ such that $\pi_{\mathcal{E}}(\sigma(m)) = m$ for any $m \in G/\mathrm{K}$. This definition is extended to $\Gamma^{(r)}(G/\mathrm{K}, \mathcal{E})$, the set of r-forms on the basis G/K of the principal bundle with values in $\mathcal{E}$. There is a canonical isomorphism

$$\Gamma^{(r)}(G/\mathrm{K}, \mathcal{E}) \simeq \Omega^r_{\mathrm{hor}}(G, W)_{\rho(\mathrm{K})} \tag{2.2}$$

from the space of $\mathcal{E}$-valued differential forms on G/K onto the space of horizontal $\rho(\mathrm{K})$-equivariant W-valued differential forms on the principal bundle (G, K, π). For $r = 0$ – with $\Gamma(G/\mathrm{K}, \mathcal{E}) \simeq \Gamma^{(0)}(G/\mathrm{K}, \mathcal{E})$ – the isomorphism gives the well known equivalence between equivariant functions of a principal bundle and sections of its associated bundle. In particular, for $W = \mathbb{R}, \mathbb{C}$ with trivial representation the isomorphism is

$$\Omega(G/\mathrm{K}) \simeq \Omega_{\mathrm{hor}}(G)_{\rho(\mathrm{K})=\mathrm{K}} = \{\phi \in \Omega(G) : i_{X_\tau}\phi = 0;\ \mathrm{r}_k^*\phi = \phi\}, \tag{2.3}$$

giving a description of the exterior algebra on the basis of the principal bundle.

A connection on a principal bundle can be given via a connection 1-form. A connection 1-form on G is an element $\omega \in \Omega(G, \mathfrak{k})$, taking values in $\mathfrak{k}$ and satisfying the two local conditions:

$$\omega(X_\tau) = \tau,$$

$$\mathrm{r}_k^*(\omega) = \mathrm{Ad}_{k^{-1}}\,\omega,$$

where the adjoint action of K is given by $(\mathrm{Ad}_{k^{-1}}\,\omega)(X) = k^{-1}\omega(X)k$ for any vector field $X \in \mathfrak{X}(G)$. At each point $g \in G$ there is on the tangent space T_gG a natural notion of vertical subspace, whose basis is given by the vectors X_τ which are tangent to the fiber group K, while the connection 1-form selects the horizontal subspace $H_g^{(\omega)}(G)$ given by the kernel of ω. Identifiying the element $\omega(X) \in \mathfrak{k}$ with the vertical vector field it generates, the expression $X^{(\omega)} = X - \omega(X)$ denotes the horizontal projection of the vector field $X \in \mathfrak{X}(G)$.

With any $\rho(\mathrm{K})$-equivariant form $\phi \in \Omega^r(G, W)_{\rho(\mathrm{K})}$, the covariant derivative is defined as the map $D : \Omega^r(G, W)_{\rho(\mathrm{K})} \to \Omega^{r+1}_{\mathrm{hor}}(G, W)_{\rho(\mathrm{K})}$ given as

$$D\phi(X_1, \ldots, X_{r+1}) = \mathrm{d}\phi(X_1^{(\omega)}, \ldots, X_{r+1}^{(\omega)}) \tag{2.4}$$

where d is the exterior derivative on G. On a $\rho(\mathrm{K})$-equivariant horizontal form $\phi \in \Omega_{\mathrm{hor}}(G, W)_{\rho(\mathrm{K})}$ the action of the covariant derivative can be written in terms of the connection 1-form as:

$$D\phi = \mathrm{d}\phi + \omega \wedge \phi. \tag{2.5}$$

The following sections describe the Hopf fibration $\pi : S^3 \mapsto S^2$, with $G \simeq SU(2)$, $\mathrm{K} \simeq U(1)$ and S^2 the space of the orbits $SU(2)/\,\mathrm{U}(1)$, and the monopole connection.

2.1. A differential calculus on the classical $SU(2)$ Lie group. The aim of this section is to describe the differential calculus on the total space of this bundle, in terms of a natural basis of global vector fields and 1-forms [**28**]. It is intended to give them an explicit expression in order to clarify the classical limit of their quantum counterparts.

Recall that a Lie group G naturally acts on itself both from the right and from the left. The left action is the smooth map $\mathrm{l} : G \times G \to G$ defined via the left multiplication $\mathrm{l}(g', g) = g'g = \mathrm{l}_{g'}(g)$: since $\mathrm{l}_{g'g''}(g) = \mathrm{l}_{g'}(\mathrm{l}_{g''}(g))$, the left action is a group homomorphism $\mathrm{l}_g : G \to \mathrm{Aut}(G)$. The right action is the smooth map $\mathrm{r} : G \times G \to G$ defined via the right multiplication $\mathrm{r}(g, g') = gg' = \mathrm{r}_{g'}(g)$; it is then immediate to see that $\mathrm{r}_{g'g''}(g) = gg'g'' = \mathrm{r}_{g''}(\mathrm{r}_{g'}(g))$: the right action is a group anti-homomorphism $\mathrm{r}_g : G \to \mathrm{Aut}(G)$. For any $T \in \mathfrak{g}$, the Lie algebra of G, it is possible to define a vector field $R_T \in \mathfrak{X}(G)$. It acts as a derivation on a smooth complex valued function defined on G, and can be written in terms of the pull-back $\mathrm{l}_g^* : C^\infty(G) \to C^\infty(G)$ induced by l_g. On $\phi \in C^\infty(G)$:

$$R_T(\phi) = \frac{\mathrm{d}}{\mathrm{d}s}\,(\mathrm{l}^*_{\exp sT}(\phi))\big|_{s=0} \tag{2.6}$$

Although defined via the left action l_g, the vector field R_T is called the right invariant vector field associated to $T \in \mathfrak{g}$; this set of fields owes its name to the fact that, given $\mathrm{r}_{g*} : \mathfrak{X}(G) \to \mathfrak{X}(G)$ the push-forward induced by the right action r_g, they satisfy a property of right invariance as $\mathrm{r}_{g*}(R_T) = R_T$. From the definition of the pull-back map $\mathrm{l}_g^* : C^\infty(G) \to C^\infty(G)$ one has:

$$\mathrm{l}^*_{g'g''}(\phi) = \phi \circ \mathrm{l}_{g'g''} = \phi \circ \mathrm{l}_{g'} \circ \mathrm{l}_{g''} = \mathrm{l}^*_{g''}(\mathrm{l}^*_{g'}(\phi))$$

for any $\phi \in C^\infty(G)$. This relation enables to prove that the map $\check{\mathrm{l}} : T \in \mathfrak{g} \to R_T \in \mathfrak{X}(G)$ is a Lie algebra anti-homomorphism, $[R_T, R_{T'}] = R_{[T',T]}$.

The analogous definitions starting from the right action naturally hold. For any $T \in \mathfrak{g}$, the vector field $L_T \in \mathfrak{X}(G)$ is defined as a derivation on $C^\infty(G)$, namely as the infinitesimal generator of the pull-back r_g^* induced by the right action r_g:

$$L_T(\phi) = \frac{\mathrm{d}}{\mathrm{d}s}\,(\mathrm{r}^*_{\exp sT}(\phi))\big|_{s=0} \tag{2.7}$$

on any $\phi \in C^\infty(G)$. Left invariant vector fields satisfy a property of left invariance given as $\mathrm{l}_g^*(L_T) = L_T$; the map $\check{\mathrm{r}} : T \in \mathfrak{g} \to L_T \in \mathfrak{X}(G)$ is a Lie algebra homomorphism, with $[L_T, L_{T'}] = L_{[T,T']}$. The sets $\{L_T\}$, $\{R_T\}$ are two basis of the left free $C^\infty(G)$-module $\mathfrak{X}(G)$.

The total space of the classical Hopf bundle is the manifold S^3, which represents the elements of the Lie group $SU(2)$. A point $g \in S^3$ can be then written via a 2×2 matrix with complex entries and unit determinant:

$$g = \begin{pmatrix} u & -\bar{v} \\ v & \bar{u} \end{pmatrix} \quad : \quad \bar{u}u + \bar{v}v = 1; \tag{2.8}$$

the left invariant vector fields $\check{\mathrm{r}}(T) = L_T$ are given, following (2.7), as the tangent vectors to the curves $g(s) = g \cdot \exp sT$. In the defining matrix representation it reads:

$$\frac{d}{ds}\begin{pmatrix} u & -\bar{v} \\ v & \bar{u} \end{pmatrix} \cdot (\exp sT)\,|_{s=0} = \begin{pmatrix} u & -\bar{v} \\ v & \bar{u} \end{pmatrix} \cdot (T) \tag{2.9}$$

Since $\exp sT$ is unitary, T is antihermitian, and the choice of a basis in terms of the Pauli matrices:

$$T_x = \frac{1}{2}\begin{pmatrix} 0 & i \\ i & 0 \end{pmatrix}, \qquad T_y = \frac{1}{2}\begin{pmatrix} 0 & -1 \\ 1 & 0 \end{pmatrix}, \qquad T_z = \frac{i}{2}\begin{pmatrix} 1 & 0 \\ 0 & -1 \end{pmatrix}, \tag{2.10}$$

gives the explicit form of the left invariant vector fields:

$$
\begin{aligned}
L_x &= -\frac{i}{2}\left(\bar{v}\frac{\partial}{\partial u} - \bar{u}\frac{\partial}{\partial v} + u\frac{\partial}{\partial \bar{v}} - v\frac{\partial}{\partial \bar{u}}\right) \\
L_y &= -\frac{1}{2}\left(\bar{v}\frac{\partial}{\partial u} - \bar{u}\frac{\partial}{\partial v} - u\frac{\partial}{\partial \bar{v}} + v\frac{\partial}{\partial \bar{u}}\right) \\
L_z &= \frac{i}{2}\left(u\frac{\partial}{\partial u} + v\frac{\partial}{\partial v} - \bar{v}\frac{\partial}{\partial \bar{v}} - \bar{u}\frac{\partial}{\partial \bar{u}}\right) \\
L_- &= L_x - iL_y \ = i\left(v\frac{\partial}{\partial \bar{u}} - u\frac{\partial}{\partial \bar{v}}\right) \\
L_+ &= L_x + iL_y \ = i\left(\bar{u}\frac{\partial}{\partial v} - \bar{v}\frac{\partial}{\partial u}\right),
\end{aligned}
\tag{2.11}
$$

satisfying the commutation relations:

$$
\begin{aligned}
[L_z; L_-] &= iL_-, \\
[L_z; L_+] &= -iL_+, \\
[L_-; L_+] &= 2iL_z.
\end{aligned}
\tag{2.12}
$$

The components of the right invariant vector fields $R_T = \check{\mathrm{I}}(T)$ are then clearly given in the defining matrix representation (2.6) as:

$$
\frac{d}{ds}(\exp sT)\cdot\begin{pmatrix} u & -\bar{v} \\ v & \bar{u} \end{pmatrix}|_{s=0} = (T)\cdot\begin{pmatrix} u & -\bar{v} \\ v & \bar{u} \end{pmatrix}
\tag{2.13}
$$

acquiring the form:

$$
\begin{aligned}
R_x &= \frac{i}{2}\left(v\frac{\partial}{\partial u} + u\frac{\partial}{\partial v} - \bar{u}\frac{\partial}{\partial \bar{v}} - \bar{v}\frac{\partial}{\partial \bar{u}}\right) \\
R_y &= -\frac{1}{2}\left(v\frac{\partial}{\partial u} - u\frac{\partial}{\partial v} - \bar{u}\frac{\partial}{\partial \bar{v}} + \bar{v}\frac{\partial}{\partial \bar{u}}\right) \\
R_z &= \frac{i}{2}\left(u\frac{\partial}{\partial u} - v\frac{\partial}{\partial v} + \bar{v}\frac{\partial}{\partial \bar{v}} - \bar{u}\frac{\partial}{\partial \bar{u}}\right) \\
R_- &= R_x - iR_y \ = i\left(v\frac{\partial}{\partial u} - \bar{u}\frac{\partial}{\partial \bar{v}}\right) \\
R_+ &= R_x + iR_y \ = i\left(u\frac{\partial}{\partial v} - \bar{v}\frac{\partial}{\partial \bar{u}}\right).
\end{aligned}
\tag{2.14}
$$

The commutation relations they satisfy are:

$$
\begin{aligned}
[R_z; R_-] &= -iR_-, \\
[R_z; R_-] &= iR_+, \\
[R_-; R_+] &= -2iR_z.
\end{aligned}
\tag{2.15}
$$

The quadratic Casimir of the Lie algebra $\mathfrak{su}(2)$ is written as

$$
C = \frac{1}{2}(L_+L_- + L_-L_+) + L_zL_z = \frac{1}{2}(R_+R_- + R_-R_+) + R_zR_z.
\tag{2.16}
$$

The set $\mathfrak{X}(S^3)$ is a free left $C^\infty(S^3)$- module. Right vector fields can be expressed in the basis of the left vector fields as $R_a = J_{ab}L_b$. The matrix J is given by:

$$\begin{pmatrix} R_- \\ R_z \\ R_+ \end{pmatrix} = \begin{pmatrix} \bar{u}^2 & 2\bar{u}v & -v^2 \\ -\bar{u}\bar{v} & u\bar{u}-v\bar{v} & -uv \\ -\bar{v}^2 & 2u\bar{v} & u^2 \end{pmatrix} \begin{pmatrix} L_- \\ L_z \\ L_+ \end{pmatrix} \tag{2.17}$$

and its inverse matrix is:

$$\begin{pmatrix} L_- \\ L_z \\ L_+ \end{pmatrix} = \begin{pmatrix} u^2 & -2uv & -v^2 \\ u\bar{v} & u\bar{u}-v\bar{v} & \bar{u}v \\ -\bar{v}^2 & -2\bar{u}\bar{v} & \bar{u}^2 \end{pmatrix} \begin{pmatrix} R_- \\ R_z \\ R_+ \end{pmatrix} \tag{2.18}$$

A similar analysis can be performed in the study of the cotangent space $\mathfrak{X}^*(G)$ of a Lie group. This is a $C^\infty(S^3)$-bimodule, with two basis of globally defined 1-forms, namely the left invariant $\{\tilde{\omega}_a\}$ dual to the set of left invariant vector fields $\{L_a\}$, and the right invariant $\{\tilde{\eta}_b\}$ dual to the set of right invariant vector fields $\{R_b\}$. They satisfy the invariance property:

$$\begin{aligned} \mathrm{l}_g^*(\tilde{\omega}_a) &= \tilde{\omega}_a, \\ \mathrm{r}_g^*(\tilde{\eta}_b) &= \tilde{\eta}_b : \end{aligned} \tag{2.19}$$

one then immediately computes:

$$R_i = J_{ij}L_j \Leftrightarrow \tilde{\eta}_s J_{sp} = \tilde{\omega}_p. \tag{2.20}$$

The left invariant 1-forms are:

$$\begin{aligned} \tilde{\omega}_z &= -2i\,(\bar{u}du + \bar{v}dv) \\ \tilde{\omega}_- &= -i\,(\bar{v}d\bar{u} - \bar{u}d\bar{v}) \\ \tilde{\omega}_+ &= -i\,(udv - vdu) \end{aligned} \tag{2.21}$$

with $\tilde{\omega}_x = (\tilde{\omega}_- + \tilde{\omega}_+)$ and $\tilde{\omega}_y = i(\tilde{\omega}_+ - \tilde{\omega}_-)$. The antilinear involution on $\Omega^1(S^3)$, compatible with the antilinear involution on $C^\infty(S^3)$, is given by $\tilde{\omega}_x^* = \tilde{\omega}_x$, $\tilde{\omega}_y^* = \tilde{\omega}_y$, $\tilde{\omega}_z^* = \tilde{\omega}_z$. The right-invariant 1-forms are:

$$\begin{aligned} \tilde{\eta}_z &= 2i\,(ud\bar{u} + \bar{v}dv) \\ \tilde{\eta}_- &= i\,(ud\bar{v} - \bar{v}du) \\ \tilde{\eta}_+ &= -i\,(\bar{u}dv - vd\bar{u})\,. \end{aligned} \tag{2.22}$$

Given a complex valued smooth function $\phi \in C^\infty(S^3)$, the exterior derivative is the map $\mathrm{d} : C^\infty(S^3) \to \Omega^1(S^3)$ defined via:

$$\mathrm{d}\phi(X) = X(\phi) \tag{2.23}$$

in terms of the Lie derivative $X(\phi)$ of ϕ along the vector field X. This map acquires the form:

$$\mathrm{d}\phi = L_a(\phi)\tilde{\omega}_a = R_b(\phi)\tilde{\eta}_b \tag{2.24}$$

where now $L_a(\phi)$ represents the Lie derivative of ϕ along the vector field L_a, while $R_b(\phi)$ represents the Lie derivative of ϕ along the vector field R_b.

From the $C^\infty(S^3)$-bimodule $\Omega^1(S^3)$ define the tensor product of forms as the $C^\infty(S^3)$-bimodule $\{\Omega^1(S^3)\}^{\otimes k} = \Omega^1(S^3) \otimes_{C^\infty(S^3)} \dots \otimes_{C^\infty(S^3)} \Omega^1(S^3)$ (k times). The exterior algebra coming from the differential calculus (2.24) is defined as the graded associative algebra $\Omega(S^3) = \left(\oplus_k \Omega^k(S^3); \wedge\right)$, with k-forms and wedge product introduced in terms of an alternation mapping $\mathfrak{A} : \{\Omega^1(S^3)\}^{\otimes k} \to \{\Omega^1(S^3)\}^{\otimes k}$

[1]. The wedge product is bilinear, and satisfies the identity $\alpha\wedge\beta=(-1)^{kl}\beta\wedge\alpha$ for any k-form α and l-form β. The complex involution is extended by requiring

$$(\alpha\wedge\beta)^*=(-1)^{kl}\beta^*\wedge\alpha^*.$$

The exterior derivative is extended to $\mathrm{d}:\Omega^k(S^3)\to\Omega^{k+1}(S^3)$ as the unique $\mathbb{C}$-linear mapping satisfying the conditions:

(1) d is a graded $\wedge$-derivation,
that is $\mathrm{d}(\alpha\wedge\beta)=(\mathrm{d}\alpha)\wedge\beta+(-1)^k\alpha\wedge\mathrm{d}\beta$ for any k-form α;
(2) $\mathrm{d}^2=\mathrm{d}\circ\mathrm{d}=0$;
(3) on $\phi\in\Omega^0(S^3)\simeq C^\infty(S^3)$, it is given by $\mathrm{d}\phi$ as in (2.24).

It is then easy to see that $\Omega^2(S^3)$ is three dimensional, with a basis given by $\{\tilde\omega_-\wedge\tilde\omega_+,\tilde\omega_+\wedge\tilde\omega_z,\tilde\omega_z\wedge\tilde\omega_-\}$: extending in a natural way via the pull back the left and right actions of the group $SU(2)$ on $\Omega^2(S^3)$, it is also clear that such basis elements are left invariant. From (2.21) one has:

$$\begin{aligned}\mathrm{d}\tilde\omega_-&=i\tilde\omega_-\wedge\tilde\omega_z,\\ \mathrm{d}\tilde\omega_+&=-i\tilde\omega_+\wedge\tilde\omega_z,\\ \mathrm{d}\tilde\omega_z&=2i\tilde\omega_-\wedge\tilde\omega_+.\end{aligned}\tag{2.25}$$

The bimodule $\Omega^3(S^3)$ is one dimensional, with again a left invariant basis 3-form given by $\{\tilde\omega_-\wedge\tilde\omega_+\wedge\tilde\omega_z\}$. A right invariant basis of the exterior algebra $\Omega(S^3)$ is analogously given in terms of the 1-forms $\tilde\eta_a$.

2.2. A Laplacian operator on the group manifold $SU(2)$. Being $SU(2)$ a semisimple Lie group, the group manifold S^3 can be equipped with the Cartan-Killing metric originated from the Cartan decomposition of the Lie algebra $\mathfrak{su}(2)$. Consider now as a riemannian metric structure on S^3 the symmetric tensor

$$g=\alpha(\tilde\omega_x\otimes\tilde\omega_x+\tilde\omega_y\otimes\tilde\omega_y)+\tilde\omega_z\otimes\tilde\omega_z,\tag{2.26}$$

with $\alpha\in\mathbb{R}^+$. For $\alpha=1$ such a metric tensor coincides with the the Cartan-Killing metric. The volume associated to the $\mathfrak{g}$-orthonormal basis and to the choice of the orientation (x,y,z) is given by $\theta=\alpha\,\tilde\omega_x\wedge\tilde\omega_y\wedge\tilde\omega_z$, so that $\theta^*=\theta$. Such a volume θ is a Haar volume, namely it is invariant with respect to both the left l_g^* and the right actions r_g^* of the Lie group $SU(2)$ on itself, since an explicit calculation gives $L_a(\theta)=R_a(\theta)=0$. The Hodge duality $\star:\Omega^k(S^3)\to\Omega^{3-k}(S^3)$ which corresponds to this volume [1] is the $C^\infty(S^3)$-linear map given on the left invariant basis of the exterior algebra $\Omega(S^3)$ by $\star(1)=\theta$, $\star(\theta)=1$, and:

$$\begin{aligned}\star(\tilde\omega_x)&=\tilde\omega_y\wedge\tilde\omega_z, & \star(\tilde\omega_y\wedge\tilde\omega_z)&=\tilde\omega_x,\\ \star(\tilde\omega_y)&=\tilde\omega_z\wedge\tilde\omega_x, & \star(\tilde\omega_z\wedge\tilde\omega_x)&=\tilde\omega_y,\\ \star(\tilde\omega_z)&=\alpha\,\tilde\omega_x\wedge\tilde\omega_y, & \star(\tilde\omega_x\wedge\tilde\omega_y)&=\alpha^{-1}\,\tilde\omega_z.\end{aligned}\tag{2.27}$$

The differential calculus on the group manifold S^3 as well as the above $\star$-Hodge duality on the exterior algebra $\Omega(S^3)$ give a Laplacian operator defined as $\square_{S^3}\phi=\star\mathrm{d}\star\mathrm{d}\phi$ on any $\phi\in C^\infty(S^3)$. It can be written as a differential operator in terms of the left invariant vector fields:

$$\square_{S^3}\phi=[\frac{1}{2\alpha}(L_-L_++L_+L_-)+L_zL_z]\phi\tag{2.28}$$

The Laplacian operator is the Casimir of the Lie algebra $\mathfrak{su}(2)$ only if $\alpha=1$, that is only if the metric from where it is originated is the Cartan-Killing metric.

The Hodge structure satisfies two identities:

$$\star^2(\xi) = (-1)^{k(3-k)}\xi = \xi \tag{2.29}$$

$$\xi \wedge (\star\xi') = \xi' \wedge (\star\xi) \tag{2.30}$$

for any $\xi, \xi' \in \Omega^k(S^3)$. This allows to define a symmetric bilinear map $\langle\ ,\ \rangle_{S^3}$: $\Omega^k(S^3) \times \Omega^k(S^3) \to C^\infty(S^3)$ $(k = 0, \dots, 3)$ as:

$$\langle \xi, \xi' \rangle_{S^3}\, \theta = \xi \wedge (\star\xi'). \tag{2.31}$$

It is clearly a symmetric tensor on $\{\mathfrak{X}^*(S^3)\}^{\otimes 2k}$, whose components can be expressed in terms of the components of the inverse metric $g^{-1} = g^{-1ab} L_a \otimes L_b \in \{\mathfrak{X}^1(S^3)\}^{\otimes 2}$ with $g^{-1ab} g_{bc} = \delta^a_c$, as

$$\langle \tilde\omega_{i_1} \wedge \dots \wedge \tilde\omega_{i_k}, \tilde\omega_{j_1} \wedge \dots \wedge \tilde\omega_{j_k} \rangle_{S^3} = \sum_\sigma \pi_\sigma g^{-1 i_1 \sigma(j_1)} \dots g^{-1 i_k \sigma(j_k)} \tag{2.32}$$

where the summation is over permutations σ of k elements, with parity π_σ. Starting from the Hodge duality a second bilinear map $\langle\ ,\ \rangle^{\sim}_{S^3} : \Omega^k(S^3) \times \Omega^k(S^3) \to C^\infty(S^3)$, can be introduced as

$$\langle \xi', \xi \rangle^{\sim}_{S^3}\, \theta = \xi^* \wedge (\star\xi') \tag{2.33}$$

for any $\xi, \xi' \in \Omega^k(S^3)$, being hermitian $((\langle \xi', \xi \rangle^{\sim}_{S^3})^* = \langle \xi, \xi' \rangle^{\sim}_{S^3}$. The Haar volume form can be used to introduce an integral on a manifold [**1**], $\int_\theta : \Omega^3(S^3) \to \mathbb{C}$; being S^3 compact, the volume of the group manifold can be normalised, setting $\int_\theta \theta = 1$. From (2.31) and (2.33) it is possible to define on the exterior algebra $\Omega(S^3)$ both a scalar product,

$$(\xi; \xi')_{S^3} = \int_\theta \xi \wedge (\star\xi') = \int_\theta \langle \xi, \xi' \rangle_{S^3}\, \theta, \tag{2.34}$$

and an hermitian inner product,

$$(\xi'; \xi)^{\sim}_{S^3} = \int_\theta \xi^* \wedge (\star\xi') = \int_\theta \langle \xi', \xi \rangle^{\sim}_{S^3}\, \theta. \tag{2.35}$$

An evaluation on a non hermitian basis in $\Omega(S^3)$ presents the differences between the non vanishing terms of two bilinear forms:

$$\begin{aligned}
&\langle 1, 1 \rangle_{S^3} = 1; \\
&\langle \tilde\omega_-, \tilde\omega_+ \rangle_{S^3} = \langle \tilde\omega_+, \tilde\omega_- \rangle_{S^3} = \frac{1}{2\alpha}, \\
&\langle \tilde\omega_z, \tilde\omega_z \rangle_{S^3} = 1; \\
&\langle \tilde\omega_+ \wedge \tilde\omega_z, \tilde\omega_- \wedge \tilde\omega_z \rangle_{S^3} = \langle \tilde\omega_- \wedge \tilde\omega_z, \tilde\omega_+ \wedge \tilde\omega_z \rangle_{S^3} = \frac{1}{2\alpha}, \\
&\langle \tilde\omega_- \wedge \tilde\omega_+, \tilde\omega_- \wedge \tilde\omega_+ \rangle_{S^3} = \frac{1}{4\alpha^2}; \\
&\langle \theta, \theta \rangle_{S^3} = 1;
\end{aligned} \tag{2.36}$$

while

$$
\begin{aligned}
&\langle 1,1\rangle^{\sim}_{S^3} = 1;\\
&\langle \tilde{\omega}_-, \tilde{\omega}_-\rangle^{\sim}_{S^3} = \langle \tilde{\omega}_+, \tilde{\omega}_+\rangle^{\sim}_{S^3} = \frac{1}{2\alpha},\\
&\langle \tilde{\omega}_z, \tilde{\omega}_z\rangle^{\sim}_{S^3} = 1;\\
&\langle \tilde{\omega}_+\wedge\tilde{\omega}_z, \tilde{\omega}_+\wedge\tilde{\omega}_z\rangle^{\sim}_{S^3} = \langle \tilde{\omega}_-\wedge\tilde{\omega}_z, \tilde{\omega}_-\wedge\tilde{\omega}_z\rangle^{\sim}_{S^3} = \frac{1}{2\alpha},\\
&\langle \tilde{\omega}_-\wedge\tilde{\omega}_+, \tilde{\omega}_-\wedge\tilde{\omega}_+\rangle^{\sim}_{S^3} = \frac{1}{4\alpha^2};\\
&\langle \theta, \theta\rangle^{\sim}_{S^3} = 1. \qquad (2.37)
\end{aligned}
$$

2.3. The principal bundle structure and the monopole connection. Consider the one parameter subgroup of $SU(2)$ given by $\gamma_{T_z}(s) = \exp sT_z$ where T_z is the generator in (2.10). In this specific matrix representation it is

$$
\gamma_{T_z}(s) = \exp\left[\frac{is}{2}\begin{pmatrix} 1 & 0 \\ 0 & -1\end{pmatrix}\right] = \begin{pmatrix} e^{is/2} & 0 \\ 0 & e^{-is/2}\end{pmatrix}, \qquad (2.38)
$$

thus proving that $\gamma_{T_z}(s) \simeq \mathrm{U}(1)$ as a subgroup in $SU(2)$. The space of left cosets $SU(2)/\mathrm{U}(1)$ is the set of the orbits of the right principal action $\check{r}_{\exp sT_z}(g) = g\exp sT_z$ which is free, and smooth; its infinitesimal generator coincides with the vector field L_z (2.9). As already mentioned the canonical projection $\pi : SU(2) \to SU(2)/\mathrm{U}(1)$ gives a principal bundle whose vertical field is L_z. A formulation for a principal bundle atlas on a homogeneous space is extensively analysed in terms of local sections [**20, 28**]. This section describes in detail how a principal bundle atlas is introduced [**15, 35**] defining suitable trivialisations.

Parametrise S^3 by

$$
\begin{aligned}
u &= \cos\theta/2\, e^{i(\varphi+\psi)/2}\\
v &= \sin\theta/2\, e^{-i(\varphi-\psi)/2},
\end{aligned}
$$

with $0 \le \theta \le \pi$ and $\phi, \psi \in \mathbb{R}$, the Hopf map $\pi : SU(2) \to S^2 \simeq SU(2)/\mathrm{U}(1)$ is defined by:

$$
\begin{aligned}
b_z &= uu^* - vv^* = \cos\theta,\\
b_y &= uv^* + vu^* = \sin\theta\cos\varphi,\\
b_x &= -i(vu^* - uv^*) = -\sin\theta\sin\varphi \qquad (2.39)
\end{aligned}
$$

with $b_z^2 + b_x^2 + b_y^2 = 1$. It is immediate to see that $\pi(u,v) = \pi(u',v')$ if and only if $u' = ue^{i\alpha}$ and $v' = ve^{i\alpha}$ with $\alpha \in \mathbb{R}$: this is also a way to recover that the projection has the standard fibre U(1). A choice for an open covering of the sphere S^2 is given by:

$$
\begin{aligned}
S^2_{(N)} &= \{S^2 : b_z \neq 1\} &\Rightarrow&\quad \pi^{-1}(S^2_{(N)}) = S^3_{(N)} = \{S^3 : v \neq 0\},\\
S^2_{(S)} &= \{S^2 : b_z \neq -1\} &\Rightarrow&\quad \pi^{-1}(S^2_{(S)}) = S^3_{(S)} = \{S^3 : u \neq 0\}, \qquad (2.40)
\end{aligned}
$$

with $S^3_{(j)} \simeq S^2_{(j)} \times \mathrm{U}(1)$ via the diffeomorphisms:

$$g \simeq (u,v) \in S^3_{(N)} \quad : \lambda_N(g) = (\pi(g); \frac{v}{|v|}) \in S^2_{(N)} \times \mathrm{U}(1),$$
$$g \simeq (u,v) \in S^3_{(S)} \quad : \lambda_S(g) = (\pi(g); \frac{u}{|u|}) \in S^2_{(S)} \times \mathrm{U}(1).$$

The set of transition functions associated with this trivialisation is given by $\lambda^{-1}_{NS} = \lambda_{SN} = \lambda_S \circ \lambda_N^{-1} : (S^2_{(N)} \cap S^2_{(S)}) \times \mathrm{U}(1) \to \mathrm{U}(1)$. Choose $b \sim (\theta, \varphi) \in S^2_{(N)} \cap S^2_{(S)}$. The element $(b, e^{i\alpha}) \in (S^2_{(N)} \cap S^2_{(S)}) \times \mathrm{U}(1)$ is mapped into

$$\lambda_N^{-1}(b, e^{i\alpha}) = (u = \frac{b_y - ib_x}{\sqrt{2(1-b_z)}}\, e^{i\alpha}; v = \sqrt{\frac{1-b_z}{2}}\, e^{i\alpha}) \in S^3_{(N)}$$
$$\Rightarrow \lambda_S \circ \lambda_N^{-1}(b; e^{i\alpha}) = (b, e^{i\varphi} e^{i\alpha}).$$

This means that $\lambda_{SN}(b) \cdot e^{i\alpha} = e^{i\varphi} e^{i\alpha}$. The transition functions describe a left action of the U(1) gauge group on itself, and trivially satisfy the cocycle conditions.

For any integer n there is a representation of the gauge group,

$$\rho_{(n)} : \mathrm{U}(1) \to \mathbb{C}^*, \qquad \rho_{(n)}(e^{i\alpha}) = e^{in\alpha} \tag{2.41}$$

so that for any $n \in \mathbb{Z}$ there is a line bundle $\mathcal{E}_n = SU(2) \times_{\rho_{(n)}} \mathbb{C}$ associated to the principal Hopf bundle. Since the representations of the gauge group given in (2.41) are defined on $\mathbb{C}$, the set $\Omega^r(S^3, \mathbb{C})_{\rho_{(n)}} \simeq \Omega^r(S^3)$ of $\rho_{(n)}(\mathrm{U}(1))$-equivariant r-forms on the Hopf bundle can be easily described in terms of the action of the vertical field of the bundle, giving the infinitesimal version of the definition in (2.1) (with $r = 0, \ldots, 3$)

$$\Omega^r(S^3)_{\rho_{(n)}} = \{\phi \in \Omega^r(S^3) : \check{\mathrm{r}}^*_k(\phi) = \rho^{-1}_{(n)}(k)\phi \Leftrightarrow L_z(\phi) = -\frac{in}{2}\phi\}. \tag{2.42}$$

The sets $\Omega^r(S^3)_{\rho_{(n)}}$ are $C^\infty(S^2)$-bimodule. The horizontal $\rho_{(n)}(\mathrm{U}(1))$-equivariant r-forms are given as:

$$\mathfrak{L}^{(r)}_n = \{\phi \in \Omega^r(S^3)_{\rho_{(n)}} : i_{L_z}(\phi) = 0\} \tag{2.43}$$

for $r > 0$: one obviously has $\mathfrak{L}^{(3)}_n = \emptyset$, while

$$\mathfrak{L}^{(0)}_n = \Omega^0(S^3)_{\rho_{(n)}} = \{\phi \in C^\infty(S^3) : \check{\mathrm{r}}^*_k(\phi) = \phi \Leftrightarrow L_z(\phi) = -(in/2)\phi\}. \tag{2.44}$$

With $\Gamma^{(r)}(S^2, \mathcal{E}_n)$ the set of $\mathcal{E}_n$-valued r-forms defined on S^2, the isomorphisms in (2.2) can be written as isomorphisms of $C^\infty(S^2)$-bimodule

$$\Gamma^{(r)}(S^2, \mathcal{E}_n) \simeq \mathfrak{L}^{(r)}_n. \tag{2.45}$$

They formalise the equivalence between r-form valued sections on each line bundle $\mathcal{E}_n$ and $\rho_{(n)}(\mathrm{U}(1))$-equivariant horizontal r-forms of the principal Hopf bundle. This equivalence can be described – as in [**29**] – using the local trivialisation (2.40). A global, algebraic description of them, naturally conceived for the generalisation to the non commutative setting, is in [**23**], and it is based on the Serre-Swan theorem[1].

[1]The theorem of Serre and Swan [**34**] constructs a complete equivalence between the category of (smooth) vector bundles over a (smooth) compact manifold $\mathcal{M}$ and bundle maps, and the category of finite projective modules over the commutative algebra $C(\mathcal{M})$ of (smooth) functions over $\mathcal{M}$ and module morphisms. The space $\Gamma(\mathcal{M}, \mathcal{E})$ of (smooth) sections of a vector bundle $\pi_{\mathcal{E}} : \mathcal{E} \to \mathcal{M}$ over a compact manifold $\mathcal{M}$ is a finite projective module over the commutative

Given $n \in \mathbb{Z}$, consider an element $\left|\tilde{\Psi}^{(n)}\right\rangle \in C^\infty(S^3)^{|n|+1}$ whose components are given by:

$$n \geq 0: \qquad \left|\tilde{\Psi}^{(n)}\right\rangle_\mu = \sqrt{\binom{n}{\mu}}\,\bar{v}^\mu \bar{u}^{n-\mu} \in \mathfrak{L}_n^{(0)},$$

$$n \leq 0: \qquad \left|\tilde{\Psi}^{(n)}\right\rangle_\mu = \sqrt{\binom{|n|}{\mu}}\,v^{|n|-\mu} u^\mu \in \mathfrak{L}_n^{(0)} \tag{2.46}$$

with $\mu = 0, \dots |n|$. Recalling the binomial expansion it is easy to compute that:

$$n \geq 0: \qquad \left\langle \tilde{\Psi}^{(n)}, \tilde{\Psi}^{(n)} \right\rangle = \sum_{\mu=0}^{n} \binom{n}{\mu} u^{n-\mu} v^\mu \bar{v}^\mu \bar{u}^{n-\mu} = (\bar{u}u + \bar{v}v)^n = 1,$$

(2.47)

$$n \leq 0: \qquad \left\langle \tilde{\Psi}^{(n)}, \tilde{\Psi}^{(n)} \right\rangle = \sum_{\mu=0}^{|n|} \binom{|n|}{\mu} \bar{u}^\mu \bar{v}^{|n|-\mu} v^{|n|-\mu} u^\mu = (\bar{u}u + \bar{v}v)^n = 1.$$

The ket-bra element $\tilde{\mathfrak{p}}^{(n)} = \left|\tilde{\Psi}^{(n)}\right\rangle \left\langle \tilde{\Psi}^{(n)}\right| \in \mathbb{M}^{|n|+1}(C^\infty(S^2))$ is then a projector in the free finitely generated module $C^\infty(S^2)^{|n|+1}$, as it satisfies the identities $(\tilde{\mathfrak{p}}^{(n)})^\dagger = \tilde{\mathfrak{p}}^{(n)}$, $(\tilde{\mathfrak{p}}^{(n)})^2 = \tilde{\mathfrak{p}}^{(n)}$. The matrix elements of the projectors are given by $\tilde{\mathfrak{p}}^{(n)}_{\mu\nu} = \left|\tilde{\Psi}^{(n)}\right\rangle_\mu \left\langle \tilde{\Psi}^{(n)}\right|_\nu$: each projector $\tilde{\mathfrak{p}}^{(n)}$ has rank 1, because its trace is the constant unit function given by

$$tr\,\tilde{\mathfrak{p}}^{(n)} = \sum_{\mu=0}^{|n|} \left|\tilde{\Psi}^{(n)}\right\rangle_\mu \left\langle \tilde{\Psi}^{(n)}\right|_\mu = 1. \tag{2.48}$$

Consider the set of $\rho_{(n)}(\mathrm{U}(1))$-equivariant map $\mathfrak{L}_n^{(0)}$ as a left module over $C^\infty(S^2) \subset C^\infty(S^3)$: any equivariant map $\phi \in \mathfrak{L}_n^{(0)}$ can be written in terms of an element $\langle f| \in C^\infty(S^2)^{|n|+1}$ as

$$\phi_f = \left\langle f, \tilde{\Psi}^{(n)} \right\rangle = \sum_{\mu=0}^{|n|} \langle f|_\mu \left|\tilde{\Psi}^{(n)}\right\rangle_\mu .$$

Given the set $\Gamma^{(0)}(S^2, \mathcal{E}_n)$ of sections of each associated line bundle $\mathcal{E}_n$, the equivalence with the set $\mathfrak{L}_n^{(0)}$ of $\rho_{(n)}(U(1))$-equivariant maps of the Hopf bundle is formalised via an isomorphism between $C^\infty(S^2)$-left modules, represented by:

$$\begin{array}{ccc} \Gamma^{(0)}(S^2, \mathcal{E}_n) & \leftrightarrow & \mathfrak{L}_n^{(0)} \\ \langle \sigma_f| = \langle f|\,\tilde{\mathfrak{p}}^{(n)} & \leftrightarrow & \left\langle f, \tilde{\Psi}^{(n)} \right\rangle \\ \langle \sigma_f| = \phi_f \left\langle \tilde{\Psi}^{(n)}\right| & \leftrightarrow & \phi_f = \left\langle \sigma_f, \tilde{\Psi}^{(n)} \right\rangle \end{array} \tag{2.49}$$

for any $\langle f| \in C^\infty(S^2)^{|n|+1}$. Since from this definition it is $\langle \sigma_f|\,\tilde{\mathfrak{p}}^{(n)} = \langle \sigma_f|$, this isomorphism enables to recover $\langle \sigma_f| \in \Gamma^{(0)}(S^2, \mathcal{E}_n) \simeq C^\infty(S^2)^{|n|+1}\tilde{\mathfrak{p}}^{(n)}$. An

algebra $C(\mathcal{M})$ and every finite projective $C(\mathcal{M})$-module can be realised as a module of sections of a vector bundle over $\mathcal{M}$.

explicit computation from (2.11) and (2.21) gives:

$$L_z(\tilde{\omega}_+) = i\tilde{\omega}_+ \quad \Rightarrow \quad \tilde{\omega}_+ \in \mathfrak{L}^{(1)}_{-2};$$
$$L_z(\tilde{\omega}_-) = -i\tilde{\omega}_- \quad \Rightarrow \quad \tilde{\omega}_- \in \mathfrak{L}^{(1)}_{2}, \tag{2.50}$$

so that for any $n \in \mathbb{Z}$ the set of $\rho_{(n)}(\mathrm{U}(1))$-equivariant horizontal 1-forms of the Hopf bundle is

$$\mathfrak{L}^{(1)}_n = \{\phi = \phi'\tilde{\omega}_- + \phi''\tilde{\omega}_+ : \phi' \in \mathfrak{L}^{(0)}_{n-2} \text{ and } \phi'' \in \mathfrak{L}^{(0)}_{n+2}\}. \tag{2.51}$$

For $n = 0$ one also recovers from (2.3) the equivalence $\mathfrak{L}^{(1)}_0 \simeq \Omega^1(S^2)$, so to have the $C^\infty(S^2)$-bimodule identification $\mathfrak{L}^{(1)}_n \simeq \Omega^1(S^2) \otimes_{C^\infty(S^2)} \mathfrak{L}^{(0)}_n$. For $r = 1$ the isomorphism in (2.45) can be written as:

$$\Gamma^{(1)}(S^2, \mathcal{E}_n) \simeq \Omega^1(S^2)^{|n|+1} \cdot \tilde{\mathfrak{p}}^{(n)} \quad \leftrightarrow \quad \mathfrak{L}^{(1)}_n \simeq \Omega^1(S^2) \otimes_{C^\infty(S^2)} \mathfrak{L}^{(0)}_n,$$
$$\langle\sigma| = \phi\left\langle\tilde{\Psi}^{(n)}\right| \quad \leftrightarrow \quad \phi = \left\langle\sigma, \tilde{\Psi}^{(n)}\right\rangle. \tag{2.52}$$

Given any $\phi \in \mathfrak{L}^{(1)}_n$, set $\langle\sigma| = \phi\left\langle\tilde{\Psi}^{(n)}\right| \in \Omega^1(S^2)^{|n|+1}$, so to have $\langle\sigma| = \langle\sigma|\,\tilde{\mathfrak{p}}^{(n)}$. To write the inverse mapping, consider $\langle\sigma| \in \Omega^1(S^2)^{|n|+1}\tilde{\mathfrak{p}}^{(n)}$ with components $\langle\sigma|_\mu \in \Omega^1(S^2)$ in the bra-vector notation, satisfying $\langle\sigma|_\mu\,\tilde{\mathfrak{p}}^{(n)}_{\mu\nu} = \langle\sigma|_\nu$. Define $\phi = \left\langle\sigma, \tilde{\Psi}^{(n)}\right\rangle$: it is then straightforward to recover that $\phi \in \mathfrak{L}^{(1)}_n$ and that $\langle\sigma|_\mu = \phi\left\langle\tilde{\Psi}^{(n)}\right|_\mu$.

The same path can be followed to analyse the higher order forms. One has $L_z(\tilde{\omega}_- \wedge \tilde{\omega}_+) = 0$, so the $C^\infty(S^2)$-bimodule of horizontal $\rho_{(n)}(\mathrm{U}(1))$-equivariant 2-forms of the Hopf bundle is given by

$$\mathfrak{L}^{(2)}_n = \{\phi = \phi'''\tilde{\omega}_- \wedge \tilde{\omega}_+ : \phi''' \in \mathfrak{L}^{(0)}_n\} \simeq \Omega^2(S^2) \otimes_{C^\infty(S^2)} \mathfrak{L}^{(0)}_n \tag{2.53}$$

for any $n \in \mathbb{Z}$. It is clear that for $r = 2$ the isomorphism in (2.45) can be written as:

$$\Gamma^{(2)}(S^2, \mathcal{E}_n) \simeq \Omega^2(S^2)^{|n|+1} \cdot \tilde{\mathfrak{p}}^{(n)} \quad \leftrightarrow \quad \mathfrak{L}^{(2)}_n \simeq \Omega^2(S^2) \otimes_{C^\infty(S^2)} \mathfrak{L}^{(0)}_n,$$
$$\langle\sigma| = \phi\left\langle\tilde{\Psi}^{(n)}\right| \quad \leftrightarrow \quad \phi = \left\langle\sigma, \tilde{\Psi}^{(n)}\right\rangle. \tag{2.54}$$

The most natural choice of a connection, compatible with the local trivialisation, is given via the definition, as a $\mathbb{C}$-valued connection 1-form, of

$$\omega = \frac{i}{2}\tilde{\omega}_z = (u^*\mathrm{d}u + v^*\mathrm{d}v). \tag{2.55}$$

It globally – i.e. trivialisation independent – selects the horizontal part of the tangent space as the left $C^\infty(S^3)$-module $H^{(\omega)}(S^3) \subset \mathfrak{X}(S^3) = \{L_\pm\}$ since $\omega(L_\pm) = 0$. On the basis of left invariant vector fields the horizontal projection acts as $L^{(\omega)}_\pm = L_\pm$, $L^{(\omega)}_z = 0$.

2.4. A Laplacian operator on the base manifold S^2. The canonical isomorphism expressed in (2.3) allows to recover the exterior algebra $\Omega(S^2)$ on the basis of the Hopf bundle as the set of horizontal forms in $\Omega(S^3)$ which are also invariant for the right principal action of the gauge group $U(1)$. Recalling the definition of the $C^\infty(S^2)$-bimodules of $\rho_{(n)}(U(1))$-equivariant forms given in (2.51) and (2.53), it is possible to identify

$$\begin{aligned}
&\Omega^0(S^2) = C^\infty(S^2) \simeq \mathfrak{L}_0^{(0)};\\
&\Omega^1(S^2) \simeq \mathfrak{L}_0^{(1)} = \{\phi = \phi'\tilde{\omega}_- + \phi''\tilde{\omega}_+ : \phi' \in \mathfrak{L}_{-2}^{(0)}, \phi'' \in \mathfrak{L}_2^{(0)}\};\\
&\Omega^2(S^2) \simeq \mathfrak{L}_0^{(2)} = \{f\tilde{\omega}_- \wedge \tilde{\omega}_+ : f \in \mathfrak{L}_0^{(0)} = C^\infty(S^2)\},
\end{aligned} \tag{2.56}$$

where all such identifications are $C^\infty(S^2)$-bimodule isomorphisms.

On the basis manifold $S^2 \simeq SU(2)/U(1) = \pi(SU(2))$, whose trivialisation is given in (2.40), consider the metric

$$\check{g} = 2\alpha\,(\tilde{\omega}_- \otimes \tilde{\omega}_+ + \tilde{\omega}_+ \otimes \tilde{\omega}_-) \tag{2.57}$$

and its associated volume $\check{\theta} = \alpha\,\tilde{\omega}_x \wedge \tilde{\omega}_y = 2i\alpha\,\tilde{\omega}_- \wedge \tilde{\omega}_+ = i_{L_z}\theta$ in terms of the volume on the group manifold S^3. The corresponding Hodge duality is the $C^\infty(S^2)$-linear map $\star : \Omega^k(S^2) \to \Omega^{2-k}(S^2)$ given by:

$$\begin{aligned}
&\star(\check{\theta}) = 1, && \star(1) = \check{\theta},\\
&\star(\phi''\tilde{\omega}_+) = i\phi''\tilde{\omega}_+, && \star(\phi'\tilde{\omega}_-) = -i\phi'\tilde{\omega}_-,
\end{aligned} \tag{2.58}$$

with $\phi' \in \mathfrak{L}_{-2}^{(0)}$ and $\phi'' \in \mathfrak{L}_2^{(0)}$. The Laplacian operator on S^2 can be now evaluated:

$$\Box_{S^2} f = \star\mathrm{d}\star\mathrm{d}f = \frac{1}{2\alpha}(L_+L_- + L_-L_+)f. \tag{2.59}$$

It corresponds to the action of the Laplacian $\Box_{S^3}$ (2.28) on the subalgebra algebra $C^\infty(S^2) \subset C^\infty(S^3)$.

REMARK 2.1. *Given the Hodge duality* (2.58), *the expression* (2.31) *defines a bilinear symmetric tensor* $\langle\ ,\ \rangle_{S^2} : \Omega^k(S^2) \times \Omega^k(S^2) \to C^\infty(S^2)$ *(with* $k = 0, 1, 2$*):*

$$\langle \xi, \xi' \rangle_{S^2}\,\check{\theta} = \xi \wedge (\star\xi'), \tag{2.60}$$

for any $\xi, \xi' \in \Omega^k(S^2)$. *Its non zero terms are given by:*

$$\begin{aligned}
&\langle 1, 1\rangle_{S^2} = 1;\\
&\langle \phi'\tilde{\omega}_-, \phi''\tilde{\omega}_+\rangle_{S^2} = \langle \phi''\tilde{\omega}_+, \phi'\tilde{\omega}_-\rangle_{S^2} = \phi'\phi''/2\alpha;\\
&\langle \check{\theta}, \check{\theta}\rangle_{S^2} = 1 :
\end{aligned} \tag{2.61}$$

such a tensor coincides with the restriction to the exterior algebra $\Omega(S^2)$ *of the analogue tensor* $\langle\ ,\ \rangle_{S^3}$. *The expression*

$$\langle \xi, \xi' \rangle_{S^2}^{\sim}\,\check{\theta} = \xi'^* \wedge (\star\xi), \tag{2.62}$$

with again $\xi, \xi' \in \Omega^k(S^2)$, *defines a bilinear map on* $\Omega(S^2)$, *which coincides with the restriction of the bilinear map* $\langle\ ,\ \rangle_{S^3}^{\sim}$ *to* $\Omega(S^2)$*:*

$$\begin{aligned}
&\langle 1, 1\rangle_{S^2}^{\sim} = 1;\\
&\langle \phi'\tilde{\omega}_-, \psi'\tilde{\omega}_-\rangle_{S^2}^{\sim} = \frac{1}{2\alpha}\psi'^*\phi' = \langle \phi'\tilde{\omega}_-, \psi'\tilde{\omega}_-\rangle_{S^3}^{\sim},\\
&\langle \phi''\tilde{\omega}_+, \psi''\tilde{\omega}_+\rangle_{S^2}^{\sim} = \frac{1}{2\alpha}\psi'^*\phi' = \langle \phi''\tilde{\omega}_+, \psi''\tilde{\omega}_+\rangle_{S^3}^{\sim};\\
&\langle \check{\theta}, \check{\theta}\rangle_{S^2}^{\sim} = 1 = \langle 2i\alpha\,\tilde{\omega}_- \wedge \tilde{\omega}_+, 2i\alpha\,\tilde{\omega}_- \wedge \tilde{\omega}_+\rangle_{S^3}^{\sim}
\end{aligned} \tag{2.63}$$

for any $\phi', \psi' \in \mathfrak{L}_{-2}^{(0)}$ *and* $\phi'', \psi'' \in \mathfrak{L}_2^{(0)}$.

REMARK 2.2. *Introducing from the volume form* $\check{\theta}$ *an integral* $\int_{\check{\theta}} : \Omega^2(S^2) \mapsto \mathbb{C}$ *with the normalisation* $\int_{\check{\theta}} \theta = 1$, *the bilinear maps in* (2.60) *and* (2.62) *give on the exterior algebra* $\Omega(S^2)$ *a symmetric scalar product and a hermitian inner product, setting:*

$$(\xi;\xi')_{S^2} = \int_{\check{\theta}} \xi \wedge (\star\xi'), \tag{2.64}$$

$$(\xi;\xi')_{\widetilde{S^2}} = \int_{\check{\theta}} \xi'^* \wedge (\star\xi). \tag{2.65}$$

It is clear that they coincide with the restrictions to $\Omega(S^2)$ *of respectively* (2.34) *and* (2.35).

3. The quantum principal Hopf bundle

This section describes a quantum formulation of a Hopf bundle. It starts with a description of the algebraic approach to the theory of differential calculi on Hopf algebras coming from [**38, 21**] and then algebraically presents the geometric structures of a principal bundle.

3.1. Algebraic approach to the theory of differential calculi on Hopf algebras. The first order differential forms on the smooth group manifold $SU(2) \simeq S^3$ have been presented as elements in the space $\mathfrak{X}^*(S^3)$, or more properly as sections of the cotangent bundle $T^*(S^3)$. The set $\Omega^1(S^3) \simeq \mathfrak{X}^*(S^3)$ of 1-forms is a bimodule over $C^\infty(S^3)$, with the exterior derivative d satisfying the basic Leibniz rule $\mathrm{d}(ff') = (\mathrm{d}f)f' + f\mathrm{d}f'$ for any $f, f' \in C^\infty(S^3)$. Moreover, being S^3 a compact manifold, any differential form $\theta \in \Omega^1(S^3)$ is necessarily of the form $\theta = f_k \mathrm{d}f'_k$ (with $k \in \mathbb{N}$).

In an algebraic setting, these properties are a definition. Given a $\mathbb{C}$-algebra with a unit $\mathcal{A}$ and Ω a bimodule over $\mathcal{A}$ with a linear map $\mathrm{d} : \mathcal{A} \to \Omega$, (Ω, d) is defined a first order differential calculus over $\mathcal{A}$ if $\mathrm{d}(ff') = (\mathrm{d}f)f' + f\mathrm{d}f'$ for any $f, f' \in \mathcal{A}$ and if any element $\theta \in \Omega$ can be written as $\theta = \sum_k f_k \mathrm{d}f'_k$ with $f_k, f'_k \in \mathcal{A}$.

For a $\mathbb{C}$-algebra with unit $\mathcal{A}$, any first order differential calculus $(\Omega^1(\mathcal{A}), \mathrm{d})$ on $\mathcal{A}$ can be obtained from the universal calculus $(\Omega^1(\mathcal{A})_{un}, \delta)$. The space of universal 1-forms is the submodule of $\mathcal{A} \otimes \mathcal{A}$ given by $\Omega^1(\mathcal{A})_{un} = \ker(m : \mathcal{A} \otimes \mathcal{A} \to \mathcal{A})$, with $m(a \otimes b) = ab$ the multiplication map. The universal differential $\delta : \mathcal{A} \to \Omega^1(\mathcal{A})_{un}$ is $\delta a = 1 \otimes a - a \otimes 1$. If $\mathcal{N}$ is any sub-bimodule of $\Omega^1(\mathcal{A})_{un}$ with projection $\pi_{\mathcal{N}} : \Omega^1(\mathcal{A})_{un} \to \Omega^1(\mathcal{A}) = \Omega^1(\mathcal{A})_{un}/\mathcal{N}$, then $(\Omega^1(\mathcal{A}), \mathrm{d})$, with $\mathrm{d} := \pi_{\mathcal{N}} \circ \delta$, is a first order differential calculus over $\mathcal{A}$ and any such a calculus can be obtained in this way. The projection $\pi_{\mathcal{N}} : \Omega^1(\mathcal{A})_{un} \to \Omega^1(\mathcal{A})$ is $\pi_{\mathcal{N}}(\sum_i a_i \otimes b_i) = \sum_i a_i \mathrm{d}b_i$ with associated subbimodule $\mathcal{N} = \ker \pi$.

The concept of action of a group on a manifold is algebraically dualised via the notion of coaction of a Hopf algebra $\mathcal{H}$ on an algebra $\mathcal{A}$: if the algebra $\mathcal{A}$ is covariant for the coaction of a quantum group $\mathcal{H} = (\mathcal{H}, \Delta, \varepsilon, S)$, one has a notion of covariant calculi on $\mathcal{A}$ as well, thus translating the idea of invariance of the differential calculus on a manifold for the action of a group. Then, let $\mathcal{A}$ be a (right, say) $\mathcal{H}$-comodule algebra, with a right coaction $\Delta_R : \mathcal{A} \to \mathcal{A} \otimes \mathcal{H}$ which is also an algebra map. In order to state the covariance of the calculus $(\Omega^1(\mathcal{A}), \mathrm{d})$ one needs to extend the coaction of $\mathcal{H}$. A map $\Delta_R^{(1)} : \Omega^1(\mathcal{A}) \to \Omega^1(\mathcal{A}) \otimes \mathcal{H}$ is defined by the requirement

$$\Delta_R^{(1)}(\mathrm{d}f) = (\mathrm{d} \otimes \mathrm{id})\Delta_R(f)$$

and bimodule structure governed by

$$\Delta_R^{(1)}(f\mathrm{d}f') = \Delta_R(f)\Delta_R^{(1)}(\mathrm{d}f'),$$
$$\Delta_R^{(1)}((\mathrm{d}f)f') = \Delta_R^{(1)}(\mathrm{d}f)\Delta_R(f').$$

The calculus is said to be right covariant if it happens that

$$(\mathrm{id} \otimes \Delta)\Delta_R^{(1)} = (\Delta_R^{(1)} \otimes \mathrm{id})\Delta_R^{(1)}$$

and

$$(\mathrm{id} \otimes \varepsilon)\Delta_R^{(1)} = 1.$$

A calculus is right covariant if and only if for the corresponding bimodule $\mathcal{N}$ it is verified that $\Delta_R^{(1)}(\mathcal{N}) \subset \mathcal{N} \otimes \mathcal{H}$, where $\Delta_R^{(1)}$ is defined on $\mathcal{N}$ by formulæ as above with the universal derivation δ replacing the derivation d:

$$\Delta_R^{(1)}(\delta f) = (\delta \otimes \mathrm{id})\Delta_R(f). \tag{3.1}$$

Differential calculi on a quantum group $\mathcal{H} = (\mathcal{H}, \Delta, \varepsilon, S)$ were studied in [**38**]. As a quantum group consider a Hopf $*$-algebra with an invertible antipode: the coproduct $\Delta : \mathcal{H} \rightarrow \mathcal{H} \otimes \mathcal{H}$ defines both a right and a left coaction of $\mathcal{H}$ on itself:

$$\Delta_R^{(1)}(\mathrm{d}h) = (\mathrm{d} \otimes 1)\Delta(h),$$
$$\Delta_L^{(1)}(\mathrm{d}h) = (1 \otimes \mathrm{d})\Delta(h). \tag{3.2}$$

Right and left covariant calculi on $\mathcal{H}$ will be defined as before. Right covariance of the calculus implies that $\Omega^1(\mathcal{H})$ has a module basis $\{\eta_a\}$ of right invariant 1-forms, that is 1-forms for which

$$\Delta_R^{(1)}(\eta_a) = \eta_a \otimes 1,$$

and left covariance of a calculus similarly implies that $\Omega^1(\mathcal{H})$ has a module basis $\{\omega_a\}$ of left invariant 1-forms, that is 1-forms for which $\Delta_L^{(1)}(\omega_a) = 1 \otimes \omega_a$. In addition one has the notion of a bicovariant calculus, namely a both left and right covariant calculus, satisfying the compatibility condition:

$$(\mathrm{id} \otimes \Delta_R^{(1)}) \circ \Delta_L^{(1)} = (\Delta_L^{(1)} \otimes \mathrm{id}) \circ \Delta_R^{(1)}.$$

Given the bijection

$$r : \mathcal{H} \otimes \mathcal{H} \rightarrow \mathcal{H} \otimes \mathcal{H}, \qquad r(h \otimes h') = (h \otimes 1)\Delta(h'), \tag{3.3}$$

one proves that $r(\Omega^1(\mathcal{H})_{un}) = \mathcal{H} \otimes \ker \varepsilon$. Then, if $\mathcal{Q} \subset \ker \varepsilon$ is a right ideal of $\ker \varepsilon$, the inverse image $\mathcal{N}_{\mathcal{Q}} = r^{-1}(\mathcal{H} \otimes \mathcal{Q})$ is a sub-bimodule contained in $\Omega^1(\mathcal{H})_{un}$. The differential calculus defined by such a bimodule, $\Omega^1(\mathcal{H}) := \Omega^1(\mathcal{H})_{un}/\mathcal{N}_{\mathcal{Q}}$, is left-covariant, and any left-covariant differential calculus can be obtained in this way. Bicovariant calculi are in one to one correspondence with right ideals $\mathcal{Q} \subset \ker \varepsilon$ which are in addition stable under the right adjoint coaction Ad of $\mathcal{H}$ onto itself, that is $\mathrm{Ad}(\mathcal{Q}) \subset \mathcal{Q} \otimes \mathcal{H}$. Explicitly, one has $\mathrm{Ad} = (\mathrm{id} \otimes m)\,(\tau \otimes \mathrm{id})\,(S \otimes \Delta)\,\Delta$, with τ the flip operator, or $\mathrm{Ad}(h) = h_{(2)} \otimes \big(S(h_{(1)})h_{(3)}\big)$ using the Sweedler notation $\Delta h =: h_{(1)} \otimes h_{(2)}$ with summation understood, and higher numbers for iterated coproducts.

Given the $*$-structure on $\mathcal{H}$, a first order differential calculus $(\Omega^1(\mathcal{H}), \mathrm{d})$ on $\mathcal{H}$ is called a $*$-calculus if there exists an anti-linear involution $* : \Omega^1(\mathcal{H}) \rightarrow \Omega^1(\mathcal{H})$ such that $(h_1(\mathrm{d}h)h_2)^* = h_2^*(d(h^*))h_1^*$ for any $h, h_1, h_2 \in \mathcal{H}$. A left covariant first order differential calculus is [38] a $*$-calculus if and only if $(S(Q))^* \in \mathcal{Q}$ for any $Q \in \mathcal{Q}$. In such a case the $*$-structure is also compatibe with the left coaction $\Delta_L^{(1)}$ of $\mathcal{H}$ on $\Omega^1(\mathcal{H})$: $\Delta_L^{(1)}(dh^*) = (\Delta^{(1)}(\mathrm{d}h))^*$.

The ideal $\mathcal{Q}$ also determines the tangent space of the calculus. This is the complex vector space of elements $\{X_a\}$ in $\mathcal{H}'$ defined by

$$\mathcal{X}_{\mathcal{Q}} := \{X \in \mathcal{H}' \ : \ X(1) = 0, \ X(Q) = 0, \ \forall Q \in \mathcal{Q}\},$$

whose dimension, which coincides with the dimension of the calculus, is given by $\dim \mathcal{X}_{\mathcal{Q}} = \dim(\ker \varepsilon_{\mathcal{H}}/\mathcal{Q})$. If the vector space $\mathcal{X}_{\mathcal{Q}}$ is finite dimensional, then [21] its elements X_a belong to the dual Hopf algebra $\mathcal{H}^o \subset \mathcal{H}'$. Given an infinite dimensional Hopf $*$-algebra $\mathcal{H}$ and the set $\mathcal{H}'$ of its linear functionals, the set $\mathcal{H}' \otimes \mathcal{H}'$ is a linear subspace of $(\mathcal{H} \otimes \mathcal{H})'$ obtained via the identification $X \otimes Y \in \mathcal{H}' \otimes \mathcal{H}'$ with the linear functional on $\mathcal{H} \otimes \mathcal{H}$ determined by $(X \otimes Y)(h_1 \otimes h_2) = X(h_1)Y(h_2)$. For any $X \in \mathcal{H}'$ consider ΔX as the element in $(\mathcal{H} \otimes \mathcal{H})'$ defined by $\Delta X(h_1 \otimes h_2) = X(h_1 h_2)$. The space $\mathcal{H}^o \subset \mathcal{H}'$ denotes the set of linear functionals $X \in \mathcal{H}'$ for which $\Delta X \in \mathcal{H}' \otimes \mathcal{H}'$, i.e. there exist functionals $\{Y_a\}, \{Z_b\} \in \mathcal{H}'$ – with $a, b = 1, \ldots, r$; $r \in \mathbb{N}$ – such that

$$X(h_1 h_2) = \sum_{i=1}^{r} Y_i(h_1) Z_i(h_2) \qquad \Leftrightarrow \qquad \Delta X = \sum_{i=1}^{r} Y_i \otimes Z_i.$$

Dualising the structure maps from $\mathcal{H}$ to $\mathcal{H}'$ via:

$$\begin{aligned} X_1 X_2(h) &= X_1(h_{(1)}) X_2(h_{(2)}), \\ \varepsilon_{\mathcal{H}'}(X) &= X(1), \\ (S_{\mathcal{H}'}(X))(h) &= X(S(h)), \\ \mathbf{1}_{\mathcal{H}'}(h) &= \varepsilon(h), \\ X^*(h) &= \overline{X(S(h)^*)} \end{aligned} \tag{3.4}$$

for any $X, X_1, X_2 \in \mathcal{H}'$ and $h, h_1, h_2 \in \mathcal{H}$, the dual $\mathcal{H}^o$ is proved to be the largest Hopf $*$-subalgebra contained in $\mathcal{H}'$. The presence of a $*$-structure on a first order left-covariant differential calculus can be translated into a condition on the quantum tangent space: $(\Omega^1(\mathcal{H}), \mathrm{d})$ is a left-covariant differential calculus if and only if $\mathcal{X}_{\mathcal{Q}}^* \subset \mathcal{X}_{\mathcal{Q}}$ with $\mathcal{H}'$ endowed by the complex structure in (3.4).

The exterior derivative can be written as:

$$\mathrm{d}h := \sum\nolimits_a (X_a \triangleright h)\, \omega_a, \tag{3.5}$$

in terms of the canonical left and right $\mathcal{H}'$-module algebra structure on $\mathcal{H}$ given by [37]:

$$\begin{aligned} X \triangleright h &:= h_{(1)}(X(h_{(2)})), \\ h \triangleleft X &:= X(h_{(1)}) h_{(2)}. \end{aligned} \tag{3.6}$$

Left and right actions mutually commute:

$$(X_1 \triangleright h) \triangleleft X_2 = X_1 \triangleright (h \triangleleft X_2),$$

and the $*$-structures are compatible with both actions:

$$\begin{aligned} X \triangleright h^* &= ((S(X))^* \triangleright h)^*, \\ h^* \triangleleft X &= (h \triangleleft (S(X))^*)^*, \qquad \forall X \in \mathcal{H}^o, \ h \in \mathcal{H}. \end{aligned}$$

Given the two Hopf $*$-algebras $\mathcal{H} = (\mathcal{H}, \Delta, \varepsilon, S)$ and $\mathcal{U} = (\mathcal{U}, \Delta_{\mathcal{U}}, \varepsilon_{\mathcal{U}}, S_{\mathcal{U}})$, they can be dually paired. This duality is expressed by the existence of a bilinear map $\langle\ ,\ \rangle : \mathcal{U} \times \mathcal{H} \to \mathbb{C}$ such that:

$$\begin{aligned}
&\langle \Delta_{\mathcal{U}}(U), h_1 \otimes h_2 \rangle = \langle U, h_1 h_2 \rangle, \\
&\langle U_1 U_2, h \rangle = \langle U_1 \otimes U_2, \Delta(h) \rangle, \\
&\langle U, 1 \rangle = \varepsilon_{\mathcal{U}}(U), \\
&\langle 1, h \rangle = \varepsilon(h)
\end{aligned} \tag{3.7}$$

for any $U_a \in \mathcal{U}(\mathcal{H})$ and $h_b \in \mathcal{H}$. The pairing is also required to be compatible with $*$-structures:

$$\begin{aligned}
&\langle U^*, h \rangle = \overline{\langle U, (S(h))^* \rangle}, \\
&\langle U, h^* \rangle = \overline{\langle (S_{\mathcal{U}}(U))^*, h \rangle}.
\end{aligned} \tag{3.8}$$

Such a dual pairing has the property that $\langle S_{\mathcal{U}}(U), h \rangle = \langle U, S(h) \rangle$. A dual pairing can be defined on the generators and then extended to the whole algebras following the relations (3.7): it is called non degenerate if the condition $\langle U, h \rangle = 0$ for any $h \in \mathcal{H}$ implies $U = 0$, and if $\langle U, h \rangle = 0$ for any $U \in \mathcal{U}$ implies $h = 0$.

It comes from this analysis out that via a non degenerate dual pairing between the two Hopf algebras $\mathcal{H}$ and $\mathcal{U}$, it is possible to regard $\mathcal{U}$ as a Hopf $*$-subalgebra of $\mathcal{H}^o$, and $\mathcal{H}$ as a Hopf $*$-subalgebra of $\mathcal{U}^o$, after the identifications $U(h) = h(U) = \langle U, h \rangle$ for any $U \in \mathcal{U}$ and $h \in \mathcal{H}$. A further comparison among relations (3.4) and (3.7) shows that $\mathcal{H}$ and $\mathcal{H}^o$ are dually paired in a natural way, with a pairing which is non degenerate if $\mathcal{H}^o$ separates the points in $\mathcal{H}$.

The derivation nature of elements in $\mathcal{X}_Q$ is expressed by their coproduct,

$$\Delta(X_a) = 1 \otimes X_a + \sum\nolimits_b X_b \otimes f_{ba},$$

with the elements $f_{ab} \in \mathcal{H}^o$ having specific properties [**38**]:

$$\begin{aligned}
&\Delta(f_{ab}) = f_{ac} \otimes f_{cb}, \\
&\varepsilon(f_{ab}) = \delta_{ab}, \\
&S(f_{ab}) f_{bc} = f_{ab} S(f_{bc}) = \delta_{ac}.
\end{aligned}$$

These elements also control the commutation relation between the basis 1-forms and elements of $\mathcal{H}$:

$$\begin{aligned}
\omega_a h &= \sum\nolimits_b (f_{ab} \triangleright h) \omega_b, \\
h \omega_a &= \sum\nolimits_b \omega_b \left((S^{-1}(f_{ab})) \triangleright h \right) \qquad \text{for} \quad h \in \mathcal{H}.
\end{aligned}$$

For a left covariant differential calculus, the elements $X_a \in \mathcal{X}_Q$ play the role which is classically played by the vectors tangent to a Lie group manifold at the group identity: the first of equations (3.6) transforms them into the analogue of left invariant derivations on the Hopf algebra of functions on the group. Their dual forms ω_a play the role of the left invariant one forms. For a bicovariant differential calculus it is possible to define a basis of the bimodule of 1-forms which are right invariant. The right coaction of $\mathcal{H}$ on $\Omega^1(\mathcal{H})$ defines a matrix:

$$\Delta_R^{(1)}(\omega_a) = \omega_b \otimes J_{ba} \tag{3.9}$$

where $J_{ab} \in \mathcal{H}$. This matrix is invertible, since $S(J_{ab})J_{bc} = \delta_{ac}$ and $J_{ab}S(J_{bc}) = \delta_{ac}$; it satisfies the properties $\Delta(J_{ab}) = J_{ac} \otimes J_{cb}$, $\varepsilon(J_{ab}) = \delta_{ab}$ and can be used to define a set of 1-forms:

$$\eta_a = \omega_b S(J_{ba}) \qquad \Leftrightarrow \qquad \eta_a J_{ab} = \omega_b \tag{3.10}$$

which are right invariant:

$$\Delta_R^{(1)}(\eta_a) = \eta_a \otimes 1. \tag{3.11}$$

On the basis of right invariant 1-forms, the exterior derivative operator acquires the form:

$$dh = \eta_a(h \triangleleft Y_a) \tag{3.12}$$

where $Y_a = -S^{-1}(X_a)$ are the analogue of the derivations associated to right invariant vector fields. Equation (2.24) is then represented, in an algebraic approach to the theory of differential calculi, by (3.5) and (3.12). The derivation nature of Y_a as well as the commutation relation between the basis of right invariant 1-forms and elements of $\mathcal{H}$ are ruled by the same elements $f_{ab} \in \mathcal{U}(\mathcal{H})$ [**2**]:

$$\begin{aligned}
&\Delta(Y_a) = Y_a \otimes 1 + \sum\nolimits_b S^{-1}(f_{ba}) \otimes Y_b \\
&\eta_a h = (h \triangleleft S^{-2}(f_{ab}))\eta_b, \\
&h\eta_a = \eta_b(h \triangleleft (S^{-1}(f_{ab})).
\end{aligned}$$

3.2. Quantum principal bundles. An algebraic formalisation of the geometric structures of a principal bundle has been introduced in [**6**] and refined in [**7**]. A slightly different formulation of such a structure is in [**12, 13**]; an interesting comparison between the two approaches is in [**14**].

Following [**6**], consider as a total space an algebra $\mathcal{P}$ (with multiplication $m : \mathcal{P} \otimes \mathcal{P} \to \mathcal{P}$) and as structure group a Hopf algebra $\mathcal{H}$. Thus $\mathcal{P}$ is a right $\mathcal{H}$-comodule algebra with coaction $\Delta_R : \mathcal{P} \to \mathcal{P} \otimes \mathcal{H}$. The subalgebra of the right coinvariant elements, $\mathcal{B} = \mathcal{P}^{\mathcal{H}} = \{p \in \mathcal{P} : \Delta_R p = p \otimes 1\}$, is the base space of the bundle. At the 'topological level' the principality of the bundle is the requirement of exactness of the sequence:

$$0 \to \mathcal{P}\left(\Omega^1(\mathcal{B})_{un}\right)\mathcal{P} \to \Omega^1(\mathcal{P})_{un} \xrightarrow{\chi} \mathcal{P} \otimes \ker \varepsilon_{\mathcal{H}} \to 0 \tag{3.13}$$

with $\Omega^1(\mathcal{P})_{un}$ and $\Omega^1(\mathcal{B})_{un}$ the universal calculi and the map χ defined by

$$\chi : \mathcal{P} \otimes \mathcal{P} \to \mathcal{P} \otimes \mathcal{H}, \qquad \chi := (m \otimes \mathrm{id})(\mathrm{id} \otimes \Delta_R), \tag{3.14}$$

or $\chi(p' \otimes p) = p'\Delta_R(p)$. The exactness of this sequence is equivalent to the requirement that the analogous 'canonical map' $\mathcal{P} \otimes_{\mathcal{B}} \mathcal{P} \to \mathcal{P} \otimes \mathcal{H}$ (defined as the formula above) is an isomorphism. This is the definition that the inclusion $\mathcal{B} \hookrightarrow \mathcal{P}$ be a Hopf-Galois extension [**33**].

REMARK 3.1. *The surjectivity of the map χ appears as the dual translation of the classical condition that the action of the structure group on the total space of the principal bundle is free. In the classical setting described in section 2, given the principal bundle $(\mathcal{P}, \mathrm{K}, [\mathcal{M}], \pi)$, the condition that the right principal action r_k is free can be written as the injectivity of the map:*

$$P \times G \to P \times_M P, \qquad (p, k) \mapsto (p, \mathrm{r}_k(p)),$$

whose dualisation is the condition of the surjectivity of the map χ.

With differential calculi on both the total algebra $\mathcal{P}$ and the structure Hopf algebra $\mathcal{H}$ one needs compatibility conditions that eventually lead to an exact sequence like in (3.13) with the calculi at hand replacing the universal ones. Then, let $(\Omega^1(\mathcal{P}), \mathrm{d})$ be a $\mathcal{H}$-covariant differential calculus on $\mathcal{P}$ given via the subbimodule $\mathcal{N}_{\mathcal{P}} \in (\Omega^1(\mathcal{P})_{un})$, and $(\Omega^1(\mathcal{H}), \mathrm{d})$ a bicovariant differential calculus on $\mathcal{H}$ given via the Ad-invariant right ideal $\mathcal{Q}_{\mathcal{H}} \in \ker \varepsilon_{\mathcal{H}}$. In order to extend the coaction Δ_R of $\mathcal{H}$ on $\mathcal{P}$ to a coaction of $\mathcal{H}$ on $\Omega^1(\mathcal{P})$, one requires $\Delta_R(\mathcal{N}_{\mathcal{P}}) \subset \mathcal{N}_{\mathcal{P}} \otimes \mathcal{H}$. The coaction Δ_R of $\mathcal{H}$ on $\mathcal{N}_{\mathcal{P}} \subset \mathcal{P} \otimes \mathcal{P}$ is understood as a usual coaction of a Hopf algebra on a tensor product of its comodule algebras, i.e.

$$\Delta_R = (\mathrm{id} \otimes \mathrm{id} \otimes \cdot) \circ (\mathrm{id} \otimes \tau\, \mathrm{id}) \circ (\Delta_R \otimes \Delta_R).$$

The condition $\Delta_R(\mathcal{N}_{\mathcal{P}}) \subset \mathcal{N}_{\mathcal{P}} \otimes \mathcal{H}$ is equivalent to the condition (3.1).

The compatibility of the calculi are then the requirements that $\chi(\mathcal{N}_{\mathcal{P}}) \subseteq \mathcal{P} \otimes \mathcal{Q}_{\mathcal{H}}$ and that the map $\sim_{\mathcal{N}_{\mathcal{P}}} : \Omega^1(\mathcal{P}) \to \mathcal{P} \otimes (\ker \varepsilon_{\mathcal{H}} / \mathcal{Q}_{\mathcal{H}})$, defined by the diagram

$$\begin{array}{ccc} \Omega^1(\mathcal{P})_{un} & \xrightarrow{\pi_{\mathcal{N}}} & \Omega^1(\mathcal{P}) \\ \downarrow \chi & & \downarrow \sim_{\mathcal{N}_{\mathcal{P}}} \\ \mathcal{P} \otimes \ker \varepsilon_{\mathcal{H}} & \xrightarrow{\mathrm{id} \otimes \pi_{\mathcal{Q}_{\mathcal{H}}}} & \mathcal{P} \otimes (\ker \varepsilon_{\mathcal{H}} / \mathcal{Q}_{\mathcal{H}}) \end{array} \tag{3.15}$$

(with $\pi_{\mathcal{N}}$ and $\pi_{\mathcal{Q}_{\mathcal{H}}}$ the natural projections) is surjective and has kernel

$$\ker \sim_{\mathcal{N}_{\mathcal{P}}} = \mathcal{P} \Omega^1(\mathcal{B}) \mathcal{P} =: \Omega^1_{\mathrm{hor}}(\mathcal{P}). \tag{3.16}$$

Here $\Omega^1(\mathcal{B}) = \mathcal{B} \mathrm{d} \mathcal{B}$ is the space of nonuniversal 1-forms on $\mathcal{B}$ associated to the bimodule $\mathcal{N}_{\mathcal{B}} := \mathcal{N}_{\mathcal{P}} \cap \Omega^1(\mathcal{B})_{un}$. These conditions ensure the exactness of the sequence:

$$0 \to \mathcal{P} \Omega^1(\mathcal{B}) \mathcal{P} \to \Omega_1(\mathcal{P}) \xrightarrow{\sim_{\mathcal{N}_{\mathcal{P}}}} \mathcal{P} \otimes (\ker \varepsilon_{\mathcal{H}} / \mathcal{Q}_{\mathcal{H}}) \to 0. \tag{3.17}$$

The condition $\chi(\mathcal{N}_{\mathcal{P}}) \subseteq \mathcal{P} \otimes \mathcal{Q}_{\mathcal{H}}$ is needed to have a well defined map $\sim_{\mathcal{N}_{\mathcal{P}}}$: with all conditions for a quantum principal bundle $(\mathcal{P}, \mathcal{B}, \mathcal{H}; \mathcal{N}_{\mathcal{P}}, \mathcal{Q}_{\mathcal{H}})$ satisfied, this inclusion implies the equality $\chi(\mathcal{N}_{\mathcal{P}}) = \mathcal{P} \otimes \mathcal{Q}_{\mathcal{H}}$. Moreover, if $(\mathcal{P}, \mathcal{B}, \mathcal{H})$ is a quantum principal bundle with the universal calculi, the equality $\chi(\mathcal{N}_{\mathcal{P}}) = \mathcal{P} \otimes \mathcal{Q}_{\mathcal{H}}$ ensures that $(\mathcal{P}, \mathcal{B}, \mathcal{H}; \mathcal{N}_{\mathcal{P}}, \mathcal{Q}_{\mathcal{H}})$ is a quantum principal bundle with the corresponding nonuniversal calculi.

Elements in the quantum tangent space $\mathcal{X}_{\mathcal{Q}_{\mathcal{H}}}(\mathcal{H})$ giving the calculus on the structure quantum group $\mathcal{H}$ act on $\ker \varepsilon_{\mathcal{H}} / \mathcal{Q}_{\mathcal{H}}$ via the pairing $\langle \cdot, \cdot \rangle$ between $\mathcal{H}^o$ and $\mathcal{H}$. Then, with each $\xi \in \mathcal{X}_{\mathcal{Q}_{\mathcal{H}}}(\mathcal{H})$ one defines a map

$$\tilde{\xi} : \Omega^1(\mathcal{P}) \to \mathcal{P}, \qquad \tilde{\xi} := (\mathrm{id} \otimes \xi) \circ (\sim_{\mathcal{N}_{\mathcal{P}}}) \tag{3.18}$$

and declare a 1-form $\omega \in \Omega^1(\mathcal{P})$ to be horizontal iff $\tilde{\xi}(\omega) = 0$, for all elements $\xi \in \mathcal{X}_{\mathcal{Q}_{\mathcal{H}}}(\mathcal{H})$. The collection of horizontal 1-forms is easily seen to coincide with $\Omega^1_{\mathrm{hor}}(\mathcal{P})$ in (3.16).

3.3. A topological quantum Hopf bundle. As a step toward a quantum formulation of the classical Hopf bundle $\pi : S^3 \to S^2$ this section will describe, following [**24**], a topological U(1)-bundle [**6**] over the standard Podleś sphere S^2_q [**30**], with total space the manifold of the quantum group $SU_q(2)$.

3.3.1. *The algebras.* The coordinate algebra $\mathcal{A}(\mathrm{SU}_q(2))$ of the quantum group $\mathrm{SU}_q(2)$ is the $*$-algebra generated by a and c, with relations

$$ac = qca \quad ac^* = qc^*a \quad cc^* = c^*c,$$
$$a^*a + c^*c = aa^* + q^2cc^* = 1. \tag{3.19}$$

The deformation parameter $q \in \mathbb{R}$ is taken in the interval $0 < q < 1$, since for $q > 1$ one gets isomorphic algebras; at $q = 1$ one recovers the commutative coordinate algebra on the group manifold SU(2). The Hopf algebra structure for $\mathcal{A}(\mathrm{SU}_q(2))$ is given by the coproduct:

$$\Delta \begin{bmatrix} a & -qc^* \\ c & a^* \end{bmatrix} = \begin{bmatrix} a & -qc^* \\ c & a^* \end{bmatrix} \otimes \begin{bmatrix} a & -qc^* \\ c & a^* \end{bmatrix},$$

antipode:

$$S \begin{bmatrix} a & -qc^* \\ c & a^* \end{bmatrix} = \begin{bmatrix} a^* & c^* \\ -qc & a \end{bmatrix},$$

and counit:

$$\epsilon \begin{bmatrix} a & -qc^* \\ c & a^* \end{bmatrix} = \begin{bmatrix} 1 & 0 \\ 0 & 1 \end{bmatrix}.$$

The quantum universal envelopping algebra $\mathcal{U}_q(\mathfrak{su}(2))$ is the Hopf $*$-algebra generated as an algebra by four elements K, K^{-1}, E, F with $KK^{-1} = 1$ and subject to relations:

$$K^{\pm}E = q^{\pm}EK^{\pm},$$
$$K^{\pm}F = q^{\mp}FK^{\pm},$$
$$[E, F] = \frac{K^2 - K^{-2}}{q - q^{-1}}. \tag{3.20}$$

The $*$-structure is

$$K^* = K, \qquad E^* = F, \qquad F^* = E,$$

and the Hopf algebra structure is provided by coproduct:

$$\Delta(K^{\pm}) = K^{\pm} \otimes K^{\pm},$$
$$\Delta(E) = E \otimes K + K^{-1} \otimes E,$$
$$\Delta(F) = F \otimes K + K^{-1} \otimes F;$$

antipode:

$$S(K) = K^{-1}, \qquad S(E) = -qE, \qquad S(F) = -q^{-1}F;$$

and a counit:

$$\varepsilon(K) = 1, \qquad \varepsilon(E) = \varepsilon(F) = 0.$$

From the relations (3.20), the quadratic quantum Casimir element:

$$C_q := \frac{qK^2 - 2 + q^{-1}K^{-2}}{(q - q^{-1})^2} + FE - \tfrac{1}{4} \tag{3.21}$$

generates the centre of $\mathcal{U}_q(\mathfrak{su}(2))$. The irreducible finite dimensional $*$-representations σ_J of $\mathcal{U}_q(\mathfrak{su}(2))$ (see e.g. [**25**]) are labelled by nonnegative half-integers $J \in \frac{1}{2}\mathbb{N}$ (the spin); they are given by[2]

$$\begin{aligned}\sigma_J(K)\,|J,m\rangle &= q^m\,|J,m\rangle\,,\\ \sigma_J(E)\,|J,m\rangle &= \sqrt{[J-m][J+m+1]}\,|J,m+1\rangle\,,\\ \sigma_J(F)\,|J,m\rangle &= \sqrt{[J-m+1][J+m]}\,|J,m-1\rangle\,,\end{aligned} \tag{3.23}$$

where the vectors $|J,m\rangle$, for $m = J, J-1, \ldots, -J+1, -J$, form an orthonormal basis for the $(2J+1)$-dimensional, irreducible $\mathcal{U}_q(\mathfrak{su}(2))$-module V_J, and the brackets denote the q-number. Moreover, σ_J is a $*$-representation of $\mathcal{U}_q(\mathfrak{su}(2))$, with respect to the hermitian scalar product on V_J for which the vectors $|J,m\rangle$ are orthonormal. In each representation V_J, the Casimir (3.21) is a multiple of the identity with constant given by:

$$C_q^{(J)} = [J+\tfrac{1}{2}]^2 - \tfrac{1}{4}. \tag{3.24}$$

The Hopf algebras $\mathcal{U}_q(\mathfrak{su}(2))$ and $\mathcal{A}(\mathrm{SU}_q(2))$ are dually paired. The bilinear mapping $\langle\cdot,\cdot\rangle : \mathcal{U}_q(\mathfrak{su}(2)) \times \mathcal{A}(\mathrm{SU}_q(2)) \mapsto \mathbb{C}$ compatible with the $*$-structures, is set on the generators by:

$$\begin{aligned}&\langle K,a\rangle = q^{-1/2}, \quad \langle K^{-1},a\rangle = q^{1/2},\\ &\langle K,a^*\rangle = q^{1/2}, \quad \langle K^{-1},a^*\rangle = q^{-1/2},\\ &\langle E,c\rangle = 1, \quad \langle F,c^*\rangle = -q^{-1},\end{aligned} \tag{3.25}$$

with all other couples of generators pairing to 0. Since the deformation parameter q runs in the real interval range $]0,1[$, this pairing is proved [**21**] to be non degenerate. The canonical left and right actions of $\mathcal{U}_q(\mathfrak{su}(2))$ on $\mathcal{A}(\mathrm{SU}_q(2))$ can be recovered by:

(3.26)

$$\begin{array}{lll} K^{\pm} \triangleright a^s = q^{\mp\frac{s}{2}} a^s & F \triangleright a^s = 0 & E \triangleright a^s = -q^{(3-s)/2}[s] a^{s-1} c^* \\ K^{\pm} \triangleright a^{*s} = q^{\pm\frac{s}{2}} a^{*s} & F \triangleright a^{*s} = q^{(1-s)/2}[s] c a^{*s-1} & E \triangleright a^{*s} = 0 \\ K^{\pm} \triangleright c^s = q^{\mp\frac{s}{2}} c^s & F \triangleright c^s = 0 & E \triangleright c^s = q^{(1-s)/2}[s] c^{s-1} a^* \\ K^{\pm} \triangleright c^{*s} = q^{\pm\frac{s}{2}} c^{*s} & F \triangleright c^{*s} = -q^{-(1+s)/2}[s] a c^{*s-1} & E \triangleright c^{*s} = 0; \end{array}$$

and:

(3.27)

$$\begin{array}{lll} a^s \triangleleft K^{\pm} = q^{\mp\frac{s}{2}} a^s & a^s \triangleleft F = q^{(s-1)/2}[s] c a^{s-1} & a^s \triangleleft E = 0 \\ a^{*s} \triangleleft K^{\pm} = q^{\pm\frac{s}{2}} a^{*s} & a^{*s} \triangleleft F = 0 & a^{*s} \triangleleft E = -q^{(3-s)/2}[s] c^* a^{*s-1} \\ c^s \triangleleft K^{\pm} = q^{\pm\frac{s}{2}} c^s & c^s \triangleleft F = 0 & c^s \triangleleft E = q^{(s-1)/2}[s] c^{s-1} a \\ c^{*s} \triangleleft K^{\pm} = q^{\mp\frac{s}{2}} c^{*s} & c^{*s} \triangleleft F = -q^{(s-3)/2}[s] a^* c^{*s-1} & c^{*s} \triangleleft E = 0. \end{array}$$

Denote $\mathcal{A}(\mathrm{U}(1)) := \mathbb{C}[z,z^*]/ < zz^*-1 >$; the map $\pi : \mathcal{A}(\mathrm{SU}_q(2)) \to \mathcal{A}(\mathrm{U}(1))$,

$$\pi \begin{bmatrix} a & -qc^* \\ c & a^* \end{bmatrix} = \begin{bmatrix} z & 0 \\ 0 & z^* \end{bmatrix} \tag{3.28}$$

[2]The 'q-number' is defined as:

$$[x] = [x]_q := \frac{q^x - q^{-x}}{q - q^{-1}}, \tag{3.22}$$

for $q \neq 1$ and any $x \in \mathbb{R}$.

is a surjective Hopf $*$-algebra homomorphism, so that $\mathcal{A}(\mathrm{U}(1))$ becomes a quantum subgroup of $SU_q(2)$ with a right coaction,

$$\Delta_R := (\mathrm{id} \otimes \pi) \circ \Delta \,:\, \mathcal{A}(\mathrm{SU}_q(2)) \mapsto \mathcal{A}(\mathrm{SU}_q(2)) \otimes \mathcal{A}(\mathrm{U}(1)). \tag{3.29}$$

The coinvariant elements for this coaction, elements $b \in \mathcal{A}(\mathrm{SU}_q(2))$ for which $\Delta_R(b) = b \otimes 1$, form a subalgebra of $\mathcal{A}(\mathrm{SU}_q(2))$ which is the coordinate algebra $\mathcal{A}(\mathrm{S}^2_q)$ of the standard Podleś sphere S^2_q. From:

$$\begin{aligned}
\Delta_R(a) &= a \otimes z, \\
\Delta_R(a^*) &= a^* \otimes z^*, \\
\Delta_R(c) &= c \otimes z, \\
\Delta_R(c^*) &= c^* \otimes z^*
\end{aligned} \tag{3.30}$$

as a set of generators for $\mathcal{A}(\mathrm{S}^2_q)$ one can choose:

$$B_- := -ac^*, \qquad B_+ := qca^*, \qquad B_0 := \frac{q^2}{1+q^2} - q^2cc^*, \tag{3.31}$$

satisfying the relations[3]:

$$\begin{aligned}
B_-B_0 &= [\frac{q^2-q^4}{1+q^2}B_- + q^2B_0B_-], \\
B_+B_0 &= [\frac{q^2-1}{q^2+1}B_+ + q^{-2}B_0B_+], \\
B_+B_- &= q\left[q^{-2}B_0 - (1+q^2)^{-1}\right]\left[q^{-2}B_0 + (1+q^{-2})^{-1}\right], \\
B_-B_+ &= q\left[B_0 + (1+q^2)^{-1}\right]\left[B_0 - (1+q^{-2})^{-1}\right],
\end{aligned}$$

and $*$-structure:

$$(B_0)^* = B_0, \qquad (B_+)^* = -qB_-.$$

The sphere S^2_q is a quantum homogeneous space of $\mathrm{SU}_q(2)$ and the coproduct of $\mathcal{A}(\mathrm{SU}_q(2))$ restricts to a left coaction of $\mathcal{A}(\mathrm{SU}_q(2))$ on $\mathcal{A}(\mathrm{S}^2_q)$ which on generators reads:

$$\begin{aligned}
\Delta(B_-) &= a^2 \otimes B_- - (1+q^{-2})B_- \otimes B_0 + c^{*2} \otimes B_+, \\
\Delta(B_0) &= q\, ac \otimes B_- + (1+q^{-2})B_0 \otimes B_0 - c^*a^* \otimes B_+, \\
\Delta(B_+) &= q^2\, c^2 \otimes B_- + (1+q^{-2})B_+ \otimes B_0 + a^{*2} \otimes B_+.
\end{aligned}$$

3.3.2. *The associated line bundles.* The left action of the group-like element K on $\mathcal{A}(\mathrm{SU}_q(2))$ allows [**27**] to give a vector basis decomposition $\mathcal{A}(\mathrm{SU}_q(2)) = \oplus_{n\in\mathbb{Z}}\mathcal{L}^{(0)}_n$, where

$$\mathcal{L}^{(0)}_n := \{x \in \mathcal{A}(\mathrm{SU}_q(2)) \,:\, K \triangleright x = q^{n/2}x\}. \tag{3.32}$$

In particular $\mathcal{A}(\mathrm{S}^2_q) = \mathcal{L}^{(0)}_0$. One also has $\mathcal{L}^{(0)*}_n \subset \mathcal{L}^{(0)}_{-n}$ and $\mathcal{L}^{(0)}_n\mathcal{L}^{(0)}_m \subset \mathcal{L}^{(0)}_{n+m}$. Each $\mathcal{L}^{(0)}_n$ is a bimodule over $\mathcal{A}(\mathrm{S}^2_q)$; relations (3.30) show that they can be equivalently characterised by the coaction Δ_R of the quantum subgroup $\mathcal{A}(U(1))$ on $\mathcal{A}(\mathrm{SU}_q(2))$:

$$\mathcal{L}^{(0)}_n = \{x \in \mathcal{A}(\mathrm{SU}_q(2)) \,:\, \Delta_R(x) = x \otimes z^{-n}\}. \tag{3.33}$$

[3]I should like to thank T.Brzezinski, who noticed that the commutation relations among the generators B_j of the algebra $\mathcal{A}(\mathrm{S}^2_q)$ written in [**24**] are not correct.

This equation appears as the natural quantum analogue of the classical relation (2.44), introducing $\mathcal{L}_n^{(0)} \subset \mathcal{A}(\mathrm{SU}_q(2))$ as $\mathcal{A}(\mathrm{S}^2_q)$-bimodule of *co*-equivariant elements with respect to the coaction (3.29) of the gauge group algebra. The relation (3.32) can then be read as an infinitesimal version of that in (3.33). The classical $\mathfrak{L}_n^{(0)}$ are recovered as rank 1 projective left $C^\infty(S^2)$-modules: the analogue property in the quantum setting was shown in [**31**]. Each $\mathcal{L}_n^{(0)}$ is isomorphic to a projective left $\mathcal{A}(\mathrm{S}^2_q)$-module of rank 1. These projective left $\mathcal{A}(\mathrm{S}^2_q)$-modules give modules of equivariant maps or of sections of line bundles over the quantum sphere S^2_q with winding numbers (monopole charge) $-n$. The corresponding projections [**8, 17**] can be explicitly written. Given $n \in \mathbb{Z}$, consider an element $\left|\Psi^{(n)}\right\rangle \in \mathcal{A}(\mathrm{SU}_q(2))^{|n|+1}$ whose components are:

$$
(3.34)\qquad
\begin{aligned}
n \geq 0: \qquad & \left|\Psi^{(n)}\right\rangle_\mu = \sqrt{\beta_{n,\mu}}\; c^{*\mu} a^{*n-\mu} \in \mathcal{L}_n^{(0)}, \\
\text{where}: \qquad & \beta_{n,0} = 1; \qquad \beta_{n,\mu} = q^{2\mu} \prod\nolimits_{j=0}^{\mu-1} \left(\frac{1-q^{-2(n-j)}}{1-q^{-2(j+1)}} \right), \qquad \mu = 1, \ldots, n
\end{aligned}
$$

$$
(3.35)\qquad
\begin{aligned}
n \leq 0: \qquad & \left|\Psi^{(n)}\right\rangle_\mu = \sqrt{\alpha_{n,\mu}}\; c^{|n|-\mu} a^{\mu} \in \mathcal{L}_n^{(0)}, \\
\text{where}: \qquad & \alpha_{n,0} = 1; \qquad \alpha_{n,\mu} = \prod\nolimits_{j=0}^{|n|-\mu-1} \left(\frac{1-q^{2(|n|-j)}}{1-q^{2(j+1)}} \right), \qquad \mu = 1, \ldots, |n|
\end{aligned}
$$

Using the commutation relations (3.19) and the explicit form of the coefficients in (3.34) and (3.35), it is possible to compute that:

$$
(3.36)\qquad
\begin{aligned}
n \geq 0: \qquad & \left\langle \Psi^{(n)}, \Psi^{(n)} \right\rangle = \sum\nolimits_{\mu=0}^{n} \beta_{n,\mu}\, a^{n-\mu} c^{\mu} c^{*\mu} a^{*n-\mu} = (aa^* + q^2 cc^*)^n = 1, \\
n \leq 0: \qquad & \left\langle \Psi^{(n)}, \Psi^{(n)} \right\rangle = \sum\nolimits_{\mu=0}^{|n|} \alpha_{n,\mu}\, a^{*\mu} c^{*|n|-\mu} c^{|n|-\mu} a^{\mu} = (a^*a + c^*c)^{|n|} = 1
\end{aligned}
$$

so that a projector $\mathfrak{p}^{(n)} \in \mathbb{M}_{|n|+1}(\mathcal{A}(S^2_q))$ can be defined as:

$$
\mathfrak{p}^{(n)} = \left|\Psi^{(n)}\right\rangle \left\langle \Psi^{(n)}\right| \tag{3.37}
$$

which is by construction an idempotent - $(\mathfrak{p}^{(n)})^2 = \mathfrak{p}^{(n)}$ - and selfadjoint operator - $(\mathfrak{p}^{(n)})^\dagger = \mathfrak{p}^{(n)}$ - whose entries are:

$$
(3.38)\qquad
\begin{aligned}
n \geq 0: \qquad & \mathfrak{p}^{(n)}_{\mu\nu} = \sqrt{\beta_{n,\mu}\beta_{n,\nu}}\; c^{*\mu} a^{*n-\mu} a^{n-\nu} c^{\nu} \in \mathcal{A}(\mathrm{S}^2_q), \\
n \leq 0: \qquad & \check{\mathfrak{p}}^{(n)}_{\mu\nu} = \sqrt{\alpha_{n,\mu}\alpha_{n,\nu}}\; c^{|n|-\mu} a^{\mu} a^{*\nu} c^{*|n|-\nu} \in \mathcal{A}(\mathrm{S}^2_q).
\end{aligned}
$$

The projections (3.37) play a central role in the description of the quantum Hopf bundle. As a first application one can prove that the algebra inclusion $\mathcal{A}(\mathrm{S}^2_q) \hookrightarrow \mathcal{A}(\mathrm{SU}_q(2))$ satisfies the topological requirements for a quantum principal bundle, when both the algebras are equipped with the universal calculus.

PROPOSITION 3.2. *The datum* $(\mathcal{A}(\mathrm{SU}_q(2)), \mathcal{A}(\mathrm{S}^2_q), \mathcal{A}(\mathrm{U}(1)))$ *is a quantum principal bundle.*

PROOF. The proof consists of showing the exactness of the sequence

$$0 \to \mathcal{A}(\mathrm{SU}_q(2))\,(\Omega^1(\mathrm{S}^2_q)_{un})\,\mathcal{A}(\mathrm{SU}_q(2)) \\ \to \Omega_1(\mathrm{SU}_q(2))_{un} \xrightarrow{\chi} \mathcal{A}(\mathrm{SU}_q(2)) \otimes \ker \varepsilon_{\mathrm{U}(1)} \to 0$$

or equivalently that the map $\chi : \Omega^1(\mathrm{SU}_q(2))_{un} \to \mathcal{A}(\mathrm{SU}_q(2)) \otimes \ker \varepsilon_{\mathrm{U}(1)}$ defined as in (3.14) – and with the $\mathcal{A}(\mathrm{U}(1))$-coaction on $\mathcal{A}(\mathrm{SU}_q(2))$ given in (3.29) – is surjective. Given an element $x \in \mathcal{L}^{(0)}_n \subset \mathcal{A}(\mathrm{SU}_q(2))$, from (3.33) the map χ acts as:

$$\chi(\delta x) = \chi(1 \otimes x - x \otimes 1) = x \otimes (z^{-n} - 1). \tag{3.39}$$

A generic element in $\mathcal{A}(\mathrm{SU}_q(2)) \otimes \ker \varepsilon_{\mathrm{U}(1)}$ is of the form $x \otimes (z^n - 1)$ with $n \in \mathbb{Z}$ and $x \in \mathcal{A}(\mathrm{SU}_q(2))$. To show surjectivity of χ the strategy is to show that $1 \otimes (z^n - 1)$ is in its image since left $\mathcal{A}(\mathrm{SU}_q(2))$-linearity of χ will give the general result: if $\gamma \in \Omega^1(\mathrm{SU}_q(2))_{un}$ is such that $\chi(\gamma) = 1 \otimes (z^n - 1)$, then $\chi(x\gamma) = x\,(1 \otimes (z^n - 1)) = x \otimes (z^n - 1)$. Fixed now $n \in \mathbb{Z}$, define an element γ in $\mathcal{A}(\mathrm{SU}_q(2))$ as $\gamma = \langle \Psi^{(-n)}, \delta \Psi^{(-n)} \rangle$ following (3.34) and (3.35). Since $|\Psi^{(-n)}\rangle \in \mathcal{L}^{(0)}_{-n}$, one computes that:

$$\chi(\gamma) = 1 \otimes (z^n - 1),$$

thus completing the proof. □

Next, it is possible to identify the spaces of equivariant maps $\mathcal{L}^{(0)}_n$ – or equivalently of *coequivariant* elements $\mathcal{L}^{(0)}_n$ – with the left $\mathcal{A}(\mathrm{S}^2_q)$-modules of sections $\mathcal{E}^{(0)}_n = (\mathcal{A}(\mathrm{S}^2_q))^{|n|+1}\mathfrak{p}^{(n)}$. For this write any element in the free module $(\mathcal{A}(\mathrm{S}^2_q))^{|n|+1}$ as $\langle f| = (f_0, f_1, \ldots, f_{|n|})$ with $f_\mu \in \mathcal{A}(\mathrm{S}^2_q)$. This allows to write equivariant maps as

$$\begin{aligned} \phi_f := \left\langle f, \Psi^{(n)} \right\rangle &= \sum\nolimits_{\mu=0}^{n} f_\mu \sqrt{\beta_{n,\mu}}\, c^{*\mu} a^{*n-\mu} && \text{for} \quad n \geq 0, \\ &= \sum\nolimits_{\mu=0}^{|n|} f_\mu \sqrt{\alpha_{n,\mu}}\, c^{|n|-\mu} a^{\mu} && \text{for} \quad n \leq 0. \end{aligned}$$

making it straightforward to establish the proposition, which generalises to the quantum setting the equivalence (2.49):

PROPOSITION 3.3. *Given* $n \in \mathbb{Z}$, *let* $\mathcal{E}^{(0)}_n := (\mathcal{A}(\mathrm{S}^2_q))^{|n|+1}\mathfrak{p}^{(n)}$. *There is a left* $\mathcal{A}(\mathrm{S}^2_q)$*-modules isomorphism:*

$$\mathcal{L}^{(0)}_n \xrightarrow{\simeq} \mathcal{E}^{(0)}_n, \quad \phi_f \mapsto \langle \sigma_f | = \phi_f \left\langle \Psi^{(n)} \right| = \langle f | \, \mathfrak{p}^{(n)},$$

with inverse

$$\mathcal{E}^{(0)}_n \xrightarrow{\simeq} \mathcal{L}^{(0)}_n, \quad \langle \sigma_f | = \langle f | \, \mathfrak{p}^{(n)} \mapsto \phi_f := \left\langle f, \Psi^{(n)} \right\rangle.$$

3.3.3. *A Peter-Weyl decomposition of* $\mathcal{A}(\mathrm{SU}_q(2))$. The aim of this section is to describe the known decomposition of the modules $\mathcal{L}^{(0)}_n$ into representation spaces under the action of $\mathcal{U}_q(\mathfrak{su}(2))$ [**21**]. From (3.32) one has a vector space decomposition $\mathcal{A}(\mathrm{SU}_q(2)) = \oplus_{n\in\mathbb{Z}} \mathcal{L}^{(0)}_n$, with

$$E \triangleright \mathcal{L}^{(0)}_n \subset \mathcal{L}^{(0)}_{n+2}, \qquad F \triangleright \mathcal{L}^{(0)}_n \subset \mathcal{L}^{(0)}_{n-2}. \tag{3.40}$$

On the other hand, commutativity of the left and right actions of $\mathcal{U}_q(\mathfrak{su}(2))$ yields that

$$\mathcal{L}_n^{(0)} \triangleleft h \subset \mathcal{L}_n^{(0)}, \qquad \forall\, h \in \mathcal{U}_q(\mathfrak{su}(2)).$$

It has already been shown in [**31**] that there is also a decomposition,

$$\mathcal{L}_n^{(0)} := \bigoplus_{J=\frac{|n|}{2}, \frac{|n|}{2}+1, \frac{|n|}{2}+2, \cdots} V_J^{(n)}, \tag{3.41}$$

with $V_J^{(n)}$ the spin J-representation space (for the right action) of $\mathcal{U}_q(\mathfrak{su}(2))$. Altogether it gives a Peter-Weyl decomposition for $\mathcal{A}(\mathrm{SU}_q(2))$ (already given in [**37**]).

More explicitly, the highest weight vector for each $V_J^{(n)}$ in (3.41) is $c^{J-n/2}a^{*J+n/2}$:

$$\begin{aligned}
K \triangleright (c^{J-n/2}a^{*J+n/2}) &= q^{n/2}(c^{J-n/2}a^{*J+n/2}), \\
(c^{J-n/2}a^{*J+n/2}) \triangleleft K &= q^{J}(c^{J-n/2}a^{*J+n/2}), \\
(c^{J-n/2}a^{*J+n/2}) \triangleleft F &= 0.
\end{aligned} \tag{3.42}$$

Analogously, the lowest weight vector for each $V_J^{(n)}$ in (3.41) is $a^{J-n/2}c^{*J+n/2}$:

$$\begin{aligned}
K \triangleright (a^{J-n/2}c^{*J+n/2}) &= q^{n/2}(a^{J-n/2}c^{*J+n/2}), \\
(a^{J-n/2}c^{*J+n/2}) \triangleleft K &= q^{-J}(a^{J-n/2}c^{*J+n/2}), \\
(a^{J-n/2}c^{*J+n/2}) \triangleleft E &= 0.
\end{aligned}$$

The elements of the vector spaces $V_J^{(n)}$ can be obtained by acting on the highest weight vectors with the lowering operator $\triangleleft E$, since clearly $\left(c^{J-n/2}a^{*J+n/2}\right) \triangleleft E \in \mathcal{L}_n^{(0)}$, or explicitly,

$$K \triangleright \left[\left(c^{J-n/2}a^{*J+n/2}\right) \triangleleft E\right] = q^{n/2}\left[\left(c^{J-n/2}a^{*J+n/2}\right) \triangleleft E\right].$$

To be definite, consider $n \geq 0$. The first admissible J is $J = n/2$; the highest weight element is a^{*n} and the vector space $V_{n/2}^{(n)}$ is spanned by $\{a^{*n} \triangleleft E^l\}$ with $l = 0, \ldots, n+1$: $V_{n/2}^{(n)} = \mathrm{span}\{a^{*n}, c^* a^{*n-1}, \ldots, c^{*n}\}$. Keeping n fixed, the other admissible values of J are $J = s + n/2$ with $s \in \mathbb{N}$. The vector spaces $V_{s+n/2}^{(n)}$ are spanned by $\{c^s a^{*s+n} \triangleleft E^l\}$ with $l = 0, \ldots, 2s+n+1$. Analogous considerations are valid when $n \leq 0$. In this cases, the admissible values of J are $J = s + |n|/2 = s - n/2$, the highest weight vector in $V_{s-n/2}^{(n)}$ is the element $c^{s-n}a^{*s}$, and a basis is given by the action of the lowering operator $\triangleleft E$, that is $V_{s-n/2}^{(n)} = \mathrm{span}\{(c^{s-n}a^{*s}) \triangleleft E^l,\ l = 0, \ldots, 2s-n+1\}$.

From (3.40) one has that the left action $F \triangleright$ maps $\mathcal{L}_n^{(0)}$ to $\mathcal{L}_{n-2}^{(0)}$. If $p \geq 0$, the element a^{*p} is the highest weight vector in $V_{p/2}^{(p)}$ and one has that $F \triangleright a^{*p} \propto c a^{*p-1}$. The element ca^{*p-1} is the highest weight vector in $V_{p/2}^{(p-2)}$ since one finds that $(ca^{*p-1}) \triangleleft F = 0$ and $(ca^{*p-1}) \triangleleft K = q^{p/2}(ca^{*p-1})$. In the same vein, the elements $F^t \triangleright a^{*p} \propto c^t a^{*p-t}$ are the highest weight elements in $V_{p/2}^{(p-2t)} \subset \mathcal{L}_{p-2t}^{(0)}$, $t = 0, \ldots, p$. Once again, a complete basis of each subspace $V_{p/2}^{(p-2t)}$ is obtained by the right action of the lowering operator $\triangleleft E$.

With these considerations, the algebra $\mathcal{A}(\mathrm{SU}_q(2))$ can be partitioned into finite dimensional blocks which are the analogues of the Wigner D-functions [**36**] for the group $SU(2)$. To illustrate the meaning of this partition, start with the element a^*, the highest weight vector of the space $V^{(1)}_{1/2}$. Representing the left action of $F\triangleright$ with a horizontal arrow and the right action of $\triangleleft E$ with a vertical one, yields the box

$$\begin{array}{ccc} a^* & \rightarrow & c \\ \downarrow & & \downarrow \\ -qc^* & \rightarrow & a \end{array},$$

where the first column is a basis of the subspace $V^{(1)}_{1/2}$, while the second column is a basis of the subspace $V^{(-1)}_{1/2}$. Starting from a^{*2} – the highest weight vector of $V^{(2)}_1$ – one gets:

$$\begin{array}{ccccc} a^{*2} & \rightarrow & q^{-1/2}\,[2]\,ca^* & \rightarrow & [2]\,c^2 \\ \downarrow & & \downarrow & & \downarrow \\ -q^{1/2}\,[2]\,c^*a^* & \rightarrow & [2]\,(aa^*-cc^*) & \rightarrow & [2]^2\,q^{1/2}ca \\ \downarrow & & \downarrow & & \downarrow \\ q^2\,[2]\,c^{*2} & \rightarrow & -q^{1/2}\,[2]^2\,ac^* & \rightarrow & [2]^2\,a^2 \end{array}.$$

The three columns of this box are bases for the subspaces $V^{(2)}_1$, $V^{(0)}_1$,$V^{(-2)}_1$, respectively. The recursive structure is clear. For a positive integer p, one has a box W_p made up of $(p+1)\times(p+1)$ elements. Without explicitly computing the coefficients, one gets:

$$\begin{array}{ccccccccccc} a^{*p} & \rightarrow & ca^{*p-1} & \rightarrow & \ldots & \rightarrow & c^ta^{*p-t} & \rightarrow & \ldots & \rightarrow & c^p \\ \downarrow & & \downarrow & & \ldots & & \downarrow & & \ldots & & \downarrow \\ c^*a^{*p-1} & \rightarrow & \ldots & \rightarrow & \ldots & \rightarrow & \ldots & \rightarrow & \ldots & \rightarrow & ac^{p-1} \\ \downarrow & & \downarrow & & \ldots & & \downarrow & & \ldots & & \downarrow \\ \ldots & \rightarrow & \ldots & \rightarrow & \ldots & \rightarrow & \ldots & \rightarrow & \ldots & \rightarrow & \ldots \\ \downarrow & & \downarrow & & \ldots & & \downarrow & & \ldots & & \downarrow \\ c^{*s}a^{*p-s} & \rightarrow & \ldots & \rightarrow & \ldots & \rightarrow & \ldots & \rightarrow & \ldots & \rightarrow & a^sc^{p-s} \\ \downarrow & & \downarrow & & \ldots & & \downarrow & & \ldots & & \downarrow \\ \ldots & \rightarrow & \ldots & \rightarrow & \ldots & \rightarrow & \ldots & \rightarrow & \ldots & \rightarrow & \ldots \\ \downarrow & & \downarrow & & \ldots & & \downarrow & & \ldots & & \downarrow \\ c^{*p} & \rightarrow & ac^{*p-1} & \rightarrow & \ldots & \rightarrow & a^tc^{*p-t} & \rightarrow & \ldots & \rightarrow & a^p \end{array}.$$

The space W_p is the direct sum of representation spaces for the right action of $\mathcal{U}_q(\mathfrak{su}(2))$,

$$W_p = \oplus_{t=0}^{p} V^{(p-2t)}_{p/2},$$

and on each W_p the quantum Casimir C_q acts is the same manner from both the right and the left, with eigenvalue (3.24), that is $C_q\triangleright w_p = w_p\triangleleft C_q = \left([\frac{p+1}{2}]^2-\frac{1}{4}\right)w_p$, for all $w_p\in W_p$. The Peter-Weyl decomposition for the algebra $\mathcal{A}(\mathrm{SU}_q(2))$ is given as

$$\mathcal{A}(\mathrm{SU}_q(2)) = \oplus_{p\in\mathbb{N}} W_p = \oplus_{p\in\mathbb{N}}\left(\oplus_{t=0}^{p} V^{(p-2t)}_{p/2}\right).$$

A compatible basis with this decomposition is given by elements

$$w_{p:t,r} := F^t\triangleright a^{*p}\triangleleft E^r \in W_p \tag{3.43}$$

for $t,r=0,1\ldots,p$. In order to get elements in the Podleś sphere subalgebra $\mathcal{A}(\mathrm{S}^2_q)\simeq\mathcal{L}^{(0)}_0$ out of a highest weight vector a^{*p} we need $p=2l$ to be even and left action of F^l: $F^l \triangleright a^{*2l} \propto c^l a^{*l} \in \mathcal{A}(\mathrm{S}^2_q)$. Then, the right action of E yields a spherical harmonic decomposition,

$$\mathcal{A}(\mathrm{S}^2_q) = \oplus_{l \in \mathbb{N}} V_l^{(0)}, \tag{3.44}$$

with a basis of $V_l^{(0)}$ given by the vectors $F^l \triangleright a^{*2l} \triangleleft E^r$, for $r = 0, 1, \ldots, 2l$.

3.4. A quantum Hopf bundle with non-universal differential calculi. Once described how the inclusion $\mathcal{A}(\mathrm{S}^2_q) \hookrightarrow \mathcal{A}(\mathrm{SU}_q(2))$ has the structure of a topological quantum principal bundle, the aim of this section is to describe non-universal differential calculi on the algebras $\mathcal{A}(\mathrm{SU}_q(2)), \mathcal{A}(\mathrm{S}^2_q), \mathcal{A}(\mathrm{U}(1))$, and to show that these are compatible [**6, 7**].

3.4.1. *The left-covariant 3D calculus on $SU_q(2)$.* The first differential calculus defined on the quantum group $\mathrm{SU}_q(2)$ is the left-covariant one developed in [**37**]. It is three dimensional with corresponding ideal $\mathcal{Q}_{\mathrm{SU}_q(2)} \subset \ker \varepsilon_{\mathrm{SU}_q(2)}$ generated by the 6 elements $\{a^* + q^2 a - (1 + q^2); c^2; c^* c; c^{*2}; (a-1)c; (a-1)c^*\}$. Its quantum tangent space turns out to be, in terms of the non degenerate pairing (3.25), the vector space over the complex $\mathcal{X}_{\mathrm{SU}_q(2)} \subset \mathcal{U}_q(\mathfrak{su}(2))$, whose basis is

$$\begin{aligned} X_- &= q^{-1/2} F K, \\ X_+ &= q^{1/2} E K, \\ X_z &= \frac{1 - K^4}{1 - q^{-2}}; \end{aligned} \tag{3.45}$$

their coproducts result:

$$\begin{aligned} \Delta X_z &= 1 \otimes X_z + X_z \otimes K^4, \\ \Delta X_\pm &= 1 \otimes X_\pm + X_\pm \otimes K^2. \end{aligned} \tag{3.46}$$

The differential $\mathrm{d} : \mathcal{A}(\mathrm{SU}_q(2)) \to \Omega^1(\mathrm{SU}_q(2))$ is

$$\mathrm{d}x = (X_+ \triangleright x)\, \omega_+ + (X_- \triangleright x)\, \omega_- + (X_z \triangleright x)\, \omega_z, \tag{3.47}$$

for all $x \in \mathcal{A}(\mathrm{SU}_q(2))$. This equation gives a basis for the dual space of 1-forms $\Omega^1(\mathcal{A}(\mathrm{SU}_q(2)))$,

$$\begin{aligned} \omega_z &= a^* \mathrm{d}a + c^* \mathrm{d}c, \\ \omega_- &= c^* \mathrm{d}a^* - q a^* \mathrm{d}c^*, \\ \omega_+ &= a \mathrm{d}c - q c \mathrm{d}a, \end{aligned} \tag{3.48}$$

of left-covariant forms, that is $\Delta^{(1)}_L(\omega_s) = 1 \otimes \omega_s$, with $\Delta^{(1)}_L$ the (left) coaction of $\mathcal{A}(\mathrm{SU}_q(2))$ onto itself extended to forms (3.2). The above relations (3.48) can be inverted to

$$\begin{aligned} \mathrm{d}a &= -q c^* \omega_+ + a \omega_z, \\ \mathrm{d}a^* &= -q^2 a^* \omega_z + c \omega_-, \\ \mathrm{d}c &= a^* \omega_+ + c \omega_z, \\ \mathrm{d}c^* &= -q^2 c^* \omega_z - q^{-1} a \omega_-. \end{aligned} \tag{3.49}$$

A direct computation shows that $(S(\mathcal{Q}_{\mathrm{SU}_q(2)}))^* \subset \mathcal{Q}_{\mathrm{SU}_q(2)}$. This differential calculus is then a $*$-calculus, with $\omega_-^* = -\omega_+$ and $\omega_z^* = -\omega_z$. The bimodule structure is:

$$\omega_z\phi = q^{2n}\phi\omega_z, \qquad \omega_\pm\phi = q^n\phi\omega_\pm \tag{3.50}$$

for any $\phi \in \mathcal{L}_n^{(0)}$. Higher dimensional forms can be defined in a natural way by requiring compatibility for commutation relations and that $\mathrm{d}^2 = 0$. Consider the tensor product $\{\Omega(\mathrm{SU}_q(2))\}^{\otimes 2} = \Omega^1(\mathrm{SU}_q(2)) \otimes_{\mathcal{A}(\mathrm{SU}_q(2))} \Omega^1(\mathrm{SU}_q(2))$. A consistent alternation mapping on $\{\Omega(\mathrm{SU}_q(2))\}^{\otimes 2}$, generalising the alternation mapping in the classical formalism, can be introduced only if the quantum differential calculus is bicovariant. The strategy to define a wedge product comes then from Lemma 15 in chapter 14 in [**21**], where it is proved that $\mathcal{S}_{\mathcal{Q}_{\mathrm{SU}_q(2)}}(x) = \sum_{a,b} \langle X_a X_b, x\rangle\, \omega_a \otimes \omega_b$ for any $x \in \mathcal{Q}_{\mathrm{SU}_q(2)}$ generates a two-sided ideal in $\{\Omega(\mathrm{SU}_q(2))\}^{\otimes 2}$. The bimodule of exterior differential 2-forms results to be the quotient

$$\Omega^2(\mathrm{SU}_q(2)) \simeq \{\Omega^1(\mathrm{SU}_q(2))\}^{\otimes 2}/\mathcal{A}(\mathrm{SU}_q(2))\{\mathcal{S}_{\mathcal{Q}}\}\mathcal{A}(\mathrm{SU}_q(2)). \tag{3.51}$$

The wedge product $\wedge : \Omega^1(\mathrm{SU}_q(2)) \times \Omega^1(\mathrm{SU}_q(2)) \to \Omega^2(\mathrm{SU}_q(2))$ embodies the commutation relations among 1-forms: from the six generators in $\mathcal{Q}_{\mathrm{SU}_q(2)}$ the elements generating $\mathcal{S}_{\mathcal{Q}}$ can be written as

$$\begin{aligned} &\omega_+ \wedge \omega_+ = \omega_- \wedge \omega_- = \omega_z \wedge \omega_z = 0, \\ &\omega_- \wedge \omega_+ + q^{-2}\omega_+ \wedge \omega_- = 0, \\ &\omega_z \wedge \omega_- + q^4\omega_- \wedge \omega_z = 0, \\ &\omega_z \wedge \omega_+ + q^{-4}\omega_+ \wedge \omega_z = 0. \end{aligned} \tag{3.52}$$

Such commutation rules also show that the bimodule $\Omega^2(\mathrm{SU}_q(2))$ is 3 dimensional, the three basis 2-forms being exact, since one has

$$\begin{aligned} \mathrm{d}\omega_z &= -\omega_- \wedge \omega_+, \\ \mathrm{d}\omega_+ &= q^2(1+q^2)\omega_z \wedge \omega_+, \\ \mathrm{d}\omega_- &= -(1+q^{-2})\omega_z \wedge \omega_-; \end{aligned} \tag{3.53}$$

the commutation relations moreover clarify that this left covariant calculus has a unique top form $\omega_- \wedge \omega_+ \wedge \omega_z$. The $*$-structure is extended to Ω^{m+n} by $(\alpha \wedge \beta)^* = (-1)^{mn}\beta^* \wedge \alpha^*$ with $\alpha \in \Omega^m$ and $\beta \in \Omega^n$. This definition is compatible with (3.52).

The left covariance of the differential calculus allows to extend to higher order forms in a natural way the left coaction $\Delta_L^{(1)}$ of $\mathcal{A}(\mathrm{SU}_q(2))$ on $\Omega^1(\mathrm{SU}_q(2))$. An element $\eta \in \{\Omega^1(\mathrm{SU}_q(2))\}^{\otimes k}$ can always be written as $\eta = x_{a_1 \ldots a_k}\, \omega_{a_1} \otimes \ldots \otimes \omega_{a_k}$ in terms of the left invariant forms ω_j in (3.48). Define

$$\Delta_L^{(k)}(\eta) = x_{a_1\ldots a_k(1)} \otimes x_{a_1\ldots a_k(2)}\omega_{a_1} \otimes \ldots \otimes \omega_{a_k},$$

from the Sweedler notation for the coproduct $\Delta(x_{a_1\ldots a_k})$. One proves that this definition is consistent on the exterior algebra $\Omega^k(\mathrm{SU}_q(2))$, as $\Delta_L^{(2)}(\mathcal{S}_{\mathcal{Q}}) \subset 1 \otimes \mathcal{S}_{\mathcal{Q}}$, and that $\Delta_L^{(k)}(\mathrm{d}\eta) = (1 \otimes \mathrm{d})\Delta_L^{(k-1)}(\eta)$ for any $\eta \in \Omega^k(\mathrm{SU}_q(2))$ with $k = 1, 2, 3$. The relations (3.53) show then that $\Omega^2(\mathrm{SU}_q(2))$ has a basis of exact left invariant forms, given by $\mathrm{d}\omega_j$; it is also clear that $\omega_- \wedge \omega_+ \wedge \omega_z$ is a left-invariant 3-form.

3.4.2. *The calculus on the structure group.* The strategy adopted in [**6**] consists in defining the calculus on U(1) via the Hopf projection π in (3.28). Out of the $\mathcal{Q}_{\mathrm{SU}_q(2)}$ which determines the left covariant calculus on $\mathrm{SU}_q(2)$, one defines a right ideal $\mathcal{Q}_{\mathrm{U}(1)} = \pi(\mathcal{Q}_{\mathrm{SU}_q(2)}) \subset \ker \varepsilon_{\mathrm{U}(1)}$ for the calculus on U(1).

This specific $\mathcal{Q}_{\mathrm{U}(1)}$ results generated by the element $\xi = (z^{-1} - 1) + q^2(z-1)$, and the differential calculus is then characterised by the quotient $\ker \varepsilon_{\mathrm{U}(1)}/\mathcal{Q}_{\mathrm{U}(1)}$. Any term in $\ker \varepsilon_{\mathrm{U}(1)}$ can be written as $\varphi = u(z-1) = \sum_{j\in\mathbb{Z}} u_j z^j(z-1)$, with $u = \sum_{j\in\mathbb{Z}} u_j z^j \in \mathcal{A}(\mathrm{U}(1))$ and $u_j \in \mathbb{C}$, so that the elements $\varphi(j) = z^j(z-1)$ define a vector space basis over $\mathbb{C}$ of $\ker \varepsilon_{\mathrm{U}(1)}$. The basis elements $\varphi(j)$ can be written in terms of the element ξ, via the two identities:

$$j \geq 0, \qquad \varphi(j) = z^j(z-1) = \xi\left(\sum_{m=1}^{j} q^{-2m} z^{j-m+1}\right) + q^{-2j}(z-1),$$

$$j \leq 0, \qquad \varphi(j) = z^{-|j|}(z-1) = -\xi\left(\sum_{m=0}^{|j|-1} q^{2m} z^{1+m-|j|}\right) + q^{2|j|}(z-1), \tag{3.54}$$

which can be proved by induction on j. Define a map $\lambda : \ker \varepsilon_{\mathrm{U}(1)} \to \ker \varepsilon_{\mathrm{U}(1)}$ setting on the basis elements $\lambda(\varphi(j)) = q^{-2j}(z-1)$, and linearly extending it to:

$$\lambda : u(z-1) = \sum_{j\in\mathbb{Z}} u_j z^j(z-1) \qquad \mapsto \qquad \sum_{j\in\mathbb{Z}} u_j q^{-2j}(z-1). \tag{3.55}$$

It is clear that λ describes the choice of a representative element out of the equivalence class $[u(z-1)] \in \ker \varepsilon_{\mathrm{U}(1)}/\mathcal{Q}_{\mathrm{U}(1)}$, since it is possible to see that $\ker \lambda = \mathcal{Q}_{\mathrm{U}(1)}$. To prove this assertion, one first directly computes that $\lambda(\xi) = 0$, then since λ is linear one recovers that $\lambda(u\xi) = \lambda(u(q^2(z-1) + (z^{-1}-1))) = q^2\lambda(u(z-1)) + \lambda(u(z^{-1}-1))$, so to have:

$$\begin{aligned} \lambda(u\xi) &= q^2\lambda(u(z-1)) + \lambda\left(\sum_{j\in\mathbb{Z}} u_j z^j(z^{-1}-1)\right) \\ &= q^2\lambda(u(z-1)) + \lambda\left(-\sum_{j\in\mathbb{Z}} u_j z^{j-1}(z^{-1})\right) \\ &= q^2\left(\sum_{j\in\mathbb{Z}} u_j q^{-2j}(z-1)\right) - \sum_{j\in\mathbb{Z}} u_j q^{-2(j-1)}(z-1) = 0, \end{aligned} \tag{3.56}$$

thus proving that $\mathcal{Q}_{\mathrm{U}(1)} \subset \ker \lambda$. To prove the inverse inclusion, consider an element $\check{u} = u(z-1) \in \ker \varepsilon_{\mathrm{U}(1)}$, and write it as:

$$\begin{aligned} u(z-1) &= \sum_{j\in\mathbb{Z}} u_j z^j(z-1) \\ &= \sum_{j\in\mathbb{N}} u_j z^j(z-1) + \sum_{j\in\mathbb{N}} u_{-j} z^{-j}(z-1) \\ &= \sum_{j\in\mathbb{N}} u_j(\alpha(j)\xi + q^{-2j}(z-1)) + \sum_{j\in\mathbb{N}} u_{-j}(\beta(-j)\xi + q^{2j}(z-1)) \end{aligned} \tag{3.57}$$

where $\alpha(j) = \sum_{m=1}^{j} q^{-2m} z^{j-m+1}$ and $\beta(-j) = \sum_{m=0}^{|j|-1} q^{2m} z^{1+m-|j|}$ are the terms proportional to ξ in (3.54) for positive and negative values of $j \in \mathbb{Z}$. The previous sum can be rewritten as:

$$u(z-1) = \xi \left(\sum_{j \in \mathbb{N}} u_j \alpha(j) + \sum_{j \in \mathbb{N}} u_{-j} \beta(-j) \right) + \sum_{j \in \mathbb{Z}} u_j q^{-2j} (z-1).$$

From the definition (3.55), it is $\lambda(\breve{u}) = 0 \leftrightarrow \sum_{j \in \mathbb{Z}} u_j q^{-2j} = 0$, so the last lines proves that $\ker \lambda \subset \mathcal{Q}_{\mathrm{U}(1)}$.

LEMMA 3.4. *Given the ideal* $\mathcal{Q}_{\mathrm{U}(1)} \subset \ker \varepsilon_{\mathrm{U}(1)}$ *generated by the element* $\xi = (z^{-1}-1) + q^2(z-1)$, *it is* $\ker \varepsilon_{\mathrm{U}(1)} / \mathcal{Q}_{\mathrm{U}(1)} \simeq \mathbb{C}$.

PROOF. Define a map $\tilde{\lambda} : \ker \varepsilon_{\mathrm{U}(1)} \to \mathbb{C}$ setting, on the basis elements $\varphi(j) \in \ker \varepsilon_{\mathrm{U}(1)}$, $\tilde{\lambda}(\varphi(j)) = q^{-2j}$ and extending it to $\ker \varepsilon_{\mathrm{U}(1)}$ by linearity. The properties of the map λ defined in (3.55) clarify that $\ker \tilde{\lambda} = \mathcal{Q}_{\mathrm{U}(1)}$, so to give a well defined map $\tilde{\lambda} : \ker \varepsilon_{\mathrm{U}(1)} / \mathcal{Q}_{\mathrm{U}(1)} \to \mathbb{C}$. It is immediate to see that $\tilde{\lambda}$ is an isomorphism of vector spaces, thus describing the equivalence: with $w \in \mathbb{C}$, the map $\tilde{\lambda}^{-1}(w) = w \in [w(z-1)] \subset \ker \varepsilon_{\mathrm{U}(1)}$ represents the inverse of the map $\tilde{\lambda}$. □

This result shows that the differential calculus generated by the specific $\mathcal{Q}_{\mathrm{U}(1)}$ is 1D, while a direct computation shows that it is bicovariant. As a basis element for its quantum tangent space one can consider

$$X = X_z = \frac{1 - K^4}{1 - q^{-2}}, \tag{3.58}$$

with dual left-invariant 1-form given by ω_z. This calculus turns out to have a $*$-structure, with $\omega_z^* = -\omega_z$. Explicitly, one has $\omega_z = z^* \mathrm{d}z$ with

$$\mathrm{d}z = z\omega_z, \qquad \mathrm{d}z^* = -q^2 z^* \omega_z;$$

and noncommutative $\mathcal{A}(\mathrm{U}(1))$-bimodule relations

$$z\mathrm{d}z = q^2 (\mathrm{d}z) z; \qquad \omega_z z = q^{-2} z \omega_z, \qquad \omega_z z^* = q^2 z^* \omega_z.$$

3.4.3. *The standard 2D calculus on* S_q^2. The restriction of the above 3D calculus to the sphere S_q^2 yields the unique left covariant 2-dimensional calculus on the latter [**26**]. An evolution of this approach has led [**32**] to a description of the unique 2D calculus of S_q^2 in term of a Dirac operator. The 'cotangent bundle' $\Omega^1(\mathrm{S}_q^2)$ is shown to be isomorphic to the direct sum $\mathcal{L}_{-2}^{(0)} \oplus \mathcal{L}_{2}^{(0)}$, that is the line bundles with winding number ± 2. Since the element K acts as the identity on $\mathcal{A}(\mathrm{S}_q^2)$, the differential (3.47) becomes, when restricted to the latter,

$$\begin{aligned} \mathrm{d}f &= (X_- \triangleright f)\,\omega_- + (X_+ \triangleright f)\,\omega_+ \\ &= (F \triangleright f)\,(q^{-1/2}\omega_-) + (E \triangleright f)\,(q^{1/2}\omega_+), \qquad \text{for } f \in \mathcal{A}(\mathrm{S}_q^2). \end{aligned}$$

These leads to break the exterior derivative into a holomorphic and an anti-holomorphic part, $\mathrm{d} = \bar{\partial} + \partial$, with:

$$\bar{\partial} f = (X_- \triangleright f)\, \omega_- = (F \triangleright f)\, (q^{-1/2} \omega_-),$$
$$\partial f = (X_+ \triangleright f)\, \omega_+ = (E \triangleright f)\, (q^{1/2} \omega_+), \qquad \text{for} \quad f \in \mathcal{A}(\mathrm{S}^2_q).$$

An explicit computation on the generators (3.31) of S^2_q yields:

$$\bar{\partial} B_- = q^{-1}\, a^2\, \omega_-, \quad \bar{\partial} B_0 = q\, ca\, \omega_-, \qquad \bar{\partial} B_+ = q\, c^2\, \omega_-,$$

$$\partial B_+ = q^2\, a^{*2}\, \omega_+, \quad \partial B_0 = -q^2\, c^* a^*\, \omega_+, \quad \partial B_- = q^2\, c^{*2}\, \omega_+.$$

The above shows that: $\Omega^1(\mathrm{S}^2_q) = \Omega^1_-(\mathrm{S}^2_q) \oplus \Omega^1_+(\mathrm{S}^2_q)$ where $\Omega^1_-(\mathrm{S}^2_q) \simeq \mathcal{L}^{(0)}_{-2} \simeq \bar{\partial}(\mathcal{A}(\mathrm{S}^2_q))$ is the $\mathcal{A}(\mathrm{S}^2_q)$-bimodule generated by:

$$\{\bar{\partial} B_-, \bar{\partial} B_0, \bar{\partial} B_+\} = \{a^2, ca, c^2\}\, \omega_- = q^2 \omega_- \{a^2, ca, c^2\}$$

and $\Omega^1_+(\mathrm{S}^2_q) \simeq \mathcal{L}^{(0)}_{+2} \simeq \partial(\mathcal{A}(\mathrm{S}^2_q))$ is the one generated by:

$$\{\partial B_+, \partial B_0, \partial B_-\} = \{a^{*2}, c^* a^*, c^{*2}\}\, \omega_+ = q^{-2} \omega_+ \{a^{*2}, c^* a^*, c^{*2}\}.$$

That these two modules of forms are not free is also expressed by the existence of relations among the differential:

$$\partial B_0 = q^{-1} B_- \partial B_+ - q^3 B_+ \partial B_-, \qquad \bar{\partial} B_0 = q B_+ \bar{\partial} B_- - q^{-3} B_- \bar{\partial} B_+.$$

Writing any 1-form as $\alpha = \phi' \omega_- + \phi'' \omega_+ \in \mathcal{L}^{(0)}_{-2} \omega_- \oplus \mathcal{L}^{(0)}_{+2} \omega_+$, the product of 1-forms is

$$(\phi' \omega_- + \phi'' \omega_+) \wedge (\psi' \omega_- + \psi'' \omega_+) = (q^{-2} \phi'' \psi' - \phi' \psi'') \omega_+ \wedge \omega_-, \tag{3.59}$$

while the exterior derivative acts as:

$$\begin{aligned} \mathrm{d}(\phi' \omega_- + \phi'' \omega_+) &= (\mathrm{d}\phi') \wedge \omega_- + \phi' \mathrm{d}\omega_- + (\mathrm{d}\phi'') \wedge \omega_+ + \phi'' \mathrm{d}\omega_+ \\ &= (X_+ \triangleright \phi') \omega_+ \wedge \omega_- + \{(X_z \triangleright \phi') \omega_z \wedge \omega_- + \phi' \mathrm{d}\omega_-\} \\ &\quad + (X_- \triangleright \phi'') \omega_- \wedge \omega_+ + \{(X_z \triangleright \phi'') \omega_z \wedge \omega_+ + \phi'' \mathrm{d}\omega_+\} \\ &= \{(X_- \triangleright \phi'') - q^2 (X_+ \triangleright \phi')\} \omega_- \wedge \omega_+, \end{aligned} \tag{3.60}$$

since the terms in curly brackets vanish: $\{(X_z \triangleright \phi') \omega_z \wedge \omega_- + \phi' \mathrm{d}\omega_-\} = \{(X_z \triangleright \phi'') \omega_z \wedge \omega_+ + \phi'' \mathrm{d}\omega_+\} = 0$ from (3.53) and (3.32). It is then clear that the calculus on the quantum sphere is 2D, and that $\Omega^2(\mathrm{S}^2_q) = \mathcal{A}(\mathrm{S}^2_q) \omega_- \wedge \omega_+ = \omega_- \wedge \omega_+ \mathcal{A}(\mathrm{S}^2_q)$, as both $\omega_\pm$ commute with elements of $\mathcal{A}(\mathrm{S}^2_q)$ and so does $\omega_- \wedge \omega_+$.

REMARK 3.5. *From (3.53) it is natural to ask that* $\mathrm{d}\omega_- = \mathrm{d}\omega_+ = 0$ *when restricted to* S^2_q. *Then, the exterior derivative of any 1-form* $\alpha = \phi' \omega_- + \phi'' \omega_+ \in \mathcal{L}^{(0)}_{-2} \omega_- \oplus \mathcal{L}^{(0)}_{+2} \omega_+$ *is given by:*

$$\begin{aligned} \mathrm{d}\alpha &= \mathrm{d}(\phi' \omega_- + \phi'' \omega_+) \\ &= \partial \phi' \wedge \omega_- + \bar{\partial} \phi'' \wedge \omega_+ \\ &= (X_+ \triangleright \phi' - q^{-2} X_- \triangleright \phi'')\, \omega_+ \wedge \omega_- \\ &= q^{-1/2} (E \triangleright \phi' - q^{-1} F \triangleright \phi'')\, \omega_+ \wedge \omega_-, \end{aligned} \tag{3.61}$$

since $K \triangleright$ *acts as* $q^{\mp}$ *on* $\mathcal{L}^{(0)}_{\mp 2}$. *Notice that in the above equality, both* $E \triangleright \phi'$ *and* $F \triangleright \phi''$ *belong to* $\mathcal{A}(\mathrm{S}^2_q)$, *as it should be.*

The above results can be summarised in the following proposition, which is the natural generalisation of the description in (2.56) of the classical exterior algebra on the sphere manifold S^2.

PROPOSITION 3.6. *The 2D differential calculus on the sphere* S^2_q *is given by:*

$$\Omega(S^2_q) = \mathcal{A}(S^2_q) \oplus \left(\mathcal{L}^{(0)}_{-2}\omega_- \oplus \mathcal{L}^{(0)}_{+2}\omega_+\right) \oplus \mathcal{A}(S^2_q)\omega_+ \wedge \omega_-,$$

with multiplication rule

$$\begin{aligned}(f_0; \phi', \phi''; f_2)\,(g_0; \psi', \psi''; g_2)& \\ = \big(f_0g_0;\, f_0\psi' + \phi' g_0,\, f_0\psi'' &+ \phi'' g_0;\, f_0 g_2 + f_2 g_0 + q^{-2}\phi''\psi' - \phi'\psi''\big),\end{aligned}$$

and exterior derivative $\mathrm{d} = \bar\partial + \partial$*:*

$$f \mapsto (q^{-1/2} F \triangleright f, q^{1/2} E \triangleright f), \qquad \text{for } f \in \mathcal{A}(S^2_q),$$

$$(\phi', \phi'') \mapsto q^{-1/2}(E \triangleright \phi' - q^{-1} F \triangleright \phi''), \quad \text{for } (\phi', \phi'') \in \mathcal{L}^{(0)}_{-2} \oplus \mathcal{L}^{(0)}_{+2}.$$

3.4.4. *The compatibility between the calculi.* Given the 3D left-covariant differential calculus on $SU_q(2)$ described in section 3.4.1, as well the 1D bicovariant differential calculus on the gauge group algebra U(1) in section 3.4.2, the 'principal bundle compatibility' of these calculi is established by showing that the sequence (3.17) is exact. For the case at hand, this sequence becomes

$$\begin{aligned}0 \to \mathcal{A}(SU_q(2))\,(\Omega^1(S^2_q))\,\mathcal{A}(SU_q(2)) &\to \\ \to \Omega^1(SU_q(2)) &\stackrel{\sim\mathcal{N}_{SU_q(2)}}{\longrightarrow} \mathcal{A}(SU_q(2)) \otimes \ker \varepsilon_{U(1)}/\mathcal{Q}_{U(1)} \to 0,\end{aligned}$$

where $\mathcal{Q}_{U(1)}$ is the ideal given in section 3.4.2 that defines the calculus on $\mathcal{A}(U(1))$ and the map $\sim\mathcal{N}_{SU_q(2)}$ is defined as in the diagram (3.15) which now acquires the form:

$$\begin{array}{ccc}\Omega^1(SU_q(2))_{un} & \stackrel{\pi\mathcal{N}_{SU_q(2)}}{\longrightarrow} & \Omega^1(SU_q(2)) \\ \downarrow\chi & & \downarrow\sim\mathcal{N}_{SU_q(2)} \\ \mathcal{A}(SU_q(2)) \otimes \ker\varepsilon_{U(1)} & \stackrel{\mathrm{id}\otimes\pi_{\mathcal{Q}_{U(1)}}}{\longrightarrow} & \mathcal{A}(SU_q(2)) \otimes (\ker\varepsilon_{U(1)}/\mathcal{Q}_{U(1)})\,.\end{array}$$

Having a quantum homogeneous bundle, that is a quantum bundle whose total space is a Hopf algebra and whose fiber is a Hopf subalgebra of it, with the differential calculus on the fiber obtained from the corresponding projection, for the above sequence to be exact it is enough [7] to check two conditions. The first one is

$$(\mathrm{id}\otimes\pi)\circ \mathrm{Ad}(\mathcal{Q}_{SU_q(2)}) \subset \mathcal{Q}_{SU_q(2)} \otimes \mathcal{A}(U(1))$$

with $\pi : \mathcal{A}(SU_q(2)) \to \mathcal{A}(U(1))$ the projection in (3.28). This is easily established by a direct calculation and using the explicit form of the elements in $\mathcal{Q}_{SU_q(2)}$. The second condition amounts to the statement that the kernel of the projection π can be written as a right $\mathcal{A}(SU_q(2))$-module of the kernel of π itself restricted to the base algebra $\mathcal{A}(S^2_q)$. Then, one needs to show that $\ker\pi \subset (\ker\pi|_{S^2_q})\mathcal{A}(SU_q(2))$, the opposite implication being obvious. With π defined in (3.28), one has that

$$\ker\pi = \{cf,\, c^*g, \quad \text{with} \quad f, g \in \mathcal{A}(SU_q(2))\}.$$

Then $cf = c(a^*a + c^*c)f = ca^*(af) + c^*c(cf)$, with both ca^* and c^*c in $\ker\pi|_{S^2_q}$. The same holds for elements of the form c^*g, and the inclusion follows.

The analysis of the map $\sim_{\mathcal{N}_{\mathrm{SU}_q(2)}}: \Omega^1(\mathrm{SU}_q(2)) \to \mathcal{A}(\mathrm{SU}_q(2)) \otimes \ker \varepsilon_{\mathrm{U}(1)}/\mathcal{Q}_{\mathrm{U}(1)}$ shows that $\omega_\pm \in \Omega^1(\mathcal{A}(\mathrm{SU}_q(2)))$ are indeed the generators of the horizontal forms of the principal bundle, being in the $\ker \sim_{\mathcal{N}_{\mathrm{SU}_q(2)}}$. From (3.39) one recovers:

$$\begin{aligned}
\chi(\delta a) &= a \otimes (z-1),\\
\chi(\delta a^*) &= a^* \otimes (z^*-1),\\
\chi(\delta c) &= c \otimes (z-1),\\
\chi(\delta c^*) &= c^* \otimes (z^*-1).
\end{aligned}$$

Given the two generators $\omega_\pm$ and the specific $\mathcal{Q}_{\mathrm{SU}_q(2)}$ which determines the 3D calculus, corresponding universal 1-forms can be taken to be:

$$\begin{aligned}
\omega_+ &= a\mathrm{d}c - qc\mathrm{d}a &\Rightarrow&\qquad (a\delta c - qc\delta a) \in [\pi_{\mathcal{N}_{\mathrm{SU}_q(2)}}]^{-1}(\omega_+),\\
\omega_- &= c^*\mathrm{d}a^* - qa^*\mathrm{d}c^* &\Rightarrow&\qquad (c^*\delta a^* - qa^*\delta c^*) \in [\pi_{\mathcal{N}_{\mathrm{SU}_q(2)}}]^{-1}(\omega_-).
\end{aligned}$$

The action of the canonical map then gives:

$$\begin{aligned}
\chi(a\delta c - qc\delta a) &= (ac - qca) \otimes (z-1) = 0,\\
\chi(c^*\delta a^* - qa^*\delta c^*) &= (c^*a^* - qa^*c^*) \otimes (z^*-1) = 0,
\end{aligned}$$

which means that

$$\sim_{\mathcal{N}_{\mathrm{SU}_q(2)}} (\omega_\pm) = 0 \tag{3.62}$$

For the third generator ω_z, one shows in a similar fashion that

$$\sim_{\mathcal{N}_{\mathrm{SU}_q(2)}} (\omega_z) = 1 \otimes (\pi_{\mathcal{Q}_{\mathrm{U}(1)}}(z-1)). \tag{3.63}$$

From these it is possible to conclude that the elements $\omega_\pm$ generate the $\mathcal{A}(\mathrm{SU}_q(2))$-bimodule of horizontal forms, while from (3.58) one has that the vector $X = X_z = (1-q^{-2})^{-1}(1-K^4)$ is the dual generator to the calculus on the structure Hopf algebra $\mathcal{A}(\mathrm{U}(1))$. For the corresponding 'vector field' $\tilde{X}$ on $\mathcal{A}(\mathrm{SU}_q(2))$ as in (3.18), one has that $\tilde{X}(\omega_\pm) = \langle X, \sim_{\mathcal{N}_{\mathrm{SU}_q(2)}} (\omega_\pm)\rangle = 0$, while $\tilde{X}(\omega_z) = \langle X, \sim_{\mathcal{N}_{\mathrm{SU}_q(2)}} (\omega_z)\rangle = 1$. These results identify $\tilde{X}$ as a vertical vector field.

4. A ⋆-Hodge duality on $\Omega(\mathrm{SU}_q(2))$ and a Laplacian on $\mathrm{SU}_q(2)$

In classical differential geometry a metric structure g on a N-dimensional manifold $\mathcal{M}$ enables to define a Hodge duality $\star : \Omega^k(\mathcal{M}) \to \Omega^{N-k}(\mathcal{M})$ on the exterior algebra $\Omega(\mathcal{M})$. The strategy is to consider the volume form $\theta \in \Omega^N(\mathcal{M})$ associated to a g-orthonormal basis; this corresponds to the choice of an orientation. Via the Hodge duality it becomes possible to introduce in $\Omega(\mathcal{M})$ both a symmetric bilinear product and a sesquilinear inner product.

The algebraic formulation of geometry of quantum groups, that has been described, presents no metric tensor. The strategy to introduce a Hodge duality on the exterior algebra $\Omega(\mathcal{H})$ coming from a N-dimensional differential calculus on a Hopf algebra $\mathcal{H}$ is then reversed with respect to the strategy used in the classical setting. The path consists in defining a suitable bilinear product on $\Omega(\mathcal{H})$ and considering a volume N-form, from which to induce a ⋆-Hodge structure, using an equation like the one in (2.33) as a definition.

The following description of the quantum formulation of a Hodge duality originates from [**22**]. Assume that $\mathcal{H}$ is a $*$-Hopf algebra equipped with a left covariant calculus $(\Omega^1(\mathcal{H}), \mathrm{d})$, with N the dimension of the calculus such that $\dim \Omega^N_{inv}(\mathcal{H}) = 1$, $\dim \Omega^k_{inv}(\mathcal{H}) = \dim^{N-k}_{inv}(\mathcal{H})$. Suppose also that $\mathcal{H}$ admits a Haar state $h : \mathcal{H} \to \mathbb{C}$, that is a unital linear functional on $\mathcal{H}$ for which $(id \otimes h)\Delta x = (h \otimes id)\Delta x = h(x)1$ for any $x \in \mathcal{H}$, where 1 is used to emphasise the unit of the algebra. Suppose further that h is positive, that is $h(x^*x) \geq 0$ for all $x \in \mathcal{H}$; it is known that the Haar state is unique and automatically faithful: if $h(x^*x) = 0$, then necessarily $x = 0$. One can endow $\mathcal{H}$ with an inner product derived from h, setting:

$$(x'; x)_{\mathcal{H}} = h(x^*x') \tag{4.1}$$

for any $x, x' \in \mathcal{H}$. The whole exterior algebra can be endowed with an inner product, defined on a left invariant basis and then extended via the requirement of left invariance,

$$(x'\omega'; x\,\omega)_{\mathcal{H}} = h(x^*x')(\omega', \omega)_{\mathcal{H}} \tag{4.2}$$

for any $x, x' \in \mathcal{H}$ and left invariant forms ω, ω' in $\Omega(\mathcal{H})$. An inner product is said graded if the spaces $\Omega^k(\mathcal{H})$ are pairwise orthogonal.

Out of $\Omega^N(\mathcal{H})$ choose a left invariant hermitian basis element $\theta = \theta^*$, which will be called the volume form of the calculus. A linear functional $\int_\theta : \Omega(\mathcal{H}) \to \mathbb{C}$ – called the integral on $\Omega(\mathcal{H})$ associated to the volume form $\theta \in \Omega^N(\mathcal{H})$ – is defined by setting $\int_\theta \eta = 0$ if η is a k-form with $k < N$, and $\int_\theta \eta = h(x)$ if $\eta = x\,\theta$ with $x \in \mathcal{H}$. The differential calculus will be said non-degenerate if, whenever $\eta \in \Omega^k(\mathcal{H})$ and $\eta' \wedge \eta = 0$ for any $\eta' \in \Omega^{N-k}(\mathcal{H})$, then necessarily $\eta = 0$. This property reflects itself in the property of left-faithfulness of the functional $\int_\theta$: starting from a non degenerate calculus, it is possible to prove that, if η is an element in $\Omega^k(\mathcal{H})$ for which $\int_\theta \eta' \wedge \eta = 0$ for all $\eta' \in \Omega^{N-k}(\mathcal{H})$, then it is $\eta = 0$.

PROPOSITION 4.1. *Given the exterior algebra $\Omega(\mathcal{H})$ coming from a left covariant, non degenerate calculus $(\Omega^1(\mathcal{H}), \mathrm{d})$, there exists a unique left $\mathcal{H}$-linear bijective operator $L : \Omega^k(\mathcal{H}) \to \Omega^{N-k}(\mathcal{H})$ for $k = 0, \ldots, N$, such that*

$$\int_\theta \eta^* \wedge L(\eta') = (\eta'; \eta)_{\mathcal{H}} \tag{4.3}$$

on any $\eta, \eta' \in \Omega^k(\mathcal{H})$.

The proof of this result is in [**22**], where the operator L is called a Hodge operator. With a left-invariant inner product which is positive definite, i.e. $(\omega, \omega)_{\mathcal{H}} > 0$ for any exterior form ω, the operator L does not yet define a $\star$-Hodge structure on $\Omega(\mathcal{H})$, since its square does not satisy the natural requirement (2.29). It is then used to define a new graded left invariant inner product setting on a basis of left invariant forms $\omega \in \Omega(\mathcal{H})$:

$$\begin{aligned} (\omega; \omega')^\natural_{\mathcal{H}} &= (\omega; \omega')_{\mathcal{H}}, && \text{on}\, \Omega^k(\mathcal{H}),\ k < N/2; \\ (\omega; \omega')^\natural_{\mathcal{H}} &= (L^{-1}(\omega); L^{-1}(\omega'))_{\mathcal{H}}, && \text{on}\, \Omega^k(\mathcal{H}),\ k > N/2. \end{aligned} \tag{4.4}$$

If N is odd, these relations completely define a new left invariant graded inner product on the exterior algebra $\Omega(\mathcal{H})$; notice also that assuming the relation (4.1) means that $(1; 1)_{\mathcal{H}} = 1$, from which one has $L(1) = \theta$, so to obtain in (4.4) that $(\theta; \theta)^\natural_{\mathcal{H}} = (1; 1)_{\mathcal{H}} = 1$.

In analogy with (4.3) define a new Hodge operator $L^{\natural} : \Omega^k(\mathcal{H}) \to \Omega^{N-k}(\mathcal{H})$ via the inner product given in (4.4) as

$$\int_\theta \eta^* \wedge L^{\natural}(\eta') = (\eta'; \eta)^{\natural}_{\mathcal{H}}. \tag{4.5}$$

Due to the left-faithfulness of the integral, it is clear that $L^{\natural}$ is a well defined bijection, which satisfies the identity $L = L^{\natural}$ when restricted to $\Omega^k(\mathcal{H})$ with $k < N/2$. Such an operator $L^{\natural}$ is also proved to satisfy $(L^{\natural})^2 = (-1)^{k(N-k)}$: this is the reason why one can define a $\star$-Hodge structure on $\Omega(\mathcal{H})$ as:

$$\star : \Omega^k(\mathcal{H}) \to \Omega^{N-k}(\mathcal{H}) \qquad \star(\eta) = L^{\natural}(\eta). \tag{4.6}$$

The relation (4.5) appears as the quantum version of the classical relation (2.35), which is now used as a definition for the Hodge duality.

If the dimension of the calculus is given by an even $N = 2m$, a more specific procedure is needed, The same procedure as before gives a $\star$-Hodge operator on $\Omega^k(\mathcal{H})$ for $k \neq m$ via the inner product (4.4). Using the volume form $\theta \in \Omega^N(\mathcal{H})$ set now a sesquilinear form

$$\langle \eta', \eta \rangle = \int_\theta \eta^* \wedge \eta', \tag{4.7}$$

which is non-degenerate by the faithfulness of the integral $\int_\theta$. The $\mathcal{H}$-bimodule $\Omega^m(\mathcal{H})$ has a basis of $\begin{pmatrix} 2m \\ m \end{pmatrix}$ left invariants elements ω_a. The restriction of (4.7) to elements ω_a defines a sesquilinear form on the vector space Ω^m_{inv}: this form is hermitian if $(-1)^{m^2} = 1$, and anti-hermitian if $(-1)^{m^2} = -1$, so it can be 'diagonalised'. There exists a basis $\breve{\omega}_j \in \Omega^m(\mathcal{H})$ such that one has $\langle \breve{\omega}_a, \breve{\omega}_b \rangle = \pm\delta_{ab}$ if it is hermitian, and $\langle \breve{\omega}_a, \breve{\omega}_b \rangle = \pm i\delta_{ab}$ if it is anti-hermitian. It is then possible to use such a basis to define a left $\mathcal{H}$-linear operator $\mathfrak{L} : \Omega^m(\mathcal{H}) \to \Omega^m(\mathcal{H})$ setting on the basis

$$\mathfrak{L}(\breve{\omega}_a) = (-1)^{m^2} \langle \breve{\omega}_a, \breve{\omega}_a \rangle \, \breve{\omega}_a. \tag{4.8}$$

(no sum on a). This map is a bijection, and satisfies $\mathfrak{L}^2 = (-1)^{m^2}$, so a $\star$-Hodge structure on $\Omega^m_{\text{inv}}(\mathcal{H})$ can be defined as:

$$\star(\breve{\omega}_a) = \mathfrak{L}(\breve{\omega}_a), \tag{4.9}$$

and extended on any $\eta \in \Omega^m(\mathcal{H})$ by the requirement of left linearity, thus giving a complete constructive procedure for a $\star$-Hodge structure on $\Omega(\mathcal{H})$. The Hodge operator $\mathfrak{L} : \Omega^m(\mathcal{H}) \to \Omega^m(\mathcal{H})$ is then used to introduce a left invariant inner product on $\Omega^m(\mathcal{H})$, defined by:

$$(\omega_a; \omega_b)^{\natural}_{\mathcal{H}} = \int_\theta \omega_b^* \wedge \mathfrak{L}(\omega_a), \tag{4.10}$$

on a basis of left invariant $\{\omega_a\}$ 2-forms, and then extended via the requirement of left invariance as in (4.2). It is easy to see that the definition eventually gives the inner product

$$(\breve{\omega}_a; \breve{\omega}_b)^{\natural}_{\mathcal{H}} = \delta_{ab}. \tag{4.11}$$

4.1. A $\star$-Hodge structure on $\Omega(\mathrm{SU}_q(2))$. This section describes how the outlined procedure yields a left invariant inner product on the exterior algebra $\Omega(SU_q(2))$ generated by the left covariant 3D calculus from section 3.4.1, and the way it gives rise to a $\star$-Hodge structure. Such a $\star$-Hodge structure will be then used to define a Laplacian operator on $\mathcal{A}(\mathrm{SU}_q(2))$, which is completely diagonalised.

The Hopf algebra $\mathcal{A}(\mathrm{SU}_q(2))$ has a Haar state $h : \mathcal{A}(\mathrm{SU}_q(2)) \to \mathbb{C}$, which is positive, unique and authomatically faithful. From the Peter-Weyl decomposition of $\mathcal{A}(\mathrm{SU}_q(2))$ in terms of the vector space basis elements $w_{p:r,t} \in W_p$ (3.43), the Haar state is determined by setting:

$$h(1) = 1 \qquad h(w_{p:r,t}) = 0 \ \forall p \geq 0.$$

The algebraic relations (3.19) among the generators of $\mathcal{A}(\mathrm{SU}_q(2))$ makes it then possible to prove that the only non trivial action of h on $\mathcal{A}(\mathrm{SU}_q(2))$ can also be written as:

$$h((cc^*)^k) = (\sum_{j=0}^{k} q^{2j})^{-1} = \frac{1}{1 + q^2 + \ldots + q^{2k}},$$

with $k \in \mathbb{N}$. One can define on $\mathcal{A}(\mathrm{SU}_q(2))$ an inner product derived from h, setting:

$$(x', x)_{\mathrm{SU}_q(2)} = h(x^* x') \tag{4.12}$$

with $x, x' \in \mathcal{A}(\mathrm{SU}_q(2))$. The differential 3D calculus being left covariant, the set of k-forms $\Omega^k(\mathrm{SU}_q(2))$ has a basis of left invariant forms. The exterior algebra $\Omega(\mathrm{SU}_q(2))$ is endowed with an inner product, defined on a left invariant basis and extended via the requirement of left invariance:

$$(x'\omega', x\,\omega)_{\mathrm{SU}_q(2)} = h(x^\star x')(\omega', \omega)_{\mathrm{SU}_q(2)}$$

for all x, x' in $\mathcal{A}(\mathrm{SU}_q(2))$ and $\omega, \omega' \in \Omega(\mathrm{SU}_q(2))$ left invariant forms. Assume the top form $\theta = \alpha' \omega_- \wedge \omega_+ \wedge \omega_z$ as volume form, with $\alpha' \in \mathbb{R}$ so that $\theta^\star = \theta$. The integral on the exterior algebra $\Omega(\mathrm{SU}_q(2))$ associated to the volume form θ is defined by $\int_\theta \eta = 0$ if η is a k-form with $k < 2$, and $\int_\theta \eta = h(x)$ if $\eta = x\,\theta$. This integral is left-faithful.

Set a left invariant graded inner product by assuming that the only non-zero products among left invariant forms are:

$$\begin{aligned} (1,1)_{\mathrm{SU}_q(2)} &= 1, \\ (\theta,\theta)_{\mathrm{SU}_q(2)} &= 1; \end{aligned} \tag{4.13}$$

while in $\Omega^1(\mathrm{SU}_q(2))$ are:

$$\begin{aligned} (\omega_-, \omega_-)_{\mathrm{SU}_q(2)} &= \beta, \\ (\omega_+, \omega_+)_{\mathrm{SU}_q(2)} &= \nu, \\ (\omega_z, \omega_z)_{\mathrm{SU}_q(2)} &= \gamma \end{aligned} \tag{4.14}$$

with $\beta, \nu, \gamma \in \mathbb{R}$, and:

$$\begin{aligned} (\omega_- \wedge \omega_+, \omega_- \wedge \omega_+)_{\mathrm{SU}_q(2)} &= 1, \\ (\omega_+ \wedge \omega_z, \omega_+ \wedge \omega_z)_{\mathrm{SU}_q(2)} &= 1, \\ (\omega_z \wedge \omega_-, \omega_z \wedge \omega_-)_{\mathrm{SU}_q(2)} &= 1 \end{aligned} \tag{4.15}$$

in $\Omega^2(\mathrm{SU}_q(2))$. This choice comes as the most natural in order to mimic the properties of the classical inner product (2.37), coming from the classical Hodge structure (2.27) originated from the metric (2.26). The Hodge operator defined in (4.3) is:

$$\begin{aligned}
&L(1) = \alpha'\,\omega_- \wedge \omega_+ \wedge \omega_z,\\
&L(\omega_-) = -\alpha'\beta q^{-6}\,\omega_z \wedge \omega_-,\\
&L(\omega_+) = -\alpha'\nu\,\omega_+ \wedge \omega_z,\\
&L(\omega_z) = -\alpha'\gamma\,\omega_- \wedge \omega_+,\\
&L(\omega_- \wedge \omega_+) = -\alpha'\,\omega_z,\\
&L(\omega_+ \wedge \omega_z) = -\alpha'\,\omega_+,\\
&L(\omega_z \wedge \omega_-) = -\alpha'\,\omega_-,\\
&L(\omega_- \wedge \omega_+ \wedge \omega_z) = \alpha'^{-1}.
\end{aligned} \tag{4.16}$$

The Hodge operator L is used to define a new graded left invariant inner product on $\Omega(\mathrm{SU}_q(2))$, as:

$$\begin{aligned}
&(\omega',\omega)^{\natural}_{\mathrm{SU}_q(2)} = (\omega',\omega)_{\mathrm{SU}_q(2)} && \text{on } \Omega^k(\mathrm{SU}_q(2)),\ k=0,1;\\
&(\omega',\omega)^{\natural}_{\mathrm{SU}_q(2)} = (L^{-1}(\omega'),L^{-1}(\omega))_{\mathrm{SU}_q(2)} && \text{on } \Omega^k(\mathrm{SU}_q(2)),\ k=2,3,
\end{aligned} \tag{4.17}$$

on the basis of left invariant forms. On $\Omega^k(\mathrm{SU}_q(2)))$ – with $k=2,3$ – one has:

$$\begin{aligned}
&(\omega_- \wedge \omega_+, \omega_- \wedge \omega_+)^{\natural}_{\mathrm{SU}_q(2)} = \alpha'^{-2}\gamma^{-1},\\
&(\omega_+ \wedge \omega_z, \omega_+ \wedge \omega_z)^{\natural}_{\mathrm{SU}_q(2)} = \alpha'^{-2}\nu^{-1},\\
&(\omega_z \wedge \omega_-, \omega_z \wedge \omega_-)^{\natural}_{\mathrm{SU}_q(2)} = q^{12}\alpha'^{-2}\beta^{-1},\\
&(\theta,\theta)^{\natural}_{\mathrm{SU}_q(2)} = 1.
\end{aligned} \tag{4.18}$$

Associated to this new inner product there is in analogy a new unique left $\mathcal{A}(\mathrm{SU}_q(2))$-linear operator $L^{\natural} : \Omega^k(\mathrm{SU}_q(2)) \to \Omega^{3-k}(\mathrm{SU}_q(2))$ defined by $\int_\theta \eta^* \wedge L^{\natural}(\eta') = (\eta',\eta)^{\natural}$, which is a bijection. This operator is such that $(L^{\natural})^2 = (-1)^{k(3-k)} = 1$, so following (4.6) one has a $\star$-Hodge structure on the exterior algebra $\Omega(\mathrm{SU}_q(2))$:

$$\star : \Omega^k(\mathrm{SU}_q(2)) \to \Omega^{3-k}(\mathrm{SU}_q(2)) \qquad \star(\eta) = L^{\natural}(\eta), \tag{4.19}$$

given by:

$$\begin{aligned}
&\star(1) = \theta = \alpha'\,\omega_- \wedge \omega_+ \wedge \omega_z,\\
&\star(\omega_-) = -\alpha'\beta q^{-6}\,\omega_z \wedge \omega_-,\\
&\star(\omega_+) = -\alpha'\nu\,\omega_+ \wedge \omega_z,\\
&\star(\omega_z) = -\alpha'\gamma\,\omega_- \wedge \omega_+,\\
&\star(\omega_- \wedge \omega_+) = -\alpha'^{-1}\gamma^{-1}\,\omega_z,\\
&\star(\omega_+ \wedge \omega_z) = -\alpha'^{-1}\nu^{-1}\,\omega_+,\\
&\star(\omega_z \wedge \omega_-) = -\alpha'^{-1}\beta^{-1}q^6\,\omega_-,\\
&\star(\omega_- \wedge \omega_+ \wedge \omega_z) = \alpha'^{-1}.
\end{aligned} \tag{4.20}$$

REMARK 4.2. *The definition of the graded left invariant inner product* $(\cdot,\cdot)^{\natural}_{\mathrm{SU}_q(2)}$ *in* (4.17) *shows that, in order to have a* $\star$*-Hodge structure on the exterior algebra*

$\Omega(\mathrm{SU}_q(2))$ *generated by the 3D calculus, it is sufficient to choice an hermitian volume form and a graded left invariant inner product* only *on* $\Omega^k(\mathrm{SU}_q(2))$ *for* $k=0,1$. *This is a general aspect: given a Hopf* $*$*-algebra* $\mathcal{H}$, *equipped with a finite odd* N *dimensional left covariant differential calculus, the formalism developed in* [**22**] *shows that what one needs is an hermitian volume form and a graded left invariant inner product on* $\Omega^k(\mathcal{H})$ *for* $k<N/2$.

4.1.1. *A Laplacian operator on* $\mathcal{A}(\mathrm{SU}_q(2))$. Given a differential calculus and a $\star$-Hodge structure on the Hopf algebra $\mathcal{A}(\mathrm{SU}_q(2))$ it is possible to define a scalar Laplacian operator $\square_{\mathrm{SU}_q(2)}:\mathcal{A}(\mathrm{SU}_q(2))\to\mathcal{A}(\mathrm{SU}_q(2))$ as $\square_{\mathrm{SU}_q(2)}\phi=\star\mathrm{d}\star\mathrm{d}\phi$ for any $\phi\in\mathcal{A}(\mathrm{SU}_q(2))$. This Laplacian can be written down by a computation on the basis of the left invariant forms of the calculus:

$$\begin{aligned}\mathrm{d}\phi&=(X_+\triangleright\phi)\omega_++(X_-\triangleright\phi)\omega_-+(X_z\triangleright\phi)\omega_z;\\ \star\mathrm{d}\phi&=-\alpha'[\nu(X_+\triangleright\phi)\omega_+\wedge\omega_z+\beta q^{-6}(X_-\triangleright\phi)\omega_z\wedge\omega_-+\gamma(X_z\triangleright\phi)\omega_-\wedge\omega_+].\end{aligned}$$

The last line comes from (4.20) and the left linearity of the $\star$-Hodge on the exterior algebra $\Omega(\mathrm{SU}_q(2))$. By (3.53) the derivative d acts on the previous 2-form as:

(4.21)
$$\begin{aligned}\mathrm{d}\star\mathrm{d}\phi&=-\alpha'[\nu(X_-X_+\triangleright\phi)(\omega_-\wedge\omega_+\wedge\omega_z)+\beta q^{-6}(X_+X_-\triangleright\phi)(\omega_+\wedge\omega_z\wedge\omega_-)\\&\quad+\gamma(X_zX_z\triangleright\phi)(\omega_z\wedge\omega_-\wedge\omega_+)]\\&=-\alpha'\{[\nu X_-X_++\beta X_+X_-+\gamma X_zX_z]\triangleright\phi\}(\omega_-\wedge\omega_+\wedge\omega_z),\end{aligned}$$

where the commutation rules (3.52) have been used. The last of (4.20) finally gives the Laplacian operator the expression:

$$\star\mathrm{d}\star\mathrm{d}\phi=-[\nu X_-X_++\beta X_+X_-+\gamma X_zX_z]\triangleright\phi \tag{4.22}$$

in terms of the left action of the quantum vector fields of the calculus. The expression (4.22) shows that $\square_{\mathrm{SU}_q(2)}:\mathcal{L}_n\to\mathcal{L}_n$. This operator can be diagonalised. One has to recall the decomposition (3.41) of the modules $\mathcal{L}_n$ for the right action of $\mathcal{U}_q(\mathfrak{su}(2))$: this right action leaves invariant the eigenspaces of the Laplacian since left and right actions of $\mathcal{U}_q(\mathfrak{su}(2))$ on $\mathcal{A}(\mathrm{SU}_q(2))$ do commute. On each irreducible subspace $V_J^{(n)}$ (3.41) for the right action of $\mathcal{U}_q(\mathfrak{su}(2))$ one has a basis $\phi_{n,J,l}=(c^{J-n/2}a^{\star J+n/2})\triangleleft E^l=w_{2J:J-\frac{n}{2},l}$ (with $l=0,\dots,2J$) of eigenvectors (3.43) for the Laplacian. The spectrum of the Laplacian does not depend on the integer l: an explicit computation shows that

$$\begin{aligned}X_zX_z\triangleright\phi_{n,J,l}&=q^{2(n+1)}[n]^2\phi_{n,J,l},\\ X_+X_-\triangleright\phi_{n,J,l}&=q^{n-1}([J-\tfrac{n}{2}][J+1+\tfrac{n}{2}]+[n])\phi_{n,J,l},\\ X_-X_+\triangleright\phi_{n,J,l}&=q^{n+1}([J-\tfrac{n}{2}][J+1+\tfrac{n}{2}])\phi_{n,J,l}.\end{aligned} \tag{4.23}$$

The spectrum of the Laplacian (4.22) is then given as $\square_{\mathrm{SU}_q(2)}\phi_{n,J,l}=\lambda_{n,J,l}\phi_{n,J,l}$ with:

$$\lambda_{n,J,l}=-q^n\{\nu q[J-\tfrac{n}{2}][J+1+\tfrac{n}{2}]+\beta q^{-1}([J-\tfrac{n}{2}][J+1+\tfrac{n}{2}]+[n])+\gamma q^{n+2}[n]^2\}. \tag{4.24}$$

5. A $\star$-Hodge structure on $\Omega(\mathrm{S}^2_q)$ and a Laplacian operator on $\mathcal{A}(\mathrm{S}^2_q)$

The way the $\star$-Hodge structure (4.20) has been introduced on $\Omega(\mathrm{SU}_q(2))$ comes from the analysis in [**22**]. The aim of this section is to extend that procedure in order to introduce a $\star$-Hodge structure on $\Omega(\mathrm{S}^2_q)$. The strategy is to directly follow the same path, and to apply to the differential calculus $\Omega(\mathrm{S}^2_q)$ the same procedure, explicitly checking its consistency in the new setting.

5.1. A $\star$-Hodge structure on $\mathcal{A}(\mathrm{S}^2_q)$. The differential calculus on the quantum sphere S^2_q has been described in section 3.4.3 and fully presented in proposition 3.6. It is a 2D left covariant calculus: as a volume form consider $\check{\theta} = i\alpha''\omega_- \wedge \omega_+$.

LEMMA 5.1. *The 2D calculus $\Omega(\mathrm{S}^2_q)$ from proposition 3.6 is non degenerate.*

PROOF. The proof of this lemma is direct. To be definite, consider a 0-form $\eta = f$ with $f \in \mathcal{A}(\mathrm{S}^2_q) \simeq \mathcal{L}^{(0)}_0$, so to have a product

$$\eta' \wedge \eta = f'(\omega_- \wedge \omega_+)f = f' f\, \omega_- \wedge \omega_+$$

from the commutation rules in (3.50), where $\eta' = f'\,\omega_- \wedge \omega_+$ with $f' \in \mathcal{L}^{(0)}_0$. One has $\eta' \wedge \eta = 0 \Leftrightarrow f' f = 0$: such a relation is satisfied for any $f' \in \mathcal{L}^{(0)}_0$ iff $f = 0$.

Consider now the 1-form $\eta = x\,\omega_-$ with $x \in \mathcal{L}^{(0)}_{-2}$, so to have a product

$$\eta' \wedge \eta = (x'\omega_- + y'\omega_+) \wedge x\,\omega_- = -y'x\,\omega_- \wedge \omega_+$$

where $(x', y') \in (\mathcal{L}^{(0)}_{-2}, \mathcal{L}^{(0)}_2)$. The relation $\eta' \wedge \eta = 0 \Leftrightarrow y'x = 0$ is satisfied for any $y' \in \mathcal{L}^{(0)}_2$ iff $x = 0$. The remaining cases can be analogously analysed, thus proving the claim. □

The restriction of the Haar state h to $\mathcal{A}(\mathrm{S}^2_q)$ yields a faithful, invariant – that is $h(f \triangleleft X) = h(f)\varepsilon(X)$ for $f \in \mathcal{A}(\mathrm{S}^2_q)$ and $X \in \mathcal{U}_q(\mathfrak{su}(2))$ – state on $\mathcal{A}(\mathrm{S}^2_q)$, allowing the definition of an integral $\int_{\check{\theta}} : \Omega(\mathrm{S}^2_q) \to \mathbb{C}$ given by:

$$\int_{\check{\theta}} f = 0, \qquad \text{on } f \in \mathcal{A}(\mathrm{S}^2_q),$$
$$\int_{\check{\theta}} \eta = 0, \qquad \text{on } \eta \in \Omega^1(\mathrm{S}^2_q),$$
$$\int_{\check{\theta}} f\,\omega_- \wedge \omega_+ = -i\alpha''^{-1}\, h(f). \tag{5.1}$$

LEMMA 5.2. *The integral $\int_{\check{\theta}}$ defined in* (5.1) *is left-faithful.*

PROOF. The proof of this result is also direct. Consider, to be definite, the 1-form $\eta = x\omega_-$ with $x \in \mathcal{L}^{(0)}_{-2}$, and a generic $\eta' = x'\omega_- + y'\omega_+ \in \Omega^1(\mathrm{S}^2_q)$. The relation $\int_{\check{\theta}} \eta' \wedge \eta = 0$ for any $\eta' \in \Omega^1(\mathrm{S}^2_q)$ is equivalent to the condition $h(y'x) = 0\ \forall\, y' \in \mathcal{L}^{(0)}_2$. Since this last equality must be valid for any $y' \in \mathcal{L}^{(0)}_2$, choosing $y' = x^*$, it results $h(x^*x) = 0$: the faithfulness of the Haar state h then gives $x = 0$. The claim of the lemma is proved by an analogous analysis on the remaining cases. □

The restriction to $\Omega(\mathrm{S}^2_q)$ of the left invariant graded product (4.17) on $\Omega(\mathrm{SU}_q(2))$, which is the one compatible with the $\star$-Hodge structure, gives a left $\mathcal{A}(\mathrm{S}^2_q)$-invariant graded inner product:

$$
\begin{aligned}
&(1,1)_{S^2_q} = 1;\\
&(x'\omega_- + y'\omega_+, x\,\omega_- + y\,\omega_+)_{S^2_q} = h(x^*x')\beta + h(y^*y')\nu;\\
&(\omega_- \wedge \omega_+, \omega_- \wedge \omega_+)_{S^2_q} = \alpha'^{-2}\gamma^{-1}, \qquad (5.2)
\end{aligned}
$$

with , $x, x' \in \mathcal{L}^{(0)}_{-2}$ and $y, y' \in \mathcal{L}^{(0)}_{2}$. Recalling proposition 4.1 – namely equation (4.3) – and the results proved in lemmas 5.1 and 5.2, a left $\mathcal{A}(S^2_q)$-linear Hodge operator $L : \Omega^k(S^2_q) \to \Omega^{2-k}(S^2_q)$ can be defined for $k = 0, 2$. From the first line in the inner product relation (5.2) one has $L(1) = \check{\theta}$, while the third gives $L(\check{\theta}) = \alpha''^2\alpha'^{-2}\gamma^{-1}$. It is evident that for such an Hodge operator it is $L^2 \neq 1$, which is a natural requirement for a $\star$-Hodge structure on $\Omega^k(S^2_q)$ for $k = 0, 2$. On the exterior algebra $\Omega(SU_q(2))$ this problem was solved by changing the inner product via the definition (4.17), and proving that the new Hodge operator does satisy all the required properties to have a consistent $\star$-Hodge. Following an analogous path, define

$$
\begin{aligned}
&(1,1)^{\natural}_{S^2_q} = 1,\\
&(x'\omega_- + y'\omega_+, x\,\omega_- + y\,\omega_+)^{\natural}_{S^2_q} = (x'\omega_- + y'\omega_+, x\,\omega_- + y\,\omega_+)_{S^2_q},\\
&(\check{\theta}, \check{\theta})^{\natural}_{S^2_q} = (L^{-1}(\check{\theta}), L^{-1}(\check{\theta}))_{S^2_q} = 1, \qquad (5.3)
\end{aligned}
$$

where the inner products on 1-forms amounts to a different labelling of the inner product in (5.2). The Hodge operator on $\Omega^k(S^2_q)$ for $k = 0, 2$ relative to such a new inner product is given by $L^{\natural}(1) = \check{\theta}$ and $L^{\natural}(\check{\theta}) = 1$. But now the inner product has changed: the requirement that the inner product $(\ ,\ )^{\natural}_{SU_q(2)}$ on the exterior algebra $\Omega(SU_q(2))$ fixed – via a restriction, as given in (5.2) – the inner product $(\ ,\ )_{S^2_q}$ on the exterior algebra $\Omega(S^2_q)$ implies that the condition

$$(\check{\theta}, \check{\theta})^{\natural}_{S^2_q} = (\check{\theta}, \check{\theta})^{\natural}_{SU_q(2)} \qquad (5.4)$$

has to be imposed, giving

$$\alpha''^2\alpha'^{-2}\gamma^{-1} = 1 \qquad (5.5)$$

as a constraint among the parameters. The constraint (5.4) can be interpreted as the quantum analogue of fixing the classical metric on the basis S^2 of the Hopf bundle as the contraction of the Cartan-Killing metric on $S^3 \sim SU(2)$, since that choice in the classical formalism, as stressed in remark 2.2, gives the equality of the inner product on $\Omega(S^2)$ defined in (2.65) with the restriction of the inner product on $\Omega(S^3)$ given in (2.35).

The differential calculus on S^2_q is even dimensional with $N = 2$, so on $\Omega^1(S^2_q)$ define a sesquilinear form:

$$\langle \eta', \eta \rangle = \int_{\check{\theta}} \eta^* \wedge \eta' = i\alpha''^{-1}\{h(y^*y') - q^2 h(x^*x')\} \qquad (5.6)$$

where $\eta = x\,\omega_- + y\,\omega_+$ and $\eta' = x'\omega_- + y'\omega_+$, with $x, x' \in \mathcal{L}^{(0)}_{-2}$ and $y, y' \in \mathcal{L}^{(0)}_{2}$. The quantum sphere S^2_q is a quantum homogeneous space and not a Hopf algebra, so there is no left-invariant basis in $\Omega^1(S^2_q)$: neverthless such a sesquilinear form can be "diagonalised", as

$$\langle x\,\omega_-, x\,\omega_-\rangle = -iq^2\alpha''^{-1}\,h(x^*x);$$
$$\langle y\,\omega_+, y\,\omega_+\rangle = i\alpha''^{-1}\,h(y^*y), \tag{5.7}$$

where the faithfulness of the Haar state ensures that the coefficients on the right hand side of these expressions never vanish. The general result from [**22**] – recalled in (4.8) – is no longer valid on a quantum homogeneous space: the diagonalisation in (5.7) suggests indeed a way to define a Hodge operator. Since α'' can be both positive or negative, define

$$x\,\theta_- = q^{-1}\left(\frac{|\alpha''|}{h(x^*x)}\right)^{1/2} x\,\omega_-,$$
$$y\,\theta_+ = \left(\frac{|\alpha''|}{h(y^*y)}\right)^{1/2} y\omega_+ \tag{5.8}$$

so to have from (5.7):

$$\langle x\,\theta_-, x\,\theta_-\rangle = -i\frac{|\alpha''|}{\alpha''},$$
$$\langle y\,\theta_+, y\,\theta_+\rangle = i\frac{|\alpha''|}{\alpha''}. \tag{5.9}$$

In the same way as in (4.8), define a left $\mathcal{A}(\mathrm{S}^2_q)$-linear operator $\mathfrak{L} : \Omega^1(\mathrm{S}^2_q) \to \Omega^1(\mathrm{S}^2_q)$ setting:

$$\mathfrak{L}(x\,\theta_-) = i\frac{|\alpha''|}{\alpha''}x\,\theta_-,$$
$$\mathfrak{L}(y\,\theta_+) = -i\frac{|\alpha''|}{\alpha''}y\,\theta_+. \tag{5.10}$$

Such an operator clearly satisfies the condition $\mathfrak{L}^2 = -1$ for any value of α''. It is not yet a consistent Hodge operator: it has to be compatible with the left invariant inner product on $\Omega^1(\mathrm{S}^2_q)$ obtained in (5.3) as a restriction of the analogue on $\Omega^1(\mathrm{SU}_q(2))$. From the relation (4.10), this compatibility must be imposed:

$$(\eta', \eta)^{\natural}_{\mathrm{S}^2_q} = \int_{\check{\theta}} \eta^* \wedge \mathfrak{L}(\eta'). \tag{5.11}$$

This condition is fulfilled if and only if the parameters in this formulation satisfy:

$$|\alpha''|\beta = q^2, \tag{5.12}$$

$$|\alpha''|\nu = 1. \tag{5.13}$$

The $\star$-Hodge structure on $\Omega(\mathrm{S}^2_q)$ is defined as a left $\mathcal{A}(\mathrm{S}^2_q)$-linear operator whose action is given by:

$$\star\,(1) = i\alpha''\,\omega_-\wedge\omega_+,$$
$$\star\,(x\,\omega_-) = i\frac{|\alpha''|}{\alpha''}(x\,\omega_-),$$
$$\star\,(y\,\omega_+) = -i\frac{|\alpha''|}{\alpha''}(y\,\omega_+),$$
$$\star\,(i\omega_-\wedge\omega_+) = \alpha''^{-1}, \tag{5.14}$$

with the parameters $\alpha', \alpha'', \beta, \nu, \gamma$ satisfying the constraints (5.5), (5.12), (5.13).

REMARK 5.3. *The $\star$-Hodge structure* (5.14) *differs from the one in* [**26**], *because in that paper the $\star$-Hodge structure was required to satisfy the relation $\star^2 = 1$, while the path followed here is to remain consistent with the requirement that $\star^2 = (-1)^{k(N-k)}$ on k-forms from a N-dimensional calculus.*

The definition (5.14) of the Hodge duality is still not complete. The constraints among the parameters involve the absolute value of α'', so one still needs to choose their relative signs. In the classical setting the only parameter was $\alpha \in \mathbb{R}$, and it has been chosen positive so to give a riemannian metric g in the analysis of section 2.2. As it is clear from (2.31) and from the definition (2.33), the positivity of the metric implies the positivity of the symmetric form $\langle\ ,\ \rangle_{S^3}$ (2.31) and of the sesquilinear inner product $\langle\ ,\ \rangle_{S^3}^{\sim}$ (2.33): the signature of the metric tensor implies the signature of both the bilinear forms

In the quantum setting, having no metric tensor, the choice of the relative signs of the parameters is equivalent to choose the signature of the left-invariant inner product (4.14) on $\Omega^1(\mathrm{SU}_q(2))$: this will encode a specific metric signature.

The natural choice for a riemannian signature is, from (4.14) and (4.18), given by $\beta, \nu, \gamma \in \mathbb{R}_+$. This choice turns out to be compatible with (5.5), (5.12) and (5.13) for every α' and α''. From (5.12) and (5.13) one also has that:

$$\beta = q^2 \nu. \tag{5.15}$$

This relation has a number of interesting and important consequences, described in the next propositions.

PROPOSITION 5.4. *The $\star$-Hodge structure given as a left $\mathcal{A}(\mathrm{S}^2_q)$-linear map $\star : \Omega^k(\mathrm{S}^2_q) \to \Omega^{2-k}(\mathrm{S}^2_q)$ for $k = 0, 1, 2$ and defined by* (5.14), *has the property*[4]

$$\star(\eta) \wedge \eta' = (-1)^{k(2-k)} \eta \wedge \star(\eta') \tag{5.16}$$

for any $\eta, \eta' \in \Omega^k(\mathrm{S}^2_q)$.

PROOF. The relation is trivially satisfied for $k = 0, 2$. Consider now the two elements $\eta = x\,\omega_- + y\,\omega_+$ and $\eta' = x'\omega_- + y'\omega_+$ in $\Omega^1(\mathrm{S}^2_q)$, which means $x, x' \in \mathcal{L}^{(0)}_{-2}$ and $y, y' \in \mathcal{L}^{(0)}_2$ by proposition 3.6. The multiplication rule from the same proposition gives:

$$\begin{aligned}(\star\eta) \wedge \eta' &= i\alpha''(\beta\, xy' + \nu\, yx')\omega_- \wedge \omega_+,\\ \eta \wedge (\star\eta') &= -i\alpha''(q^{-2}\beta\, yx' + q^2\nu\, xy')\omega_- \wedge \omega_+.\end{aligned} \tag{5.17}$$

The two expression are equal – up to the sign, which is the claim of the proposition – from (5.15). □

PROPOSITION 5.5. *The left $\mathcal{A}(\mathrm{S}^2_q)$-linear $\star$-Hodge map defined by* (5.14) *is right $\mathcal{A}(\mathrm{S}^2_q)$-linear: given $\eta \in \Omega(\mathrm{S}^2_q)$, it is $\star(\eta f) = \star(\eta) f$ for any $f \in \mathcal{A}(\mathrm{S}^2_q)$.*

PROOF. The 2D differential calculus on the quantum sphere S^2_q has the specific property, coming from the bimodule structure (3.50) of $\Omega^1(\mathrm{SU}_q(2))$ – where one has $\omega_\pm \phi = q^n \phi\, \omega_\pm$ for any $\phi \in \mathcal{L}^{(0)}_n$ – that $\omega_\pm f = f \omega_\pm$ with $f \in \mathcal{L}^{(0)}_0 \simeq \mathcal{A}(\mathrm{S}^2_q)$.

[4] In the classical formalism, the $\star$-Hodge structure on an exterior algebra coming from a N dimensional differential calculus $\star : \Omega^k(\mathcal{H}) \mapsto \Omega^{N-k}(\mathcal{H})$ satisfies the identity (2.30):

$$\eta \wedge (\star\eta') = \eta' \wedge (\star\eta)$$

to which the identity (5.16) reduces in the classical limit.

The claim of the proposition is trivial for $\eta \in \Omega^0(S^2_q) \simeq \mathcal{A}(S^2_q)$. For a 1-form $\eta = x\omega_- + y\omega_+$ in $\Omega^1(S^2_q)$, one has:

$$\star(\eta f) = \star((x\omega_- + y\omega_+)f) = \star(xf\omega_- + yf\omega_+) = i\alpha''\nu(xf\omega_- - yf\omega_+)$$
$$= i\alpha''\nu(x\omega_- - y\omega_+)f = \star(\eta)f.$$

An analogue chain of equalities is valid for $\eta = f'\omega_- \wedge \omega_+ \in \Omega^2(S^2_q)$, with $f' \in \mathcal{A}(S^2_q)$. □

In the same way it is possible to prove the following identities, which will be explicitly used in the analysis of the gauged Laplacian operator, and which slightly generalise the last proposition.

LEMMA 5.6. *Given the left $\mathcal{A}(S^2_q)$-linear $\star$-Hodge map defined by (5.14), with $\phi \in \mathcal{L}_n^{(0)}$, $\phi' \in \mathcal{L}_{-n}^{(0)}$ and $\eta \in \Omega^1(S^2_q)$ one has:*

$$\star\,(\phi'\eta\phi) = \phi'(\star\eta)\phi,$$
$$\star\,(\phi'(\omega_- \wedge \omega_+)\phi) = q^{2n}\phi'\{\star(\omega_- \wedge \omega_+)\}\phi.$$

PROOF. With $\phi'\eta\phi \in \Omega^1(S^2_q)$, and again $\eta = x\omega_- + y\omega_+$, it is explicitly:

$$\star(\phi'\eta\phi) = \star(\phi' q^n(y\phi\omega_+ + x\phi\omega_-)) = -iq^n\alpha''\nu\,\phi'(y\phi\omega_+ - x\phi\omega_-)$$
$$= -i\alpha''\nu\,\phi'(y\omega_+ - x\omega_-)\phi = \phi'(\star\eta)\phi.$$

$$\star(\phi'(\omega_-\wedge\omega_+)\phi) = q^{2n}\star(\phi'\phi(\omega_-\wedge\omega_+)) = q^{2n}\phi'\phi\star(\omega_-\wedge\omega_+) = q^{2n}\phi'\{\star(\omega_-\wedge\omega_+)\}\phi,$$

where the last equality is evident, since $\star(\omega_- \wedge \omega_+) \in \mathbb{C}$. □

5.2. A Laplacian operator on $\mathcal{A}(S^2_q)$. Using the 2D differential calculus on the Podleś sphere S^2_q and the $\star$-Hodge structure on $\Omega(S^2_q)$ it is natural to define a Laplacian operator $\Box_{S^2_q} : \mathcal{A}(S^2_q) \to \mathcal{A}(S^2_q)$ as $\Box_{S^2_q} f = \star d \star df$ on any $f \in \mathcal{A}(S^2_q)$. An explicit computation using the properties of the exterior algebra $\Omega(S^2_q)$ represented in proposition 3.6 gives:

$$\begin{aligned} df &= (X_+\triangleright f)\omega_+ + (X_-\triangleright f)\omega_-,\\ \star df &= -i\alpha''[\nu(X_+\triangleright f)\omega_+ - q^{-2}\beta(X_-\triangleright f)\omega_-],\\ d\star df &= -i\alpha''[\nu X_-X_+ + \beta X_+X_-]\triangleright f\,(\omega_- \wedge \omega_+),\\ \star d\star df &= -[\nu X_-X_+ + \beta X_+X_-]\triangleright f. \end{aligned} \tag{5.18}$$

The relation (3.32) shows that such a Laplacian operator can be seen as an operator $\Box_{S^2_q} : \mathcal{L}_0^{(0)} \to \mathcal{L}_0^{(0)}$. In particular, from (4.22), the Laplacian $\Box_{S^2_q}$ is the restriction of the Laplacian $\Box_{SU_q(2)}$ to the subalgebra $\mathcal{A}(S^2_q) \subset \mathcal{A}(SU_q(2))$. A basis of the eigenvector spaces $\mathcal{L}_0^{(0)} = \oplus_{J\in\mathbb{N}} V_J^{(0)}$ coming from (3.41) is given by elements $\phi_{0,J,l} = c^J a^{*J} \triangleleft E^l = w_{2J:J,l}$, so that formulas (4.23) drive to a spectrum of this Laplacian on S^2_q as:

$$\begin{aligned} \Box_{S^2_q}\phi_{0,J,l} &= -(q\nu + q^{-1}\beta)\{[J][J+1]\}\phi_{0,J,l}\\ &= -2q\nu\{[J][J+1]\}\phi_{0,J,l}. \end{aligned} \tag{5.19}$$

REMARK 5.7. *Equations* (4.22) *and* (5.18) *show that the classical relations between the Laplacians* $\square_{SU(2)}$ *and* $\square_{S^2}$*, coming from the Hodge duality associated to the metric tensor* g (2.26) *related to the Cartan-Killing metric, is then reproduced in the quantum formalism, in the specific realisation of the quantum Hopf bundle that has been described. The constraints among the 5 real parameters used in the analysis of the Hodge duality can be written as:*

$$\begin{aligned} \gamma &= \alpha''^2 \alpha'^{-2}, \\ \nu &= |\alpha''|^{-1}, \\ \beta &= q^2 \nu. \end{aligned} \tag{5.20}$$

The parameters α', α'' *are the coefficients of the volume forms. The analysis of the classical limit of this formulation is in section 8. The choice:*

$$\begin{aligned} \lim_{q\to 1} \alpha' &= -4\alpha, \\ \lim_{q\to 1} \alpha'' &= -2\alpha \end{aligned} \tag{5.21}$$

gives (4.22) *and* (5.18) *in the classical limit. Being* α *a positive real number, it seems natural to assume* α' *and* α'' *negative real numbers. This also gives* $\nu = -\alpha''^{-1}$ *from the second relation in* (5.20)*, so to have a Hodge duality* (5.14) *which is now:*

$$\begin{aligned} \star\,(1) &= \check{\theta} = i\alpha''\, \omega_- \wedge \omega_+, \\ \star\,(x\,\omega_-) &= -ix\,\omega_-, \\ \star\,(y\,\omega_+) &= iy\,\omega_+, \\ \star\,(i\omega_- \wedge \omega_+) &= \alpha''^{-1}, \end{aligned} \tag{5.22}$$

giving, if (5.21) *is satisfied, the Hodge duality* (2.58) *in the classical limit.*

6. Connections on the Hopf bundle

The structure of a quantum principal bundle $(\mathcal{P}, \mathcal{B}, \mathcal{H}; \mathcal{N}_{\mathcal{P}}, \mathcal{Q}_{\mathcal{H}})$ with compatible differential calculi, given the total space algebra $\mathcal{P}$ on which the gauge group Hopf algebra $\mathcal{H}$ coacts, has been described in section 3.2. The compatibility conditions ensure the exactness of the sequence (3.17):

$$0 \to \mathcal{P}\Omega^1(\mathcal{B})\mathcal{P} \to \Omega_1(\mathcal{P}) \stackrel{\sim_{\mathcal{N}_{\mathcal{P}}}}{\longrightarrow} \mathcal{P} \otimes (\ker \varepsilon_{\mathcal{H}}/\mathcal{Q}_{\mathcal{H}}) \to 0. \tag{6.1}$$

with the map $\sim_{\mathcal{N}_{\mathcal{P}}}$ defined via the commutative diagram (3.15). Among the compatibility conditions, the requirement that $\Delta_R \mathcal{N}_{\mathcal{P}} \subset \mathcal{N}_{\mathcal{P}} \otimes \mathcal{H}$ – giving a right covariance of the differential structure on $\mathcal{P}$ – allows to extend the coaction Δ_R of $\mathcal{H}$ on $\mathcal{P}$ to a coaction of $\mathcal{H}$ on 1-forms, $\Delta_R^{(1)} : \Omega^1(\mathcal{P}) \to \Omega^1(\mathcal{P}) \otimes \mathcal{H}$, defining $\Delta_R^{(1)} \circ \mathrm{d} = (\mathrm{d} \otimes 1) \circ \Delta_R$.

Note that $\mathrm{Ad}(\ker \varepsilon_{\mathcal{H}}) \subset (\ker \varepsilon_{\mathcal{H}}) \otimes \mathcal{H}$. If the right ideal $\mathcal{Q}_{\mathcal{H}}$ is Ad-invariant (which is equivalent to say that the differential calculus on $\mathcal{H}$ is bicovariant), it is possible to define a right-adjoint coaction $\mathrm{Ad}^{(R)} : \ker\varepsilon_{\mathcal{H}}/\mathcal{Q}_{\mathcal{H}} \to \ker\varepsilon_{\mathcal{H}}/\mathcal{Q}_{\mathcal{H}} \otimes \mathcal{H}$ by the commutative diagram

$$\begin{array}{ccc} \ker\varepsilon_{\mathcal{H}} & \xrightarrow{\pi_{\mathcal{Q}_{\mathcal{H}}}} & \ker\varepsilon_{\mathcal{H}}/\mathcal{Q}_{\mathcal{H}} \\ \downarrow \mathrm{Ad} & & \downarrow \mathrm{Ad}^{(R)} \\ \ker\varepsilon_{\mathcal{H}} \otimes \mathcal{H} & \xrightarrow{\pi_{\mathcal{Q}_{\mathcal{H}}}\otimes\mathrm{id}} & (\ker\varepsilon_{\mathcal{H}}/\mathcal{Q}_{\mathcal{H}}) \otimes \mathcal{H} \end{array}$$

Together with the right coaction Δ_R of $\mathcal{H}$ on $\mathcal{P}$, such a right-adjoint coaction $\mathrm{Ad}^{(R)}$ allows to define a right coaction $\Delta_R^{(\mathrm{Ad})}$ of $\mathcal{H}$ on $\mathcal{P} \otimes \ker\varepsilon_{\mathcal{H}}/\mathcal{Q}_{\mathcal{H}}$ as a coaction of a Hopf algebra on the tensor product of its comodules. This coaction is explicitly given by the relation:

$$\Delta_R^{(\mathrm{Ad})}(p \otimes \pi_{\mathcal{Q}_{\mathcal{H}}}(h)) = p_{(0)} \otimes \pi_{\mathcal{Q}_{\mathcal{H}}}(h_{(2)}) \otimes p_{(1)}(Sh_{(1)})h_{(3)}, \tag{6.2}$$

adopting the Sweedler notation for the coaction as $\Delta_R(p) = p_{(0)} \otimes p_{(1)}$.

It is now possible to define a connection on the quantum principal bundle as a right invariant splitting of the sequence (6.1). Given a left $\mathcal{P}$-linear map $\sigma : \mathcal{P} \otimes (\ker\varepsilon_{\mathcal{H}}/\mathcal{Q}_{\mathcal{H}}) \to \Omega^1(\mathcal{P})$ such that

$$\begin{aligned} \Delta_R^{(1)} \circ \sigma &= (\sigma \otimes id)\Delta_R^{(\mathrm{Ad})}, \\ \sim_{\mathcal{N}_{\mathcal{P}}} \circ\sigma &= id, \end{aligned} \tag{6.3}$$

then the map $\Pi : \Omega^1(\mathcal{P}) \to \Omega^1(\mathcal{P})$ defined by $\Pi = \sigma\circ \sim_{\mathcal{N}_{\mathcal{P}}}$ is a right invariant left $\mathcal{P}$-linear projection, whose kernel coincides with the horizontal forms $\mathcal{P}\Omega^1(\mathcal{B})\mathcal{P}$:

$$\begin{aligned} \Pi^2 &= \Pi, \\ \Pi(\mathcal{P}\Omega^1(\mathcal{B})\mathcal{P}) &= 0, \\ \Delta_R^{(1)} \circ \Pi &= (\Pi \otimes id) \circ \Delta_R^{(1)}. \end{aligned} \tag{6.4}$$

The image of the projection Π is the set of vertical 1-forms of the principal bundle. A connection on a principal bundle can also be written in terms of a connection 1-form, which is a map $\omega : \mathcal{H} \to \Omega^1(\mathcal{P})$. Given a right invariant splitting σ of the exact sequence (6.1), define the connection 1-form as $\omega(h) = \sigma(1 \otimes \pi_{\mathcal{Q}_{\mathcal{H}}}(h - \varepsilon_{\mathcal{H}}(h)))$ on $h \in \mathcal{H}$. Such a connection 1-form has the following properties:

$$\begin{aligned} \omega(\mathcal{Q}_{\mathcal{H}}) &= 0, \\ \sim_{\mathcal{N}_{\mathcal{P}}} (\omega(h)) &= 1 \otimes \pi_{\mathcal{Q}_{\mathcal{H}}}(h - \varepsilon_{\mathcal{H}}(h)) \qquad \forall\, h \in \mathcal{H}, \\ \Delta_R^{(1)} \circ \omega &= (\omega \otimes \mathrm{id}) \circ \mathrm{Ad}, \\ \Pi(\mathrm{d}p) &= (id \otimes \omega)\Delta_R(p) \qquad \forall\, p \in \mathcal{P}. \end{aligned} \tag{6.5}$$

Conversely if ω is a linear map $\ker\varepsilon_{\mathcal{H}} \to \Omega^1(\mathcal{P})$ that satisfies the first three conditions in (6.5), then there exists a unique connection on the principal bundle, such that ω is its connection 1-form. In this case, the splitting of the sequence (6.1) is given by:

$$\sigma(p \otimes [h]) = p\omega([h]) \tag{6.6}$$

with $[h]$ in $\ker\varepsilon_{\mathcal{H}}/\mathcal{Q}_{\mathcal{H}}$, while the projection Π is given by:

$$\Pi = m \circ (id \otimes \omega)\circ \sim_{\mathcal{N}_{\mathcal{P}}} \tag{6.7}$$

The general proof of these results is in [**6**]. This section explicitly describes the connections on the quantum Hopf bundle with the compatible differential calculi presented in sections 3.4.1 and 3.4.2.

6.1. Vertical subspaces on the quantum Hopf bundle. The right coaction $\Delta_R^{(1)} : \Omega^1(\mathrm{SU}_q(2)) \to \Omega^1(\mathrm{SU}_q(2)) \otimes \mathcal{A}(\mathrm{U}(1))$ of the gauge group algebra $\mathcal{A}(\mathrm{U}(1))$ on the set of 1-forms on the total space algebra of the bundle, whose consistency is allowed by the compatibility conditions between the 3D left covariant calculus on $\mathcal{A}(\mathrm{SU}_q(2))$ and the 1D bicovariant calculus on $\mathcal{A}(\mathrm{U}(1))$, gives:

$$\begin{aligned} \Delta_R^{(1)} \omega_z &= \omega_z \otimes 1, \\ \Delta_R^{(1)} \omega_\pm &= \omega_\pm \otimes z^{\pm 2}. \end{aligned} \tag{6.8}$$

From the analysis on the 1D calculus on $\mathcal{A}(\mathrm{U}(1))$ performed in section 3.4.4 and the result of lemma 3.4, a connection on the quantum Hopf bundle is given via a splitting map $\sigma : \mathcal{A}(\mathrm{SU}_q(2)) \otimes (\ker \varepsilon_{\mathrm{U}(1)}/\mathcal{Q}_{\mathrm{U}(1)}) \to \Omega^1(\mathrm{SU}_q(2))$, which can be defined recalling the isomorphism $\tilde{\lambda} : \ker \varepsilon_{\mathrm{U}(1)}/\mathcal{Q}_{\mathrm{U}(1)} \to \mathbb{C}$. Given $w \in \mathbb{C}$ set:

$$\sigma(1 \otimes w) = \sigma(w \otimes 1) = w(\omega_z + U\omega_+ + V\omega_-); \tag{6.9}$$

and extend by the requirement of left $\mathcal{A}(\mathrm{SU}_q(2))$-linearity, so to have:

$$\begin{aligned} \sigma(1 \otimes [\varphi(j)]) &= q^{-2j}(\omega_z + U\omega_+ + V\omega_-), \\ \sigma(\phi \otimes [\varphi(j)]) &= q^{-2j}\phi(\omega_z + U\omega_+ + V\omega_-), \end{aligned} \tag{6.10}$$

where $\phi \in \mathcal{A}(\mathrm{SU}_q(2))$ and the requirement of right covariance (6.3) selects – from (6.8) – $U \in \mathcal{L}_2^{(0)}$ and $V \in \mathcal{L}_{-2}^{(0)}$. The projection Π associated to this connection is easily seen to be:

$$\begin{aligned} \Pi(\omega_\pm) &= \sigma(\sim_{\mathcal{N}_{\mathrm{SU}_q(2)}} (\omega_\pm)) = 0, \\ \Pi(\omega_z) &= \sigma(\sim_{\mathcal{N}_{\mathrm{SU}_q(2)}} (\omega_z)) = \sigma(1 \otimes [\varphi(0)]) = \omega_z + U\omega_+ + V\omega_-. \end{aligned} \tag{6.11}$$

In this expression the 1-forms $\omega_\pm$ are recovered as horizontal (3.62), a notion depending only on the compatibility conditions between the differential calculi, while a choice of a connection is equivalent to the choice of the vertical part of $\Omega^1(\mathrm{SU}_q(2))$. The set of connections for the quantum Hopf bundle corresponds to the set of the possible choices of 1-forms on the basis of the bundle as $\mathrm{a} = U\omega_+ + V\omega_- \in \Omega^1(\mathrm{S}_q^2)$, so that the second line in (6.11) can be written as

$$\Pi(\omega_z) = \omega_z + \mathrm{a}. \tag{6.12}$$

The connection one form (6.5) $\omega : U(1) \to \Omega^1(\mathrm{SU}_q(2))$ is given by:

$$\begin{aligned} \omega(z^j) &= \sigma(1 \otimes [z^j - 1]) \\ &= \left(\frac{1 - q^{-2j}}{1 - q^{-2}}\right)(\omega_z + U\omega_+ + V\omega_-) = \left(\frac{1 - q^{-2j}}{1 - q^{-2}}\right)(\omega_z + \mathrm{a}). \end{aligned} \tag{6.13}$$

Given the projection Π and the connection 1-form ω, it is possible to compute the lhs and the rhs of the last line in (6.5). On the basis of left invariant differential forms and using the explicit form of the quantum vector fields in (3.45), with $\phi \in \mathcal{L}_j^{(0)}$ one has:

$$\begin{aligned} \Pi(\mathrm{d}\phi) = \Pi((X_j \triangleright \phi)\omega_j) &= (X_j \triangleright \phi)\Pi(\omega_j) \\ &= \left(\frac{1 - q^{2j}}{1 - q^{-2}}\right)\phi(\omega_z + U\omega_+ + V\omega_-); \end{aligned} \tag{6.14}$$

and also:

$$(id \otimes \omega)\Delta_R(\phi) = (id \otimes \omega)(\phi \otimes z^{-j})$$
$$= \left(\frac{1-q^{2j}}{1-q^{-2}}\right)\phi(\omega_z + U\omega_+ + V\omega_-) = \Pi(\mathrm{d}\phi). \tag{6.15}$$

The monopole connection corresponds to the choice $U = V = 0 \Leftrightarrow \mathrm{a} = 0$, so to have $\Pi_0(\omega_z) = \omega_z$ and the monopole connection 1-form $\omega_0(z^j) = [(1-q^{-2j})/(1-q^{-2})]\omega_z$ [**5, 9**]. With a connection, one has the notion of covariant derivative $D : \mathcal{A}(\mathrm{SU}_q(2)) \to \Omega^1(\mathcal{A}(\mathrm{SU}_q(2)))$ of equivariant maps. Given $\phi \in \mathcal{L}_n^{(0)}$, define

$$D\phi = (1 - \Pi)\mathrm{d}\phi. \tag{6.16}$$

The covariant derivative $D\phi$ is clearly an horizontal 1-form: the adjective "covariant" refers to the behaviour under the coaction of the gauge group algebra, as one directly (3.33) shows that:

$$\Delta_R\phi = \phi \otimes z^{-j} \qquad \Leftrightarrow \qquad \Delta_R^{(1)}(D\phi) = D\phi \otimes z^{-j}, \tag{6.17}$$

from the right invariance (6.4) of the projection Π. In terms of the connection 1-form the covariant derivative can be written, using (6.15), as :

$$D\phi = (1-\Pi)\mathrm{d}\phi = \mathrm{d}\phi - \Pi(\mathrm{d}\phi)$$
$$= \mathrm{d}\phi - \phi \wedge \omega(z^{-j}) \tag{6.18}$$

on a $\phi \in \mathcal{L}_j^{(0)}$. It is then immediate to recover that, for any $f \in \mathcal{L}_0^{(0)} \simeq \mathcal{A}(\mathrm{S}_q^2)$, one has $Df = \mathrm{d}f$.

REMARK 6.1. *Given any $\phi \in \mathcal{L}_n^{(0)}$, from* (6.18) *and* (6.12), *the covariant derivative can be written as:*

$$D\phi = \{(X_+ \triangleright \phi) - (X_z \triangleright \phi)U\}\omega_+ + \{(X_- \triangleright \phi) - (X_z \triangleright \phi)V\}\omega_-.$$

It is an easy computation using the $\mathcal{A}(\mathrm{SU}_q(2))$-bimodule properties (3.50) *of $\Omega^1(\mathrm{SU}_q(2))$ to prove that $D\phi \simeq \Omega^1(\mathrm{S}_q^2) \cdot \mathcal{A}(\mathrm{SU}_q(2))$ for any connection represented by* $\mathrm{a} \in \Omega^1(\mathrm{S}_q^2)$. *This means that any connection on this quantum Hopf bundle is a strong connection, following the analysis in* [**16**].

6.2. Covariant derivative on the associated line bundles. A covariant derivative, or a connection, on the left $\mathcal{A}(\mathrm{S}_q^2)$-module $\mathcal{E}_n^{(0)}$ is a $\mathbb{C}$-linear map

$$\nabla : \Omega^k(\mathrm{S}_q^2) \otimes_{\mathcal{A}(\mathrm{S}_q^2)} \mathcal{E}_n^{(0)} \to \Omega^{k+1}(\mathrm{S}_q^2) \otimes_{\mathcal{A}(\mathrm{S}_q^2)} \mathcal{E}_n^{(0)}, \tag{6.19}$$

defined for any $k \geq 0$ and satisfying a left Leibniz rule:

$$\nabla(\alpha \langle\sigma|) = \mathrm{d}\alpha \wedge \langle\sigma| + (-1)^m \alpha \wedge (\nabla\langle\sigma|)$$

for any $\alpha \in \Omega^m(\mathrm{S}_q^2)$ and $\langle\sigma| \in \Omega^k(\mathrm{S}_q^2) \otimes_{\mathcal{A}(\mathrm{S}_q^2)} \mathcal{E}_n^{(0)}$. A connection is completely determined by its restriction $\nabla : \mathcal{E}_n^{(0)} \to \Omega^1(\mathrm{S}_q^2) \otimes_{\mathcal{A}(\mathrm{S}_q^2)} \mathcal{E}_n^{(0)}$ and then extended by the Leibniz rule. Connections always exist on projective modules: the canonical (Levi-Civita, or Grassmann) connection on a left projective $\mathcal{A}(\mathrm{S}_q^2)$-module $\mathcal{E}_n^{(0)}$ is given as

$$\nabla_0\langle\sigma| = (\mathrm{d}\langle\sigma|)\mathfrak{p}^{(n)}; \tag{6.20}$$

the space $C(\mathcal{E}_n^{(0)})$ of all connections on $\mathcal{E}_n^{(0)}$ is an affine space modelled on

$$\mathrm{Hom}_{\mathcal{A}(\mathrm{S}_q^2)}(\mathcal{E}_n^{(0)}, \mathcal{E}_n^{(0)} \otimes_{\mathcal{A}(\mathrm{S}_q^2)} \Omega^1(\mathrm{S}_q^2)),$$

so that any connection can be written as:

$$\nabla \langle\sigma| = (\mathrm{d}\langle\sigma|)\mathfrak{p}^{(n)} + (-1)^k \langle\sigma| \wedge \mathrm{A}^{(\mathrm{n})} \tag{6.21}$$

with $\langle\sigma| \in \Omega^k(\mathrm{S}^2_q)\otimes_{\mathcal{A}(\mathrm{S}^2_q)}\mathcal{E}^{(0)}_n$ and $\mathrm{A}^{(\mathrm{n})} \in \mathbb{M}_{|n|+1}\otimes_{\mathcal{A}(\mathrm{S}^2_q)}\Omega^1(\mathrm{S}^2_q)$ – which is called the gauge potential of the connection ∇ – subject to the condition $\mathrm{A}^{(\mathrm{n})} = \mathrm{A}^{(\mathrm{n})}\mathfrak{p}^{(n)} = \mathfrak{p}^{(n)}\mathrm{A}^{(\mathrm{n})}$. The composition

$$\nabla^2 = \nabla\circ\nabla \quad : \quad \Omega^k(\mathrm{S}^2_q)\otimes_{\mathcal{A}(\mathrm{S}^2_q)}\mathcal{E}^{(0)}_n \to \Omega^{k+2}(\mathrm{S}^2_q)\otimes_{\mathcal{A}(\mathrm{S}^2_q)}\mathcal{E}^{(0)}_n$$

is $\Omega(\mathrm{S}^2_q)$-linear. This map can be explicitly calculated: given $\langle\sigma| \in \Omega^k(\mathrm{S}^2_q)\otimes_{\mathcal{A}(\mathrm{S}^2_q)}\mathcal{E}^{(0)}_n$, from (6.21) one has

$$\begin{aligned}
\nabla^2\langle\sigma| &= \mathrm{d}(\nabla\langle\sigma|)\mathfrak{p}^{(n)} + (-1)^{k+1}(\nabla\langle\sigma|)\wedge\mathrm{A}^{(\mathrm{n})} \\
&= \mathrm{d}\{(\mathrm{d}\sigma)\mathfrak{p}^{(n)} + (-1)^k\langle\sigma|\wedge\mathrm{A}^{(\mathrm{n})}\}\mathfrak{p}^{(n)} + (-1)^{k+1}\{(\mathrm{d}\langle\sigma|)\mathfrak{p}^{(n)} \qquad (6.22)\\
&\quad + (-1)^k\langle\sigma|\wedge\mathrm{A}^{(\mathrm{n})}\}\wedge\mathrm{A}^{(\mathrm{n})} \\
&= \mathrm{d}\{(\mathrm{d}\langle\sigma|)\mathfrak{p}^{(n)}\}\mathfrak{p}^{(n)} + (-1)^k(\mathrm{d}\langle\sigma|\wedge\mathrm{A}^{(\mathrm{n})})\mathfrak{p}^{(n)} + (\langle\sigma|\wedge\mathrm{d}\mathrm{A}^{(\mathrm{n})})\mathfrak{p}^{(n)} \\
&\qquad + (-1)^{k+1}\{(\mathrm{d}\langle\sigma|)\mathfrak{p}^{(n)}\wedge\mathrm{A}^{(\mathrm{n})}\} - \langle\sigma|\wedge\mathrm{A}^{(\mathrm{n})}\wedge\mathrm{A}^{(\mathrm{n})} \\
&= \langle\sigma|\{-(\mathrm{d}\mathfrak{p}^{(n)}\wedge\mathrm{d}\mathfrak{p}^{(n)})\mathfrak{p}^{(n)} + (\mathrm{d}\mathrm{A}^{(\mathrm{n})})\mathfrak{p}^{(n)} - \mathrm{A}^{(\mathrm{n})}\wedge\mathrm{A}^{(\mathrm{n})}\}. \qquad (6.23)
\end{aligned}$$

The restriction of the map ∇^2 to $\mathcal{E}^{(0)}_n$, seen as an element in $\Omega^2(\mathrm{S}^2_q)\otimes_{\mathcal{A}(\mathrm{S}^2_q)}\mathcal{E}^{(0)}_n$, is the curvature F_∇ of the given connection.

The left $\mathcal{A}(\mathrm{S}^2_q)$-module isomorphism between $\mathcal{L}^{(0)}_n$ and $\mathcal{E}^{(0)}_n$ described in proposition 3.3 allows for the definition of an hermitian structure on each projective left module $\mathcal{E}^{(0)}_n$, $\{\ ;\ \} : \mathcal{E}^{(0)}_n \times \mathcal{E}^{(0)}_n \to \mathcal{A}(\mathrm{S}^2_q)$ given as:

$$\{\langle\sigma|_\phi\,;\langle\sigma|_{\phi'}\} = \phi\phi'^*, \tag{6.24}$$

with $\phi, \phi' \in \mathcal{L}^{(0)}_n$. Such an hermitian structure satisfies the relations:

$$\begin{aligned}
&\{f\langle\sigma|_\phi\,; f'\langle\sigma|_{\phi'}\} = f\phi(f'\phi')^*, \\
&\{\langle\sigma|_\phi\,,\langle\sigma|_\phi\} \geq 0, \quad \{\langle\sigma|_\phi\,,\langle\sigma|_\phi\} = 0 \Leftrightarrow \langle\sigma| = 0.
\end{aligned}$$

The left $\mathcal{A}(\mathrm{S}^2_q)$-module isomorphism between $\mathcal{L}^{(0)}_n$ and $\mathcal{E}^{(0)}_n$ also enables to relate the concept of connection on the quantum Hopf bundle to that of covariant derivative on the associated line bundles. As first step, define the $\mathcal{A}(\mathrm{S}^2_q)$-bimodule:
(6.25)

$$\mathcal{L}^{(1)}_n = \{\phi \in \Omega^1_{\mathrm{hor}}(\mathrm{SU}_q(2)) \simeq \mathcal{A}(\mathrm{SU}_q(2))\Omega^1(\mathrm{S}^2_q)\mathcal{A}(\mathrm{SU}_q(2))\ :\ \Delta^{(1)}_R\phi = \phi\otimes z^{-n}\}$$

and introduce the notations:

$$\mathcal{E}^{(k)}_n = \Omega^k(\mathrm{S}^2_q)\otimes_{\mathcal{A}(\mathrm{S}^2_q)}\mathcal{E}^{(0)}_n.$$

The maps:

$$\begin{aligned}
&\mathcal{L}^{(1)}_n \xrightarrow{\simeq} \mathcal{E}^{(1)}_n : \qquad \phi \mapsto \langle\sigma|_\phi = \phi\left\langle\Psi^{(n)}\right|, \\
&\mathcal{E}^{(1)}_n \xrightarrow{\simeq} \mathcal{L}^{(1)}_n : \qquad \langle\sigma| \mapsto \phi = \left\langle\sigma, \Psi^{(n)}\right\rangle \qquad (6.26)
\end{aligned}$$

give left $\mathcal{A}(\mathrm{S}^2_q)$-module isomorphisms (in this notation the explicit dependence on $\langle f| \in \mathcal{A}(\mathrm{S}^2_q)^{|n|+1}$ as in proposition 3.3 has been dropped). Via this isomorphism, any connection on the quantum Hopf bundle – represented by a projection Π (6.11) or by a connection 1-form (6.13) – induces a gauge potential $\mathrm{A}^{(\mathrm{n})}$ on any associated line bundle $\mathcal{E}_n^{(0)}$.

PROPOSITION 6.2. *Given the left $\mathcal{A}(\mathrm{S}^2_q)$-isomorphism $\mathcal{L}_n^{(0)} \simeq \mathcal{E}_n^{(0)}$ described in proposition 3.3, as well as the analogue left $\mathcal{A}(\mathrm{S}^2_q)$-module isomorphism $\mathcal{L}_n^{(1)} \simeq \mathcal{E}_n^{(1)}$ described in* (6.26)*, there is an equivalence between the set of connections on the quantum Hopf bundle via a projection Π in $\Omega(\mathrm{SU}_q(2))$ as in* (6.11)*, and the set of covariant derivative $\nabla \in C(\mathcal{E}_n^{(0)})$ on any associated line bundle. With $\phi \in \mathcal{L}_n^{(0)}$ so that $\langle\sigma_\phi| = \phi\left\langle\Psi^{(n)}\right| \in \mathcal{E}_n^{(0)}$, the equivalence is given by $D\phi = (\nabla\langle\sigma|_\phi)\left|\Psi^{(n)}\right\rangle$.*

PROOF. Choose $\phi \in \mathcal{L}_n^{(0)}$, so to have $\sigma_\phi = \phi\left\langle\Psi^{(n)}\right|$ and from the definition in (6.21) express a covariant derivative on $\mathcal{E}_n^{(0)}$ via a gauge potential as:

$$\nabla\langle\sigma|_\phi = \mathrm{d}\left(\phi\left\langle\Psi^{(n)}\right|\right)\left|\Psi^{(n)}\right\rangle\left\langle\Psi^{(n)}\right| + \phi\left\langle\Psi^{(n)}\right|\mathrm{A}^{(\mathrm{n})} \tag{6.27}$$

$$= \{\mathrm{d}\phi - \phi[\left\langle\Psi^{(n)},\mathrm{d}\Psi^{(n)}\right\rangle - \left\langle\Psi^{(n)}\right|\mathrm{A}^{(\mathrm{n})}\left|\Psi^{(n)}\right\rangle]\}\left\langle\Psi^{(n)}\right| \tag{6.28}$$

since $\mathrm{A}^{(\mathrm{n})} = \mathrm{A}^{(\mathrm{n})}\mathfrak{p}^{(n)}$. On the other hand, being $\phi \in \mathcal{L}_n^{(0)}$ one has:

$$\begin{aligned} D\phi &= (1-\Pi)\mathrm{d}\phi = \mathrm{d}\phi - (X_z \triangleright \phi)\Pi(\omega_z) \\ &= \mathrm{d}\phi - \left(\frac{1-q^{2n}}{1-q^{-2}}\right)\phi\,\Pi(\omega_z), \end{aligned}$$

with $D\phi \in \mathcal{L}_n^{(1)}$ from (6.17). By the isomorphism (6.26), equating

$$D\phi = (\nabla\langle\sigma|_\phi)\left|\Psi^{(n)}\right\rangle$$

defines the gauge potential $\mathrm{A}^{(\mathrm{n})}$ as:

$$\begin{aligned} \left\langle\Psi^{(n)},\mathrm{d}\Psi^{(n)}\right\rangle - \left\langle\Psi^{(n)}\right|\mathrm{A}^{(\mathrm{n})}\left|\Psi^{(n)}\right\rangle &= \frac{1-q^{2n}}{1-q^{-2}}(\omega_z + U\omega_+ + V\omega_-) \\ &= \frac{1-q^{2n}}{1-q^{-2}}(\omega_z + \mathrm{a}) = \omega(z^{-n}) : \end{aligned} \tag{6.29}$$

an explicit calculation shows that $\left\langle\Psi^{(n)},\mathrm{d}\Psi^{(n)}\right\rangle = [(1-q^{2n})/(1-q^{-2})]\omega_z$, so the previous expression becomes:

$$\left\langle\Psi^{(n)}\right|\mathrm{A}^{(\mathrm{n})}\left|\Psi^{(n)}\right\rangle = -\frac{1-q^{2n}}{1-q^{-2}}(U\omega_+ + V\omega_-), \tag{6.30}$$

which is solved by

$$\begin{aligned} \mathrm{A}^{(\mathrm{n})} &= -\frac{1-q^{2n}}{1-q^{-2}}\left|\Psi^{(n)}\right\rangle(U\omega_+ + V\omega_-)\left\langle\Psi^{(n)}\right| \\ &= -\frac{1-q^{2n}}{1-q^{-2}}\left|\Psi^{(n)}\right\rangle \mathrm{a}\left\langle\Psi^{(n)}\right|. \end{aligned} \tag{6.31}$$

This solution is unique. Being the set of connection an affine space, any different gauge potential, solution of equation (6.30), should be $\check{\mathrm{A}}^{(n)} = \mathrm{A}^{(\mathrm{n})} + \mathrm{A}'^{(\mathrm{n})}$ where $\mathrm{A}^{(\mathrm{n})}$ is given in (6.31) and $\mathrm{A}'^{(\mathrm{n})}$ must satisfy $\left\langle\Psi^{(n)}\right|\mathrm{A}'^{(\mathrm{n})}\left|\Psi^{(n)}\right\rangle = 0$, with $\mathrm{A}'^{(\mathrm{n})} =$

$\mathfrak{p}^{(n)}\mathrm{A}'^{(\mathrm{n})}\mathfrak{p}^{(n)} = \mathfrak{p}^{(n)}\mathrm{A}'^{(\mathrm{n})} = \mathrm{A}'^{(\mathrm{n})}\mathfrak{p}^{(n)} = \mathrm{A}'^{(\mathrm{n})}$. One directly has:

$$\left\langle \Psi^{(n)} \middle| \mathrm{A}'^{(\mathrm{n})} \middle| \Psi^{(n)} \right\rangle = 0$$
$$\Rightarrow\ 0 = \left|\Psi^{(n)}\right\rangle \left\langle \Psi^{(n)} \middle| \mathrm{A}'^{(\mathrm{n})} \middle| \Psi^{(n)} \right\rangle \left\langle \Psi^{(n)}\right| = \mathfrak{p}^{(n)}\mathrm{A}'^{(\mathrm{n})}\mathfrak{p}^{(n)} = \mathrm{A}'^{(\mathrm{n})}.$$

The complete equivalence claimed in the proposition comes by (6.30), which gives for any gauge potential $\mathrm{A}^{(\mathrm{n})}$ a 1-form $\mathrm{a} \in \Omega^1(\mathrm{S}^2_q)$, suitable to define a connection as in (6.12). □

The form of the gauge potential (6.31) shows that the monopole connection $\Pi_0(\omega_z) = \omega_z$ corresponds to the Grassmann, or canonical covariant derivative $\nabla_0 \langle\sigma| = (\mathrm{d}\langle\sigma|)\mathfrak{p}^{(n)}$ on the line bundles $\mathcal{E}^{(0)}_n$, having $\mathrm{A}^{(\mathrm{n})} = 0$ for any $n \in \mathbb{Z}$. A connection on the quantum Hopf bundle is defined compatible with the hermitian structure (6.24) on each module of sections of the associated line bundle if

$$\mathrm{d}\{\langle\sigma|_\phi ; \langle\sigma|_{\phi'}\} = \{\nabla\langle\sigma|_\phi ; \langle\sigma|_{\phi'}\} + \{\langle\sigma|_\phi ; \nabla\langle\sigma|_{\phi'}\}.$$

It is easy to compute that this condition amounts to have a connection (6.12) satisfying the condition

$$\mathrm{a}^* = -\mathrm{a}.$$

The compatibility between the differential calculi allows to extend the concept of right coaction of the gauge group algebra on the whole exterior algebra $\Omega(\mathrm{SU}_q(2))$, introducing a right coaction $\Delta^{(k)}_R : \Omega^k(\mathrm{SU}_q(2)) \to \Omega^k(\mathrm{SU}_q(2)) \otimes \mathcal{A}(\mathrm{U}(1))$ by induction as

$$\Delta^{(k)}_R \circ \mathrm{d} = (\mathrm{d} \otimes \mathrm{id}) \circ \Delta^{(k-1)}_R. \tag{6.32}$$

It becomes now natural to define the $\mathcal{A}(\mathrm{S}^2_q)$-bimodule:

$$\mathcal{L}^{(2)}_n = \{\phi \in \mathcal{A}(\mathrm{SU}_q(2))\Omega^2(\mathrm{S}^2_q)\mathcal{A}(\mathrm{SU}_q(2)) \ :\ \Delta^{(2)}_R\phi = \phi \otimes z^{-n}\}; \tag{6.33}$$

so that the maps:

$$\begin{aligned} \mathcal{L}^{(2)}_n \xrightarrow{\simeq} \mathcal{E}^{(2)}_n : &\qquad \phi \mapsto \langle\sigma|_\phi = \phi\left\langle \Psi^{(n)}\right|, \\ \mathcal{E}^{(2)}_n \xrightarrow{\simeq} \mathcal{L}^{(2)}_n : &\qquad \langle\sigma| \mapsto \phi = \left\langle \sigma, \Psi^{(n)}\right\rangle \end{aligned} \tag{6.34}$$

are left $\mathcal{A}(\mathrm{S}^2_q)$-module isomorphisms, generalising the isomorphisms given in proposition 3.3 and in (6.26). In the formulation of [**6**], the elements in $\mathcal{L}^{(k)}_n$ are strongly tensorial forms.

Recall that the covariant derivative ∇ is defined in (6.19) as an operator $\nabla : \mathcal{E}^{(k)}_n \to \mathcal{E}^{(k+1)}_n$ for $k = 0, 1, 2$, since the differential calculus on $\mathcal{A}(\mathrm{S}^2_q)$ is 2 dimensional; the covariant derivative D has been defined by (6.16) only on the $\mathcal{A}(\mathrm{S}^2_q)$-bimodule $\mathcal{L}^{(0)}_n$, while the proposition 6.2 shows the equivalence between $D : \mathcal{L}^{(0)}_n \to \mathcal{L}^{(1)}_n$ and $\nabla : \mathcal{E}^{(0)}_n \to \mathcal{E}^{(1)}_n$. The isomorphism (6.34) allows then to extend the covariant derivative to $D : \mathcal{L}^{(1)}_n \to \mathcal{L}^{(2)}_n$, defining:

$$D\phi = (\nabla\langle\sigma|_\phi)\left|\Psi^{(n)}\right\rangle \tag{6.35}$$

for any $\phi \in \mathcal{L}_n^{(1)}$ with $\langle\sigma|_\phi = \phi\left\langle\Psi^{(n)}\right| \in \mathcal{E}_n^{(1)} = \Omega^1(\mathrm{S}_q^2) \otimes_{\mathcal{A}(\mathrm{S}_q^2)} \mathcal{E}_n^{(0)}$. Such an operator can be represented in terms of the connection (6.13) 1-form ω. From the Leibniz rule one has:

$$\mathrm{d}(\phi\left\langle\Psi^{(n)}\right|) = (\mathrm{d}\phi)\left\langle\Psi^{(n)}\right| + (-1)^k\phi\,\mathrm{d}\left\langle\Psi^{(n)}\right|,$$

with $\phi \in \mathcal{L}_n^{(k)}$. This identity gives the next proposition.

PROPOSITION 6.3. *Given $\phi \in \mathcal{L}_n^{(1)}$, so that $\langle\sigma|_\phi = \phi\left\langle\Psi^{(n)}\right| \in \mathcal{E}_n^{(1)}$, the action of the operator $D : \mathcal{L}_n^{(1)} \to \mathcal{L}_n^{(2)}$ defined by* (6.35) *can be written as:*

$$D\phi = \mathrm{d}\phi + \phi \wedge \omega(z^{-n}) \tag{6.36}$$

PROOF. The proposition is proved by a direct computation. Start from $\phi \in \mathcal{L}_n^{(1)}$, so that from(6.21) one has $\nabla\langle\sigma|_\phi = (\mathrm{d}\langle\sigma|_\phi)\mathfrak{p}^{(n)} - \langle\sigma|_\phi \wedge \mathrm{A}^{(\mathrm{n})}$, so that :

$$\begin{aligned} D\phi &= (\nabla\langle\sigma|_\phi)\left|\Psi^{(n)}\right\rangle \\ &= (\mathrm{d}\langle\sigma|_\phi)\left|\Psi^{(n)}\right\rangle - \langle\sigma|_\phi \wedge \mathrm{A}^{(\mathrm{n})}\left|\Psi^{(n)}\right\rangle \\ &= \mathrm{d}(\phi\left\langle\Psi^{(n)}\right|)\left|\Psi^{(n)}\right\rangle - \phi\wedge\left\langle\Psi^{(n)}\right|\mathrm{A}^{(\mathrm{n})}\left|\Psi^{(n)}\right\rangle \\ &= \mathrm{d}\phi + \phi\wedge\left\langle\Psi^{(n)}, \mathrm{d}\Psi^{(n)}\right\rangle - \phi\wedge\left\langle\Psi^{(n)}\right|\mathrm{A}^{(\mathrm{n})}\left|\Psi^{(n)}\right\rangle = \mathrm{d}\phi + \phi\wedge\omega(z^{-n}), \end{aligned} \tag{6.37}$$

where the last equality comes from (6.29), expressing the gauge potential $\mathrm{A}^{(\mathrm{n})}$ in terms of the connection 1-form ω. □

To give the curvature F_∇ of the given connection (6.23) a more explicit form, one can make use of two further relations. The first one, involving the projectors $\mathfrak{p}^{(n)}$ only, comes from [**24**], while the second is proved again by direct calculation.

LEMMA 6.4. *Let $\mathfrak{p}^{(n)}$ denote the projection given in* (3.37). *With the 2D calculus on S_q^2 of section 3.4.3 one finds:*

$$\begin{aligned} \mathrm{d}\mathfrak{p}^{(n)} \wedge \mathrm{d}\mathfrak{p}^{(n)}\,\mathfrak{p}^{(n)} &= -q^{-n-1}[n]\;\mathfrak{p}^{(n)}\,\omega_+ \wedge \omega_-, \\ \mathfrak{p}^{(n)}\,\mathrm{d}\mathfrak{p}^{(n)} \wedge \mathrm{d}\mathfrak{p}^{(n)} &= -q^{-n-1}[n]\;\mathfrak{p}^{(n)}\,\omega_+ \wedge \omega_-. \end{aligned}$$

LEMMA 6.5. *Given for any $n \in \mathbb{Z}$ the projectors $\mathfrak{p}^{(n)}$ as in* (3.37) *and the expression of the gauge potential $\mathrm{A}^{(\mathrm{n})}$ as in* (6.31), *one has:*

$$\mathfrak{p}^{(n)}\mathrm{d}\mathrm{A}^{(\mathrm{n})}\mathfrak{p}^{(n)} = -\left(\frac{1-q^{2n}}{1-q^{-2}}\right)\left|\Psi^{(n)}\right\rangle \mathrm{d}(U\omega_+ + V\omega_-)\left\langle\Psi^{(n)}\right|. \tag{6.38}$$

PROOF. Setting

$$\mathrm{a}^{(n)} = \left\langle\Psi^{(n)}\right|\mathrm{A}^{(\mathrm{n})}\left|\Psi^{(n)}\right\rangle = -\{(1-q^{2n})/(1-q^{-2})\}(U\omega_+ + V\omega_-) = -\frac{1-q^{2n}}{1-q^{-2}}\mathrm{a},$$

the expression (6.38) can be written as the sum of three terms, from the Leibniz rule satisfied by the exterior derivation d:

$$\begin{aligned}
&\mathfrak{p}^{(n)}\mathrm{d}\mathrm{A}^{(\mathrm{n})}\mathfrak{p}^{(n)} \\
&\quad= \Big[\big|\Psi^{(n)}\big\rangle \big\langle\Psi^{(n)}, \mathrm{d}\Psi^{(n)}\big\rangle \mathrm{a}^{(n)} \big\langle\Psi^{(n)}\big|\Big] + \Big[\big|\Psi^{(n)}\big\rangle (\mathrm{d}\mathrm{a}^{(n)}) \big\langle\Psi^{(n)}\big|\Big] \\
&\quad- \Big[\big|\Psi^{(n)}\big\rangle \mathrm{a}^{(n)} \big\langle\mathrm{d}\Psi^{(n)}, \Psi^{(n)}\big\rangle \big\langle\Psi^{(n)}\big|\Big] \\
&\quad= \left(\frac{1-q^{2n}}{1-q^{-2}}\right) \big|\Psi^{(n)}\big\rangle \Big[\omega_z \wedge \mathrm{a}^{(n)} + \mathrm{a}^{(n)} \wedge \omega_z\Big] \big\langle\Psi^{(n)}\big| + \Big[\big|\Psi^{(n)}\big\rangle \mathrm{d}\mathrm{a}^{(n)} \big\langle\Psi^{(n)}\big|\Big],
\end{aligned}$$

where the second equality comes from the identities

$$\big\langle\Psi^{(n)}, \mathrm{d}\Psi^{(n)}\big\rangle = -\big\langle\mathrm{d}\Psi^{(n)}, \Psi^{(n)}\big\rangle = \{(1-q^{2n})/(1-q^{-2})\}\omega_z,$$

while the $\mathcal{A}(\mathrm{SU}_q(2))$-bimodule relations (3.50) of 1-forms in $\Omega^1(\mathrm{SU}_q(2))$, as well as commutation relations among them (3.52), give:

$$\omega_z \wedge (U\omega_+ + V\omega_-) = q^4 U\omega_z \wedge \omega_+ + q^{-4} V\omega_z \wedge \omega_- = -(U\omega_+ + V\omega_-) \wedge \omega_z,$$

so that $\omega_z \wedge \mathrm{a}^{(n)} + \mathrm{a}^{(n)} \wedge \omega_z = 0$ and the identity claimed in (6.38) is verified. □

REMARK 6.6. *The identity* $\omega_z \wedge \mathrm{a}^{(n)} + \mathrm{a}^{(n)} \wedge \omega_z = 0$ *also shows that the 1-form* ω_z *anti-commutes with every 1-form in* $\Omega^1(\mathrm{S}^2_q)$.

PROPOSITION 6.7. *Given the covariant derivative* $\nabla : \mathcal{E}_n^{(k)} \to \mathcal{E}_n^{(k+1)}$ *from* (6.21) *with a gauge potential* (6.31) $\mathrm{A}^{(\mathrm{n})} = -(1-q^{2n})(1-q^{-2})^{-1} \big|\Psi^{(n)}\big\rangle \mathrm{a} \big\langle\Psi^{(n)}\big|$, *the operator* $\nabla^2 : \mathcal{E}_n^{(0)} \to \mathcal{E}_n^{(2)}$ *can be written as:*

(6.39)
$$\nabla^2 \langle\sigma| = \langle\sigma| \wedge F_\nabla = -\langle\sigma| \wedge \{\big|\Psi^{(n)}\big\rangle q^{n+1}[n](\omega_- \wedge \omega_+ - \mathrm{da} + q^{n+1}[n]\mathrm{a} \wedge \mathrm{a}) \big\langle\Psi^{(n)}\big|\}.$$

PROOF. From the general expression (6.23), the action of the operator ∇^2 on a $\langle\sigma| \in \mathcal{E}_n^{(0)}$ is linear, and given by the sum of three terms. The first one, recalling the result of the lemma 6.4 and the commutation rules (3.50) and (3.52), is:

$$\begin{aligned}
-(\mathrm{d}\mathfrak{p}^{(n)} \wedge \mathrm{d}\mathfrak{p}^{(n)})\mathfrak{p}^{(n)} &= q^{-n-1}[n]\mathfrak{p}^{(n)}\omega_+ \wedge \omega_- \\
&= -q^{1-n}[n] \big|\Psi^{(n)}\big\rangle \big\langle\Psi^{(n)}\big| \omega_- \wedge \omega_+ \\
&= -q^{n+1}[n] \big|\Psi^{(n)}\big\rangle \omega_- \wedge \omega_+ \big\langle\Psi^{(n)}\big|. \qquad (6.40)
\end{aligned}$$

Since one has $\langle\sigma| \mathfrak{p}^{(n)} = \langle\sigma|$, being elements in the projective modules $\mathcal{E}_n^{(0)}$, the other two terms in (6.21) are:

$$\begin{aligned}
\mathfrak{p}^{(n)}\mathrm{d}\mathrm{A}^{(\mathrm{n})}\mathfrak{p}^{(n)} &= -\left(\frac{1-q^{2n}}{1-q^{-2}}\right) \big|\Psi^{(n)}\big\rangle \mathrm{da} \big\langle\Psi^{(n)}\big| \\
&= q^{n+1}[n] \big|\Psi^{(n)}\big\rangle \mathrm{da} \big\langle\Psi^{(n)}\big|, \\
-\mathrm{A}^{(\mathrm{n})} \wedge \mathrm{A}^{(\mathrm{n})} &= -\left(\frac{1-q^{2n}}{1-q^{-2}}\right)^2 \big|\Psi^{(n)}\big\rangle \mathrm{a} \wedge \mathrm{a} \big\langle\Psi^{(n)}\big| \\
&= -q^{2(n+1)}[n]^2 \big|\Psi^{(n)}\big\rangle \mathrm{a} \wedge \mathrm{a} \big\langle\Psi^{(n)}\big|.
\end{aligned}$$

The sum of these three lines gives the curvature $F_\nabla \in \mathbb{M}_{|n|+1} \otimes_{\mathcal{A}(S^2_q)} \Omega^2(S^2_q)$ the expression:

$$F_\nabla = -\left|\Psi^{(n)}\right\rangle q^{n+1}[n](\omega_- \wedge \omega_+ - \mathrm{da} + q^{n+1}[n]\mathrm{a} \wedge \mathrm{a}) \left\langle \Psi^{(n)}\right|. \tag{6.41}$$

□

The isomorphism (6.34) allows to formulate the curvature as a linear map $D^2 : \mathcal{L}^{(0)}_n \to \mathcal{L}^{(2)}_n$, defined by:

$$D^2\phi = (\nabla^2 \langle\sigma|_\phi) \left|\Psi^{(n)}\right\rangle \tag{6.42}$$

for a given $\phi = \langle \sigma, \Psi^{(n)}\rangle$. This operator can also be written in terms of the connection 1-form ω.

PROPOSITION 6.8. *The operator $D^2 : \mathcal{L}^{(0)}_n \to \mathcal{L}^{(2)}_n$ defined in* (6.42) *can be written as*

$$D^2\phi = -\phi \wedge \{\mathrm{d}\omega(z^{-n}) + \omega(z^{-n}) \wedge \omega(z^{-n})\} = \phi \wedge \left(\left\langle \Psi^{(n)}\right| F_\nabla \left|\Psi^{(n)}\right\rangle\right) \tag{6.43}$$

on any $\phi \in \mathcal{L}^{(0)}_n$.

PROOF. The proof is a direct application of the result in propositions 6.18 and 6.3. It is $D\phi = \mathrm{d}\phi - \phi \wedge \omega(z^{-n})$ with $\phi \in \mathcal{L}^{(0)}_n$, so that:

$$\begin{aligned} D^2\phi &= \mathrm{d}(D\phi) + (D\phi) \wedge \omega(z^{-n}) \\ &= -\mathrm{d}(\phi \wedge \omega(z^{-n})) + (\mathrm{d}\phi - \phi \wedge \omega(z^{-n})) \wedge \omega(z^{-n}) \\ &= -\phi \wedge (\mathrm{d}\omega(z^{-n}) + \omega(z^{-n}) \wedge \omega(z^{-n})). \end{aligned}$$

The relation (6.29) can be rewritten as $\omega(z^{-n}) = -q^{1+n}[n](\omega_z + \mathrm{a})$, so to have:

$$\begin{aligned} \mathrm{d}\omega(z^{-n}) &= -q^{1+n}[n](\mathrm{d}\omega_z + \mathrm{da}) = q^{1+n}[n](\omega_- \wedge \omega_+ - \mathrm{da}), \\ \omega(z^{-n}) \wedge \omega(z^{-n}) &= \{q^{1+n}[n]\}^2(\omega_z + \mathrm{a}) \wedge (\omega_z + \mathrm{a}) = q^{2(1+n)}[n]^2\mathrm{a} \wedge \mathrm{a}, \end{aligned}$$

where the last equality in the second line comes from the remark 6.6. It becomes then clear to recover from (6.39)

$$D^2\phi = -\phi \wedge q^{1+n}[n]\{\omega_- \wedge \omega_+ - \mathrm{da} + q^{1+n}[n]\mathrm{a} \wedge \mathrm{a}\} = \phi \wedge \left(\left\langle \Psi^{(n)}\right| F_\nabla \left|\Psi^{(n)}\right\rangle\right),$$

meaning that the action of the operator D^2 can be represented by the 2-form $(\langle\Psi^{(n)}| F_\nabla |\Psi^{(n)}\rangle) \in \mathcal{L}^{(2)}_0$. □

REMARK 6.9. *Recall from* (6.16) *that, given $\phi \in \mathcal{L}^{(0)}_n$, the covariant derivative $D : \mathcal{L}^{(0)}_n \to \mathcal{L}^{(1)}_n$ has been defined in terms of the projector Π associated to the connection as:*

$$D\phi = (1 - \Pi)\mathrm{d}\phi.$$

Given the left $\mathcal{A}(S^2_q)$-module isomorphisms $\mathcal{L}^{(k)}_n \simeq \Omega^k(S^2_q) \otimes_{\mathcal{A}(S^2_q)} \mathcal{E}^{(0)}_n = \mathcal{E}^{(k)}_n$, the proposition 6.2 shows that any connection written as a projector Π as in (6.11) *induces a gauge potential $\mathrm{A}^{(n)}$, so to have a covariant derivative $\nabla : \Omega^k(S^2_q) \otimes_{\mathcal{A}(S^2_q)} \mathcal{E}_n \to \Omega^{k+1}(S^2_q) \otimes_{\mathcal{A}(S^2_q)} \mathcal{E}_n$. The operator D is then extended in* (6.35) *as $D : \mathcal{L}^{(1)}_n \to \mathcal{L}^{(2)}_n$ in terms of the operator ∇, without using the projector Π. This definition is perfectly consistent, but it seems natural to understand whether it is possible to define $D : \mathcal{L}^{(1)}_n \to \mathcal{L}^{(2)}_n$ via the projector Π, and even whether it is possible to extend* *the domain of such a covariant derivative operator D from the set of horizontal forms $\mathcal{L}_n^{(k)}$ to the whole exterior algebra $\Omega(\mathrm{SU}_q(2))$, in analogy to the classical case* (2.4).

Given $\phi \in \mathcal{L}_n^{(1)}$, the most natural definition of a covariant derivative seems to be:

$$\check{D}\phi = (1 - \Pi)\mathrm{d}\phi, \tag{6.44}$$

with the horizontal projector $(1 - \Pi)$ *extended to* $\Omega^2(\mathrm{SU}_q(2))$ *by assuming a compatibility with the wedge product*

$$\Omega^2(\mathrm{SU}_q(2)) = \{\Omega^1(\mathrm{SU}_q(2)) \otimes_{\mathcal{A}(\mathrm{SU}_q(2))} \Omega^1(\mathrm{SU}_q(2))\}/\mathcal{S}_\mathcal{Q} = \Omega^1(\mathrm{SU}_q(2)) \wedge \Omega^1(\mathrm{SU}_q(2))$$

so to have:

$$(1 - \Pi)\Omega^2(\mathrm{SU}_q(2)) = \{(1 - \Pi)\Omega^1(\mathrm{SU}_q(2))\} \wedge \{(1 - \Pi)\Omega^1(\mathrm{SU}_q(2))\}. \tag{6.45}$$

It is easy to see that such a compatibility does not *exist. To be definite, consider an example. Choose $\omega_+ \in \mathcal{L}_{-2}^{(1)}$, so that* $\mathrm{d}\omega_+ = q^2(1+q^2)\omega_z \wedge \omega_+ = -(1+q^{-2})\omega_+ \wedge \omega_z$ *by the commutation properties of the* $\wedge$ *product* (3.52). *Compute now:*

$$\begin{aligned} q^2(1+q^2)(1-\Pi)\{\omega_z \wedge \omega_+\} &= q^2(1+q^2)\{(1-\Pi)\omega_z\} \wedge \{(1-\Pi)\omega_+\} \\ &= q^2(1+q^2)V\omega_- \wedge \omega_+, \\ -(1+q^{-2})(1-\Pi)\{\omega_+ \wedge \omega_z\} &= -(1+q^{-2})\{(1-\Pi)\omega_+\} \wedge \{(1-\Pi)\omega_z\} \\ &= (1+q^{-2})V\omega_- \wedge \omega_+, \end{aligned}$$

The two expressions are different: the problem is that, for the given 3D calculus on $\mathcal{A}(\mathrm{SU}_q(2))$, *one has*

$$(1 - \Pi)\mathcal{S}_\mathcal{Q} \not\subseteq \mathcal{S}_\mathcal{Q}. \tag{6.46}$$

Consider the 6 relations (3.52) *generating $\mathcal{S}_\mathcal{Q}$. An explicit calculation shows that, from the three of them not involving ω_z, one has:*

$$\begin{aligned} &\{(1-\Pi)\omega_+\} \wedge \{(1-\Pi)\omega_+\} = 0, \\ &\{(1-\Pi)\omega_-\} \wedge \{(1-\Pi)\omega_-\} = 0, \\ &\{(1-\Pi)\omega_-\} \wedge \{(1-\Pi)\omega_+\} + q^{-2}\{(1-\Pi)\omega_+\} \wedge \{(1-\Pi)\omega_-\} = 0, \end{aligned}$$

while from the remaining terms:

$$\begin{aligned} &\{(1-\Pi)\omega_z\} \wedge \{(1-\Pi)\omega_-\} + q^4\{(1-\Pi)\omega_-\} \wedge \{(1-\Pi)\omega_z\} = (1-q^4)U\omega_+ \wedge \omega_-, \\ &\{(1-\Pi)\omega_z\} \wedge \{(1-\Pi)\omega_+\} + q^{-4}\{(1-\Pi)\omega_+\} \wedge \{(1-\Pi)\omega_z\} = (1-q^{-4})V\omega_- \wedge \omega_+, \\ &\{(1-\Pi)\omega_z\} \wedge \{(1-\Pi)\omega_z\} = \mathrm{a} \wedge \mathrm{a}. \end{aligned}$$

These computations show that only in the case of the monopole connection – that is $\mathrm{a} = 0$ – *it is*

$$(1 - \Pi_0)\mathcal{S}_\mathcal{Q} \subseteq \mathcal{S}_\mathcal{Q} : \tag{6.47}$$

only in the case of the monopole connection it is consistent to set

$$(1 - \Pi_0)\Omega^2(\mathrm{SU}_q(2)) = \{(1 - \Pi_0)\Omega^1(\mathrm{SU}_q(2))\} \wedge \{(1 - \Pi_0)\Omega^1(\mathrm{SU}_q(2))\}$$

and to define

$$D_0 : \Omega^k(\mathrm{SU}_q(2)) \mapsto \Omega^{k+1}(\mathrm{SU}_q(2)), \qquad D_0\phi = (1 - \Pi_0)\mathrm{d}\phi \tag{6.48}$$

The operator D_0 is a 'covariant' operator: given $\phi \in \Omega^k(\mathrm{SU}_q(2))$ such that $\Delta_R^{(k)}\phi = \phi \otimes z^{-n}$, it is $\Delta_R^{(k+1)}(D_0\phi) = D_0\phi \otimes z^{-n}$, and moreover $D_0\phi \in \mathcal{L}_n^{(k)}$: $D_0\phi$ is horizontal. Note that $\mathcal{L}_n^{(3)} = \emptyset$, as the calculus on S_q^2 is 2D. It becomes an easy computation to prove that the restriction $D_0 : \mathcal{L}_n^{(k)} \to \mathcal{L}_n^{(k+1)}$ acquires the form:

$$D_0\phi = (1 - \Pi_0)\mathrm{d}\phi = \mathrm{d}\phi - (-1)^k\phi \wedge \omega_0(z^{-n}). \tag{6.49}$$

This relation is the quantum analogue of the classical (2.5)*. The classical covariant derivative of an equivariant differential form ϕ can be expressed in terms of the connection 1-form ω only if such ϕ is horizontal. In this quantum formulation, the classical condition that ϕ is horizontal and equivariant has been translated into the condition $\phi \in \mathcal{L}_n^{(k)}$.*

7. A gauged Laplacian on the quantum Hopf bundle

With a covariant derivative ∇ acting on the left $\mathcal{A}(\mathrm{S}_q^2)$-projective modules $\mathcal{E}_n^{(k)} = \Omega^k(\mathrm{S}_q^2) \otimes_{\mathcal{A}(\mathrm{S}_q^2)} \mathcal{E}_n$ and the $\star$-Hodge structure on the exterior algebra $\Omega(\mathrm{S}_q^2)$ introduced in section 5 it is possible to define a gauged Laplacian operator $\Box_\nabla : \mathcal{E}_n^{(0)} \to \mathcal{E}_n^{(0)}$ as:

$$\Box_\nabla \langle\sigma| = \star\nabla\star\nabla\langle\sigma| \tag{7.1}$$

on any $\langle\sigma| \in \mathcal{E}_n^{(0)}$. From the left $\mathcal{A}(\mathrm{S}_q^2)$-linearity of the $\star$-Hodge map, and the relation (6.21), one has:

$$\begin{aligned}\nabla\star\nabla\langle\sigma| &= \mathrm{d}\{\star(\nabla\langle\sigma|)\}\mathfrak{p}^{(n)} - (\star\nabla\langle\sigma|) \wedge \mathrm{A}^{(\mathrm{n})} \\ &= \mathrm{d}\{\star[(\mathrm{d}\langle\sigma|)\mathfrak{p}^{(n)}] + \langle\sigma| \wedge (\star\mathrm{A}^{(\mathrm{n})})\}\mathfrak{p}^{(n)} - \{(\star[(\mathrm{d}\langle\sigma|)\mathfrak{p}^{(n)}] \wedge \mathrm{A}^{(\mathrm{n})} \\ &\quad + \langle\sigma| \wedge (\star\mathrm{A}^{(\mathrm{n})}) \wedge \mathrm{A}^{(\mathrm{n})}\} \\ &= \mathrm{d}\{\star[(\mathrm{d}\langle\sigma|)\mathfrak{p}^{(n)}]\}\mathfrak{p}^{(n)} + \mathrm{d}\{\langle\sigma| \wedge (\star\mathrm{A}^{(\mathrm{n})})\}\mathfrak{p}^{(n)} - \star\{(\mathrm{d}\langle\sigma|)\mathfrak{p}^{(n)}\} \wedge \mathrm{A}^{(\mathrm{n})} \\ &\quad - \langle\sigma| \wedge (\star\mathrm{A}^{(\mathrm{n})}) \wedge \mathrm{A}^{(\mathrm{n})}\end{aligned} \tag{7.2}$$

The second term in the last line can be written as:

$$\begin{aligned}\mathrm{d}\{\langle\sigma| \wedge (\star\mathrm{A}^{(\mathrm{n})})\}\mathfrak{p}^{(n)} &= \mathrm{d}\langle\sigma| \wedge (\star\mathrm{A}^{(\mathrm{n})})\mathfrak{p}^{(n)} + \langle\sigma| \wedge \{\mathrm{d}(\star\mathrm{A}^{(\mathrm{n})})\}\mathfrak{p}^{(n)} \\ &= \mathrm{d}\langle\sigma| \wedge (\star\mathrm{A}^{(\mathrm{n})}) + \langle\sigma| \wedge \{\mathrm{d}(\star\mathrm{A}^{(\mathrm{n})})\}\mathfrak{p}^{(n)},\end{aligned} \tag{7.3}$$

while the third term in (7.2) is:

$$\begin{aligned}-\star\{(\mathrm{d}\langle\sigma|)\mathfrak{p}^{(n)}\} \wedge \mathrm{A}^{(\mathrm{n})} &= -\star(\mathrm{d}\langle\sigma|)\mathfrak{p}^{(n)} \wedge \mathrm{A}^{(\mathrm{n})} \\ &= -(\star\mathrm{d}\langle\sigma|) \wedge \mathrm{A}^{(\mathrm{n})} :\end{aligned} \tag{7.4}$$

in both the relations (7.3) and (7.4) the specific property of right $\mathcal{A}(\mathrm{S}_q^2)$-linearity of the $\star$-Hodge map has been used, namely as $\star(\mathrm{A}^{(\mathrm{n})})\mathfrak{p}^{(n)} = \star(\mathrm{A}^{(\mathrm{n})}\mathfrak{p}^{(n)}) = \star\mathrm{A}^{(\mathrm{n})}$ in (7.3) and as $\star\{(\mathrm{d}\langle\sigma|)\mathfrak{p}^{(n)}\} = \star(\mathrm{d}\langle\sigma|)\mathfrak{p}^{(n)}$ in (7.4). Moreover, from the proposition 5.4 one has $\mathrm{d}\langle\sigma| \wedge (\star\mathrm{A}^{(\mathrm{n})}) = -(\star\mathrm{d}\langle\sigma|) \wedge \mathrm{A}^{(\mathrm{n})}$, so that

$$\begin{aligned}\star\nabla\star\nabla\langle\sigma| &= \star\,\mathrm{d}\{\star(\mathrm{d}\langle\sigma|)\mathfrak{p}^{(n)}\}\mathfrak{p}^{(n)} - 2\star\{(\star\mathrm{d}\langle\sigma|) \wedge \mathrm{A}^{(\mathrm{n})}\} + \langle\sigma| \wedge \{\star\mathrm{d}\star\mathrm{A}^{(\mathrm{n})}\}\mathfrak{p}^{(n)} \\ &\quad - \langle\sigma| \wedge \star\{(\star\mathrm{A}^{(\mathrm{n})}) \wedge \mathrm{A}^{(\mathrm{n})}\}\end{aligned} \tag{7.5}$$

The four terms componing the gauged Laplacian can be individually studied.

- Recalling the result of lemma 5.6, one has:

$$\begin{aligned}\star\mathrm{A}^{(\mathrm{n})} &= q^{n+1}[n]\star\{\left|\Psi^{(n)}\right\rangle \mathrm{a}\left\langle\Psi^{(n)}\right|\} \\ &= q^{n+1}[n]\left|\Psi^{(n)}\right\rangle(\star\mathrm{a})\left\langle\Psi^{(n)}\right|.\end{aligned} \tag{7.6}$$

The fourth term in (7.5) is, using once more the result of lemma 5.6 with $\left\langle\Psi^{(n)}\right| \in \mathcal{L}^{(0)}_{-n}$:

$$\begin{aligned}-\langle\sigma|\wedge\star\{(\star\mathrm{A}^{(\mathrm{n})})\wedge\mathrm{A}^{(\mathrm{n})}\} &= -\langle\sigma|\wedge q^{2(1+n)}[n]\star\{\left|\Psi^{(n)}\right\rangle(\star\mathrm{a})\wedge\mathrm{a}\left\langle\Psi^{(n)}\right|\} \\ &= -q^2[n]\langle\sigma|\wedge\left|\Psi^{(n)}\right\rangle(\star\{(\star\mathrm{a})\wedge\mathrm{a}\})\left\langle\Psi^{(n)}\right|.\end{aligned} \tag{7.7}$$

- From (7.6) the third term in the expression (7.5) of the gauged Laplacian is:

$$\begin{aligned}\langle\sigma|\wedge\{\star\mathrm{d}\star\mathrm{A}^{(\mathrm{n})}\}\mathfrak{p}^{(n)} &= \langle\sigma|\wedge q^{1+n}[n]\star\{\mathrm{d}\left(\left|\Psi^{(n)}\right\rangle(\star\mathrm{a})\left\langle\Psi^{(n)}\right|\right)\}\mathfrak{p}^{(n)} \\ &= \langle\sigma|\wedge q^{1+n}[n]\star\{\mathfrak{p}^{(n)}\mathrm{d}\left(\left|\Psi^{(n)}\right\rangle(\star\mathrm{A}^{(\mathrm{n})})\left\langle\Psi^{(n)}\right|\right)\mathfrak{p}^{(n)}\}.\end{aligned} \tag{7.8}$$

The last term in curly bracket is, by the derivation property of d:

$$\begin{aligned}&\mathfrak{p}^{(n)}\mathrm{d}\left(\left|\Psi^{(n)}\right\rangle(\star\mathrm{A}^{(\mathrm{n})})\left\langle\Psi^{(n)}\right|\right)\mathfrak{p}^{(n)} = \\ &= \left|\Psi^{(n)}\right\rangle\left(\left\langle\Psi^{(n)},\mathrm{d}\Psi^{(n)}\right\rangle(\star\mathrm{a})\right)\left\langle\Psi^{(n)}\right| - \left|\Psi^{(n)}\right\rangle\left((\star\mathrm{a})\left\langle\mathrm{d}\Psi^{(n)},\Psi^{(n)}\right\rangle\right)\left\langle\Psi^{(n)}\right| \\ &\quad + \left|\Psi^{(n)}\right\rangle(\mathrm{d}(\star\mathrm{a}))\left\langle\Psi^{(n)}\right| \\ &= \left|\Psi^{(n)}\right\rangle\{-q^{1+n}[n]\omega_z\wedge(\star\mathrm{a}) - q^{1+n}[n](\star\mathrm{a})\wedge\omega_z + \mathrm{d}(\star\mathrm{a})\}\left\langle\Psi^{(n)}\right|,\end{aligned} \tag{7.9}$$

where the last equality comes from $\left\langle\Psi^{(n)},\mathrm{d}\Psi^{(n)}\right\rangle = -q^{1+n}[n]\omega_z$. Recalling the remark 6.6, and using the commutation rules (3.50) as they were used in (7.7), the expression (7.8) becomes:

$$\begin{aligned}\langle\sigma|\wedge\{\star\mathrm{d}\star\mathrm{A}^{(\mathrm{n})}\}\mathfrak{p}^{(n)} &= q^{1+n}[n]\langle\sigma|\wedge\star\{\left|\Psi^{(n)}\right\rangle\mathrm{d}(\star\mathrm{a})\left\langle\Psi^{(n)}\right|\} \\ &= q^{1-n}[n]\langle\sigma|\wedge\left|\Psi^{(n)}\right\rangle\{\star\mathrm{d}\star\mathrm{a}\}\left\langle\Psi^{(n)}\right|.\end{aligned} \tag{7.10}$$

- It is now straightforward to analyse the second term in the expression (7.5) of the gauged Laplacian. From the definition (6.31) and the Hodge duality (5.14), with again $\mathrm{a} = U\omega_+ + V\omega_-$, $U \in \mathcal{L}^{(0)}_2$ and $V \in \mathcal{L}^{(0)}_{-2}$:

$$2\star\{\mathrm{d}\langle\sigma|\wedge(\star\mathrm{A}^{(\mathrm{n})})\} \tag{7.11}$$

$$\begin{aligned}&= 2i\alpha''\nu\, q^{n+1}[n]\star\{(X_+\triangleright\langle\sigma|)\omega_+\left|\Psi^{(n)}\right\rangle\wedge\mathrm{a}\left\langle\Psi^{(n)}\right| \\ &\qquad - (X_-\triangleright\langle\sigma|)\omega_-\left|\Psi^{(n)}\right\rangle\wedge\mathrm{a}\left\langle\Psi^{(n)}\right|\} \\ &= -2i\alpha''\nu\, q[n]\{(X_+\triangleright\langle\sigma|\left|\Psi^{(n)}\right\rangle V\left\langle\Psi^{(n)}\right| \\ &\qquad + q^2(X_-\triangleright\langle\sigma|)\left|\Psi^{(n)}\right\rangle U\left\langle\Psi^{(n)}\right|\}\star(\omega_-\wedge\omega_+)\end{aligned}$$

$$= -2q[n]\{\nu(X_+\triangleright\langle\sigma|\left|\Psi^{(n)}\right\rangle V\left\langle\Psi^{(n)}\right| + \beta(X_-\triangleright\langle\sigma|)\left|\Psi^{(n)}\right\rangle U\left\langle\Psi^{(n)}\right|\} \tag{7.12}$$

- To analyse the first term in (7.5), which is the only one not depending on the gauge potential a, start with:

$$\begin{aligned}\star\{(\mathrm{d}\,\langle\sigma|)\mathfrak{p}^{(n)}\} &= \star\left(\{(X_+\triangleright\langle\sigma|)\omega_+ + (X_-\triangleright\langle\sigma|)\omega_-\}\mathfrak{p}^{(n)}\right)\\ &= \star\{(X_+\triangleright\langle\sigma|)\mathfrak{p}^{(n)}\omega_+ + (X_-\triangleright\langle\sigma|)\mathfrak{p}^{(n)}\omega_-\}\\ &= -i\alpha''\nu\{(X_+\triangleright\langle\sigma|)\mathfrak{p}^{(n)}\omega_+ - (X_-\triangleright\langle\sigma|)\mathfrak{p}^{(n)}\omega_-\}\end{aligned} \tag{7.13}$$

so to have:

$$\begin{aligned}&\mathrm{d}\star\{(\mathrm{d}\,\langle\sigma|)\mathfrak{p}^{(n)}\}\\ &\quad= -i\alpha''\nu\left(X_-\triangleright\left[\{X_+\triangleright\langle\sigma|\}\mathfrak{p}^{(n)}\right]\omega_-\wedge\omega_+ - X_-\triangleright\left[\{X_-\triangleright\langle\sigma|\}\mathfrak{p}^{(n)}\right]\omega_+\wedge\omega_-\right)\\ &\quad= -i\alpha''\nu\left(X_-\triangleright\left[\{X_+\triangleright\langle\sigma|\}\mathfrak{p}^{(n)}\right] + q^2X_+\triangleright\left[\{X_-\triangleright\langle\sigma|\}\mathfrak{p}^{(n)}\right]\right)\omega_-\wedge\omega_+.\\ &\star\left(\mathrm{d}\star\{(\mathrm{d}\,\langle\sigma|)\mathfrak{p}^{(n)}\}\right)\\ &\quad= -i\alpha''\left(\nu X_-\triangleright\left[\{X_+\triangleright\langle\sigma|\}\mathfrak{p}^{(n)}\right] + \beta X_+\triangleright\left[\{X_-\triangleright\langle\sigma|\}\mathfrak{p}^{(n)}\right]\right)\star(\omega_-\wedge\omega_+)\\ &\quad= -\left(\nu X_-\triangleright\left[\{X_+\triangleright\langle\sigma|\}\mathfrak{p}^{(n)}\right] + \beta X_+\triangleright\left[\{X_-\triangleright\langle\sigma|\}\mathfrak{p}^{(n)}\right]\right)\end{aligned} \tag{7.14}$$

The gauged Laplacian can be seen as an operator $\square_D : \mathcal{L}_n^{(0)} \to \mathcal{L}_n^{(0)}$ via the equivalence between equivariant maps $\phi \in \mathcal{L}_n^{(0)}$ and section of the associated line bundles $\sigma \in \mathcal{E}_n^{(0)}$, represented by the isomorphism in proposition 3.3:

$$\square_D\phi = (\square_\nabla\,\langle\sigma|)\left|\Psi^{(n)}\right\rangle \tag{7.15}$$

on any equivariant $\phi = \langle\sigma, \Psi^{(n)}\rangle$. The terms $(X_\pm\triangleright\langle\sigma|)\left|\Psi^{(n)}\right\rangle$ in (7.12) and(7.14) need a specific analysis. Given the coproduct $\Delta X_\pm = 1\otimes X_\pm + X_\pm\otimes K^2$, one has:

$$\begin{aligned}(X_\pm\triangleright\langle\sigma|)\left|\Psi^{(n)}\right\rangle &= (X_\pm\triangleright\{\phi\left\langle\Psi^{(n)}\right|\})\left|\Psi^{(n)}\right\rangle\\ &= \phi(X_\pm\triangleright\left\langle\Psi^{(n)}\right|)\left|\Psi^{(n)}\right\rangle + q^{-n}(X_\pm\triangleright\phi)\\ &= q^{-n}(X_\pm\triangleright\phi).\end{aligned} \tag{7.16}$$

This last equality is clear from (3.26) with X_+ and $n < 0$, and with X_- and $n > 0$. In the other two cases, it is possible to apply once more the deformed Leibniz rule to products of elements in $\mathcal{A}(\mathrm{SU}_q(2))$, having:

$$\begin{aligned}q^n(X_\pm\triangleright\left\langle\Psi^{(n)}\right|)\left|\Psi^{(n)}\right\rangle &= X_\pm\triangleright\left\langle\Psi^{(n)},\Psi^{(n)}\right\rangle - \left\langle\Psi^{(n)}\right|(X_\pm\triangleright\left|\Psi^{(n)}\right\rangle)\\ &= X_\pm\triangleright(1) - \left\langle\Psi^{(n)}\right|(X_\pm\triangleright\left|\Psi^{(n)}\right\rangle)\\ &= -\left\langle\Psi^{(n)}\right|(X_\pm\triangleright\left|\Psi^{(n)}\right\rangle) = 0;\end{aligned} \tag{7.17}$$

since again from (3.26) one has $X_+\triangleright\left|\Psi^{(n)}\right\rangle = 0$ with $n > 0$, and $X_-\triangleright\left|\Psi^{(n)}\right\rangle = 0$ with $n < 0$.

Recollecting the four terms from (7.5) and making use of the relation (7.16), one has:

$$
\begin{aligned}
-\sigma\wedge\star\{(\star \mathrm{A}^{(\mathrm{n})})\wedge \mathrm{A}^{(\mathrm{n})}\}\left|\Psi^{(n)}\right\rangle &= -q^2[n]\phi\wedge\star\{(\star \mathrm{a})\wedge \mathrm{a}\},\\
\sigma\wedge\{\star \mathrm{d}\star \mathrm{A}^{(\mathrm{n})}\}\left|\Psi^{(n)}\right\rangle &= q^{1-n}[n]\phi\wedge\{\star \mathrm{d}\star \mathrm{a}\},\\
2\star\{\mathrm{d}\sigma\wedge(\star \mathrm{a})\}\left|\Psi^{(n)}\right\rangle &= -2q^{1-n}[n]\left(\nu(X_+\triangleright\phi)V+\beta(X_-\triangleright\phi)U\right),\\
\left[\star\left(\mathrm{d}\star\{(\mathrm{d}\sigma)\mathfrak{p}^{(n)}\}\right)\right]\left|\Psi^{(n)}\right\rangle &= -q^{-2n}\left(\nu X_-X_+ + \beta X_+X_-\right)\triangleright\phi.
\end{aligned}
\tag{7.18}
$$

It is clear that the gauged Laplacian operator can be completely diagonalised only if one chooses the gauge potential $\mathrm{a}=0$, that is if one gauges the Laplacian by the monopole connection. Such a gauged Laplacian $\square_{D_0}:\mathcal{L}_n^{(0)}\to\mathcal{L}_n^{(0)}$ can be written as:

$$
\square_{D_0}\phi = -q^{-2n}\left(\nu X_-X_+ + \beta X_+X_-\right)\triangleright\phi, \qquad \text{for } \phi\in\mathcal{L}_n^{(0)}. \tag{7.19}
$$

The diagonalisation is straightforward, following (4.23). One has:

$$
\begin{aligned}
\square_{D_0}\phi_{n,J,l} &= -q^{1-n}\nu\{[J-\tfrac{n}{2}][J+1+\tfrac{n}{2}]\} - q^{-1-n}\beta\{[J-\tfrac{n}{2}][J+1+\tfrac{n}{2}]+[n]\}\phi_{n,J,l}\\
&= -q^{1-n}\nu\{2[J-\tfrac{n}{2}][J+1+\tfrac{n}{2}]+[n]\}\phi_{n,J,l}.
\end{aligned}
\tag{7.20}
$$

Recall the Laplacian operators on $\mathcal{A}(\mathrm{SU}_q(2))$ and on $\mathcal{A}(\mathrm{S}^2_q)$ from equations (4.22) and (5.18):

$$
\begin{aligned}
&\square_{\mathrm{SU}_q(2)}\phi = -(\nu X_-X_+ + \beta X_+X_- + \gamma X_zX_z)\triangleright\phi, && \phi\in\mathcal{L}_n^{(0)},\\
&\square_{\mathrm{S}^2_q} f = -(\nu X_-X_+ + \beta X_+X_-)\triangleright f, && f\in\mathcal{A}(\mathrm{S}^2_q)\simeq\mathcal{L}_0^{(0)},\\
&\square_{D_0}\phi = -q^{-2n}\left(\nu X_-X_+ + \beta X_+X_-\right)\triangleright\phi, && \phi\in\mathcal{L}_n^{(0)}.
\end{aligned}
\tag{7.21}
$$

One has that the restriction of $\square_{D_0}$ to $\phi\in\mathcal{L}_0^{(0)}$ coincides with the operator $\square_{\mathrm{S}^2_q}$. Moreover it is now possible to generalise to the quantum Hopf bundle with the specific differential calculi studied so far, the classical relation (1.1), from which this analysis started:

$$
q^{2n}\square_{D_0}\triangleright\phi = \left(\square_{\mathrm{SU}_q(2)} + \gamma X_zX_z\right)\triangleright\phi, \qquad \phi\in\mathcal{L}_n^{(0)}. \tag{7.22}
$$

This relation appears as the natural generalisation of the classical relation (1.1) to this specific quantum setting. The quantum Casimir operator (3.21) can not be written as a polynomial in the basis derivations X_j (3.45) of the 3D left covariant calculus from Woronowicz, so its role is played by the Laplacian $\square_{\mathrm{SU}_q(2)}$. Its quantum vertical part can still be written as a quadratic operator in the vertical field X_z of the quantum Hopf fibration.

8. An algebraic formulation of the classical Hopf bundle

The aim of this section is to apply the formalism developed to study the quantum Hopf bundle to the case when all the space algebras are commutative, in order to recover the standard formulation of the classical Hopf bundle described at the beginning of the paper, from a dual viewpoint.

8.1. An algebraic description of the differential calculus on the group manifold $SU(2)$. Rephrasing the relations (2.8) which define the matrix Lie group $SU(2)$, the coordinate algebra $\mathcal{A}(SU(2))$ of the simple Lie group $SU(2)$ is the commutative $*$-algebra generated by u and v, satisfying the spherical relation $u^*u+v^*v=1$. The Hopf algebra structure is given by the coproduct:

$$\Delta \begin{bmatrix} u & -v^* \\ v & u^* \end{bmatrix} = \begin{bmatrix} u & -v^* \\ v & u^* \end{bmatrix} \otimes \begin{bmatrix} u & -v^* \\ v & u^* \end{bmatrix}, \tag{8.1}$$

antipode:

$$S \begin{bmatrix} u & -v^* \\ v & u^* \end{bmatrix} = \begin{bmatrix} u^* & v^* \\ -v & u \end{bmatrix}, \tag{8.2}$$

and counit:

$$\epsilon \begin{bmatrix} u & -v^* \\ v & u^* \end{bmatrix} = \begin{bmatrix} 1 & 0 \\ 0 & 1 \end{bmatrix}. \tag{8.3}$$

The universal envelopping algebra $\mathcal{U}(\mathfrak{su}(2))$ is the Hopf $*$-algebra generated by the three elements $\mathrm{e}, \mathrm{f}, \mathrm{h}$ which satisfy the algebraic relations (2.12) coming from the Lie algebra structure in $\mathfrak{su}(2)$:

$$\begin{aligned} [\mathrm{e}, \mathrm{f}] &= 2\mathrm{h}, \\ [\mathrm{f}, \mathrm{h}] &= \mathrm{f}, \\ [\mathrm{e}, \mathrm{h}] &= -\mathrm{e}. \end{aligned} \tag{8.4}$$

The $*$-structure is:

$$\mathrm{h}^* = \mathrm{h}, \qquad \mathrm{e}^* = \mathrm{f}, \qquad \mathrm{f}^* = \mathrm{e},$$

and the Hopf algebra structure is provided by the coproduct:

$$\begin{aligned} \Delta(\mathrm{e}) &= \mathrm{e} \otimes 1 + 1 \otimes \mathrm{e}, \\ \Delta(\mathrm{f}) &= \mathrm{f} \otimes 1 + 1 \otimes \mathrm{f}, \\ \Delta(\mathrm{h}) &= \mathrm{h} \otimes 1 + 1 \otimes \mathrm{h}; \end{aligned}$$

antipode:

$$S(\mathrm{e}) = -\mathrm{e}, \qquad S(\mathrm{f}) = -\mathrm{f}, \qquad S(\mathrm{h}) = -\mathrm{h};$$

and a counit which is trivial:

$$\varepsilon(\mathrm{e}) = \varepsilon(\mathrm{f}) = \varepsilon(\mathrm{h}) = 0.$$

The centre of the algebra $\mathcal{U}(\mathfrak{su}(2))$ is generated by the Casimir element:

$$C = \mathrm{h}^2 + \frac{1}{2}(\mathrm{ef} + \mathrm{fe}) \tag{8.5}$$

The irreducible finite dimensional $*$-representations σ_j of $\mathcal{U}(\mathfrak{su}(2))$ are well known and labelled by nonnegative half-integers $j \in \frac{1}{2}\mathbb{N}$. They are given by:

$$\begin{aligned} \sigma_j(\mathrm{h}) \left|j, m\right\rangle &= m \left|j, m\right\rangle, \\ \sigma_j(\mathrm{e}) \left|j, m\right\rangle &= \sqrt{(j-m)(j+m+1)} \left|j, m+1\right\rangle, \\ \sigma_j(\mathrm{f}) \left|j, m\right\rangle &= \sqrt{(j-m+1)(j+m)} \left|j, m-1\right\rangle. \end{aligned} \tag{8.6}$$

The algebras $\mathcal{A}(SU(2))$ and $\mathcal{U}(\mathfrak{su}(2))$ are dually paired. The bilinear (3.7) mapping $\langle\ ,\ \rangle : \mathcal{U}(\mathfrak{su}(2)) \times \mathcal{A}(SU(2)) \to \mathbb{C}$, compatible with the $*$-structures, is set by:

$$\begin{aligned}
&\langle \mathrm{h}, u\rangle = -1/2, \\
&\langle \mathrm{h}, u^*\rangle = 1/2, \\
&\langle \mathrm{e}, v\rangle = 1, \\
&\langle \mathrm{f}, v^*\rangle = -1;
\end{aligned} \tag{8.7}$$

all other couples of generators pairing to 0. This pairing is non degenerate: the condition $\langle l, x\rangle = 0\ \forall l \in \mathcal{U}(\mathfrak{su}(2))$ implies $x = 0$, while $\langle l, x\rangle = 0\ \forall x \in \mathcal{A}(SU(2))$ implies $h = 0$.

It is possible to prove [**21**] that a finite dimensional vector space $\mathcal{X} \subset \mathcal{H}'$ of linear functionals on a Hopf algebra $\mathcal{H}$ is a tangent space of a finite dimensional left covariant first order differential calculus $(\Omega^1(\mathcal{H}), \mathrm{d})$ if and only if $X(1) = 0$ and $(\Delta(X) - \varepsilon \otimes X) \in \mathcal{X} \otimes \mathcal{H}^o$, for any $X \in \mathcal{X}$, where $\mathcal{H}^o \subset \mathcal{H}'$ is the dual Hopf algebra to $\mathcal{H}$. The ideal $\mathcal{Q} = \{x \in \ker \varepsilon_{\mathcal{H}} : X(x) = 0 \forall X \in \mathcal{X}\}$ characterises the calculus, the bimodule of 1-forms being isomorphic to $\Omega^1(\mathcal{H}) = \Omega^1_{un}(\mathcal{H})/\mathcal{N}_{\mathcal{Q}}$ with $\mathcal{N}_{\mathcal{Q}} = r^{-1}(\mathcal{H} \otimes \mathcal{Q})$. This result shows the path to prove the following proposition.

PROPOSITION 8.1. *Given the nondegenerate bilinear pairing* $\langle\ ,\ \rangle : \mathcal{U}(\mathfrak{su}(2)) \times \mathcal{A}(SU(2)) \to \mathbb{C}$ *as in* (8.7), *the set* $\{\mathrm{e}, \mathrm{f}, \mathrm{h}\}$ *of generators in* $\mathcal{U}(\mathfrak{su}(2))$ *defines a basis of the tangent space* $\mathcal{X}_{SU(2)}$ *for a bicovariant differential $*$-calculus on* $\mathcal{A}(SU(2))$. *Such a differential calculus is isomorphic to the differential calculus* (2.24), *once the algebra* $C^\infty(S^3)$ *is restricted to the polynomial algebra* $\mathcal{A}(SU(2))$.

PROOF. The definition of counit in the Hopf algebra $\mathcal{U}(\mathfrak{su}(2))$ shows that the generators $l_a = \{\mathrm{e}, \mathrm{f}, \mathrm{h}\}$, seen as linear functionals on $\mathcal{A}(SU(2))$ via the pairing, are such that:

$$\begin{aligned}
\mathrm{e}(1) &= \langle \mathrm{e}, 1\rangle = \varepsilon(\mathrm{e}) = 0, \\
\mathrm{f}(1) &= \langle \mathrm{f}, 1\rangle = \varepsilon(\mathrm{f}) = 0, \\
\mathrm{h}(1) &= \langle \mathrm{h}, 1\rangle = \varepsilon(\mathrm{h}) = 0;
\end{aligned}$$

while the coproduct relations can be cast in the form:

$$\begin{aligned}
&\Delta(\mathrm{e}) - 1 \otimes \mathrm{e} = \mathrm{e} \otimes 1, \\
&\Delta(\mathrm{f}) - 1 \otimes \mathrm{f} = \mathrm{f} \otimes 1, \\
&\Delta(\mathrm{h}) - 1 \otimes \mathrm{h} = \mathrm{h} \otimes 1;
\end{aligned} \tag{8.8}$$

thus proving that the set $\{\mathrm{e}, \mathrm{f}, \mathrm{h}\}$ in $\mathcal{U}(\mathfrak{su}(2))$ defines a complex vector space basis of a tangent space $\mathcal{X}_{SU(2)}$ for a left covariant differential calculus. The obvious inclusion $\mathcal{X}^*_{SU(2)} \subset \mathcal{X}_{SU(2)}$ proves, as described in section 3, that such a calculus admits a $*$ structure.

In order to recover the ideal $\mathcal{Q}_{SU(2)} \subset \ker \varepsilon_{SU(2)}$ for this specific calculus, consider a generic element $x \in \ker \varepsilon_{SU(2)}$. It must necessarily be written as $x = \{(u-1)x_1, (u^*-1)x_2, vx_3, v^*x_4\}$ with $x_j \in \mathcal{A}(SU(2))$. Such an element x will belong to $\mathcal{Q}_{SU(2)}$ if $\langle l_a, x\rangle = 0$ for any of the generators $l_a \in \mathcal{U}(\mathfrak{su}(2))$, since they form a vector space basis for the tangent space $\mathcal{X}_{SU(2)}$ relative to this calculus. For the element $x = (u-1)x_1$ the three conditions are:

$$
\begin{aligned}
\langle \mathrm{e}, (u-1)x_1 \rangle &= \langle \mathrm{e}, u-1 \rangle \langle 1, x_1 \rangle + \langle 1, u-1 \rangle \langle \mathrm{e}, x_1 \rangle = 0, \\
\langle \mathrm{f}, (u-1)x_1 \rangle &= \langle \mathrm{f}, u-1 \rangle \langle 1, x_1 \rangle + \langle 1, u-1 \rangle \langle \mathrm{f}, x_1 \rangle = 0, \\
\langle \mathrm{h}, (u-1)x_1 \rangle &= \langle \mathrm{h}, u-1 \rangle \langle 1, x_1 \rangle + \langle 1, u-1 \rangle \langle \mathrm{h}, x_1 \rangle = -\frac{1}{2} \langle 1, x_1 \rangle = -\frac{1}{2}\varepsilon(x_1),
\end{aligned} \tag{8.9}
$$

where, in each of the three lines, the first equality comes from the general properties of dual pairing and from the specific coproduct in $\mathcal{U}(\mathfrak{su}(2))$, while the final result depends on the specific form of the pairing. This means that $x = (u-1)x_1$ belongs to $\mathcal{Q}_{SU(2)}$ if and only if $x_1 \in \ker \varepsilon_{SU(2)}$. The analysis is similar for the other three elements $x = \{(u^*-1)x_2, vx_3, v^*x_4\}$. It is then proved that this left covariant differential calculus on $\mathcal{A}(SU(2))$ - whose tangent space is 3 dimensional - can be characterised by the ideal $\mathcal{Q}_{SU(2)} = \{\ker \varepsilon_{SU(2)}\}^2 \subset \ker \varepsilon_{SU(2)}$, which is generated by the ten elements: $\mathcal{Q}_{SU(2)} = \{(u-1)^2, (u-1)(u^*-1), (u-1)v, (u-1)v^*, (u^*-1)^2, (u^*-1)v, (u^*-1)v^*, v^2, vv^*, v^{*2}\}$. The equation (3.5) allows then to write the exterior derivative for this calculus as:

$$
\mathrm{d}x = (\mathrm{e} \triangleright x)\omega_{\mathrm{e}} + (\mathrm{f} \triangleright x)\omega_{\mathrm{f}} + (\mathrm{h} \triangleright x)\omega_{\mathrm{h}} \tag{8.10}
$$

The commutation properties between the left invariant forms $\{\omega_{\mathrm{e}}, \omega_{\mathrm{f}}, \omega_{\mathrm{h}}\}$ and elements of the algebra $\mathcal{A}(SU(2))$ depend on the functionals f_{ab} defined as $\Delta(l_a) = 1 \otimes l_a + l_b \otimes f_{ba}$. From (8.8) one has $f_{ab} = \delta_{ab}$, so 1-forms do commute with elements of the algebra $\mathcal{A}(SU(2))$, $\omega_a x = x\omega_a$.

The ideal $\mathcal{Q}_{SU(2)}$ is in addition stable under the right coaction Ad of the algebra $\mathcal{A}(SU(2))$ onto itself: $\mathrm{Ad}(\mathcal{Q}_{SU(2)}) \subset \mathcal{Q}_{SU(2)} \otimes \mathcal{A}(SU(2))$. The proof of this result consists of a direct computation. The stability of the ideal $\mathcal{Q}_{SU(2)}$ under the right coaction Ad means that this differential calculus is bicovariant.

The explicit form of the left action of the generators of $\mathcal{U}(\mathfrak{su}(2))$ on the generators of the coordinate algebra $\mathcal{A}(SU(2))$ is:

$$
\begin{array}{lll}
\mathrm{h} \triangleright u = -\frac{1}{2}u & \mathrm{e} \triangleright u = -v^* & \mathrm{f} \triangleright u = 0 \\
\mathrm{h} \triangleright u^* = \frac{1}{2}u^* & \mathrm{e} \triangleright u^* = 0 & \mathrm{f} \triangleright u^* = v \\
\mathrm{h} \triangleright v = -\frac{1}{2}v & \mathrm{e} \triangleright v = u^* & \mathrm{f} \triangleright v = 0 \\
\mathrm{h} \triangleright v^* = \frac{1}{2}v^* & \mathrm{e} \triangleright v^* = 0 & \mathrm{f} \triangleright v^* = -u
\end{array} \tag{8.11}
$$

Starting from these relations it is immediate to see that the left action of the generators $l_a \in \mathcal{U}(\mathfrak{su}(2))$ is equivalent to the Lie derivative along the left invariant vector fields L_a (2.11). This equivalence can now be written as:

$$
\begin{aligned}
\mathrm{e} \triangleright (x) &= -iL_+(x), \\
\mathrm{f} \triangleright (x) &= -iL_-(x), \\
\mathrm{h} \triangleright (x) &= iL_z(x),
\end{aligned} \tag{8.12}
$$

and it is valid for any $x \in \mathcal{A}(SU(2))$, as the Leibniz rule for the action of the derivations L_a is encoded in the definition of the left action (3.6) and the properties of the functionals $f_{ab} = \delta_{ab}$. From relation (8.10) it is possible to recover:

$$\begin{aligned} \mathrm{d}u &= -v^*\omega_{\mathrm{e}} - \frac{1}{2}u\omega_{\mathrm{h}}, \\ \mathrm{d}u^* &= v\omega_{\mathrm{f}} + \frac{1}{2}u^*\omega_{\mathrm{h}}, \\ \mathrm{d}v &= u^*\omega_{\mathrm{e}} - \frac{1}{2}v\omega_{\mathrm{h}}, \\ \mathrm{d}v^* &= -u\omega_{\mathrm{f}} + \frac{1}{2}v^*\omega_{\mathrm{h}}. \end{aligned}$$

These relations can be inverted, so that left invariant 1-forms $\{\omega_{\mathrm{e}}, \omega_{\mathrm{f}}, \omega_{\mathrm{h}}\}$ can be compared to (2.21):

$$\begin{aligned} \omega_{\mathrm{e}} &= u\mathrm{d}v - v\mathrm{d}u = i\tilde{\omega}_+, \\ \omega_{\mathrm{f}} &= v^*\mathrm{d}u^* - u^*\mathrm{d}v^* = i\tilde{\omega}_-, \\ \omega_{\mathrm{h}} &= -2(u^*\mathrm{d}u + v^*\mathrm{d}v) = -i\tilde{\omega}_z. \end{aligned} \tag{8.13}$$

The $*$-structure is given, on the basis of left-invariant generators, as $\omega_{\mathrm{e}}^* = -\omega_{\mathrm{f}}$, $\omega_{\mathrm{h}}^* = -\omega_{\mathrm{h}}$. The equalities (8.13), which are dual to (8.12), represent the isomorphism between the first order differential calculus introduced via the action of the exterior derivative in (8.10), and the one analysed in section 2.1.

□

It is now straightforward to recover this bicovariant calculus as the classical limit of the quantum 3D left covariant calculus $(\Omega(\mathrm{SU}_q(2)), \mathrm{d})$ described in section 3.4.1. In the classical limit $\mathcal{A}(\mathrm{SU}_q(2)) \to \mathcal{A}(SU(2))$ as $q \to 1$, with $\phi \to x$, one has:

$$\begin{aligned} &\omega_+ \to \omega_{\mathrm{e}}, && (X_+ \triangleright \phi) \to (\mathrm{e} \triangleright x), \\ &\omega_- \to \omega_{\mathrm{f}}, && (X_- \triangleright \phi) \to (\mathrm{f} \triangleright x), \\ &\omega_z \to -\tfrac{1}{2}\,\omega_{\mathrm{h}}, && (X_z \triangleright \phi) \to (-2\mathrm{h} \triangleright x). \end{aligned}$$

The coaction $\Delta_R^{(1)}$ of $\mathcal{A}(SU(2))$ on the basis of left invariant forms defines the matrix $\Delta_R^{(1)}(\omega_a) = \omega_b \otimes J_{ba}$:

$$\begin{aligned} \Delta_R^{(1)}(\omega_{\mathrm{f}}) &= \omega_{\mathrm{f}} \otimes u^{*2} + \omega_{\mathrm{h}} \otimes u^* v^* - \omega_{\mathrm{e}} \otimes v^{*2}, \\ \Delta_R^{(1)}(\omega_{\mathrm{h}}) &= -\omega_{\mathrm{f}} \otimes 2u^* v + \omega_{\mathrm{h}} \otimes (u^* u - v^* v) - \omega_{\mathrm{e}} \otimes 2uv^*, \\ \Delta_R^{(1)}(\omega_{\mathrm{e}}) &= -\omega_{\mathrm{f}} \otimes v^2 + \omega_{\mathrm{h}} \otimes uv + \omega_{\mathrm{e}} \otimes u^2, \end{aligned} \tag{8.14}$$

which is used to define a basis of right invariant one forms $\eta_a = \omega_b S(J_{ba})$:

$$\begin{aligned} \eta_{\mathrm{f}} &= u^2\omega_{\mathrm{f}} - uv^*\omega_{\mathrm{h}} - v^{*2}\omega_{\mathrm{e}} = v^*\mathrm{d}u - u\mathrm{d}v^*, \\ \eta_{\mathrm{h}} &= 2uv\omega_{\mathrm{f}} + (uu^* - vv^*)\omega_{\mathrm{h}} + 2u^*v^*\omega_{\mathrm{e}} = 2(u\mathrm{d}u^* + v^*\mathrm{d}v), \\ \eta_{\mathrm{e}} &= -v^2\omega_{\mathrm{f}} - u^*v\omega_{\mathrm{h}} + u^{*2}\omega_{\mathrm{e}} = u^*\mathrm{d}v - v\mathrm{d}u^*; \end{aligned} \tag{8.15}$$

- note that it has been made explicit use of the commutativity between forms ω_a and elements of the algebra $\mathcal{A}(SU(2))$. The right acting derivation associated to this basis are given by (3.12) as

$$\mathrm{d}x = \eta_a \triangleleft (-S^{-1}(l_a)) = \eta_a \triangleleft l_a$$

for any $x \in \mathcal{A}(SU(2))$, since an immediate evaluation gives $S^{-1}(l_a) = -l_a$ for the three vector basis elements of the tangent space $l_a \in \mathcal{X}$. Using again the commutativity of the right invariant one forms η_a with element of $\mathcal{A}(SU(2))$, the action of the exterior derivation (8.10) can be written as:

$$\mathrm{d}x = (x \triangleleft \mathrm{f})\eta_{\mathrm{f}} + (x \triangleleft \mathrm{h})\eta_{\mathrm{h}} + (x \triangleleft \mathrm{e})\eta_{\mathrm{e}}. \tag{8.16}$$

Comparing (8.15) to (2.22) one has:

$$\begin{aligned} \eta_{\mathrm{f}} &= i\tilde{\eta}_-, \\ \eta_{\mathrm{h}} &= -i\tilde{\eta}_z, \\ \eta_{\mathrm{e}} &= i\tilde{\eta}_+, \end{aligned} \tag{8.17}$$

while for the right action of the generators l_a on $\mathcal{A}(SU(2))$ one computes:

$$\begin{array}{lll} u \triangleleft \mathrm{h} = -\frac{1}{2}u & u \triangleleft \mathrm{e} = 0 & u \triangleleft \mathrm{f} = v \\ u^* \triangleleft \mathrm{h} = \frac{1}{2}u^* & u^* \triangleleft \mathrm{e} = -v^* & u^* \triangleleft \mathrm{f} = 0 \\ v \triangleleft \mathrm{h} = \frac{1}{2}v & v \triangleleft \mathrm{e} = u & v \triangleleft \mathrm{f} = 0 \\ v^* \triangleleft \mathrm{h} = -\frac{1}{2}v^* & v^* \triangleleft \mathrm{e} = 0 & v^* \triangleleft \mathrm{f} = -u^*; \end{array} \tag{8.18}$$

so that the identification with the action of the right invariant vector fields (2.14) can be recovered as:

$$\begin{aligned} (x) \triangleleft \mathrm{f} &= -iR_-(x), \\ (x) \triangleleft \mathrm{e} &= -iR_+(x), \\ (x) \triangleleft \mathrm{h} &= iR_z(x), \end{aligned} \tag{8.19}$$

being dual to the identification (8.17). It is also evident that relations (8.17) and (8.19) define a different realisation of the isomorphism between the differential calculus introduced in this section (8.16) and the differential calculus from section 2.1.

REMARK 8.2. *The identification* (8.12) *can be read as a Lie algebra isomorphism between the Lie algebra* $\{\mathrm{e}, \mathrm{f}, \mathrm{h}\}$ *given in* (8.4) *and the Lie algebra of the left invariant vector fields* $\{L_a\}$ (2.12)*:*

$$\mathrm{e} = -iL_+, \qquad \mathrm{f} = -iL_-, \qquad \mathrm{h} = iL_z. \tag{8.20}$$

The notion of pairing between the algebras $\mathcal{U}(\mathfrak{su}(2))$ *and* $\mathcal{A}(SU(2))$ *can be recovered as the Lie derivative of the coordinate functions along the vector fields* L_a*, evaluated at the identity of the group manifold. The terms in* (8.7) *giving the nonzero terms of the pairing are:*

$$\begin{array}{lll} L_z(u)|_{\mathrm{id}} = \frac{i}{2} & \Rightarrow & \langle \mathrm{h}, u \rangle = -\frac{1}{2} \\ L_z(u^*)|_{\mathrm{id}} = -\frac{i}{2} & \Rightarrow & \langle \mathrm{h}, u^* \rangle = \frac{1}{2} \\ L_+(v)|_{\mathrm{id}} = i & \Rightarrow & \langle \mathrm{e}, v \rangle = 1 \\ L_-(v^*)|_{\mathrm{id}} = -i & \Rightarrow & \langle \mathrm{f}, v^* \rangle = -1 \end{array}$$

The whole exterior algebra $\Omega(SU(2))$ can now be constructed from the differential calculus (8.10). Any 1-form $\theta \in \Omega^1(SU(2))$ can be written on the basis of left invariant forms as $\theta = \sum_k x_k \omega_k = \omega_k x_k$ with $x_k \in \mathcal{A}(SU(2))$. Higher dimensional forms can be defined by requiring their total antisimmetry, and that $\mathrm{d}^2 = 0$. One has then $\omega_a \wedge \omega_b + \omega_b \wedge \omega_a = 0$ and:

$$\begin{aligned} \mathrm{d}\omega_{\mathrm{f}} &= \omega_{\mathrm{h}} \wedge \omega_{\mathrm{f}}, \\ \mathrm{d}\omega_{\mathrm{e}} &= \omega_{\mathrm{e}} \wedge \omega_{\mathrm{h}}, \\ \mathrm{d}\omega_{\mathrm{h}} &= 2\omega_{\mathrm{f}} \wedge \omega_{\mathrm{e}}. \end{aligned} \tag{8.21}$$

Finally, there is a unique volume top form $\omega_{\mathrm{f}} \wedge \omega_{\mathrm{e}} \wedge \omega_{\mathrm{h}}$.

The algebra $\mathcal{A}(SU(2))$ can be partitioned into finite dimensional blocks, whose elements are related to the Wigner D-functions [**36**] for the group $SU(2)$. Considering all the unitary irreducible representations of $SU(2)$, their matrix elements will give a Peter-Weyl basis for the Hilbert space $\mathcal{L}^2(SU(2), \mu)$ of complex valued functions defined on the group manifold with respect to the Haar invariant measure. The Wigner D-function $D^J_{ks}(g)$ is defined to be the matrix element (k, s are the matrix indices) representing the element $g \simeq (u, v)$ in $SU(2)$ (2.8) in the representation of weight J. They are known:

$$D^J_{ks} = (-i)^{s+k}[(J+s)!(J-s)!(J+k)!(J-k)!]^{1/2} \cdot \sum_l (-1)^{k+l} \frac{u^{*l} v^{*J-k-l} v^{J-s-l} u^{*k+s+l}}{l!(J-k-l)!(J-s-l)!(s+k+l)!} \tag{8.22}$$

with $J = 0, 1/2, 1, \ldots$ and $k = -J, \ldots, +J$, $s = -J, \ldots, +J$. In (8.22) the index l runs over the set of natural numbers such that all the arguments of the factorial are non negative. To illustrate the meaning of this partition, proceed as in the quantum setting, and consider the element $u^* \in \mathcal{A}(SU(2))$. Representing the left action $\mathrm{f}\triangleright$ with a horizontal arrow and the right action $\triangleleft\mathrm{e}$ with a vertical one yields the box:

$$\begin{array}{ccc} u^* & \rightarrow & v \\ \downarrow & & \downarrow \\ -v^* & \rightarrow & u \end{array} \tag{8.23}$$

while starting from $u^{*2} \in \mathcal{A}(SU(2))$ yields the box:

$$\begin{array}{ccccc} u^{*2} & \rightarrow & 2u^*v & \rightarrow & 2v^2 \\ \downarrow & & \downarrow & & \downarrow \\ -2u^*v^* & \rightarrow & 2(u^*u - v^*v) & \rightarrow & 4uv \\ \downarrow & & \downarrow & & \downarrow \\ 2v^{*2} & \rightarrow & -4v^*u & \rightarrow & 4u^2 \end{array} \tag{8.24}$$

A recursive structure emerges now clear. For each positive integer p one has a box W_p made up of the $(p+1) \times (p+1)$ elements $w_{p:t,r} = \mathrm{f}^t\triangleright u^{*p} \triangleleft \mathrm{e}^r$. An explicit calculation proves that:

$$\mathrm{f}^t\triangleright u^{*p} \triangleleft \mathrm{e}^r = i^{t+r} j! \left[\frac{t!r!}{(p-t)!(p-r)!}\right]^{1/2} D^{p/2}_{t-p/2, r-p/2} \tag{8.25}$$

with $t \leq p, r \leq p$. As an element in $\mathcal{U}(\mathfrak{su}(2))$, the quadratic Casimir C (8.5) of the Lie algebra $\mathfrak{su}(2)$ acts on $x \in \mathcal{A}(SU(2))$ as $C\triangleright x = x \triangleleft C$, and its action clearly commutes with the actions $\mathrm{f}\triangleright$ and $\triangleleft\mathrm{e}$. This means that the decomposition $\mathcal{A}(SU(2)) = \oplus_{j\in\mathbb{N}} W_p$ gives the spectral resolution of the action of C:

$$C\triangleright w_{p:t,r} = \frac{p}{2}(\frac{p}{2} + 1) w_{p:t,r}. \tag{8.26}$$

8.2. The bundle structure.

8.2.1. *The base algebra of the bundle.* Given the abelian $*$-algebra $\mathcal{A}(\mathrm{U}(1)) = \mathbb{C}[z, z^*]/ < zz^* - 1 >$, the map $\breve{\pi} : \mathcal{A}(SU(2)) \rightarrow (\mathrm{U}(1))$

$$\breve{\pi} \begin{bmatrix} u & -v^* \\ v & u^* \end{bmatrix} = \begin{bmatrix} z & 0 \\ 0 & z^* \end{bmatrix}, \tag{8.27}$$

is a surjective Hopf $*$-algebra homomorphism, so that $\mathcal{A}(\mathrm{U}(1))$ can be seen as a $*$-subalgebra of $\mathcal{A}(SU(2))$, with a right coaction:

$$\check{\Delta}_R = (1 \otimes \check{\pi}) \circ \Delta, \qquad \mathcal{A}(SU(2) \to \mathcal{A}(SU(2)) \otimes \mathcal{A}(\mathrm{U}(1)). \tag{8.28}$$

The coinvariant elements for this coaction, that is elements $b \in \mathcal{A}(SU(2))$ for which $\check{\Delta}_R(b) = b \otimes 1$, form the subalgebra $\mathcal{A}(S^2) \subset \mathcal{A}(SU(2))$, which is the coordinate subalgebra of the sphere S^2. From:

$$\begin{aligned} \check{\Delta}_R(u) &= u \otimes z, \\ \check{\Delta}_R(u^*) &= u^* \otimes z^*, \\ \check{\Delta}_R(v) &= v \otimes z, \\ \check{\Delta}_R(v^*) &= v^* \otimes z^*, \end{aligned} \tag{8.29}$$

one has that a set of generators for $\mathcal{A}(S^2)$ is given by (2.39):

$$\begin{aligned} b_z &= uu^* - vv^*, \\ b_y &= uv^* + vu^*, \\ b_x &= -i(vu^* - uv^*) \end{aligned} \tag{8.30}$$

The comparison with section 2.3 shows that $\check{\pi}$ dually describes the choice of the gauge group $\mathrm{U}(1)$ as a subgroup of $SU(2)$, whose right principal pull-back action $\check{\mathfrak{r}}_k^*$ is now replaced by the right $\mathcal{A}(\mathrm{U}(1))$- coaction $\check{\Delta}_R$. The basis of the principal Hopf bundle $S^2 \simeq SU(2)/\mathrm{U}(1)$ will be given as the algebra $\mathcal{A}(S^2)$ of right coinvariant elements $b_a \in \mathcal{A}(SU(2))$, which is a homogeneous space algebra. The coproduct Δ of $\mathcal{A}(SU(2))$ restricts to a left coaction $\Delta : \mathcal{A}(SU(2)) \mapsto \mathcal{A}(SU(2)) \otimes \mathcal{A}(S^2)$ as:

$$\begin{aligned} \Delta(b_{\mathrm{f}}) &= u^2 \otimes b_{\mathrm{f}} - v^* u \otimes b_{\mathrm{h}} - v^{*2} \otimes b_{\mathrm{e}}, \\ \Delta(b_{\mathrm{h}}) &= 2uv \otimes b_{\mathrm{f}} + (u^*u - v^*v) \otimes b_{\mathrm{h}} + 2u^*v^* \otimes b_{\mathrm{e}}, \\ \Delta(b_{\mathrm{e}}) &= -v^2 \otimes b_{\mathrm{f}} - u^*v \otimes b_{\mathrm{h}} + u^{*2} \otimes b_{\mathrm{e}}. \end{aligned} \tag{8.31}$$

with $b_{\mathrm{f}} = 1/2(b_y - ib_x) = uv^*$, $b_{\mathrm{e}} = 1/2(b_y + ib_x) = vu^*$, $b_{\mathrm{h}} = b_z$. The choice of this specific basis shows that $\Delta(b_a) = S(J_{ka}) \otimes b_k$ where the matrix J is exactly the one defined in (8.14) as $\Delta_R^{(1)}(\omega_a) = \omega_b \otimes J_{ba}$.

The identification (8.12) between the left action $\mathrm{h} \triangleright x$ – given the generator $\mathrm{h} \in \mathcal{U}(\mathfrak{su}(2))$ on any $x \in \mathcal{A}(SU(2))$ – and the action $iL_z(x)$ – given the left invariant vector field L_z – as well as the definition of the $\mathcal{A}(\mathrm{U}(1))$-right coaction $\check{\Delta}_R$ on $\mathcal{A}(SU(2))$ (8.29), allow to recover the set of the $\mathrm{U}(1)$-equivariant functions $\mathfrak{L}_n^{(0)} \subset \mathcal{A}(SU(2))$ in (2.44) as:

$$\mathfrak{L}_n^{(0)} = \{\phi \in \mathcal{A}(SU(2)) \, : \, \mathrm{h} \triangleright \phi = \frac{n}{2}\phi \Leftrightarrow \check{\Delta}_R(\phi) = \phi \otimes z^{-n}\}. \tag{8.32}$$

8.2.2. *A differential calculus on the gauge group algebra.* The strategy underlining the proof of the proposition 8.1 brings also to the definition of a differential calculus on the gauge group algebra $\mathcal{A}(\mathrm{U}(1))$. The bilinear pairing $\langle \cdot, \cdot \rangle$: $\mathcal{U}(\mathfrak{su}(2)) \times \mathcal{A}(SU(2)) \to \mathbb{C}$ (8.7) is restricted via the surjection $\check{\pi}$ (8.27) to a bilinear pairing $\langle \cdot, \cdot \rangle : \mathcal{U}\{\mathrm{h}\} \times \mathcal{A}(\mathrm{U}(1)) \to \mathbb{C}$, which is still compatible with the $*$-structure, given on generators as:

$$\langle \mathrm{h}, z\rangle = -\frac{1}{2},$$
$$\langle \mathrm{h}, z^{-1}\rangle = \frac{1}{2}.$$

The set $\mathcal{X}_{\mathrm{U}(1)} = \{\mathrm{h}\}$ is proved to be the basis of the tangent space for a 1-dimensional bicovariant commutative calculus on $\mathcal{A}(\mathrm{U}(1))$. The ideal $\mathcal{Q}_{\mathrm{U}(1)} \subset \ker\varepsilon_{\mathrm{U}(1)}$ turns out again to be $\mathcal{A}_{\mathrm{U}(1)} = (\ker\varepsilon_{\mathrm{U}(1)})^2$ generated by $\{(z-1)^2, (z-1)(z^{-1}-1), (z^{-1}-1)^2\}$, which can also be recovered as $\mathcal{Q}_{\mathrm{U}(1)} = \breve{\pi}((\ker\varepsilon_{SU(2)})^2)$. From:

$$\mathrm{h}\triangleright z = -\frac{1}{2}z,$$
$$\mathrm{h}\triangleright z^{-1} = \frac{1}{2}z^{-1},$$

one has that:

$$\mathrm{d}z = -\frac{1}{2}z\breve{\omega},$$
$$\mathrm{d}z^{-1} = \frac{1}{2}z^{-1}\breve{\omega} \tag{8.33}$$

with $z\mathrm{d}z = (\mathrm{d}z)z$. The only left invariant 1-form is

$$\breve{\omega} = -2z^{-1}\mathrm{d}z = 2z\mathrm{d}z^{-1},$$

while the role of the right invariant derivation associated to $\mathrm{h} \in \mathcal{U}\{\mathrm{h}\}$ is played by $-S^{-1}(\mathrm{h}) = \mathrm{h}$, so that the right invariant form generating this calculus is:

$$\begin{array}{lcl} \mathrm{d}z = \breve{\eta}(z\triangleleft\mathrm{h}) = \breve{\eta}(-\frac{1}{2}z) & \Rightarrow & \breve{\eta} = -2z^{-1}\mathrm{d}z, \\ \mathrm{d}z^{-1} = \breve{\eta}(z^{-1}\triangleleft\mathrm{h}) = \breve{\eta}(\frac{1}{2}z^{-1}) & \Rightarrow & \breve{\eta} = 2z\mathrm{d}z^{-1} \end{array}$$

so that one obtains $\breve{\eta} = \breve{\omega}$.

It is possible to characterise the quotient $\ker\varepsilon_{\mathrm{U}(1)}/\mathcal{Q}_{\mathrm{U}(1)} = \ker\varepsilon_{\mathrm{U}(1)}/(\ker\varepsilon_{\mathrm{U}(1)})^2$. The three elements generating the ideal $\mathcal{Q}_{\mathrm{U}(1)} = (\ker\varepsilon_{\mathrm{U}(1)})^2$ can be written as:

$$\xi = (z-1)(z^{-1}-1) = (z-1) + (z^{-1}-1),$$
$$\xi' = (z-1)(z-1) = \xi + \xi(z-1),$$
$$\xi'' = (z^{-1}-1)(z^{-1}-1) = \xi + \xi(z^{-1}-1),$$

so that $\mathcal{Q}_{\mathrm{U}(1)}$ can be seen generated by $\xi = (z-1) + (z^{-1}-1)$. Set a map $\lambda : \ker\varepsilon_{\mathrm{U}(1)} \to \mathbb{C}$ by $\lambda(u(z-1)) = \sum_{j\in\mathbb{Z}} u_j$, where $u = \sum_{j\in\mathbb{Z}} u_j z^j$ is generic element in $\mathcal{A}(\mathrm{U}(1))$. The techniques outlined in lemma 3.4 in the quantum setting enable to prove that λ can be used to define a complex vector space isomorphism between $\ker\varepsilon_{\mathrm{U}(1)}/(\ker\varepsilon_{\mathrm{U}(1)})^2$ and $\mathbb{C}$, whose inverse is given by $\lambda^{-1} : w \in \mathbb{C} \mapsto \lambda^{-1}(w) = w(z-1) \in \ker\varepsilon_{\mathrm{U}(1)}$. It is evident that such a map λ gives the projection $\pi_{\mathcal{Q}_{\mathrm{U}(1)}} : \ker\varepsilon_{\mathrm{U}(1)} \to \ker\varepsilon_{\mathrm{U}(1)}/\mathcal{Q}_{\mathrm{U}(1)} \simeq \mathbb{C}$, since it chooses a representative in each equivalence class in the quotient $\ker\varepsilon_{\mathrm{U}(1)}/\mathcal{Q}_{\mathrm{U}(1)}$.

8.2.3. *The Hopf bundle structure.* With the 3D bicovariant calculus on the total space algebra $\mathcal{A}(SU(2))$ and the 1D bicovariant calculus on the gauge group algebra $\mathcal{A}(\mathrm{U}(1))$, one needs to prove the compatibility conditions that lead to the exact sequence:

$$0 \to \mathcal{A}(SU(2))\,(\Omega^1(S^2))\,\mathcal{A}(SU(2)) \to$$
$$\to \Omega^1(\mathcal{A}(SU(2)) \stackrel{\sim_{\mathcal{N}_{SU(2)}}}{\longrightarrow} \mathcal{A}(SU(2)) \otimes \ker \varepsilon_{\mathrm{U}(1)}/\mathcal{Q}_{\mathrm{U}(1)} \to 0,$$

where the map $\sim_{\mathcal{N}_{SU(2)}}$ is defined as in the diagram (3.15) which now acquires the form:

$$(8.34)\qquad \begin{array}{ccc} \Omega^1(SU(2))_{un} & \stackrel{\pi_{\mathcal{Q}_{SU(2)}}}{\longrightarrow} & \Omega^1(\mathcal{A}(SU(2)) \\ \downarrow \chi & & \downarrow\sim_{\mathcal{N}_{SU(2))}} \\ \mathcal{A}(SU(2)) \otimes \ker \varepsilon_{\mathrm{U}(1)} & \stackrel{\mathrm{id}\otimes\pi_{\mathcal{Q}_{\mathrm{U}(1)}}}{\longrightarrow} & \mathcal{A}(SU(2)) \otimes (\ker \varepsilon_{\mathrm{U}(1)}/\mathcal{Q}_{\mathrm{U}(1)})\,. \end{array}$$

The proof of the compatibility conditions is in the following lemmas. The first one analyses the right covariance of the differential structure on $\mathcal{A}(SU(2))$.

LEMMA 8.3. *From the 3D bicovariant calculus on $\mathcal{A}(SU(2))$ generated by the ideal $\mathcal{Q}_{SU(2)} = (\ker \varepsilon_{SU(2)})^2 \subset \ker \varepsilon_{SU(2)}$ given in proposition 8.1, one has*

$$\check{\Delta}_R \mathcal{N}_{SU(2)} \subset \mathcal{N}_{SU(2)} \otimes \mathcal{A}(\mathrm{U}(1)).$$

PROOF. Using the bijection given in (3.3), it is $\Omega^1(SU(2)) \simeq \Omega^1(SU(2))/\mathcal{N}_{SU(2)}$ with $\mathcal{N}_{SU(2)} = r^{-1}(\mathcal{A}(SU(2)) \otimes \mathcal{Q}_{SU(2)})$. For this specific calculus one has that $\mathcal{N}_{SU(2)}$ is the sub-bimodule generated by $\{\delta\phi\,\delta\psi\}$ for any $\phi, \psi \in \mathcal{A}(SU(2))$, where $\delta\phi = (1 \otimes \phi - \phi \otimes 1) \in \Omega^1(SU(2))_{un}$. Choose $\phi \in \mathfrak{L}^{(0)}_n$ and $\psi \in \mathfrak{L}^{(0)}_m$ so to have $\check{\Delta}_R\phi = \phi \otimes z^{-n}$ and $\check{\Delta}_R\psi = \psi \otimes z^{-m}$. Extending the coaction $\check{\Delta}_R$ to a coaction $\check{\Delta}_R : \mathcal{A}(SU(2)) \otimes \mathcal{A}(SU(2)) \to \mathcal{A}(SU(2)) \otimes \mathcal{A}(SU(2)) \otimes \mathcal{A}(U(1))$ as $\check{\Delta}_R = (\mathrm{id}\otimes\mathrm{id}\otimes m) \circ (\mathrm{id}\otimes\tau\otimes\mathrm{id}) \circ (\check{\Delta}_R \otimes \check{\Delta}_R)$ in terms of the flip operator τ, it becomes an easy calculation to find:

$$\begin{aligned} \check{\Delta}_R(\delta\phi\,\delta\psi) &= (1 \otimes \phi\psi + \phi\psi \otimes 1 - \phi \otimes \psi - \psi \otimes \phi) \\ &= (1 \otimes \phi\psi + \phi\psi \otimes 1 - \phi \otimes \psi - \psi \otimes \phi) \otimes z^{-m-n} = (\delta\phi\,\delta\psi) \otimes z^{-m-n}. \end{aligned}$$

□

LEMMA 8.4. *The map $\chi : \Omega^1(SU(2))_{un} \to \mathcal{A}(SU(2)) \otimes \mathcal{A}(\mathrm{U}(1))$ defined in (3.14) as $\chi = (m \otimes \mathrm{id}) \circ (\mathrm{id}\otimes\check{\Delta}_R)$ is surjecive.*

PROOF. The proof of this result closely follows the proof of the proposition 3.2. From the spherical relation $1 = (u^*u + v^*v)^n = \sum_{a=0}^n \begin{pmatrix} n \\ a \end{pmatrix} u^{*a}v^{*n-a}v^{n-a}u^a$ it is possible to set $\left|\Psi^{(n)}\right\rangle_a \in \mathfrak{L}^{(0)}_n$ for $a = 0, \ldots, |n|$ with $\langle\Psi^{(n)}, \Psi^{(n)}\rangle = 1$ as:

$$n > 0 : \left|\Psi^{(n)}\right\rangle_a = \sqrt{\begin{pmatrix} n \\ a \end{pmatrix}}\, v^{*a}u^{*n-a},$$

$$n < 0 : \left|\Psi^{(n)}\right\rangle_a = \sqrt{\begin{pmatrix} |n| \\ a \end{pmatrix}}\, v^{*|n|-a}u^{a}.$$

Fixed $n \in \mathbb{Z}$, define $\gamma = \langle\Psi^{(-n)}, \delta\Psi^{(-n)}\rangle$. Since $\left|\Psi^{(-n)}\right\rangle \in \mathfrak{L}^{(0)}_{-n}$, one computes that $\chi(\gamma) = 1 \otimes (z^n - 1)$, and this sufficient to prove the surjectivity of the map χ, being χ left $\mathcal{A}(SU(2))$-linear and $\ker \varepsilon_{U(1)}$ is a complex vector space with a basis $(z^n - 1)$.

□

LEMMA 8.5. *Given the map χ as in the previous lemma, it is $\chi(\mathcal{N}_{SU(2)}) \subset \mathcal{A}(SU(2)) \otimes \mathcal{Q}_{\mathrm{U}(1)}$, where $\mathcal{N}_{SU(2)}$ is as in lemma 8.3 and $\mathcal{Q}_{\mathrm{U}(1)} = (\ker \varepsilon_{\mathrm{U}(1)})^2$.*

PROOF. To be definite, consider $\phi \in \mathfrak{L}_n^{(0)}$ and $\psi \in \mathfrak{L}_m^{(0)}$. One has:

$$\begin{aligned}\chi(\delta\phi\,\delta\psi) &= \phi\psi \otimes \{z^{-n-m} + 1 - z^{-n} - z^{-m}\}\\ &= \phi\psi \otimes \{(1 - z^{-n})(1 - z^{-m})\} \subset \mathcal{A}(SU(2)) \otimes (\ker \varepsilon_{\mathrm{U}(1)})^2.\end{aligned}$$

□

The results of these lemmas allow to define the map $\sim_{\mathcal{N}_{SU(2)}}: \Omega^1(SU(2)) \to \mathcal{A}(SU(2)) \otimes \ker \varepsilon_{\mathrm{U}(1)}/\mathcal{Q}_{\mathrm{U}(1)}$ from the diagram (8.34). Using the isomorphism $\lambda : \ker \varepsilon_{\mathrm{U}(1)}/\mathcal{Q}_{\mathrm{U}(1)} \to \mathbb{C}$ described in section 8.2.2, one has:

$$\begin{aligned}&\sim_{\mathcal{N}_{SU(2)}} (\omega_{\mathrm{e}}) = 0\\ &\sim_{\mathcal{N}_{SU(2)}} (\omega_{\mathrm{f}}) = 0\\ &\sim_{\mathcal{N}_{SU(2)}} (\omega_{\mathrm{h}}) = -2 \otimes \pi_{\mathcal{Q}_{\mathrm{U}(1)}}(z-1) = -2 \otimes 1.\end{aligned} \tag{8.35}$$

The next lemma completes the analysis of the compatibility conditions between the differential structures on $\mathcal{A}(SU(2))$ and on $\mathcal{A}(\mathrm{U}(1))$. The horizontal part of the set of k-forms out of $\Omega^k(SU(2))$ is defined as $\Omega^k_{\mathrm{hor}}(SU(2)) = \Omega^k(S^2)\mathcal{A}(SU(2)) = \mathcal{A}(SU(2))\Omega^k(S^2)$.

LEMMA 8.6. *Given the differential calculus on the basis $\Omega^1(S^2) = \Omega^1(S^2)_{un}/\mathcal{N}_{S^2}$ with $\mathcal{N}_{S^2} = \mathcal{N}_{SU(2)} \cap \Omega^1(S^2)_{un}$, it is*

$$\ker \sim_{\mathcal{N}_{SU(2)}} = \Omega^1(S^2)\mathcal{A}(SU(2)) = \mathcal{A}(SU(2))\Omega^1(S^2) = \Omega^1_{\mathrm{hor}}(SU(2)).$$

PROOF. Consider a 1-form $[\eta] \in \Omega^1(SU(2))$ and choose the element $\eta = \psi\,\delta\phi \in \Omega^1(SU(2))_{un}$ as a representative of $[\eta]$, with $\phi \in \mathfrak{L}_n^{(0)}$ and $\psi \in \mathfrak{L}_m^{(0)}$. One finds:

$$\begin{aligned}&\chi(\psi\,\delta\phi) = \psi\phi \otimes (z^{-n} - 1),\\ &\sim_{\mathcal{N}_{SU(2)}} (\eta) = \psi\phi \otimes \pi_{\mathcal{Q}_{\mathrm{U}(1)}}(z^{-n} - 1).\end{aligned}$$

Recalling once more the isomorphism $\lambda : \ker \varepsilon_{\mathrm{U}(1)}/\mathcal{Q}_{\mathrm{U}(1)} \to \mathbb{C}$, it is $\lambda(z^{-n} - 1) = 0$ if and only if $n = 0$, so to have $\eta = \psi\,\delta\phi$ with $\delta\phi \in \Omega^1(S^2)_{un}$ and then $\eta \in \Omega^1(S^2)_{un}\mathcal{A}(\mathrm{U}(1))$. It is clear that the condition $\chi(\mathcal{N}_{SU(2)}) \subset \mathcal{A}(SU(2)) \otimes \mathcal{Q}_{\mathrm{U}(1)}$ proved in lemma 8.5 ensures that the map $\sim_{\mathcal{N}_{SU(2)}}$ is well-defined: its image does not depend on the specific choice of the representative $\eta \in [\eta] \subset \Omega^1(SU(2))$.

□

The property of right covariance of the calculus on $\mathcal{A}(SU(2))$ – proved in lemma 8.3 – allows to extend the coaction $\check{\Delta}_R$ to a coaction $\check{\Delta}_R^{(k)} : \Omega^k(SU(2)) \to \Omega^k(SU(2)) \otimes \mathcal{A}(\mathrm{U}(1))$ via $\check{\Delta}_R^{(k)} \circ \mathrm{d} = (\mathrm{d} \otimes id) \circ \check{\Delta}_R^{(k-1)}$. Via such a coaction it is possible to recover (2.42) the set $\Omega^k(SU(2))_{\rho_{(n)}}$ as the $\rho_{(n)}(\mathrm{U}(1))$-equivariant k-forms on the Hopf bundle:

$$\Omega^k(SU(2))_{\rho_{(n)}} = \{\phi \in \Omega^k(SU(2)) : \check{\Delta}_R^{(k)}(\phi) = \phi \otimes z^{-n}\}.$$

as well as the $\mathcal{A}(S^2)$-bimodule $\mathfrak{L}_n^{(k)}$ of horizontal elements in $\Omega^k(SU(2))_{\rho_{(n)}}$.

8.2.4. *Connections and covariant derivative on the classical Hopf bundle.* The compatibility conditions bring the exactness of the sequence:

$$0 \longrightarrow \Omega^1_{\mathrm{hor}}(SU(2)) \longrightarrow \Omega^1(SU(2)) \stackrel{\sim\mathcal{N}_{SU(2)}}{\longrightarrow} \mathcal{A}(SU(2)) \otimes \ker \varepsilon_{\mathrm{U}(1)}/\mathcal{Q}_{\mathrm{U}(1)}, \tag{8.36}$$

whose every right invariant splitting $\sigma : \mathcal{A}(SU(2)) \otimes \ker \varepsilon_{\mathrm{U}(1)}/\mathcal{Q}_{\mathrm{U}(1)} \to \Omega^1(SU(2))$ represents a connection (6.3). With $w \in \mathbb{C} \simeq \ker \varepsilon_{\mathrm{U}(1)}/\mathcal{Q}_{\mathrm{U}(1)}$, one has:

$$\begin{aligned} \sigma(1 \otimes w) &= -\frac{w}{2}(\omega_{\mathrm{h}} + U\omega_{\mathrm{e}} + V\omega_{\mathrm{f}}), \\ \sigma(\phi \otimes w) &= -\frac{w}{2}\phi(\omega_{\mathrm{h}} + U\omega_{\mathrm{e}} + V\omega_{\mathrm{f}}) \end{aligned} \tag{8.37}$$

where $\phi \in \mathcal{A}(SU(2))$, and $U \in \mathfrak{L}^{(0)}_2$, $V \in \mathfrak{L}^{(0)}_{-2}$. The right invariant projection defined in(6.4) $\Pi : \Omega^1(SU(2)) \to \Omega^1(SU(2))$ associated to this splitting is, from (8.35):

$$\begin{aligned} \Pi(\omega_{\mathrm{e}}) &= \Pi(\omega_{\mathrm{f}}) = 0, \\ \Pi(\omega_{\mathrm{h}}) &= \omega_{\mathrm{h}} + U\omega_{\mathrm{e}} + V\omega_{\mathrm{f}}. \end{aligned} \tag{8.38}$$

The connection one form $\omega : \mathcal{A}(\mathrm{U}(1)) \mapsto \Omega^1(SU(2))$ defined in (6.5) is:

$$\omega(z^n) = \sigma(1 \otimes [z^n - 1]) = -\frac{n}{2}(\omega_{\mathrm{h}} + U\omega_{\mathrm{e}} + V\omega_{\mathrm{f}}). \tag{8.39}$$

The horizontal projector $(1 - \Pi) : \Omega^1(SU(2)) \to \Omega^1_{\mathrm{hor}}(SU(2))$ can be extended to whole exterior algebra $\Omega(SU(2))$, since it is compatible with the wedge product: one finds that $\{(1-\Pi)\omega_a \wedge (1-\Pi)\omega_b\} + \{(1-\Pi)\omega_b \wedge (1-\Pi)\omega_a\} = 0$ or any pair of 1-forms. This property, which is *not* valid in the quantum setting for a general connection – recall the remark 6.9 –, allows to define an operator of covariant derivative $D : \Omega^k(SU(2)) \mapsto \Omega^{k+1}(SU(2))$ as:

$$D\phi = (1 - \Pi)\mathrm{d}\phi, \qquad \forall \phi \in \Omega^k(SU(2)). \tag{8.40}$$

This definition is the dual counterpart of definition (2.4). It is not difficult to prove the main properties of such an operator of covariant derivative D:

- For any $\phi \in \Omega^k(SU(2))$, $D\phi \in \Omega^{k+1}_{\mathrm{hor}}(SU(2))$.
- The operator D is 'covariant'. One has $\check{\Delta}^{(k)}_R \phi = \phi \otimes z^n \Leftrightarrow \check{\Delta}^{(k+1)}_R(D\phi) = D\phi \otimes z^n$.
- Given $\phi \in \mathfrak{L}^{(k)}_n$, that is $\phi \in \Omega^k_{\mathrm{hor}}(SU(2))$ such that $\check{\Delta}^{(k)}_R \phi = \phi \otimes z^n$, it is $D\phi = \mathrm{d}\phi + \omega(z^n) \wedge \phi$. This last property recovers the relation (2.5).

9. Back on a covariant derivative on the exterior algebra $\Omega(\mathrm{SU}_q(2))$

The analysis in section 6 presents the formalism of connections on a quantum principal bundle [**6**] and explicitly describes both the set of connections on a quantum Hopf bundle and the corresponding set of covariant derivative operators $\nabla : \mathcal{E}^{(k)}_n \to \mathcal{E}^{(k+1)}_n$ acting on k-form valued sections of the associated quantum line bundles. The left $\mathcal{A}(\mathrm{S}^2_q)$-module equivalence between $\mathcal{E}^{(k)}_n$ and horizontal elements $\mathcal{L}^{(k)}_n \subset \Omega^k_{\mathrm{hor}}(\mathrm{SU}_q(2))$ allows then for the definition of a covariant derivative operator $D : \mathcal{L}^{(k)}_n \to \mathcal{L}^{(k+1)}_n$ with $k = 0, 1, 2$.

The equation (6.46) in remark 6.9 clarifies the reasons why, presenting a connection via the projector (6.7) $\Pi : \Omega^1(\mathrm{SU}_q(2)) \to \Omega^1(\mathrm{SU}_q(2))$ given in (6.12), the operator $\check{D} = (1 - \Pi)\mathrm{d} : \Omega^1(\mathrm{SU}_q(2)) \to \Omega^2_{\mathrm{hor}}(\mathrm{SU}_q(2))$ as in (6.44) defined a consistent covariant derivative on the whole exterior algebra $\Omega(\mathrm{SU}_q(2))$ only in the case of the monopole connection: the operator $(1 - \Pi) : \Omega^1(\mathrm{SU}_q(2)) \to \Omega^1_{\mathrm{hor}}(\mathrm{SU}_q(2))$ is a covariant projector compatible with the properties of the wedge product (6.47) in the exterior algebra $\Omega(\mathrm{SU}_q(2))$ only if the connection is the monopole connection.

The problem of defining, for any connection on a principal quantum bundle, a consistent covariant projection operator on the whole exterior algebra on the total space of the bundle whose range is given by the horizontal exterior forms has been studied in [**12, 13**]. The aim of this section is, from one side, to describe the properties of the horizontal projector arising from that analysis, and then to show that in such a formulation of the Hopf bundle more than one horizontal covariant projector can be consistently introduced.

As already mentioned, the formulation presented in [**12, 13**] of the geometrical structures of a quantum principal bundle slightly differs from that described in section 3.2 and a comparison between them is in [**14**]. This formalism will not be explicitly reviewed: the main results concerning how to define an horizontal covariant projector will be translated into the language extensively described in the previous sections.

The differential $*$-calculus $(\Omega(\mathrm{U}(1)), \mathrm{d})$ on the gauge group algebra $\mathrm{U}(1)$ is described in section 3.4.2. It canonically corresponds to the right $\mathcal{A}(\mathrm{U}(1))$-ideal $\mathcal{Q}_{\mathrm{U}(1)} \subset \ker \varepsilon_{\mathrm{U}(1)}$ generated by the element $\{(z^* - 1) + q^2(z - 1)\}$, so that by lemma 3.4 it is $\Omega^1(\mathrm{U}(1))_{\mathrm{inv}} \simeq \ker \varepsilon_{\mathrm{U}(1)}/\mathcal{Q}_{\mathrm{U}(1)} \simeq \mathbb{C}$. Such a calculus is bicovariant: given the left and right coactions (3.2) of the $*$-Hopf algebra $\mathcal{A}(\mathrm{U}(1))$ on $\Omega^1(\mathrm{U}(1))$ one has that the 1-form ω_z is both left and right invariant,

$$\begin{aligned} &\Delta^{(1)}_{\ell} : \Omega^1(\mathrm{U}(1)) \to \mathcal{A}(\mathrm{U}(1)) \otimes \Omega^1(\mathrm{U}(1)), && \Delta^{(1)}_{\ell}(\omega_z) = 1 \otimes \omega_z; \\ &\Delta^{(1)}_{\wp} : \Omega^1(\mathrm{U}(1)) \to \Omega^1(\mathrm{U}(1)) \otimes \mathcal{A}(\mathrm{U}(1)), && \Delta^{(1)}_{\wp}(\omega_z) = \omega_z \otimes 1. \end{aligned} \tag{9.1}$$

The exterior algebra on this differential calculus is built following [**21**], as explained in section (3.4.1), where the same procedure has been applied to the analysis of the 3D left-covariant calculus on $\mathrm{SU}_q(2)$. It results $S_{\mathcal{Q}_{\mathrm{U}(1)}} = (\Omega^1(\mathrm{U}(1)))^{\otimes 2}$, so that

$$\Omega(\mathrm{U}(1)) = \sum_{k \geq 0} {}^{\oplus}\Omega(\mathrm{U}(1))^{\wedge k} = \mathcal{A}(\mathrm{U}(1)) \oplus \Omega^1(\mathrm{U}(1)). \tag{9.2}$$

The coproduct map in the Hopf $*$-algebra $\mathcal{A}(\mathrm{U}(1))$ can be extended to a homomorphism $\hat{\Delta}_{\mathrm{U}(1)} : \Omega(\mathrm{U}(1)) \to \Omega(\mathrm{U}(1)) \otimes \Omega(\mathrm{U}(1))$ given by

$$\begin{aligned} &\hat{\Delta}_{\mathrm{U}(1)}(\varphi) = \Delta(\varphi) = \varphi \otimes \varphi, \\ &\hat{\Delta}_{\mathrm{U}(1)}(\varphi\,\omega_z) = \Delta^{(1)}_{\ell}(\varphi\,\omega_z) + \Delta^{(1)}_{\wp}(\varphi\,\omega_z) = \varphi(1 \otimes \varphi\,\omega_z + \omega_z \otimes \varphi), \end{aligned} \tag{9.3}$$

for any $\varphi \in \mathcal{A}(\mathrm{U}(1))$. Given the principal bundle structure, the compatibility conditions among calculi on the total space algebra and the gauge group algebra allow to prove that there exists a unique extension of the coaction (3.29) of the gauge group $\mathrm{U}(1)$ on the total space $\mathrm{SU}_q(2)$ to a left $\mathcal{A}(\mathrm{SU}_q(2))$-module homomorphism $\mathfrak{F} : \Omega(\mathrm{SU}_q(2)) \to \Omega(\mathrm{SU}_q(2)) \otimes \Omega(\mathrm{U}(1))$ implicitly defined by:

$$\begin{aligned} &(\mathfrak{F} \otimes \mathrm{id})\mathfrak{F} = (\mathrm{id} \otimes \hat{\Delta}_{\mathrm{U}(1)})\mathfrak{F}, \\ &\mathfrak{F} *_{\mathrm{SU}_q(2)} = (*_{\mathrm{SU}_q(2)} \otimes *_{\mathrm{U}(1)})\mathfrak{F} : \end{aligned}$$

where the second condition expresses a compatiblity between the map $\mathfrak{F}$ and the $*$-structures on the exterior algebras built over the calculi on $\mathrm{SU}_q(2)$ and $\mathrm{U}(1)$. One has

$$\begin{aligned}
&\mathfrak{F}(x) = \Delta_R(x) = x \otimes z^{-n},\\
&\mathfrak{F}(x\,\omega_-) = x\,\omega_- \otimes z^{-2-n},\\
&\mathfrak{F}(x\,\omega_+) = x\,\omega_+ \otimes z^{2-n},\\
&\mathfrak{F}(x\,\omega_z) = (x \otimes z^{-n}\omega_z) + (x\,\omega_z \otimes z^{-n}),\\
&\mathfrak{F}(x\,\omega_- \wedge \omega_+) = x\,\omega_- \wedge \omega_+ \otimes z^{-n},\\
&\mathfrak{F}(x\,\omega_+ \wedge \omega_z) = (x\,\omega_+ \otimes z^{2-n}\omega_z) + (x\,\omega_+ \wedge \omega_z \otimes z^{2-n}),\\
&\mathfrak{F}(x\,\omega_z \wedge \omega_-) = (x\,\omega_- \otimes z^{-2-n}\omega_z) + (x\,\omega_z \wedge \omega_- \otimes z^{-2-n}),
\end{aligned}$$

$$\mathfrak{F}(x\,\omega_- \wedge \omega_+ \wedge \omega_z) = (x\,\omega_- \wedge \omega_+ \otimes z^{-n}\omega_z) + (x\,\omega_- \wedge \omega_+ \wedge \omega_z \otimes z^{-n}), \tag{9.4}$$

with $x \in \mathcal{A}(\mathrm{SU}_q(2))$, such that $\Delta_R(x) = x \otimes z^{-n} \Leftrightarrow x \in \mathcal{L}_n^{(0)}$. The homomorphism $\mathfrak{F}$ can be restricted to the right coaction $\Delta_R^{(k)} : \Omega^k(\mathrm{SU}_q(2)) \to \Omega^k(\mathrm{SU}_q(2)) \otimes \mathcal{A}(\mathrm{U}(1))$ given in (6.32):

$$\Delta_R^{(k)}(\phi) = (\mathrm{id} \otimes p_0)\mathfrak{F}(\phi)$$

with $\phi \in \Omega^k(\mathrm{SU}_q(2))$ and p_0 the projection $\Omega(\mathrm{U}(1)) \to \mathcal{A}(\mathrm{U}(1))$ coming from (9.2). The horizontal subset of the exterior algebra $\Omega(\mathrm{SU}_q(2))$ can be defined via:

$$\Omega_{\mathrm{hor}}(\mathrm{SU}_q(2)) = \{\phi \in \Omega(\mathrm{SU}_q(2)) \quad : \quad \mathfrak{F}(\phi) = (id \otimes p_0)\mathfrak{F}(\phi)\}, \tag{9.5}$$

while the exterior algebra $\Omega(\mathrm{S}_q^2)$ described in section 3.4.3 can be recovered as

$$\Omega(\mathrm{S}_q^2) = \{\phi \in \Omega(\mathrm{SU}_q(2)) \quad : \quad \mathfrak{F}(\phi) = \phi \otimes 1\}.$$

From the analysis in section 6 one has that a connection 1-form is given via a map $\tilde{\omega} : \Omega^1(\mathrm{U}(1))_{\mathrm{inv}} \to \Omega(\mathrm{SU}_q(2))$ satisfying the conditions (6.5). The equation (6.13) shows that any connection can be written as:

$$\tilde{\omega}(\omega_z) = \omega_z + \mathrm{a},$$

with $\mathrm{a} \in \Omega^1(\mathrm{S}_q^2)$. Given a connection, one can define a map

$$m_\omega : \Omega_{\mathrm{hor}}(\mathrm{SU}_q(2)) \otimes \Omega(\mathrm{U}(1))_{\mathrm{inv}} \to \Omega(\mathrm{SU}_q(2)), \tag{9.6}$$

where the relation (9.2) enables to recover $\Omega(\mathrm{U}(1))_{\mathrm{inv}} \simeq \{\mathbb{C} \oplus \Omega^1(\mathrm{U}(1))_{\mathrm{inv}}\}$: given $\psi \in \mathfrak{hor}(\mathrm{SU}_q(2))$ and $\theta = \lambda + \mu\,\omega_z \in \Omega(\mathrm{U}(1))_{\mathrm{inv}}$ (with $\lambda, \mu \in \mathbb{C}$) set:

$$m_\omega(\psi \otimes \theta) = \psi \wedge (\mu + \lambda\tilde{\omega}(\omega_z)). \tag{9.7}$$

The map m_ω is proved to be bijective, and the operator

$$h_\omega = (\mathrm{id} \otimes p_0)m_\omega^{-1} \tag{9.8}$$

a covariant horizontal projector $h_\omega : \Omega(\mathrm{SU}_q(2)) \to \mathfrak{hor}(\mathrm{SU}_q(2))$. Given an element $\phi \in \Omega^k(\mathrm{SU}_q(2))$, define its covariant derivative:

$$\mathfrak{D}\phi = h_\omega \mathrm{d}\phi. \tag{9.9}$$

In the formulation developed in [**12, 13**] this definition is meant to be the quantum analogue of the classical relation (8.40).

The previous analysis allows for a complete study of this quantum horizontal projector. Consider a connection 1-form $\tilde{\omega}(\omega_z) = \omega_z + U\omega_- + V\omega_+ = \omega_z + \mathrm{a}$ with

$U \in \mathcal{L}_2^{(0)}$ and $V \in \mathcal{L}_{-2}^{(0)}$ as in equation (6.11). The inverse of the multiplicative map m_ω – the map $m_\omega^{-1} : \Omega(\mathrm{SU}_q(2)) \to \mathfrak{hor}(\mathrm{SU}_q(2)) \otimes \Omega(\mathrm{U}(1))_{\mathrm{inv}}$ – as well as the horizontal projector are given on 0-forms and 1-forms by:

$$\begin{aligned} m_\omega^{-1}(x) &= x \otimes 1 && \Rightarrow \quad h_\omega(x) = x; \\ m_\omega^{-1}(x\,\omega_\pm) &= x\,\omega_\pm \otimes 1 && \Rightarrow \quad h_\omega(x\,\omega_\pm) = x\,\omega_\pm, \\ m_\omega^{-1}(x\,\omega_z) &= (-x\,\mathrm{a} \otimes 1) + (x \otimes \omega_z) && \Rightarrow \quad h_\omega(x\,\omega_z) = -x\,\mathrm{a} \end{aligned} \tag{9.10}$$

with $x \in \mathcal{A}(\mathrm{SU}_q(2))$. This means that one has $\mathfrak{D}\phi = D\phi$ where $\phi \in \mathcal{A}(\mathrm{SU}_q(2))$ with respect to the covariant derivative defined in (6.18). On higher order exterior forms one has:

$$\begin{aligned} m_\omega^{-1}(x\,\omega_- \wedge \omega_+) &= x\,\omega_- \wedge \omega_+ \otimes 1 \\ &\Rightarrow \quad h_\omega(x\,\omega_- \wedge \omega_+) = x\,\omega_- \wedge \omega_+, \\ m_\omega^{-1}(x\,\omega_+ \wedge \omega_z) &= (-x\omega_+ \wedge \mathrm{a} \otimes 1) + (x\,\omega_+ \otimes \omega_z) \\ &\Rightarrow \quad h_\omega(x\,\omega_+ \wedge \omega_z) = -x\,\omega_+ \wedge \mathrm{a} = x\,U\,\omega_- \wedge \omega_+, \\ m_\omega^{-1}(x\,\omega_- \wedge \omega_z) &= (-x\,\omega_- \wedge \mathrm{a} \otimes 1) + (x\,\omega_- \otimes \omega_z) \\ &\Rightarrow \quad h_\omega(x\,\omega_- \wedge \omega_z) = -x\,\omega_- \wedge \mathrm{a} = -q^2 x\,V\,\omega_- \wedge \omega_+, \\ m_\omega^{-1}(x\,\omega_- \wedge \omega_+ \wedge \omega_z) &= x\,\omega_- \wedge \omega_+ \otimes \omega_z \\ &\Rightarrow \quad h_\omega(x\,\omega_- \wedge \omega_+ \wedge \omega_z) = 0. \end{aligned} \tag{9.11}$$

Recalling the analysis in remark 6.9, it is important to stress that the projector h_ω from (9.8) is well defined on the exterior algebra $\Omega(\mathrm{SU}_q(2))$ for any choice of the connection, and defines a covariant derivative $\mathfrak{D} : \Omega^k(\mathrm{SU}_q(2)) \to \Omega^{k+1}_{\mathrm{hor}}(\mathrm{SU}_q(2))$ which reduces to the operators (6.18) on 0-forms and (6.35) on 1-forms. The last equation out of (9.11) shows also that $\mathfrak{D} : \Omega^2(\mathrm{SU}_q(2)) \to 0$.

REMARK 9.1. *Is the horizontal projector h_ω defined in* (9.8) *the only well-defined horizontal covariant projector operator whose domain coincides with $\Omega(\mathrm{SU}_q(2))$ and whose range is $\mathfrak{hor}(\mathrm{SU}_q(2)) \subset \Omega(\mathrm{SU}_q(2))$, such that the associated horizontal projection of the exterior derivative* (9.9) *reduces to the well established operator $D : \mathcal{L}_n^{(k)} \to \mathcal{L}_n^{(k+1)}$ given in* (6.16),(6.35)*? The answer is no. To be definite, consider the operator $h'_\omega : \Omega(\mathrm{SU}_q(2)) \to \Omega_{\mathrm{hor}}(\mathrm{SU}_q(2))$ given by:*

$$\begin{aligned} h'_\omega(x) &= x; \\ h'_\omega(x\,\omega_\pm) &= x\,\omega_\pm, \\ h'_\omega(x\,\omega_z) &= -x\,\mathrm{a}, \end{aligned} \tag{9.12}$$

so to coincide with the projector h_ω (9.10) *on 0-forms and 1-forms, and:*

$$\begin{aligned} h'_\omega(x\,\omega_- \wedge \omega_+) &= x\,\omega_- \wedge \omega_+, \\ h'_\omega(x\,\omega_+ \wedge \omega_z) &= q^4 x\,\mathrm{a} \wedge \omega_+ = q^4 xU\,\omega_- \wedge \omega_+, \\ h'_\omega(x\,\omega_- \wedge \omega_z) &= q^{-4} x\,\mathrm{a} \wedge \omega_- = -q^{-2} xV\,\omega_- \wedge \omega_+; \\ h'_\omega(x\,\omega_- \wedge \omega_+ \wedge \omega_z) &= 0. \end{aligned} \tag{9.13}$$

It is clear that the operator $\mathfrak{D}' = h'_\omega \mathrm{d} : \Omega^k(\mathrm{SU}_q(2)) \to \Omega^{k+1}_{\mathrm{hor}}(\mathrm{SU}_q(2))$ defines a consistent covariant derivative on the whole exterior algebra on the total space algebra of the quantum Hopf bundle, which reduces to the operator $\mathfrak{D}$ from (9.9) *when* *restricted to horizontal elements* $\mathcal{L}_n^{(k)} \subset \Omega^k(\mathrm{SU}_q(2))$. *Both the operators* $\mathfrak{D}, \mathfrak{D}'$ *coincide in the classical limit with the covariant derivative on the classical Hopf bundle* (8.40) *presented in section 8.*

The last step is to understand from where it is possible to trace the origin of such a projector h'_ω *back. It is easy to see that the isomorphism* m_ω^{-1} *coming from* (9.7) *can be recovered as the choice of a specific left* $\mathcal{A}(\mathrm{SU}_q(2))$*-module basis for the exterior algebra* $\Omega(\mathrm{SU}_q(2))$, *namely*

$$\Omega(\mathrm{SU}_q(2)) \simeq \mathcal{A}(\mathrm{SU}_q(2))\{1 \oplus \omega_- \oplus \omega_+ \oplus \tilde{\omega}(\omega_z)\}$$

(9.14)

$$\oplus \mathcal{A}(\mathrm{SU}_q(2))\{(\omega_- \wedge \omega_+) \oplus (\omega_- \wedge \tilde{\omega}(\omega_z)) \oplus (\omega_+ \wedge \tilde{\omega}(\omega_z)) \oplus (\omega_- \wedge \omega_+ \wedge \tilde{\omega}(\omega_z))\},$$

while the horizontal projection obviously annihilates all the coefficients associated to exterior forms having the connection 1-form $\tilde{\omega}(\omega_z)$ *as a term. The projector* h'_ω *in* (9.12),(9.13) *comes from the choice of a different left* $\mathcal{A}(\mathrm{SU}_q(2))$*-module basis of* $\Omega(\mathrm{SU}_q(2))$, *that is setting – as analogue of* (9.14) *– the isomorphism*

$$\Omega(\mathrm{SU}_q(2)) \simeq \mathcal{A}(\mathrm{SU}_q(2))\{1 \oplus \omega_- \oplus \omega_+ \oplus \tilde{\omega}(\omega_z)\}$$

(9.15)

$$\oplus \mathcal{A}(\mathrm{SU}_q(2))\{(\omega_- \wedge \omega_+) \oplus (\tilde{\omega}(\omega_z) \wedge \omega_-) \oplus (\tilde{\omega}(\omega_z) \wedge \omega_+) \oplus (\omega_- \wedge \omega_+ \wedge \tilde{\omega}(\omega_z))\}.$$

and then defining h'_ω *as the projector whose nucleus is given as the left* $\mathcal{A}(\mathrm{SU}_q(2))$*-module spanned by* $\{\tilde{\omega}(\omega_z), \tilde{\omega}(\omega_z) \wedge \omega_\pm, \omega_- \wedge \omega_+ \wedge \tilde{\omega}(\omega_z)\}$. *An explicit computation shows that*

$$\omega_- \wedge \tilde{\omega}(\omega_z) = (q^2 - q^{-2})V\omega_- \wedge \omega_+ - q^{-4}\tilde{\omega}(\omega_z) \wedge \omega_- \quad \Rightarrow \quad \ker h_\omega \neq \ker h'_\omega :$$

the two projectors are not equivalent, being equivalent if and only if the connection is the monopole connection.

Acknowledgements. Some days ago, reading once more this manuscript, I felt that for almost any sentence of it I am indebted to one and to all of the travelmates I had during the last year. This paper has been originated and developed as a part of a more general research project with G.Landi: I want to thank him for his support, guidance and feedback. I should like to thank M.Marcolli, who suggested me to write it, and S.Albeverio: with them I often discussed about many of the themes here described. I should like to thank G.Marmo, L.Cirio and C.Pagani, who read a draft of the paper and made my understanding of many important details clearer, and G.Dell'Antonio, who encouraged me to write it. And I should like to thank the referee: his report was precious and helped me to improve it.

It is a pleasure to thank the Max-Planck-Institut für Mathematik in Bonn and the Hausdorff Center for Mathematics at the University Bonn for their invitation, the Foundation Blanceflor Boncompagni-Ludovisi née Bildt (Stockholm) for the support, the Joint Institut Research for Nuclear Physics in Dubna-Moscow: A.Motovilov was a wonderful host.

References

[1] R. Abraham, J.E. Marsden, T. Ratiu, *Manifolds, tensor analysis and applications*, App. Math. Scie. 75, Springer 1988.

[2] P. Aschieri, L. Castellani, *An introduction to non-commutative differential geometry on quantum groups*, Int.J.Mod.Phys. A8 (1993) 1667-1706.

[3] P. Baum, P.M. Hajac, R. Matthes, W. Szymanski, *Noncommutative geometry approach to principal and associated bundles*, arXiv:math/0701033.

[4] N. Berline, E. Getzler, M. Vergne, *Heat Kernels and Dirac Operators*, Springer 1991.

[5] T. Brzezinski, *Quantum fibre bundles. An introduction*, Banach Center Publications, Warsaw (1995), arXiv:q-alg/9508008.

[6] T. Brzezinski, S. Majid, *Quantum group gauge theory on quantum spaces*, Comm. Math. Phys. 157 (1993) 591–638; Erratum 167 (1995) 235.

[7] T. Brzezinski, S. Majid, *Quantum differential and the q-monopole revisited*, Acta Appl. Math. 54 (1998) 185-233.

[8] T. Brzezinski, S. Majid, *Line bundles on quantum spheres*, AIP Conf. Proc. 345 (1998) 3–8.

[9] C.-S. Chu, P.-M. Ho, H. Steinacker, *q-deformed Dirac monopole with arbitrary charge*, Z. Phys. C71 (1996) 171-177.

[10] P.A.M. Dirac, Proc.R.Soc A 133 (1931) 60.

[11] F. D'Andrea, G. Landi, *Anti-self dual connections on the quantum projective plane: monopole*, arXiv:0903.3551.

[12] M. Durdevic, *Geometry of quantum principal bundles I*, Comm.Math.Phys. 175 (1996), 457-520.

[13] M. Durdevic, *Geometry of quantum principal bundles II*, Rev.Math.Phys 9 (1997) 531-607.

[14] M. Durdevic, *Quantum principal bundles as Hopf-Galois extensions*, arXiv:q-alg/9507022.

[15] M. Göckeler, T. Schücker, *Differential geometry, gauge theories and gravity*, Cambridge University Press 1987.

[16] P.M.Hajac, *Strong connections on quantum principal bundles*, Comm.Math.Phys. 182 (1996) 579-617.

[17] P.M. Hajac, S. Majid, *Projective module description of the q-monopole*, Comm. Math. Phys. 206 (1999) 247–264.

[18] H. Hopf, *Über die Abbildungen der dreidimensionalen Sphäre auf die Kugelfläche*, Mathematische Annalen 104 (1931) 637-635.

[19] H. Hopf, *Über die Abbildungen von Sphären auf Sphären niedrigerer Dimension*, Fundamenta Mathematicae 25 (1935) 427-440

[20] D. Husemoller, *Principal Bundles*, Springer-Verlag - New York 1998.

[21] A. Klimyk, K. Schmüdgen, *Quantum Groups and Their Representations*, Springer 1997.

[22] J.Kustermans, G.J.Murphy, L.Tuset, *Quantum groups, differential calculi and the eigenvalues of the Laplacian*, Trans.Amer.Math.Soc. 357 (2005) 4681-4717.

[23] G. Landi, *Projective modules of finite type and monopoles over* S^2, J.Geom.Phys. 37 (2001) 47-62.

[24] G.Landi, C.Reina, A.Zampini, *Gauged Laplacians on quantum Hopf bundles*, Comm.Math.Phys. 287 (2009) 179-209.

[25] S. Majid, *Foundations of Quantum Group Theory*, Cambridge Univ. Press, 1995.

[26] S. Majid, *Noncommutative Riemannian and spin geometry of the standard q-sphere*, Comm. Math. Phys. 256 (2005) 255–285.

[27] T. Masuda, K. Mimachi, Y. Nakagami, M. Noumi, K. Ueno, *Representations of the Quantum Group* $\mathrm{SU}_q(2)$ *and the Little q-Jacobi Polynomials*, J. Funct. Anal. 99 (1991) 357–387.

[28] P.W.Michor, *Topics in differential geometry*, Graduate Studies in Mathematics vol 93, AMS 2008.

[29] J.A. Mignaco, C. Sigaud, A.R. da Silva, F.J. Vanhecke, *The Connes-Lott program on the sphere*, Rev.Math.Phys. 9 (1997) 689-718.

[30] P. Podleś, *Quantum spheres*, Lett. Math. Phys. 14 (1987) 193–202.

[31] K. Schmüdgen, E. Wagner, *Representations of cross product algebras of Podleś quantum spheres*, J. Lie Theory 17 (2007) 751–790.

[32] K. Schmüdgen, E. Wagner, *Dirac operator and a twisted cyclic cocycle on the standard Podleś quantum sphere*, J. reine angew. Math. 574 (2004) 219–235.

[33] H. Schneider, *Principal homogeneous spaces for arbitrary Hopf algebras*, Israel J. Math. 72 (1990) 167–195.

[34] R.G.Swan, *Vector bundles and projective modules*, Trans.Am.Math.Soc. 105 (1962) 264-277.

[35] A. Trautman, *Solutions of the Maxwell and Yang-Mills equations associated with Hopf fibrings*, Int.J.Theor.Phys. 16, 8 (1977) 561-565.

[36] D.A. Varshalovic, A.N. Moskalev, V.K. Khersonskii, *Quantum theory of angular momentum*, World Scientific 1988.

[37] S.L. Woronowicz, *Twisted* $\mathrm{SU}_q(2)$ *group. An example of a noncommutative differential calculus*, Publ. Rest. Inst. Math.Sci., Kyoto Univ. 23 (1987) 117–181.

[38] S.L. Woronowicz, *Differential calculus on compact matrix pseudogroups (quantum groups)*, Comm. Math. Phys. 122 (1989) 125–170.

Max Planck Institut für Mathematik - Bonn, Vivatsgasse 7, D-53111 Bonn, Germany

Hausdorff Zentrum für Mathematik der Universität Bonn, Endenicher Allee 62, D-53115 Bonn, Germany
E-mail address: zampini@mpim-bonn.mpg.de